JACARANDA
MATHS QUEST

SPECIALIST MATHEMATICS 11
FOR QUEENSLAND
UNITS 1 & 2 | SECOND EDITION

JACARANDA MATHS QUEST

SPECIALIST MATHEMATICS 11 FOR QUEENSLAND

UNITS 1 & 2 | SECOND EDITION

CATHERINE SMITH

RAYMOND ROZEN

CONTRIBUTING AUTHORS

Judi Dau | Robyna Martinovic | Lilian Jian

REVIEWED BY

Graham Stevenson | Grace Zhang
Andrew Johnston | Fiona Swan

jacaranda
A Wiley Brand

Second edition published 2024 by
John Wiley & Sons Australia, Ltd
Level 4, 600 Bourke Street, Melbourne, Vic 3000

First edition published 2018

Typeset in 10.5/13 pt TimesLTStd

ISBN: 978-1-394-26979-2

Front and back cover images: © TWINS DESIGN STUDIO/Adobe Stock Photos, veekicl/Adobe Stock Photos, mrhighsky/Adobe Stock Photos, sapunkele/Adobe Stock Photos, NARANAT STUDIO/Adobe Stock Photos, Kullaya/Adobe Stock Photos, vectorplus/Adobe Stock Photos, WinWin/Adobe Stock Photos, valeriya_dor/Adobe Stock Photos, Ludmila/Adobe Stock Photos, Анастасия Трофимова/Adobe Stock Photos, katarinalas/Adobe Stock Photos, izzul fikry (ijjul)/Adobe Stock Photos, Tatsiana/Adobe Stock Photos, nadiinko/Adobe Stock Photos

Illustrated by various artists, diacriTech and Wiley Composition Services

Typeset in India by diacriTech

A catalogue record for this book is available from the National Library of Australia

Printed in Singapore
M130193_120824

The publisher of this series acknowledges and pays its respects to Aboriginal Peoples and Torres Strait Islander Peoples as the traditional custodians of the land on which this resource was produced.

This suite of resources may include references to (including names, images, footage or voices of) people of Aboriginal and/or Torres Strait Islander heritage who are deceased. These images and references have been included to help Australian students from all cultural backgrounds develop a better understanding of Aboriginal and Torres Strait Islander Peoples' history, culture and lived experience.

It is strongly recommended that teachers examine resources on topics related to Aboriginal and/or Torres Strait Islander Cultures and Peoples to assess their suitability for their own specific class and school context. It is also recommended that teachers know and follow the guidelines laid down by the relevant educational authorities and local Elders or community advisors regarding content about all First Nations Peoples.

All activities in this resource have been written with the safety of both teacher and student in mind. Some, however, involve physical activity or the use of equipment or tools. **All due care should be taken when performing such activities**. To the maximum extent permitted by law, the authors and publisher disclaim all responsibility and liability for any injury or loss that may be sustained when completing activities described in this resource.

The publisher acknowledges ongoing discussions related to gender-based population data. At the time of publishing, there was insufficient data available to allow for the meaningful analysis of trends and patterns to broaden our discussion of demographics beyond male and female gender identification.

Contents

online only

Problem-solving and modelling task guide

---

on**line**only

PRACTICE ASSESSMENT 1
Problem-solving and Modelling Task

PRACTICE ASSESSMENT 2
Unit 1 Examination

UNIT 2 COMPLEX NUMBERS, FURTHER PROOF, TRIGONOMETRY, FUNCTIONS AND TRANSFORMATIONS   267

---

on**line**only

PRACTICE ASSESSMENT 3
Unit 2 Examination

PRACTICE ASSESSMENT 4
Units 1&2 Examination

Learning with learnON

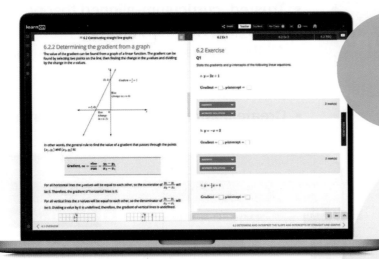

JACARANDA MATHS QUEST
SPECIALIST MATHEMATICS 11
UNITS 1 AND 2 FOR QUEENSLAND | SECOND EDITION

Developed by expert teachers for students

Tried, tested and trusted. The completely revised and updated second edition of *Jacaranda Maths Quest Specialist Mathematics 11 Units 1&2 for Queensland* continues to focus on helping teachers achieve learning success for every student — ensuring no student is left behind and no student held back.

Because both what and how students learn matter

Learning is personal

Whether students need a challenge or a helping hand, you'll find what you need to create engaging lessons.

Whether in class or at home, students can get unstuck and progress! Scaffolded lessons, with detailed worked examples, including both TI and Casio calculator support. Automatically marked, differentiated question sets are all supported by detailed worked solutions. And brand-new exam-style questions support in-depth skill acquisition in every lesson.

Learning is effortful

Learning happens when students push themselves. With learnON, Australia's most powerful online learning platform, students can challenge themselves, build confidence and ultimately achieve success.

Learning is rewarding

Through real-time results data, students can track and monitor their own progress and easily identify areas of strength and weakness.

And for teachers, Learning Analytics provide valuable insights to support student growth and drive informed intervention strategies.

Learn online with Australia's most

- Trusted, curriculum-aligned theory
- Engaging, rich multimedia
- All the teacher support resources you need
- Deep insights into progress
- Immediate feedback for students
- Create custom assignments in just a few clicks.

Practical teaching advice and ideas for each lesson provided in teachON

Each lesson linked to content points from the QCAA Specialist Mathematics 2025 General senior syllabus

Reading content and rich media including interactivities and calculator support.

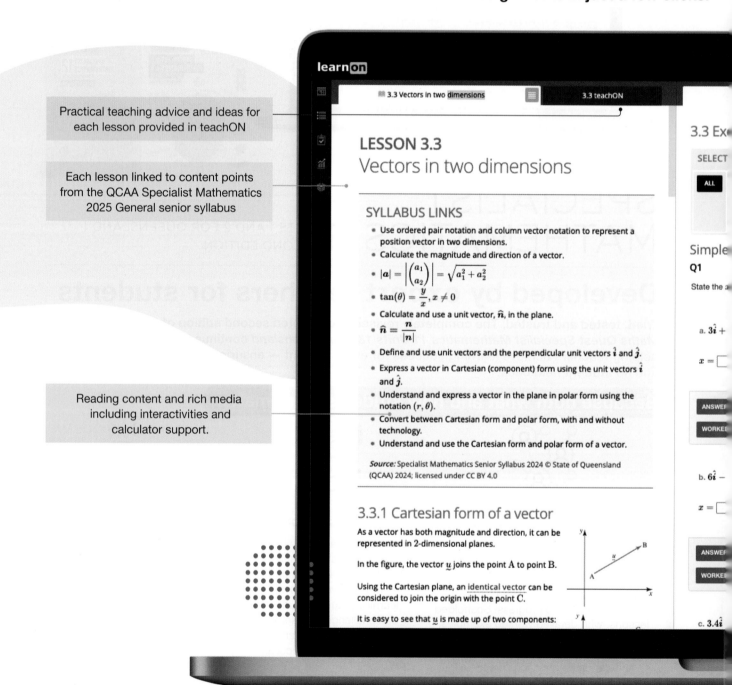

learnon

3.3 Vectors in two dimensions   3.3 teachON

3.3 Ex

SELECT

ALL

LESSON 3.3
Vectors in two dimensions

SYLLABUS LINKS

- Use ordered pair notation and column vector notation to represent a position vector in two dimensions.
- Calculate the magnitude and direction of a vector.
- $|\boldsymbol{a}| = \left|\begin{pmatrix} a_1 \\ a_2 \end{pmatrix}\right| = \sqrt{a_1^2 + a_2^2}$
- $\tan(\theta) = \dfrac{y}{x}, x \neq 0$
- Calculate and use a unit vector, $\hat{\boldsymbol{n}}$, in the plane.
- $\hat{\boldsymbol{n}} = \dfrac{\boldsymbol{n}}{|\boldsymbol{n}|}$
- Define and use unit vectors and the perpendicular unit vectors $\hat{\boldsymbol{i}}$ and $\hat{\boldsymbol{j}}$.
- Express a vector in Cartesian (component) form using the unit vectors $\hat{\boldsymbol{i}}$ and $\hat{\boldsymbol{j}}$.
- Understand and express a vector in the plane in polar form using the notation (r, θ).
- Convert between Cartesian form and polar form, with and without technology.
- Understand and use the Cartesian form and polar form of a vector.

Source: Specialist Mathematics Senior Syllabus 2024 © State of Queensland (QCAA) 2024; licensed under CC BY 4.0

3.3.1 Cartesian form of a vector

As a vector has both magnitude and direction, it can be represented in 2-dimensional planes.

In the figure, the vector $\underset{\sim}{u}$ joins the point A to point B.

Using the Cartesian plane, an identical vector can be considered to join the origin with the point C.

It is easy to see that $\underset{\sim}{u}$ is made up of two components:

Simple

Q1

State the a

a. $3\hat{\boldsymbol{i}} +$

$x =$ ☐

ANSWER

WORKE

b. $6\hat{\boldsymbol{i}} -$

$x =$ ☐

ANSWER

WORKE

c. $3.4\hat{\boldsymbol{i}}$

powerful learning tool, learnON

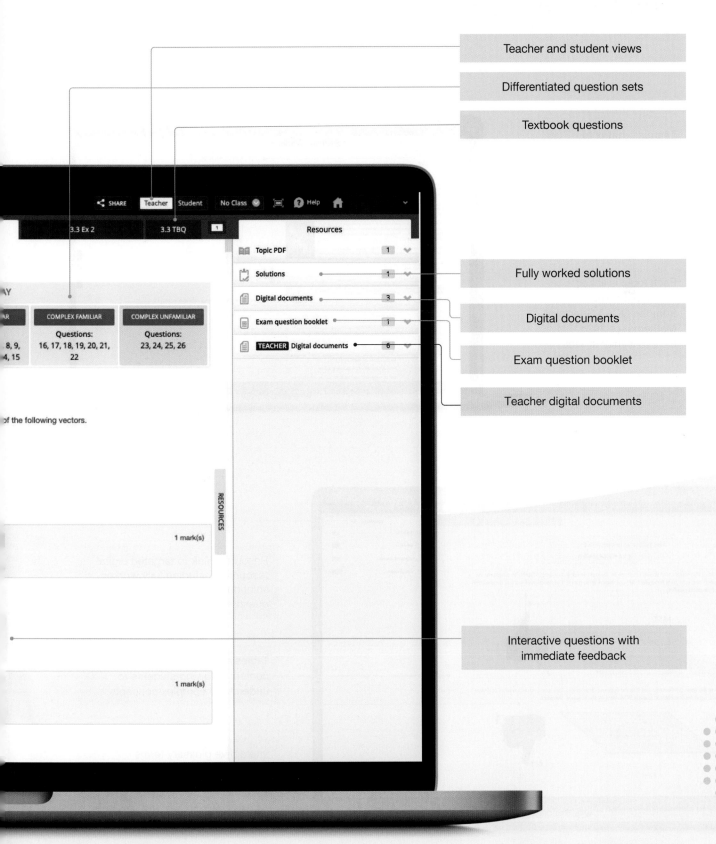

Teacher and student views

Differentiated question sets

Textbook questions

Fully worked solutions

Digital documents

Exam question booklet

Teacher digital documents

Interactive questions with immediate feedback

Online, these new editions are the complete package

Trusted Jacaranda theory, plus tools to support teaching and make learning more engaging, personalised and visible.

Learning matrix to monitor student's confidence level throughout topics

Each topic is linked to content points from the QCAA Specialist Mathematics 2025 General senior syllabus

Resources link to targeted digital resources including fully worked solutions, interactivities and exam question booklets.

Tables and images break down content, allowing students to understand complex concepts.

Interactive glossary terms help develop and support mathematical literacy.

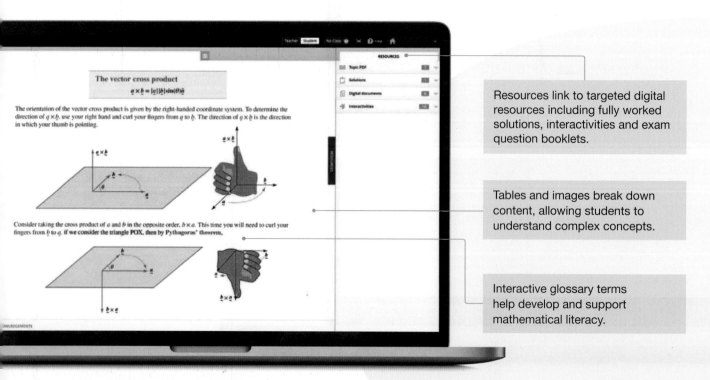

Pink highlight boxes summarise key information.

The diagram shows the line L_1 passing through the points A and B and the line L_2 passing through the points A and D, with the angle BAD being a right angle.

Taking AC as 1 unit, the sides in the diagram are labelled with their lengths. The side CB has length m_1. Because lengths must be positive, the side CD is labelled as $-m_2$, since $m_2 < 0$.

From the triangle ABC in the diagram, $\tan\theta = \frac{m_1}{1} = m_1$, and from the triangle ACD in the diagram, $\tan\theta = \frac{1}{-m_2}$.

Hence,

$$m_1 = \frac{1}{-m_2}$$
$$\therefore m_1 m_2 = -1$$

Gradients of perpendicular lines

$$m_1 m_2 = -1 \text{ or } m_2 = -\frac{1}{m_1}$$

- If two lines with gradients m_1 and m_2 are perpendicular, then the product of their gradients is -1. One gradient is the negative reciprocal of the other.
- It follows that if $m_1 m_2 = -1$, then the two lines are perpendicular. This can be used to test for perpendicularity.

Worked examples, supported by teacher-led videos, break down the process of answering questions using a think/write format, including calculator support.

Each unit has practice assessments (Problem solving and modelling task or examination) with sample responses to help build skills

- Online and offline question sets contain practice questions and exam style questions with exemplary responses and marking guides.
- Every question has immediate, corrective feedback to help students to overcome misconceptions as they occur and to study independently — in class and at home.

Topic and Unit reviews

Topic and Unit reviews include online summaries and topic-level and unit review exercises that cover multiple concepts.

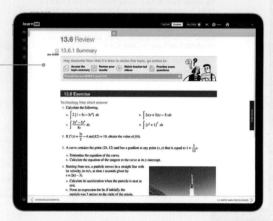

Get exam-ready!

Topic-level exam questions are structured just like the exams.

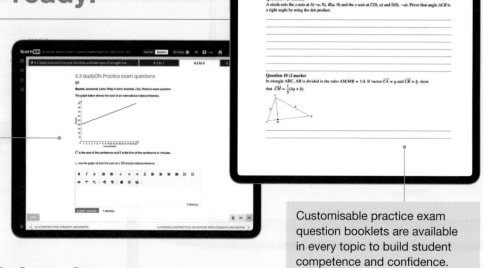

Customisable practice exam question booklets are available in every topic to build student competence and confidence.

Expert advice for problem solving and modelling tasks

Step by step guide on how to complete problem solving and modelling tasks with tips for teachers on how to create good assessments

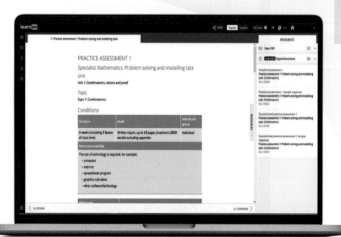

Teaching with learnON

Enhanced teacher support resources, including:

- work programs and curriculum grids
- teaching advice and additional activities
- quarantined topic tests (with solutions)
- Units reviews
- Quarantined PSMTs and examinations
- Custom exam-builder with question differentiation (SF/CF/CU) and tech-active/Tech-free question filters

Customise and assign

A testmaker enables you to create custom tests from the complete bank of thousands of questions.

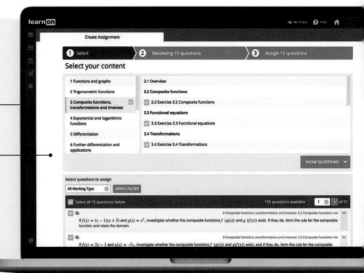

Reports and results

Data analytics and instant reports provide data-driven insights into performance across the entire course.

Show students (and their parents or carers) their own assessment data in fine detail. You can filter their results to identify areas of strength and weakness.

Acknowledgements

The authors and publisher would like to thank the following copyright holders, organisations and individuals for their assistance and for permission to reproduce copyright material in this book.

The full list of acknowledgements can be found here:

www.jacaranda.com.au/acknowledgements/#2024

Every effort has been made to trace the ownership of copyright material. Information that will enable the publisher to rectify any error or omission in subsequent reprints will be welcome. In such cases, please contact the Permissions Section of John Wiley & Sons Australia, Ltd.

UNIT

1 Combinatorics, proof, vectors and matrices

1 Combinatorics

LESSON SEQUENCE

Fully worked solutions for this chapter are available online.

EXAM PREPARATION

Access exam-style questions in every lesson, available online.

on Resources

Solutions	Solutions — Chapter 1 (sol-0393)
Exam questions	Exam question booklet — Chapter 1 (eqb-0279)
Digital documents	Learning matrix — Chapter 1 (doc-41920)
	Chapter summary — Chapter 1 (doc-41921)

LESSON
1.1 Overview

1.1.1 Introduction

Ever since you were a small child, even before you started school, you have had the experience of counting. These simple counting techniques are built upon to develop very sophisticated ways of counting and arranging, in a field of mathematics known as combinatorics. Permutations and combinations allow us to calculate the number of ways objects belonging to a finite set can be arranged. This is particularly useful in computer science.

Do you know how many three digit numbers can be formed from the digits $0 - 9$? How many of these are divisible by 9? How about if there can be no repetition of a digit, or if the first digit cannot be zero?

The techniques provided in combinatorics are useful in various areas of mathematics, such as algebra and probability.

In this chapter you will apply combinatorics to determine, for example, the number of ways a team of five players can be chosen from a group of ten. Consider its usefulness in developing rosters for staff or flow charts for projects. Combinatorics spans industries as varied as gambling, internet information transfer and security, communication networks, computer chip architecture, logistics and DNA modelling. It has applications in any field where different choices mean different efficiencies.

1.1.2 Syllabus links

Lesson	Lesson title	Syllabus links
1.2	**Counting techniques**	○ Use the inclusion-exclusion principle formulas to determine the number of elements in the union of two and the union of three sets. • $\lvert A \cup B \rvert = \lvert A \rvert + \lvert B \rvert - \lvert A \cap B \rvert$ • $\lvert A \cup B \cup C \rvert = \lvert A \rvert + \lvert B \rvert + \lvert C \rvert - \lvert A \cap B \rvert - \lvert A \cap C \rvert - \lvert B \cap C \rvert + \lvert A \cap B \cap C \rvert$
		○ Use the multiplication principle.
		○ Use the addition principle.
1.3	**Factorials and permutations**	○ Define and use permutations.
		○ Use factorial notation.
		○ Use the notation $^{n}\mathrm{P}_r$ to represent the number of ways of selecting r objects from n distinct objects where order is important. • $^{n}\mathrm{P}_r = \dfrac{n!}{(n-r)!} = n \times (n-1) \times (n-2) \times \cdots \times (n-r+1)$
		○ Solve problems that involve permutations.
1.4	**Permutations with restrictions**	○ Solve problems that involve permutations with restrictions including repeated objects, specific objects grouped together and selection from multiple groups.
1.5	**Combinations**	○ Define and use combinations
		○ Use the notation $\dbinom{n}{r}$ and $^{n}\mathrm{C}_r$ to represent the number of ways of selecting r objects from n distinct objects where order is not important. • $^{n}\mathrm{C}_r = \dfrac{n!}{r!\,(n-r)!}$
		○ Solve problems that involve combinations.
1.6	**Applications of permutations and combinations**	○ Solve problems that involve combinations with restrictions including specific objects grouped together and selection from multiple groups.
		○ Model and solve problems that involve permutations and combinations including probability problems, with and without technology.

Source: Specialist Mathematics Senior Syllabus 2024 © State of Queensland (QCAA) 2024; licensed under CC BY 4.0

LESSON
1.2 Counting techniques

SYLLABUS LINKS

- Use the inclusion-exclusion principle formulas to determine the number of elements in the union of two and the union of three sets.
  - $|A \cup B| = |A| + |B| - |A \cap B|$
  - $|A \cup B \cup C| = |A| + |B| + |C| - |A \cap B| - |A \cap C| - |B \cap C| + |A \cap B \cap C|$
- Use the multiplication principle.
- Use the addition principle.

Source: Specialist Mathematics Senior Syllabus 2024 © State of Queensland (QCAA) 2024; licensed under CC BY 4.0

1.2.1 Review of set notation

A **set**, S, is a collection of objects. The objects in a set are referred to as the **elements** of the set.

A set can be written in a variety of ways. Curly brackets are generically used to denote a set. Consider the following. Let the sample space be the set of natural numbers between 1 and 20, that is $\{1, 2, 3, 4, 5, 6, 7, 8, 9, 10, 11, 12, 13, 14, 15, 16, 17, 18, 19, 20\}$.

Let the set S be the set of even numbers between 1 and 20 inclusive.

S can be:
- written as a list: $S = \{2, 4, 6, 8, 10, 2, 14, 16, 18, 20\}$ (in any order)
- written as a rule: $S = \{n : n = 2r \text{ for } 1 \leq r \leq 10\}$
- shown in a Venn diagram.

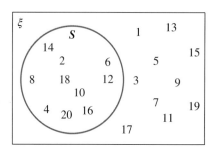

The complement of S, written as S', is the set of all things NOT in S. In this example, $S' = \{1, 3, 5, 7, 9, 11, 13, 15, 17, 19\}$.

The complete set of objects being considered is called the **universal set**, ξ. It is represented by the rectangle in the Venn diagram, and is abbreviated with Greek letter ξ (pronounced 'ksi'). In this example, $\xi = \{1, 2, 3, 4, 5, 6, 7, 8, 9, 10, 11, 12, 13, 14, 15, 16, 17, 18, 19, 20\}$.

Now consider a second set, T, which is the set of numbers that are multiples of 3 between 1 and 20; that is, $T = \{3, 6, 9, 12, 15, 18\}$. The sets S and T can be combined in various ways.

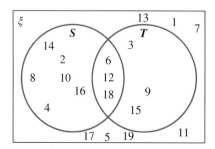

The **union** of S and T is all the elements in either S or T or both. It is shown as follows.

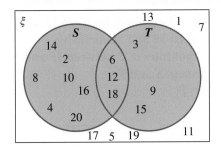

$$S \cup T = \{2, 3, 4, 6, 8, 9, 10, 12, 14, 15, 16, 18, 20\}$$

The **intersection** of S and T is all the elements that are in both S and T.

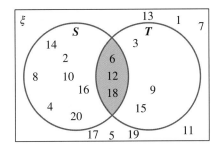

$$S \cap T = \{6, 12, 18\}$$

The size of a set S, which is the number of elements in the set, is noted $|S|$ or $n(S)$.

1.2.2 The inclusion – exclusion principle

Using the previous example with the two sets $S = \{2, 4, 6, 8, 10, 12, 14, 16, 18, 20\}$ and $T = \{3, 6, 9, 12, 15, 18\}$, and their union $S \cup T = \{2, 3, 4, 6, 8, 9, 10, 12, 14, 15, 16, 18, 20\}$ and their intersection $S \cap T = \{6, 12, 18\}$, we can determine the size of each set by counting the number of elements in each set: $|S| = 10$, $|T| = 6$, $|S \cup T| = 13$ and $|S \cap T| = 3$.

We can observe that the size of the union of two sets S and T is equal to the sum of their individual sizes, minus the size of their intersection, the number of elements in both S and T, as these have already been counted in sets S and T.

$$13 = 10 + 6 - 3$$
$$|S \cup T| = |S| + |T| - |S \cap T|$$

This is known as the **inclusion–exclusion principle**.

> ### Inclusion – exclusion principle for two sets
>
> **The size of the union of two sets is equal to the sum of their individual sizes, minus the size of their intersection**
>
> $$|A \cup B| = |A| + |B| - |A \cap B|$$

Let the sample space be the set of natural numbers between 15 and 30 inclusive.
Let S be the set of even numbers between 15 and 30 inclusive.
Let T be the set of numbers that are multiples of 3 between 15 and 30 inclusive.
a. Construct a Venn diagram to represent S and T.
b. Show that $|S \cup T| = |S| + |T| - |S \cap T|$

THINK

WRITE

a. 1. S is the set of natural numbers between 15 and 30 divisible by 2 so: $S = \{16, 18, 20, 22, 24, 26, 28, 30\}$.

a.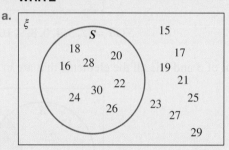

2. T is the set of natural numbers between 10 and 30 divisible by 3 so: $T = \{15, 18, 21, 24, 27, 30\}$.
$S \cap T$ are the numbers which occur in both S and T, so: $S \cap T = \{18, 24, 30\}$.

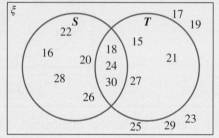

b. 1. $|S|$ is the number of elements in set S and $|T|$ is the number of elements in set T.
$|S \cap T|$ is the number of elements in the intersection of both circles.

b. $|S| = 8$
$|T| = 6$
$|S \cap T| = 3$

2. Evaluate $|S \cup T|$ and $|S| + |T| - |S \cap T|$.
$|S \cup T|$ is the number of elements in sets S and T. Count the number of elements in each portion of each circle.

$|S| + |T| - |S \cap T| = 8 + 6 - 3 = 11$
$|S \cup T| = 11$
$\therefore |S \cup T| = |S| + |T| - |S \cap T|$

3. Conclude.

The inclusion – exclusion principles can be generalised to more than two sets.

> ### Inclusion – exclusion principle for three sets
>
> **If three sets are involved, the inclusion – exclusion principle becomes:**
> $$|A \cup B \cup C| = |A| + |B| + |C| - |A \cap B| - |A \cap C| - |B \cap C| + |A \cap B \cap C|$$

An example of the inclusion – exclusion principle with three sets is shown in Worked example 2.

WORKED EXAMPLE 2 The inclusion – exclusion principle for three sets

Let the sample space be the set of natural numbers between 15 and 30 inclusive.
Let R be the set of natural numbers between 15 and 30 inclusive that are divisible by 2.
Let S be the set of natural numbers between 15 and 30 inclusive that are divisible by 3.
Let T be the set of natural numbers between 15 and 30 inclusive that are divisible by 5.
a. Construct a Venn diagram to represent R, S and T.
b. Use this diagram to evaluate $|R \cup S \cup T|$.
c. Recall the inclusion – exclusion principle to compute $|R \cup S \cup T|$.

THINK

a. 1. R is the set of natural numbers between 15 and 30 divisible by 2, so:
$R = \{16, 18, 20, \ldots 30\}$

2. S is the set of natural numbers between 15 and 30 divisible by 3, so:
$S = \{15, 18, 21, 24, 27, 30\}$
$R \cap S$ are the numbers which occur in both S and R, so:
$R \cap S = \{18, 24, 30\}$

3. T is the set of natural numbers between 15 and 30 divisible by 5, so:
$T = \{15, 20, 25, 30\}$
$R \cap T = \{20, 30\}$
$S \cap T = \{15, 30\}$
$R \cap S \cap T = \{30\}$

WRITE

a.

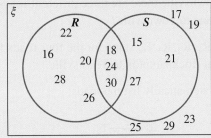

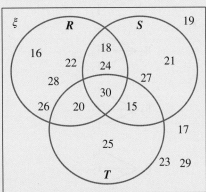

b. $|R \cup S \cup T|$ is the number of elements in sets R, S and T. Count the number of elements within each portion of each circle.

b. $|R \cup S \cup T| = 4 + 2 + 1 + 2 + 1 + 1 + 1$
$= 12$

c. 1. Recall the inclusion-exclusion principle formula.

c. $|S \cup T \cup R| = |S| + |T| + |R| - |S \cap T|$
$- |T \cap R| - |S \cap R| + |S \cap T \cap R|$

2. The number of elements in each set can be substituted into the formula.

$|S \cup T \cup R| = 6 + 4 + 8 - 2 - 2 - 3 + 1$
$= 12$
This is in agreement with $S \cup T \cup R =$
$\{15, 16, 18, 20, 21, 22, 14, 25, 26, 27, 28, 30\}$
as represented in the Venn diagram in part **a**.

1.2.3 Types of counting techniques

Counting techniques allow us to determine the number of ways an activity can occur. This in turn allows us to calculate the **probability** of an event. Recall from your earlier probability studies that the probability of event $A, P(A)$, can be determined by counting the number of elements in A and dividing by the total number in the sample space, ξ, according to the formula

$$P(A) = \frac{n(A)}{n(\xi)}.$$

Different types of counting techniques are employed depending on whether order is important. When order is important, this is called an **arrangement** or a **permutation**; when it is not important, it is called a **selection** or a **combination**. Permutations and combinations are defined more formally in sections 1.3 and 1.5.

1.2.4 The multiplication principle

To count the number of ways in which an activity can occur, first make a list. Let each outcome be represented by a letter and then systematically list all the possibilities.

Consider the following question:

In driving from Brisbane to Rockhampton I can take any one of four different roads *and* in driving from Rockhampton to Townsville there are three different roads I can take. How many different routes can I take in driving from Brisbane to Townsville?

To answer this, let R_1, R_2, R_3, R_4 stand for the four roads from Brisbane to Rockhampton and T_1, T_2, T_3 stand for the three roads from Rockhampton to Townsville.

Use the figure to systematically list the roads:

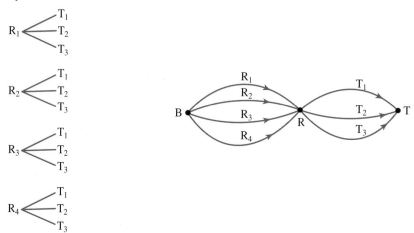

Hence, there are 12 different ways I can drive from Brisbane to Townsville.

In the above example it can be argued logically that if there are four ways of getting from Brisbane to Rockhampton and three ways of getting from Rockhampton to Townsville then there are 4×3 ways of getting from Brisbane to Townsville.

This idea is formalised in the **multiplication principle**.

The multiplication principle should be used when there are operations or events (say, A and B), where one event is followed by the other – that is, when order is important (the events are **dependent**).

The multiplication principle

If there are *n* ways of performing operation A and *m* ways of performing operation B, then there are *n* × *m* ways of performing A *and* B in the order AB.

Note: **In this case '*and*' means to multiply.**

A useful technique for solving problems based on the multiplication principle is to use boxes. In the example above we would write:

1st	2nd
4	3

The value in the '1st' column represents the number of ways the first operation — the trip from Brisbane to Rockhampton — can be performed.

The value in the '2nd' column stands for the number of ways the second operation — the trip from Rockhampton to Townsville — can be performed.

To apply the multiplication principle you multiply the numbers in the lower row of boxes.

WORKED EXAMPLE 3 Application of the multiplication principle

Two letters are to be chosen from the set of five letters. A, B, C, D and E, where order is important and the same letter is not repeated in a pair.
a. Recall how to list all the different ways that this may be done.
b. Use the multiplication principle to calculate the number of ways that this may be done.
c. Determine the probability the first letter will be a C.

THINK	WRITE
a. 1. Begin with A in first place and make a list of each of the possible pairs.	a. AB AC AD AE
2. Make a list of each of the possible pairs with B in the first position.	BA BC BD BE
3. Make a list of each of the possible pairs with C in the first position.	CA CB CD CE
4. Make a list of each of the possible pairs with D in the first position.	DA DB DC DE
5. Make a list of each of the possible pairs with E in the first position. *Note:* AB and BA need to be listed separately as order is important.	EA EB EC ED

b. The multiplication principle could have been used to determine the number of ordered pairs.

b.

1st	2nd
5	4

1. Rule up two boxes which represent the pair.
2. Write down the number of letters which may be selected for the first box. That is, in first place any of the five letters may be used.

3. Write down the number of letters which may be selected for the second box. That is, in second place, any of the four remaining letters may be used.
   Note: One less letter is used to avoid repetition.
4. Evaluate.
5. Write the answer.

$5 \times 4 = 20$ ways

There are 20 ways in which two letters may be selected from a group of five where order is important and the same letter is not repeated in a pair.

c. 1. Recall the probability formula. The total number in the set $n(\xi)$ was determined in part **b**.

c. $P(A) = \dfrac{n(A)}{n(\xi)}$

$n(\xi) = 20$

2. Let A be the event that the pair starts with a C. Draw a table showing the requirement imposed by the first letter to be C.

1	

3. Complete the table. Once the first letter has been completed, there are four choices for the second letter. Use the multiplication principle to determine the number of combinations starting with C.

1	4

There are $1 \times 4 = 4$ possible combinations beginning with C.
So, $n(A) = 4$.

4. Use the probability formula to answer the question.

$P(A) = \dfrac{n(A)}{n(\xi)}$

$= \dfrac{4}{20}$

$= \dfrac{1}{5}$

This is confirmed by examining the answer to part **a**.

WORKED EXAMPLE 4 Multiplication principle and probability

a. Use the multiplication principle to calculate how many ways an arrangement of five numbers can be chosen from $\{1, 2, 3, 4, 5, 6\}$.

b. Determine the probability of the number ending with 4.

THINK

WRITE

a. 1. Instead of listing all possibilities, draw five boxes to represent the five numbers chosen.
   Label each box on the top row as 1st, 2nd, 3rd, 4th and 5th.
   Note: The word arrangement implies order is important.

1st	2nd	3rd	4th	5th

2. Fill in each of the boxes showing the number of ways a number may be chosen.

   a. In the first box there are six choices for the first number.

   b. In the second box there are five choices for the second number as one number has already been used.

   c. In the third box there are four choices for the third number as two numbers have already been used.

   d. Continue this process until each of the five boxes is filled.

a.

1st	2nd	3rd	4th	5th
6	5	4	3	2

3. Use the multiplication principle as this is an *'and'* situation.

No. of ways $= 6 \times 5 \times 4 \times 3 \times 2$
$= 720$

4. Write the answer.

An arrangement of five numbers may be chosen 720 ways.

b. 1. Recall the probability formula. The total number of arrangements, $n(\xi)$, was determined in part **a**.

b. $P(A) = \dfrac{n(A)}{n(\xi)}$

$n(\xi) = 720$

2. Let A be the event that the number ends with 4. Draw a table showing the requirement imposed by the last digit to be 4.

1st	2nd	3rd	4th	5th
				1

3. Complete the table. Once the last number has been completed, there are five choices for the number in the first position, four choices for the next number. Continue this process until each of the five columns has been filled. Use the multiplication principle to determine the number of combinations ending with 4.

1st	2nd	3rd	4th	5th
5	4	3	2	1

There are $5 \times 4 \times 3 \times 2 \times 1 = 120$ possible combinations ending with 4.
So, $n(A) = 120$.

4. Use the probability formula to answer the question.

$P(A) = \dfrac{n(A)}{n(\xi)}$

$= \dfrac{120}{720}$

$= \dfrac{1}{6}$

This is confirmed by examining the answer to part **a**: there are 720 ways to arrange five numbers, all have the same probability, there are six digits possible for the last digit thus the probability the number ends with a specific digit is $\dfrac{1}{6}$.

1.2.5 The addition principle

Now consider a different situation, one in which the two operations do not occur one after the other.

I am going to travel from Brisbane to either Sydney *or* Adelaide. There are four ways of travelling from Brisbane to Sydney and three ways of travelling from Brisbane to Adelaide. Travelling to Brisbane and travelling to Sydney are mutually exclusive here.

How many different ways can I travel to either Sydney *or* Adelaide?

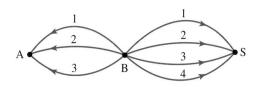

It can be seen from the figure that there are $4 + 3 = 7$ ways of completing the journey. This idea is summarised in the **addition principle**.

The addition principle should be used when trying to count the total ways/probabilities of doing several actions that cannot be done at the same time (mutually exclusive).

The addition principle

If there are *n* ways of performing operation A and *m* ways of performing operation B, then there are *n* + *m* ways of performing A *or* B.

Note: **In this case 'or' means to add.**

WORKED EXAMPLE 5 The multiplication and the addition principles

One or two letters are to be chosen from the set of six letters A, B, C, D, E, F. Assuming order is important, use the multiplication principle and the addition principle to calculate:
a. the number of ways to choose two letters.
b. the number of ways to choose one or two letters.

THINK	WRITE
a. 1. Determine the number of ways of choosing one letter.	**a.** Number of ways of choosing one letter = six.
2. Rule up two boxes for the first and second letters.	
3. Determine the number of ways of choosing two letters from six. In the first box there are six choices for the first letter. In the second box there are five choices for the second letter as one letter has already been used.	
4. Use the multiplication principle (as this is an '*and*' situation) to evaluate the number of ways of choosing two letters from six.	Number of ways of choosing two letters $= 6 \times 5$ $= 30$
5. Write the answer.	There are 30 ways of choosing two letters and there are six ways of choosing one letter.
b. 1. Determine the number of ways of choosing one or two letters from six letters. Use the addition principle as this is an '*or*' situation.	**b.** The number of ways of choosing one or two letters is $6 + 30 = 36$.
2. Write the answer.	There are 36 ways of choosing one or two letters from six.

For part a, the boxes table:

1st	2nd
6	5

The multiplication and addition principles can be used to count the number of elements in the union of two or three sets in the same way as for one set. Remember, the multiplication principle is used with '*and*' situations, and the addition principle is used with '*or*' situations.

WORKED EXAMPLE 6 Application of the multiplication and the addition principles

Oscar's cafe offers a choice of three starters, nine main courses and four desserts.
a. How many choices of three-course meals (starter, main, dessert) are available?
b. How many choices of starter and main course meals are offered?
c. How many choices of meals comprising a main course and dessert are offered?
d. How many choices of two- or three-course meals are available (assuming that a main course is always ordered)?

e. If one of the starter options is chicken wings and one of the mains is grilled fish, determine the probability of choosing chicken wings and grilled fish in a three-course meal.

THINK	WRITE
a. 1. Consider each course as separate sets and rule up three boxes to represent each course — starter, main, dessert. Label each box on the top row as S, M and D.	**a.** S M D 3 9 4
2. Determine the number of ways of choosing each meal: starter = three, main = nine, dessert = four.	
3. Use the multiplication principle (as this is an '*and*' situation) to evaluate the number of choices of three-course meals.	Number of choices $= 3 \times 9 \times 4$ $\qquad\qquad\qquad = 108$
4. Write the answer.	There are 108 choices of three-course meals.
b. 1. Rule up two boxes to represent each course — starter, main. Label each box on the top row as S and M.	**b.** S M 3 9
2. Determine the number of ways of choosing each meal: starter = three, main = nine.	
3. Use the multiplication principle (as this is an '*and*' situation) to evaluate the number of choices of starter and main courses.	Number of choices $= 3 \times 9$ $\qquad\qquad\qquad = 27$
4. Write the answer.	There are 27 choices of starter and main course.
c. 1. Rule up two boxes to represent each course — main and dessert. Label each box on the top row as M and D.	**c.** M D 9 4
2. Determine the number of ways of choosing each meal: main = nine, dessert = four.	

3. Use the multiplication principle (as this is an *'and'* situation) to evaluate the number of choices of main course and dessert.

Number of choices $= 9 \times 4$
$= 36$

4. Write the answer.

There are 36 choices of main course and dessert.

d. 1. Determine the number of ways of choosing two- or three-course meals, assuming that a main course is always ordered.

Use the addition principle as this is an *'or'* situation.

d. The number of ways of choosing two- or three-course meals, assuming that a main course is always ordered, is:
$108 + 27 + 36 = 171$

2. Write the answer.

There are 171 ways of choosing two- or three-course meals, assuming that a main course is always ordered.

e. 1. Recall the probability formula.
The total number in the set $n(\xi)$ was determined in part **a.**

e. $P(A) = \dfrac{n(A)}{n(\xi)}$

$n(\xi) = 108$

2. Let A be the event that the meal contains chicken wings and grilled fish.
Draw a table showing the requirement imposed that the starter and main be these two dishes.

S	M	D
1	1	

3. Complete the table — there are four dessert options. Use the multiplication principle to determine the number of combinations of the meal.

S	M	D
1	1	4

There are $1 \times 1 \times 4 = 4$ possible combinations for the meal.
So, $n(A) = 4$.

4. Use the probability formula to answer the question.

$P(A) = \dfrac{n(A)}{n(\xi)}$

$= \dfrac{4}{108}$

$= \dfrac{1}{27}$

The probability of choosing chicken wings and grilled fish as part of the three-course meal is $\dfrac{1}{27}$.

Simple familiar	Complex familiar	Complex unfamiliar
1, 2, 3, 4, 5, 6, 7, 8, 9, 10, 11, 12, 13, 14, 15, 16, 17, 18	19, 20, 21	22

Simple familiar

1. **WE1** Let the sample space be the set of natural numbers between 20 and 40 inclusive.
   Let M be the set of odd numbers between 20 and 40 inclusive.
   Let N be the set of numbers that are multiples of 5 between 20 and 40 inclusive.

   a. Construct a Venn diagram to represent M and N.
   b. Show that $|M \cup N| = |M| + |N| - |M \cap N|$

2. **WE2** Let the sample space be the set of natural numbers between 30 and 45 inclusive.
   Let R be the set of natural numbers between 30 and 45 inclusive that are divisible by 2.
   Let S be the set of natural numbers between 30 and 45 inclusive that are divisible by 3.
   Let T be the set of natural numbers between 30 and 45 inclusive that are divisible by 5.

   a. Construct a Venn diagram to represent R, S and T.
   b. Use this diagram to evaluate $|R \cup S \cup T|$.
   c. Recall the inclusion–exclusion principle to compute $|R \cup S \cup T|$.

3. Recall the inclusion–exclusion principle to calculate the number of cards in a deck of 52 that are either red or even or a 4.

4. Student Services has the following data on Year 11 students and their sport commitments:
   - 18 play no sport.
   - 16 play netball (and possibly other sports).
   - 24 play football.
   - 20 are involved in a gym program.
   - seven play netball and football.
   - six play netball and are in the gym program.
   - 15 play football and are involved in the gym program.
   - five students do all three activities.
     How many students are there in Year 11?

5. **WE3** Two letters are to be chosen from A, B and C, where order is important.

   a. Recall how to list all the different ways that this may be done.
   b. Use multiplication principle to calculate the number of ways that this may be done.
   c. Determine the probability the last letter will be a B.

6. a. List all the different arrangements possible for a group of two colours to be chosen from B (blue), G (green), Y (yellow) and R (red).
   b. Use the multiplication principle to confirm the answer to part a.

7. a. List all the different arrangements possible for a group of three letters to be chosen from A, B and C.
    b. Use the multiplication principle to confirm the answer to part a.

8. a. **WE4** Use the multiplication principle to calculate how many ways can an arrangement of two letters be chosen from A, B, C, D, E, F and G?
    b. In how many ways can an arrangement of three letters be chosen from seven different letters?
    c. In how many ways can an arrangement of four letters be chosen from seven different letters?
    d. How many different arrangements of five letters can be made from seven letters?
    e. Determine the probability of the letters starting with an E.

9. a. A teddy bear's wardrobe consists of three different hats, four different shirts and two different trousers. How many different outfits can the teddy bear wear?
    b. A surfboard is to have one colour on its top and a different colour on its bottom. The three possible colours are red, blue and green. In how many different ways can the surfboard be coloured?
    c. A new phone comes with a choice of three cases, two different sized screens and two different storage capacities. With these choices, determine how many different arrangements are possible.
    d. Messages can be sent by placing three different coloured flags in order on a pole. If the flags come in four colours, determine how many different messages can be sent.

10. a. **WE5** One or two letters are to be chosen in order from the letters A, B, C, D, E, F and G. Use the multiplication principle and the addition principle to calculate the number of ways can this be done.
    b. Two or three letters are to be chosen in order from the letters A, B, C, D, E, F and G. In how many ways can this be done?

11. Manish is in a race with seven other runners. If we are concerned only with the first, second and third placings, in how many ways can Manish finish first or second or third?

12. **WE6** Hani and Mary's restaurant offers its patrons a choice of four entrees, ten main courses and five desserts.

    a. How many choices of three-course meals (entree, main, dessert) are available?
    b. How many choices of entree and main course are offered?
    c. How many choices of meals comprising a main course and dessert are offered?
    d. How many choices of two- or three-course meals are available (assuming that a main course is always ordered)?
    e. Determine the probability of choosing vegetable soup for entree and roast for main in a three-course meal.

13. Jake is able to choose his work outfits from the following items of clothing: three jackets, seven shirts, six ties, five pairs of trousers, seven pairs of socks and three pairs of shoes.

    a. Calculate how many different outfits are possible if he wears one of each of the above items. (He wears matching socks and matching shoes.)
    b. If Jake has the option of wearing a jacket or not, but he must wear one of each of the above items, determine how many different outfits are possible. Justify your answer.

14. **MC** There are 12 people on the committee at the local football club. In how many ways can a president and a secretary be chosen from this committee?

    A. 2   B. 23   C. 132   D. 144

15. **MC** A TV station runs a cricket competition called *Classic Catches*. Six catches, A to F, are chosen and viewers are asked to rank them in the same order as the judges. The number of ways in which the six catches can be ranked:

    A. 1   **B.** 6   **C.** 30   **D.** 720

16. How many different four-digit numbers can be made from the numbers 1, 3, 5 and 7 if the numbers can be repeated (that is 3355 and 7777 are valid)?

17. How many four-digit numbers can be made from the numbers 1, 3, 5, 7, 9 and 0 if the numbers can be repeated? (Remember — a number cannot start with 0.)

18. Jasmin has a phone that has a four-digit security code. She remembers that the first number in the code was 9 and that the others were 3, 4 and 7 but forgets the order of the last three digits. How many different trials must she make to be sure of unlocking the phone?

Complex familiar

19. The local soccer team sells 'doubles' at each of their games to raise money. A 'double' is a card with two digits on it representing the score at full time. The card with the actual full time score on it wins a prize. If the digits on the cards run from 00 to 99, how many different tickets are there?

20. Julia has a banking app that has two four-digit codes. She remembers that she used the digits 1, 3, 5 and 7 on the first code and 2, 4, 6 and 8 on the second code, but cannot remember the order. What is the maximum number of trials she would need to make before she has opened both codes? (Assume that she can try an unlimited number of times and once the first code is correct, she can try the second code.)

21. A combination lock has three digits each from 0 to 9.

    a. How many combinations are possible?
    The lock mechanism becomes loose and will open if the digits are within one either side of the correct digit. For example if the true combination is 382 then the lock will open on 271, 272, 371, 493 and so on.
    b. How many combinations would unlock the safe?
    c. List the possible combinations that would open the lock if the true combination is 382.

Complex unfamiliar

22. Determine how many integers between 1 and 315 included are divisible by 3 or by 5 or by 7.

Fully worked solutions for this chapter are available online.

LESSON
1.3 Factorials and permutations

SYLLABUS LINKS

- Define and use permutations.
- Use factorial notation.
- Use the notation ^nP_r to represent the number of ways of selecting r objects from n distinct objects where order is important.
  - $^nP_r = \dfrac{n!}{(n-r)!} = n \times (n-1) \times (n-2) \times \cdots \times (n-r+1)$
- Solve problems that involve permutations.

Source: Specialist Mathematics Senior Syllabus 2024 © State of Queensland (QCAA) 2024; licensed under CC BY 4.0

1.3.1 Factorials

The Physical Education department is to display five new trophies along a shelf in the school foyer and wishes to know in how many ways this can be done.

Using the multiplication principle from the previous section, the display may be done in the following way:

Position 1	Position 2	Position 3	Position 4	Position 5
5	4	3	2	1

That is, there are $5 \times 4 \times 3 \times 2 \times 1 = 120$ ways.

Depending on the number of items we have, the notation of the working out of this method could become quite time consuming.

In general when we need to multiply each of the integers from a particular number, n, down to 1, we write $n!$, which is read as n **factorial**.

Hence:

$$\begin{aligned}
5! &= 5 \times 4 \times 3 \times 2 \times 1 \\
&= 120 \\
8! &= 8 \times 7 \times 6 \times 5 \times 4 \times 3 \times 2 \times 1 \\
&= 40\,320 \\
n! &= n \times (n-1) \times (n-2) \times (n-3) \times \ldots \times 3 \times 2 \times 1
\end{aligned}$$

Factorials

The number of ways n distinct objects may be arranged is $n!$ (n factorial) where:

$$n! = n \times (n-1) \times (n-2) \times (n-3) \times \ldots \times 3 \times 2 \times 1$$

That is, $n!$ is the product of each of the integers from n down to 1.

A special case of the factorial function is: $0! = 1$.

Evaluate the following factorials.

a. $7!$ b. $13!$ c. $\dfrac{8!}{5!}$ d. $\dfrac{(n-1)!}{(n-3)!}$

THINK

a. 1. Write 7! in its expanded form and evaluate.

   2. Verify the answer obtained using the factorial function on a calculator.

b. 1. Write 13! in its expanded form and evaluate.

   2. Verify the answer obtained using the factorial function on a calculator.

c. 1. Write each factorial term in its expanded form.
   2. Cancel down like terms.

   3. Evaluate.
   4. Verify the answer obtained using the factorial function on a calculator.

d. 1. Write each factorial term in its expanded form.

   2. Cancel like terms.

WRITE

a. $7! = 7 \times 6 \times 5 \times 4 \times 3 \times 2 \times 1$
$ = 5040$

7!
5040

b. $13! = 13 \times 12 \times 11 \times 10 \times 9 \times 8 \times 7 \times 6 \times 5 \times 4 \times 3 \times 2 \times 1$
$ = 6\,227\,020\,800$

13!
6227020800

c. $\dfrac{8!}{5!} = \dfrac{8 \times 7 \times 6 \times \cancel{5} \times \cancel{4} \times \cancel{3} \times \cancel{2} \times \cancel{1}}{\cancel{5} \times \cancel{4} \times \cancel{3} \times \cancel{2} \times \cancel{1}}$
$\phantom{\dfrac{8!}{5!}} = 8 \times 7 \times 6$
$\phantom{\dfrac{8!}{5!}} = 336$

OR

$\dfrac{8!}{5!} = \dfrac{8 \times 7 \times 6 \times \cancel{5!}}{\cancel{5!}}$
$\phantom{\dfrac{8!}{5!}} = 336$

$\frac{8!}{5!}$
336

d. $\dfrac{(n-1)!}{(n-3)!} = \dfrac{(n-1)(n-2)\cancel{(n-3)}\cancel{(n-4)} \times \cdots \times \cancel{3} \times \cancel{2} \times \cancel{1}}{\cancel{(n-3)}\cancel{(n-4)} \times \cdots \times \cancel{3} \times \cancel{2} \times \cancel{1}}$

$\phantom{\dfrac{(n-1)!}{(n-3)!}} = (n-1)(n-2)$

OR

$\dfrac{(n-1)!}{(n-3)!} = \dfrac{(n-1) \times (n-2) \times \cancel{(n-3)!}}{\cancel{(n-3)!}}$
$\phantom{\dfrac{(n-1)!}{(n-3)!}} = (n-1)(n-2)$

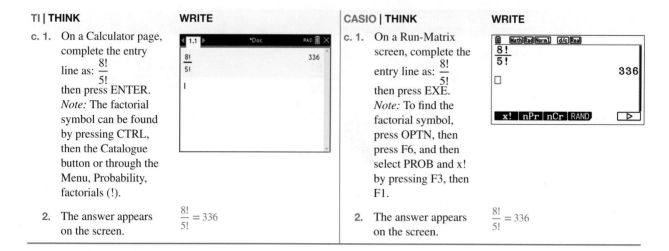

TI | THINK	WRITE	CASIO | THINK	WRITE
c. 1. On a Calculator page, complete the entry line as: $\dfrac{8!}{5!}$ then press ENTER. *Note:* The factorial symbol can be found by pressing CTRL, then the Catalogue button or through the Menu, Probability, factorials (!).		c. 1. On a Run-Matrix screen, complete the entry line as: $\dfrac{8!}{5!}$ then press EXE. *Note:* To find the factorial symbol, press OPTN, then press F6, and then select PROB and x! by pressing F3, then F1.	
2. The answer appears on the screen.	$\dfrac{8!}{5!} = 336$	2. The answer appears on the screen.	$\dfrac{8!}{5!} = 336$

1.3.2 Permutations

The term permutation is often used instead of the term arrangement, and in this section we begin by giving a formal definition of permutation.

Previously, we learned that if you want to create a three-letter word from seven letters (that is, CAT is a different word from ACT), this means that *order is important*. In these cases, the number of arrangement (or number of permutations) is:

1st	2nd	3rd
7	6	5

$$\text{The number of arrangements} = 7 \times 6 \times 5$$
$$= 210$$

This value may also be expressed in factorial form: $7 \times 6 \times 5 = \dfrac{7 \times 6 \times 5 \times 4!}{4!} = \dfrac{7!}{4!}$

Using more formal terminology we say that in choosing three things from seven things where order is important, the number of permutations is $^7P_3 = 7 \times 6 \times 5$. The letter P is used to remind us that we are finding permutations.

Ordered arrangements

The number of ways of choosing r things from n distinct things is given by the rule:

$$^nP_r = n \times (n-1) \times \ldots \times (n-r+1)$$
$$= \dfrac{n \times (n-1) \times \ldots \times (n-r+1)(n-r)!}{(n-r)!}$$
$$^nP_r = \dfrac{n!}{(n-r)!}$$

The definition of ^nP_r may be extended to the cases of ^nP_n and $^nP_0$.

^nP_n represents the number of ways of choosing n objects from n distinct things. Logically, we know that there are $n!$ ways of choosing n objects from n distinct things.

$$^nP_n = n \times (n-1) \times (n-2) \times \ldots \times (n-n+1)$$
$$= n \times (n-1) \times (n-2) \times \ldots \times 1$$
$$= n!$$

From the definition:

$$^{n}P_{n} = \frac{n!}{(n-n)!}$$
$$= \frac{n!}{0!}$$

Therefore, equating both sides, we obtain: $n! = \frac{n!}{0!}$.

This can occur only if $0! = 1$.

$$^{n}P_{0} = \frac{n!}{(n-0)!}$$
$$= \frac{n!}{n!}$$
$$= 1$$

Permutations special cases

The two special cases are:

$$^{n}P_{n} = n!$$

$$^{n}P_{0} = 1$$

WORKED EXAMPLE 8 Calculating the number of permutations

a. Calculate the number of permutations for $^{6}P_{4}$ by expressing it in expanded form.
b. Write $^{8}P_{3}$ as a quotient of factorials and hence evaluate.

THINK	WRITE
a. 1. Write down the first four terms beginning with 6.	a. $^{6}P_{4} = 6 \times 5 \times 4 \times 3$
2. Evaluate.	$= 360$
b. 1. Recall the rule for permutations.	b. $^{n}P_{r} = \frac{n!}{(n-r)!}$
2. Substitute the given values of n and r into the permutation formula.	$^{8}P_{3} = \frac{8!}{(8-3)!}$ $= \frac{8!}{5!}$
3. Use a calculator to evaluate 8! and 5!	$= \frac{40\,320}{120}$
4. Evaluate.	$= 336$

TI | THINK	WRITE	CASIO | THINK	WRITE
a. 1. On a Calculator page, press MENU, then select: 5: Probability 2: Permutations. Complete the entry line as: nPr (6, 4) then press ENTER.		a. 1. On a Run-Matrix screen, press OPTN, then F6. Select PROB by pressing F3, then select nPr by pressing F2. Complete the entry line as: 6P4 then press EXE.	
2. The answer appears on the screen.	$^6P_4 = 360$	2. The answer appears on the screen.	$^6P_4 = 360$

WORKED EXAMPLE 9 Solving problems using permutations

The netball club needs to appoint a president, secretary and treasurer. From the committee seven people have volunteered for these positions. Each of the seven nominees is happy to fill any one of the three positions.

a. Determine how many different ways these positions can be filled.

b. For three years, the same seven people volunteer for these positions. Determine the probability one of them is president three years in a row.

THINK

a. 1. The three roles are different, thus the order matters, thus use permutations here. Recall the rule for permutations.

2. Substitute the given values of n and r into the permutation formula.

3. Use a calculator to evaluate 7! and 4!.

4. Evaluate.

5. Write the answer.

b. 1. In three years of the president's position, each year this could be awarded to one of the seven people.

2. Let A be the event that the same person fills the position three years in a row.

WRITE

a. $^nP_r = \dfrac{n!}{(n-r)!}$

$^7P_3 = \dfrac{7!}{(7-3)!}$

$= \dfrac{7!}{4!}$

$= \dfrac{5040}{24}$

$= 210$

There are 210 different ways of filling the positions of president, secretary and treasurer.

b.

7	7	7

The total number of ways the president's position could be filled is $7 \times 7 \times 7$.

$n(\xi) = 7 \times 7 \times 7$

$= 343$

There are seven choices of the same person to fill the positions three years in a row. $n(A) = 7$.

3. Calculate the probability that the same person fills the president's position three years in a row.

$$P(A) = \frac{n(A)}{n(\xi)}$$

$$= \frac{7}{7 \times 7 \times 7}$$

$$= \frac{1}{49}$$

The probability that one of the seven volunteers, fills the president's position three years in a row is $\frac{1}{49}$.

Note that when the order does not matter (for instance if the three roles to fill are the same), then combinations are used (see lesson 1.5).

1.3.3 Permutations in a circle

Consider the following situation: you want to arrange four objects in a circle.
There are 4! ways for you to arrange those four objects in a circle. However, you can cyclically rearrange them in four ways that are not considered different arrangements, as it does not matter where the circle starts.

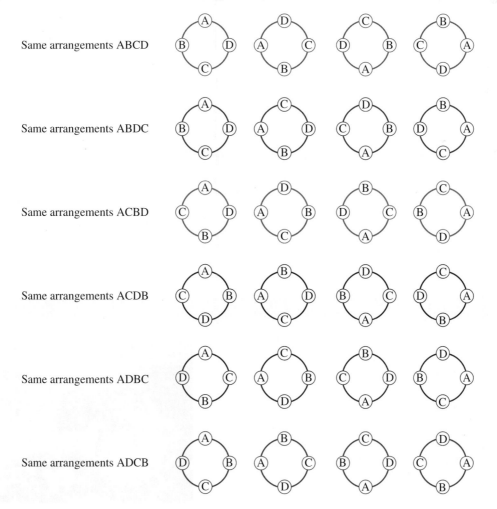

Thus the number of arrangements of four objects in a circle is $\frac{4!}{4} = (4-1)!$

This can be generalised to n objects.

Circular permutations

The number of arrangements of n objects in a circle is $\dfrac{n!}{n} = (n-1)!$

Now consider this problem: In how many different ways can seven people be seated, four at a time, on a bench?

By now you should quickly see the answer: $^7P_4 = 840$.

Let us change the problem slightly: In how many different ways can seven people be seated, four at a time, at a circular table?

The solution must recognise that when people are seated on a bench, each of the following represents a different arrangement:

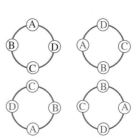

$$ABCD \quad BCDA \quad CDAB \quad DABC$$

However, when sitting in a circle, each represents the *same* arrangement. It is important to note that in a circle arrangement we do not consider positions on the circle as different — remember, it does not matter where the circle starts.

In each case B has A on the left and C on the right.

We conclude that the number $^7P_4$ gives four times the number of arrangements of seven people in a circle four at a time. Therefore, the number of arrangements is $\dfrac{^7P_4}{4} = 210$.

Permutations in a circle, n objects among r at a time

In general, the number of different ways n objects can be arranged, r at a time, in a circle is:

$$\frac{^nP_r}{r}$$

WORKED EXAMPLE 10 Calculating the number of permutations in a circle

a. **By recalling the appropriate formula, give an expression for the number of different arrangements if, from a group of eight people, five are to be seated at a round table.**

b. **Evaluate this expression.**

c. **Each table receives one lucky door prize and one lucky seat prize. Determine the probability of the same person at one table winning both.**

THINK	WRITE

THINK

a. 1. Write down the rule for the number of arrangements in a circle.

2. Substitute the given values of n and r into the formula.

3. Write the answer.

b. 1. Use a calculator to evaluate $^8P_5$.

2. Evaluate.

3. Write the answer.

c. 1. There are two prizes, and each prize can be won by any one of the five people.

2. Let A be the event that the same person wins both prizes.

3. Calculate the probability that the same person wins both.

WRITE

a. $\dfrac{^nP_r}{r}$

$= \dfrac{^8P_5}{5}$

The number of ways of seating five people from a group of eight people at a round table is given by the expression $\dfrac{^8P_5}{5}$.

b. $^8P_5 = \dfrac{6720}{5}$

$= 1344$

The number of ways of seating five from a group of eight people at a round table is 1344.

c.

5	5

The total number of ways the prizes could be won is 5×5
$n(\xi) = 5 \times 5$
$ = 25$

There are five choices of the same person to win both prizes.
$n(A) = 5$

$P(A) = \dfrac{n(A)}{n(\xi)}$

$= \dfrac{5}{5 \times 5}$

$= \dfrac{1}{5}$

The probability of the same person winning both prizes is $\dfrac{1}{5}$.

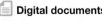 Resources

Exercise 1.3 Factorials and permutations

Simple familiar

1. **WE7a, b** Recall the definition of $n!$ and write each of the following in expanded form.
   a. $4!$
   b. $5!$
   c. $6!$
   d. $7!$

2. Evaluate the following factorials.
   a. $4!$
   b. $5!$
   c. $6!$
   d. $10!$
   e. $14!$
   f. $9!$
   g. $7!$
   h. $3!$

3. **WE7c** Evaluate the following factorials.
   a. $\dfrac{9!}{5!}$
   b. $\dfrac{10!}{4!}$
   c. $\dfrac{7!}{3!}$
   d. $\dfrac{6!}{0!}$

4. **WE7d** Evaluate the following factorials.
   a. $\dfrac{n!}{(n-5)!}$
   b. $\dfrac{(n+3)!}{(n+1)!}$
   c. $\dfrac{(n-3)!}{n!}$
   d. $\dfrac{(n-2)!}{(n+2)!}$

5. **WE8a** Calculate each of the following by expressing it in expanded form.
   a. $^8P_2$
   b. $^7P_5$
   c. $^8P_7$

6. **WE8b** Write each of the following as a quotient of factorials and hence evaluate.
   a. $^9P_6$
   b. $^5P_2$
   c. $^{18}P_5$

7. Use your calculator to determine the value of:
   a. $^{20}P_6$
   b. $^{800}P_2$
   c. $^{18}P_5$

8. The school musical needs a producer, director, musical director and script coach. Nine people have volunteered for any of these positions. In how many different ways can the positions be filled? (*Note:* One person cannot take on more than 1 position.)

DIRECTOR

9. There are 14 swimmers in a race. In how many different ways can the 1st, 2nd and 3rd positions be filled?

10. **WE10** a. By recalling the appropriate formula, give an expression for the number of different arrangements if, from a group of 15 people, four are to be seated at a round table.
    b. Evaluate this expression.
    c. Each table receives a lucky door prize, a lucky seat prize and a best-dressed prize. Determine the probability of the same person at one table winning all three prizes.

11. A round table seats six people. From a group of eight people, give an expression for, and hence calculate, the number of ways six people can be seated at the table.

12. At a dinner party for ten people all the guests were seated at a circular table. How many different arrangements were possible?

13. At one stage in the court of Camelot, King Arthur and 12 knights would sit at the round table. If each person could sit anywhere determine how many different arrangements were possible.

14. In how many ways can the letters of the word TODAY be arranged if they are used once only and taken:
    a. three at a time?
    b. four at a time?
    c. five at a time?
    Show your answers in the form $^{n}P_{r}$ and then evaluate.

Complex familiar

15. **WE9** A soccer club will appoint a president and a vice-president each year. Eight people have volunteered for either of the two positions.
    a. In how many different ways can these positions be filled?
    b. For two consecutive years the same eight people volunteered for these positions. What is the probability one of them is president for both years?

16. There are 26 players in an online game. How many different results for 1st, 2nd, 3rd and 4th can occur?

17. A rowing crew consists of four rowers who sit in a definite order. How many different crews are possible if five people try out for selection?

18. **MC** Which one of the following permutations cannot be calculated?

    A. $^{1000}P_{100}$ **B.** $^{1}P_{0}$ **C.** $^{8}P_{8}$ **D.** $^{4}P_{8}$

19. **MC** The result of 100! is greater than 94!.
    Which of the following gives the best comparison between these two numbers?

    A. 100! is 6 more than 94!
    B. 100! is 6 times bigger than 94!
    C. 100! is about 10 000 more than 94!
    D. 100! is $^{100}P_{6}$ times bigger than 94!

Complex unfamiliar

20. Twelve teachers (six men and six women) from the same school are attending a conference together.
    They notice that there are twelve empty seats in the front row and decide to sit there. Five of these teachers are maths teachers, four are humanities teacher and three are science teachers. One of the maths teacher says that there are approximately 20 more ways to sit if all the science teachers sit together than if all the men (or women) sit together, and that there are nearly 500 million ways for the group to sit without any specific order.
    Justify whether this teacher's estimations are reasonable.

Fully worked solutions for this chapter are available online.

LESSON
1.4 Permutations with restrictions

SYLLABUS LINKS

- Solve problems that involve permutations with restrictions including repeated objects, specific objects grouped together and selection from multiple groups.

Source: Specialist Mathematics Senior Syllabus 2024 © State of Queensland (QCAA) 2024; licensed under CC BY 4.0

1.4.1 Like objects

A five-letter word is to be made from three As and two Bs. How many different permutations or arrangements can be made?

If the five letters were all different, it would be easy to calculate the number of arrangements. It would be $5! = 120$. Perhaps you can see that when letters are repeated, the number of different arrangements will be less than 120. To analyse the situation let us imagine that we can distinguish one A from another. We will write A_1, A_2, A_3, B_1 and B_2 to represent the five letters.

As we list some of the possible arrangements we notice that some are actually the same, as shown in the table.

$A_1A_2B_1A_3B_2$	$A_1A_2B_2A_3B_1$	Each of these 12 arrangements
$A_1A_3B_1A_2B_2$	$A_1A_3B_2A_2B_1$	is the same — AABAB — if
$A_2A_1B_1A_3B_2$	$A_2A_1B_2A_3B_1$	$A_1 = A_2 = A_3$ and $B_1 = B_2$.
$A_2A_3B_1A_1B_2$	$A_2A_3B_2A_1B_1$	
$A_3A_1B_1A_2B_2$	$A_3A_1B_2A_2B_1$	
$A_3A_2B_1A_1B_2$	$A_3A_2B_2A_1B_1$	
$B_2A_1A_2B_1A_3$	$B_1A_1A_2B_2A_3$	Each of these 12 arrangements
$B_2A_1A_3B_1A_2$	$B_1A_1A_3B_2A_2$	is the same — BAABA — if
$B_2A_2A_1B_1A_3$	$B_1A_2A_1B_2A_3$	$A_1 = A_2 = A_3$ and $B_1 = B_2$.
$B_2A_2A_3B_1A_1$	$B_1A_2A_3B_2A_1$	
$B_2A_3A_1B_1A_2$	$B_1A_3A_1B_2A_2$	
$B_2A_3A_2B_1A_1$	$B_1A_3A_2B_2A_1$	

The number of repetitions is $3!$ for the As and $2!$ for the Bs. Thus, the number of different arrangements in choosing five letters from three As and two Bs is $\dfrac{5!}{3! \times 2!}$.

> **Permutations with restrictions**
>
> **The number of different ways of arranging n objects made up of groups of indistinguishable objects, n_1 in the first group, n_2 in the second group and so on, is:**
>
> $$\frac{n!}{n_1! \, n_2! \, n_3! \ldots n_r!}.$$

Note: If there are elements of the group which are not duplicated, then they can be considered as a group of one. It is not usual to divide by $1!$; it is more common to show only those groups which have duplications.

Determine how many different permutations of seven counters can be made from four black and three white counters.

THINK	WRITE
1. Write down the total number of counters.	There are seven counters in all; therefore, $n = 7$.
2. Write down the number of times any of the coloured counters are repeated.	There are three white counters; therefore, $n_1 = 3$. There are four black counters; therefore, $n_2 = 4$.
3. Write down the rule for arranging groups of like things.	$$\frac{n!}{n_1! \, n_2! \, n_3! \dots n_r!}$$
4. Substitute the values of n, n_1 and n_2 into the rule.	$$= \frac{7!}{3! \times 4!}$$
5. Expand each of the factorials.	$$= \frac{7 \times 6 \times 5 \times 4 \times 3 \times 2 \times 1}{3 \times 2 \times 1 \times 4 \times 3 \times 2 \times 1}$$
6. Simplify the fraction.	$$= \frac{7 \times 6 \times 5}{6}$$
7. Evaluate.	$= 35$
8. Write the answer.	35 different arrangements can be made from seven counters, of which three are white and four are black.

1.4.2 Restrictions

Sometimes restrictions are introduced so that a smaller number of objects from the original group need to be considered. This results in limiting the numbers of possible permutations.

A rowing crew of four rowers is to be selected, in order from the first seat to the fourth seat, from eight candidates. Determine how many different arrangements are possible if:
a. **there are no restrictions**
b. **Jason or Kris must row in the first seat**
c. **Jason must be in the crew, but he can row anywhere in the boat**
d. **Jason is not in the crew.**

THINK	WRITE
a. 1. Write down the permutation formula. *Note:* four rowers are to be selected from eight and the order is important.	a. $${}^nP_r = \frac{n!}{(n-r)!}$$

2. Substitute the given values of n and r into the permutation formula.

$$^8P_4 = \frac{8!}{(8-4)!}$$
$$= \frac{8!}{4!}$$

3. Expand the factorials or use a calculator to evaluate 8! and 4!.

$$= \frac{8 \times 7 \times 6 \times 5 \times \cancel{4} \times \cancel{3} \times \cancel{2} \times \cancel{1}}{\cancel{4} \times \cancel{3} \times \cancel{2} \times \cancel{1}}$$
$$= 8 \times 7 \times 6 \times 5$$

4. Evaluate.

$$= 1680$$

5. Write the answer.

There are 1680 ways of arranging four rowers from a group of eight.

b. 1. Apply the multiplication principle since two events will follow each other; that is, Jason will fill the first seat and the remaining three seats will be filled in $7 \times 6 \times 5$ ways or Kris will fill the first seat and the remaining three seats will be filled in $7 \times 6 \times 5$ ways.

b. No. of arrangements

= no. of ways of filling the first seat × no. of ways of filling the remaining three seats.

$$= 2 \times {}^nP_r$$

| J | 7 | 6 | 5 | $\rightarrow {}^7P_3$ |

or

| K | 7 | 6 | 5 | $\rightarrow {}^7P_3$ |

2. Substitute the values of n and r into the formula and evaluate.

$$= 2 \times {}^7P_3$$
$$= 2 \times 210$$
$$= 420$$

3. Write the answer.

There are 420 ways of arranging the four rowers if Jason or Kris must row in the first seat.

c. 1. Apply the addition principle, since Jason must be in either the first, second, third or fourth seat. The remaining three seats will be filled in $7 \times 6 \times 5$ ways each time.

c. No. of arrangements

= no. of arrangements with Jason in seat 1

+ No. of arrangements with Jason in seat 2

+ No. of arrangements with Jason in seat 3

+ No. of arrangements with Jason in seat 4.

| J | 7 | 6 | 5 | + | 7 | J | 6 | 5 |
| 7 | 6 | J | 5 | + | 7 | 6 | 5 | J |

Alternatively, select Jason (1 way of selecting Jason), select the other three seats in order ($^7P_3$ ways) and then slot Jason in any of the four spaces between or on the end of them (four ways).

No. of arrangements $= 1 \times {}^7P_3 \times 4$
$$= 4 \times 120$$
$$= 840$$

2. Substitute the values of n and r into the formula.

No. of arrangements
$$= 1 \times {}^7P_3 + 1 \times {}^7P_3 + 1 \times {}^7P_3 + 1 \times {}^7P_3$$
$$= 4 \times {}^7P_3$$
$$= 4 \times 210$$
$$= 840$$

3. Evaluate.

4. Write the answer.

There are 840 ways of arranging the four rowers if Jason must be in the crew of four.

d. As Jason is not in the crew, there are only seven candidates. Four rowers are to be chosen from seven and order is important.

d. $^{7}P_4 = \dfrac{7!}{(7-4)!}$

$= \dfrac{7!}{3!}$

$= \dfrac{7 \times 6 \times 5 \times 4 \times 3 \times 2 \times 1}{3 \times 2 \times 1}$

$= 7 \times 6 \times 5 \times 4$

$= 840$

There are 840 ways of arranging the crew when Jason is not included.

WORKED EXAMPLE 13 Permutations with objects grouped together

a. Calculate the number of permutations of the letters in the word COUNTER.
b. In how many of these do the letters C and N appear side by side?
c. In how many permutations do the letters C and N appear apart?
d. Determine the probability of the letters C and N appearing side by side.

THINK	WRITE
a. 1. Count the number of letters in the given word.	**a.** There are seven letters in the word COUNTER.
2. Determine the number of ways the seven letters may be arranged.	The seven letters may be arranged $7! = 5040$ ways.
3. Write the answer.	There are 5040 permutations of letters in the word COUNTER.
b. 1. Imagine the C and N are 'tied' together and are therefore considered as one unit. Determine the number of ways C and N may be arranged: CN and NC.	**b.** Let C and N represent one unit. The 'CN' unit may be arranged $2! = 2$ ways.
2. Determine the number of ways six things can be arranged. *Note:* There are now six letters: the 'CN' unit along with O, U, T, E and R.	Six units may be arranged $6! = 720$ ways.
3. Determine the number of permutations in which the letters C and N appear together.	The number of permutations $= 2 \times 6!$ $= 2 \times 720$ $= 1440$
4. Write the answer.	There are 1440 permutations in which the letters C and N appear together.
c. 1. Determine the total number of arrangements of the seven letters.	**c.** Total number of arrangements $= 7!$ $= 5040$
2. Write down the number of arrangements in which the letters C and N appear together, as obtained in **a**.	Arrangements with C and N together $= 1440$

CN	NC

3. Determine the difference between the values obtained in steps 1 and 2.

   Note: The number of arrangements in which C and N are apart is the total number of arrangements less the number of times they are together.

The number of arrangements $= 5040 - 1440$
$$= 3600$$

4. Write the answer.

The letters C and N appear apart 3600 times.

d. 1. Recall the probability formula.

d. $P(A) = \dfrac{n(\xi)}{n(A)}$

2. State the number of elements in the set ξ, that is $n(\xi)$.

From part **a**, there are 5040 permutations of letters in the word COUNTER.
$n(\xi) = 5040$

3. Determine the number of elements in the set A, that is the number of arrangements in which the letters C and N appear side by side.

From part **b**, there are 1440 permutations in which the letters C and N appear side by side.
$n(A) = 1440$

4. Calculate the answer.

$P(A) = \dfrac{n(A)}{n(\xi)}$
$$= \dfrac{1440}{5040}$$
$$= \dfrac{2}{7}$$

The probability of the letters C and N appearing side by side is $\dfrac{2}{7}$.

WORKED EXAMPLE 14 Solving problems involving permutations with restrictions

Consider the two words 'PARALLEL' and 'LINES'.
a. How many arrangements of the letters of the word LINES have the vowels grouped together?
b. How many arrangements of the letters of the word LINES have the vowels separated?
c. How many arrangements of the letters of the word PARALLEL are possible?
d. Determine the probability that in a randomly chosen arrangement of the word PARALLEL, the letters A are together.

THINK

a. 1. Group the required letters together.

2. Arrange the unit of letters together with the remaining letters.

3. Use the multiplication principle to allow for any internal rearrangements.

WRITE

a. There are two vowels in the word LINES. Treat these letters, I and E, as one unit.

Now there are four groups to arrange: (IE), L, N, S. These arrange in 4! ways.

The unit (IE) can internally rearrange in 2! ways. Hence, the total number of arrangements is:
$4! \times 2! = 24 \times 2$
$$= 48$$

b. 1. State the method of approach to the problem.

b. The number of arrangements with the vowels separated is equal to the total number of arrangements minus the number of arrangements with the vowels together.

2. State the total number of arrangements.

The five letters of the word LINES can be arranged in $5! = 120$ ways.

3. Calculate the answer.

From part **a**, there are 48 arrangements with the two vowels together. Therefore, there are $120 - 48 = 72$ arrangements in which the two vowels are separated.

c. 1. Count the letters, stating any identical letters.

The word PARALLEL contains eight letters of which there are two As and three Ls.

2. Recall the rule $\dfrac{n!}{n_1! \, n_2! \, ...}$ and state the number of distinct arrangements.

There are $\dfrac{8!}{2! \times 3!}$ arrangements of the word PARALLEL.

3. Calculate the answer.

$$\frac{8!}{2! \times 3!} = \frac{8 \times 7 \times 6 \times 5 \times \cancel{4}^2 \times \cancel{3!}}{\cancel{2} \times \cancel{3!}}$$
$$= 3360$$

There are 3360 arrangements.

d. 1. State the number of elements in the sample space.

d. There are 3360 total arrangements of the word PARALLEL, so $n(\xi) = 3360$ or $\dfrac{8!}{2! \times 3!}$.

2. Group the required letters together.

For the letters A to be together, treat these two letters as one unit. This creates seven groups: (AA), P, R, L, L, E, L, of which three are identical L s.

3. Calculate the number of elements in the event.

The seven groups arrange in $\dfrac{7!}{3!}$ ways. As the unit (AA) contains two identical letters, there are no distinct internal rearrangements of this unit that need to be taken into account. Hence, $\dfrac{7!}{3!}$ is the number of elements in the event.

4. Calculate the required probability.
Note: It helps to use factorial notation in the calculations.

The probability that the As are together

$$= \frac{\text{number of arrangements with the As together}}{\text{total number of arrangements}}$$

$$= \frac{7!}{3!} \div \frac{8!}{2! \times 3!}$$

$$= \frac{7!}{3!} \div \frac{2! \times 3!}{8 \times 7!}$$

$$= \frac{2}{8}$$

$$= \frac{1}{4}$$

Exercise 1.4 Permutations with restrictions

1.4 Exercise	1.4 Exam questions on

Simple familiar	Complex familiar	Complex unfamiliar
1, 2, 3, 4, 5, 6, 7, 8, 9, 10, 11, 12	13, 14, 15, 16	17, 18

These questions are even better in jacPLUS!
- Receive immediate feedback
- Access sample responses
- Track results and progress

Find all this and MORE in jacPLUS ▶

Simple familiar

1. **WE11** Determine how many different arrangements of five counters can be made using three red and two blue counters.

2. Determine how many different arrangements of nine counters can be made using four black, three red and two blue counters.

3. Recall the appropriate formula and calculate the number of different arrangements can be made using the six letters of the word NEWTON, assuming:

   a. the first N is distinct from the second N   **b.** there is no distinction between the two Ns.

4. How many different permutations can be made using the 11 letters of the word ABRACADABRA?

5. A collection of 12 books is to be arranged on a shelf. The books consist of three copies of *Great Expectations*, five copies of *Catcher in the Rye* and four copies of *Huntin', Fishin' and Shootin'*. How many different arrangements of these books are possible?

6. A shelf holding 24 cans of dog food is to be stacked using nine cans of Yummy and 15 cans of Ruff for Dogs. Determine how many different ways the shelf can be stocked.

7. **WE12** A cricket team of 11 players is to be selected, in batting order, from 15. Determine how many different arrangements are possible if:

   a. there are no restrictions
   b. Arjun must be in the team at number 1
   c. Arjun must be in the team but he can be anywhere from 1 to 11
   d. Arjun is not in the team.

8. The Student Council needs to fill the positions of president, secretary and treasurer from six candidates. Each candidate can fill only one of the positions. Determine how many ways can this be done if:

   a. there are no restrictions   **b.** Tan must be secretary
   c. Tan must have one of the three positions   **d.** Tan is not in any of the three positions.

9. The starting five in a basketball team is to be picked, in order, from the ten players in the squad. Determine how many ways can this be done if:

   a. there are no restrictions
   b. Jamahl needs to be player number five
   c. Jamahl and Anfernee must be in the first five players (starting five)
   d. Jamahl is not in the team.

10. **WE13** **a.** Calculate the number of permutations of the letters in the word MATHS.

    b. In how many of these do the letters M and A appear together?
    c. In how many permutations do the letters M and A appear apart?
    d. Determine the probability of the letters M and A appearing apart.

11. A rowing team of four rowers is to be selected in order from eight rowers.

   a. In how many different ways can this be done?
   b. In how many of these ways do two rowers, Jane and Lee, sit together in the boat?
   c. In how many ways can the crew be formed without using Jane or Lee?
   d. In how many ways can the crew be formed if it does not contain Jane?

12. A decathlon has 12 runners.

   a. In how many ways can 1st, 2nd and 3rd be filled?
   b. In how many ways can 1st, 2nd and 3rd be filled if Najim finishes first?

Complex familiar

13. **WE14** Consider the words SIMULTANEOUS and EQUATIONS.

   a. How many arrangements of the letters of the word EQUATIONS have the letters Q and U grouped together.
   b. How many arrangements of the letters of the word EQUATIONS have the letters Q and U separated?
   c. How many arrangements of the letters of the word SIMULTANEOUS are possible?
   d. Determine the probability that in a randomly chosen arrangement of the word SIMULTANEOUS, both the letters U are together.

14. **MC** If the answer is 10, which of the following options best matches this answer?

   A. The number of ways 1st and 2nd can occur in a race with five entrants
   B. The number of distinct arrangements of the letters in NANNA
   C. The number of permutations of the letters in POCKET where P and O are together
   D. $^{10}P_2 \div {}^4P_2$

15. **MC** If the answer is 480, which of the following options best matches this answer?

   A. The number of ways 1st and 2nd can occur in a race with five entrants
   B. The number of distinct arrangements of the letters in NANNA
   C. The number of permutations of the letters in POCKET where P and O are apart
   D. $^{10}P_2 \div {}^4P_2$

16. The clue in a crossword puzzle says that a particular answer is an anagram of STOREY. An anagram is another word that can be obtained by rearranging the letters of the given word.

   a. Determine the number of possible arrangements of the letters of STOREY.
   b. The other words in the crossword puzzle indicate that the correct answer is O__T__. How many arrangements are now possible? Can you see the word?

Complex unfamiliar

17. In web design, hex colours are expressed as a six-digit combination of numbers and letters from A to F, b95d74 or 51de4c for instance. There are 16^6 different codes for hex colours.
Determine the percentage of hex code that contains two identical letters grouped together with no other repetition of this letter. Give your answer to one decimal place.

18. Twelve musicians (six violinists, two cellists and four pianists) sit around a circular table. The six violinists always sit next to one another, and the four pianists do the same, but no violinist sit next to a pianist. One of the violinists and one of the cellists are siblings. Determine the probability that they sit next to each other.

Fully worked solutions for this chapter are available online.

LESSON
1.5 Combinations

SYLLABUS LINKS

- Define and use combinations
- Use the notation $\binom{n}{r}$ and ^nC_r to represent the number of ways of selecting r objects from n distinct objects where order is not important.
  - $^nC_r = \dfrac{n!}{r!\,(n-r)!}$
- Solve problems that involve combinations.

Source: Specialist Mathematics Senior Syllabus 2024 © State of Queensland (QCAA) 2024; licensed under CC BY 4.0

1.5.1 When order does not matter

A group of things chosen from a larger group where order is not important is called a combination. In previous sections we performed calculations of the number of ways a task could be done where order is important — permutations or arrangements. We now examine situations where *order does not matter*.

For instance, going back to Worked Example 9, if instead of selecting a president, a secretary and a treasurer among seven volunteers, the netball club selected three treasurers, then as the roles are the same, the order is not important, and combinations, instead of permutations, would be used.

Suppose five people have nominated for a committee consisting of three members. It does not matter in what order the candidates are placed on the committee; it matters only whether they are there or not. If order was important we know there would be $^5P_3$, or 60, ways in which this could be done. Here are the possibilities:

ABC	ACB	BAC	BCA	CAB	CBA
ABD	ADB	BAD	BDA	DAB	DBA
ABE	AEB	BAE	BEA	EAB	ABA
ACE	AEC	CAE	CEA	EAC	ECA
ACD	ADC	CAD	CDA	DAC	DCA
ADE	AED	DAE	DEA	EAD	EDA
BCD	BDC	CBD	CDB	DBC	DCB
BCE	BEC	CBE	CEB	EBC	ECB
BDE	BED	DBE	DEB	EBD	EDB
CDE	CED	DCE	DEC	ECD	EDC

The 60 arrangements are different only if we take order into account; that is, ABC is different from CAB and so on. You will notice in this table that there are 10 distinct committees corresponding to the 10 distinct rows. Each row merely repeats, in a different order, the committee in the first column. This result (10 distinct committees) can be arrived at logically:

1. There are $^5P_3$ ways of choosing or selecting three from five in order.
2. Each choice of three is repeated 3! times.
3. The number of distinct selections or combinations is $^5P_3 \div 3! = 10$.

This leads to the general rule of selecting r objects from n objects.

Unordered selections

The number of ways of choosing or selecting r objects from n distinct objects, where order is not important, is given by ^nC_r:

$$^nC_r = \frac{^nP_r}{r!}$$

C is used to represent combinations.

The formula we use to determine the number of ways of selecting r objects from n distinct objects, where order is not important, is useful but needs to be simplified.

$$^nC_r = \frac{^nP_r}{r!}$$
$$= \frac{\frac{n!}{(n-r)!}}{r!}$$
$$= \frac{n!}{r!\,(n-r)!}$$

$$^nC_r = \frac{n!}{r!\,(n-r)!} = \binom{n}{r}$$

$0 \leq r \leq n$ where r and n are non-negative integers.

WORKED EXAMPLE 15 Calculating combinations

Apply the concept of ^nC_r to calculate the number ways a basketball team of five players can be selected from a squad of nine if the order in which they are selected does not matter.

THINK

1. Recall the rule for ^nC_r. *Note:* Since order does not matter, use the ^nC_r rule.

2. Substitute the values of n and r into the formula.

WRITE

$$^nC_r = \frac{n!}{r!\,(n-r)!}$$

$$^9C_5 = \frac{9!}{5!\,(9-5)!}$$

3. Simplify the fraction.

$$= \frac{9!}{5! \ 4!}$$

$$= \frac{9 \times 8 \times 7 \times 6 \times 5!}{4 \times 3 \times 2 \times 1 \times 5!}$$

$$= \frac{3024}{24}$$

4. Evaluate.

$$= 126$$

WORKED EXAMPLE 16 Using the combination notations

Determine the value of the following.

a. $^{12}C_5$

b. $\dbinom{10}{2}$

THINK

WRITE

a. 1. Recall the formula for ^nC_r.

a. $^nC_r = \dfrac{n!}{(n-r)! \ r!}$

2. Substitute the given values of n and r into the combination formula.

$^{12}C_5 = \dfrac{12!}{(12-5)! \ 5!}$

$= \dfrac{12!}{7! \ 5!}$

3. Simplify the fraction.

$= \dfrac{12 \times 11 \times 10 \times 9 \times 8 \times 7!}{7! \times 5 \times 4 \times 3 \times 2 \times 1}$

$= \dfrac{12 \times 11 \times \cancel{10} \times \cancel{9}^{3} \times \cancel{8}^{2}}{\cancel{5} \times \cancel{4} \times \cancel{3} \times \cancel{2} \times 1}$

4. Evaluate.

$= 12 \times 11 \times 3 \times 2$

$= 792$

b. 1. Recall the rule for $\dbinom{n}{r}$.

b. $\dbinom{n}{r} = {}^nC_r$

$= \dfrac{n!}{(n-r)! \ r!}$

2. Substitute the given values of n and r into the combination formula.

$\dbinom{10}{2} = \dfrac{10!}{(10-2)! \ 2!}$

$= \dfrac{10!}{8! \ 2!}$

3. Simplify the fraction.

$= \dfrac{10 \times 9 \times 8!}{8! \times 2 \times 1}$

$= \dfrac{10 \times 9}{2 \times 1}$

4. Evaluate.

$= \dfrac{90}{2}$

$= 45$

1.5.2 Probability calculations

The combination formula is always used in selection problems. Most calculators have a ^nC_r key to assist with the evaluation when the figures become large.

Both the multiplication and addition principles apply and are used in the same way as for permutations.

The calculation of probabilities from the rule $P(A) = \dfrac{n(A)}{n(\xi)}$ requires that the same counting technique is used for the numerator and denominator. We have seen for permutations that it can assist calculation to express numerator and denominator in terms of factorials and then simplify. Similarly for combinations, express the numerator and denominator in terms of the appropriate combinatoric coefficients and then carry out the calculations.

WORKED EXAMPLE 17 Solving problems using combinations

A committee of five students is to be chosen from seven boys and six girls. Use ^nC_r and the multiplication and addition principles to answer the following.
a. Calculate how many committees can be formed.
b. Calculate how many of the committees contain exactly two boys and three girls.
c. Calculate how many committees have at least four girls.
d. Determine the probability of the oldest and youngest students both being on the committee in part a.

THINK

a. 1. As there is no restriction, choose the committee from the total number of students.

2. Use the formula $^nC_r = \dfrac{n!}{r! \times (n-r)!}$ to calculate the answer.

WRITE

a. There are 13 students in total from whom five students are to be chosen. This can be done in $^{13}C_5$ ways.

$$^{13}C_5 = \frac{13!}{5!\,(13-5)!}$$
$$= \frac{13!}{5!\,8!}$$
$$= \frac{13 \times 12 \times 11 \times 10 \times 9 \times 8!}{5! \times 8!}$$
$$= \frac{13 \times 12 \times 11 \times 10 \times 9}{5 \times 4 \times 3 \times 2 \times 1}$$
$$= 13 \times 11 \times 9$$
$$= 1287$$

There are 1287 possible committees.

b. 1. Select the committee to satisfy the given restriction.

2. Use the multiplication principle to form the total number of committees.
Note: The upper numbers on the combinatoric coefficients sum to the total available, $7 + 6 = 13$, while the lower numbers sum to the number that must be on the committee, $2 + 3 = 5$.

b. The two boys can be chosen from the seven boys available in $^7C_2$ ways. The three girls can be chosen from the six girls available in $^6C_3$ ways.

The total number of committees which contain two boys and three girls is $^7C_2 \times {^6C_3}$.

3. Calculate the answer.

$$^7C_2 \times {}^6C_3 = \frac{7!}{2!\,5!} \times \frac{6!}{3!\,3!}$$
$$= \frac{7 \times 6}{2} \times \frac{6 \times 5 \times 4}{3 \times 2}$$
$$= 21 \times 20$$
$$= 420$$

There are 420 committees possible with the given restriction.

c. 1. List the possible committees which satisfy the given restriction.

c. As there are six girls available, at least four girls means either four or five girls. The committees of five students which satisfy this restriction have either four girls and one boy, or they have five girls and zero boys.

2. Write the number of committees in terms of combinatoric coefficients.

Four girls and one boy are chosen in $^6C_4 \times {}^7C_1$ ways. Five girls and zero boys are chosen in $^6C_5 \times {}^7C_0$ ways.

3. Use the addition principle to state the total number of committees.

The number of committees with at least four girls is $^6C_4 \times {}^7C_1 = 15 \times 7 = 105$

4. Calculate the answer.

$$^6C_5 \times {}^7C_0 = 6 \times 1 = 6$$
$$^6C_4 \times {}^7C_1 + {}^6C_5 \times {}^7C_0 = 105 + 6$$
$$= 111$$

There are 111 committees with at least four girls.

d. 1. State the number in the sample space.

d. The total number of committees of five students is $^{13}C_5 = 1287$ from part **a**.

2. Form the number of ways the given event can occur.

Each committee must have five students. If the oldest and youngest students are placed on the committee, then three more students need to be selected from the remaining 11 students to form the committee of five. This can be done in $^{11}C_3$ ways.

3. State the probability in terms of combinatoric coefficients.

Let A be the event the oldest and the youngest students are on the committee.

4. Calculate the answer.

$$P(A) = \frac{n(A)}{n(\xi)}$$
$$= \frac{^{11}C_3}{^{13}C_5}$$
$$= \frac{11!}{3!\,8!} \div \frac{13!}{5!\,8!}$$
$$= \frac{5 \times 4}{13 \times 12}$$
$$= \frac{5}{39}$$

The probability of the committee containing the youngest and the oldest students is $\frac{5}{39}$.

WORKED EXAMPLE 18 Comparing the binomial coefficients at the same distance from the beginning to the end

Evaluate the following using your calculator and comment on your results.

a. $^9C_3$ b. $^9C_6$ c. $^{15}C_5$ d. $^{15}C_{10}$ e. $^{12}C_7$ f. $^{12}C_5$

THINK	WRITE
a–f. Use your calculator to evaluate the listed combinations.	a. $^9C_3 = 84$ b. $^9C_6 = 84$ c. $^{15}C_5 = 3003$ d. $^{15}C_{10} = 3003$ e. $^{12}C_7 = 792$ f. $^{12}C_5 = 792$
Comment on your results.	So $^9C_3 = {}^9C_6$, $^{15}C_5 = {}^{15}C_{10}$ and $^{12}C_7 = {}^{12}C_5$. It appears that when, for example, $^{12}C_p = {}^{12}C_q$ $p + q = 12$.

TI	THINK	WRITE	CASIO	THINK	WRITE
a. 1. On a Calculator page, press MENU, then select: 5: Probability 3: Combinations. Complete the entry line as: nCr(9, 3) then press ENTER.		a. 1. On a Run-Matrix screen, press OPTN then F6. Select PROB by pressing F3, then select nCr by pressing F3. Complete the entry line as: 9**C**3 then press EXE.	
2. The answer appears on the screen.	$^9C_3 = 84$	2. The answer appears on the screen.	$^9C_3 = 84$

For each of the preceding examples, it can be seen that $^nC_r = {}^nC_{n-r}$.

Drawing on our understanding of combinations, we have:
- $^nC_r = {}^nC_{n-r}$, as choosing r objects must leave behind $(n - r)$ objects and vice versa
- $^nC_0 = 1 = {}^nC_n$, as there is only one way to choose none or all of the n objects
- $^nC_1 = n$, as there are n ways of choosing 1 object from a group of n objects.

The formula for ^nC_r, also noted $\binom{n}{r}$ is exactly that for the binomial coefficients used in the binomial theorem, or binomial expansion, which is used to describe the algebraic expansion of powers of a binomial, in the mathematical methods course.

Combinations produce the elements of any row in Pascal's triangle, which can be used to calculate ^nC_r by hand. Each new row in Pascal's triangle is obtained by first placing a 1 at the beginning and end of the row and then adding adjacent entries from the previous row.

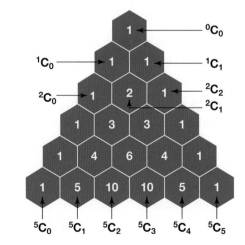

The elements of the 6th row for instance are 1, 1 + 5, 5 + 10, 10 + 10, 10 + 5, 5 + 1 and 1. Those elements can be used to expand $(x + y)^6$:

$$(x + y)^6 = \binom{6}{0} x^6 y^0 + \binom{6}{1} x^5 y^1 + \binom{6}{2} x^4 y^2 + \binom{6}{3} x^3 y^3 + \binom{6}{4} x^2 y^4 + \binom{6}{5} x^1 y^5 + \binom{6}{6} x^0 y^6$$

$$= x^6 + 6x^5 y + 15x^4 y^2 + 20x^3 y^3 + 15x^2 y^4 + 6xy^5 + y^6$$

 Resources

Digital documents SkillSHEET Listing possibilities (doc-26826)
SpreadSHEET Combinations (doc-26827)

Interactivity Counting techniques (int-6293)

Exercise 1.5 Combinations

learn on

1.5 Exercise	1.5 Exam questions on

These questions are even better in jacPLUS!
- Receive immediate feedback
- Access sample responses
- Track results and progress

Find all this and MORE in jacPLUS

Simple familiar	Complex familiar	Complex unfamiliar
1, 2, 3, 4, 5, 6, 7, 8, 9, 10, 11, 12, 13, 14, 15	16, 17, 18, 19	20

Simple familiar

1. Write each of the following as statements in terms of permutations.
   a. $^8C_3$
   b. $^{19}C_2$
   c. $^1C_1$
   d. $^5C_0$

2. Write each of the following using the notation ^nC_r.
   a. $\dfrac{^8P_2}{2!}$
   b. $\dfrac{^9P_3}{3!}$
   c. $\dfrac{^8P_0}{0!}$
   d. $\dfrac{^{10}P_4}{4!}$

3. **WE15** Apply the concept of ^nC_r to calculate the number of ways three types of ice-cream can be chosen in any order from a supermarket freezer if the freezer contains:
   a. three types
   b. six types
   c. ten types
   d. 12 types.

4. A mixed netball team must have three women and four men in the side. If the squad has six women and five men wanting to play, determine how many different teams are possible.

5. A *quinella* is a bet made on a horse race which pays a win if the punter selects the first two horses in any order. Determine how many different quinellas are possible in a race that has:
   a. eight horses
   b. 16 horses.

6. **MC** At a party there are 40 guests and they decide to have a toast. Each guest 'clinks' glasses with every other guest. How many clinks are there in all?
   A. 39
   B. 40
   C. 40!
   D. 780

7. **MC** On a bookshelf there are 15 books — seven geography books and eight law books. Abena selects five books from the shelf — two geography books and three law books. How many different ways can she make this selection?
   A. $^{15}C_2 \times {}^{15}C_3$
   B. $^{15}C_7 \times {}^{15}C_8$
   C. $^7C_2 \times {}^8C_3$
   D. $^7C_2 + {}^8C_3$

8. A cricket team of 11 players is to be chosen from a squad of 15 players. Determine how many ways can this be done.

9. A basketball team of five players is to be chosen from a squad of ten players. Determine how many ways can this be done.

10. **WE16** Determine the value of the following:

   a. $^{12}C_4$
   b. $^{11}C_1$
   c. $^{12}C_{12}$
   d. $\binom{21}{15}$
   e. $\binom{100}{1}$
   f. $\binom{17}{14}$

11. From a pack of 52 cards, a hand of five cards is dealt.

   a. How many different hands are there?
   b. How many of these hands contain only red cards?
   c. How many of these hands contain only black cards?
   d. How many of these hands contain at least one red and at least one black card?

12. **WE17** A committee of five students is to be chosen from six boys and eight girls. Use ^nC_r and the multiplication principle to answer the following.

   a. Calculate how many committees can be formed.
   b. Calculate how many of the committees contain exactly two boys and three girls.
   c. Calculate how many committees have at least four boys.
   d. Determine the probability of neither the oldest nor the youngest student being on the committee.

13. A rugby union squad has 12 forwards and ten backs in training. A team consists of eight forwards and seven backs. Determine how many different teams can be chosen from the squad.

14. A music collection contains 32 albums. Determine how many ways five albums can be chosen from the collection.

15. a. **WE18** Calculate the value of:

   i. $^{12}C_3$ and $^{12}C_9$
   ii. $^{15}C_8$ and $^{15}C_7$
   iii. $^{10}C_1$ and $^{10}C_9$
   iv. $^8C_3$ and $^8C_5$

   b. What do you notice? Give your answer as a general statement such as 'The value of ^nC_r is ...'.

Complex familiar

Questions **16**, **17** and **18** refer to the following information. A tennis tournament involve 16 players. The organisers plan to use three courts and assume that each match will last on average two hours and that no more than four matches will be played on any court per day.

16. In a 'round robin' each player plays every other player once.

   a. If the organisers use a round robin format, determine how many games will be played in all.
   b. For how many days would the tournament last?

17. The organisers split the 16 players into two pools of eight players each. After a 'round robin' within each pool, a final is played between the winners of each pool.

   a. Determine how many matches are played in the tournament.
   b. How long does the tournament last?

18. A 'knock out' format is one in which the loser of every match drops out and the winners proceed to the next round until there is only one winner left.

   a. If the game starts with 16 players, determine how many matches are needed before a winner is obtained.
   b. How long would the tournament last?

19. Lotto is a gambling game played by choosing six numbers from 45. Gamblers try to match their choice with those numbers chosen at the official draw. No number can be drawn more than once and the order in which the numbers are selected does not matter.

   a. Calculate how many different selections of six numbers can be made from 45.
   b. Suppose the first numbers drawn at the official draw are 42, 3 and 18. How many selections of six numbers will contain these three numbers?

   Note: This question ignores supplementary numbers. Lotto is discussed further in the next section.

Complex unfamiliar

20. Consider n distinct points in the plane. Determine how many different k-sided polygons ($3 \leq k \leq n$) can be created from those n points. Comment on the reasonableness of your answer using five points and triangles.

Fully worked solutions for this chapter are available online.

LESSON
1.6 Applications of permutations and combinations

SYLLABUS LINKS

- Solve problems that involve combinations with restrictions including specific objects grouped together and selection from multiple groups.
- Model and solve problems that involve permutations and combinations including probability problems, with and without technology.

Source: Specialist Mathematics Senior Syllabus 2024 © State of Queensland (QCAA) 2024; licensed under CC BY 4.0

1.6.1 Permutations and combinations in the real world

Counting techniques, particularly those involving permutations and combinations, can be applied in gambling, logistics and various forms of market research. In this section we investigate when to use permutations and when to use combinations as well as examining problems associated with these techniques.

Permutations are used to count when order is important. Some examples are:
- the number of ways the positions of president, secretary and treasurer can be filled
- the number of ways a team can be chosen from a squad *in distinctly different positions*
- the number of ways the first three positions of a race can be filled.

Combinations are used to count when order is not important. Some examples are:
- the number of ways a committee can be chosen
- the number of ways a team can be chosen from a squad
- the number of ways a hand of five cards can be dealt from a deck.

These relatively simple applications of permutations and combinations are explored in the worked examples that follow. However, it is important to be mindful that the modern world relies on combinatorial algorithms. These algorithms are important for any system that benefits from finding the fastest ways to operate. Examples include communication networks, molecular biology, enhancing security and protecting privacy in internet information transfer, data base queries and data mining, computer chip design, simulations and scheduling.

WORKED EXAMPLE 19 Solving problems using combinations and permutations

a. **Ten points are marked on a page and no three of these points are in a straight line. Determine how many triangles can be drawn joining these points.**
b. **Determine how many different three-digit numbers can be made using the digits $1, 3, 5, 7$ and 9 without repetition.**

THINK	WRITE
a. 1. *Note:* A triangle is made by choosing three points. It does not matter in what order the points are chosen, so ^nC_r is used. Recall the rule for ^nC_r.	a. $^nC_r = \dfrac{n!}{(n-r)!\, r!}$
2. Substitute the given values of n and r into the combination formula.	$^{10}C_3 = \dfrac{10!}{(10-3)!\ 3!}$ $= \dfrac{10!}{7!\ 3!}$
3. Simplify the fraction.	$= \dfrac{10 \times 9 \times 8 \times 7!}{7! \times 3 \times 2 \times 1}$ $= \dfrac{10 \times 9^3 \times 8^4}{3 \times 2 \times 1}$
4. Evaluate.	$= 10 \times 3 \times 4$ $= 120$
5. Write the answer.	120 triangles may be drawn by joining three points.
6. Verify the answer obtained by using the combination function on a calculator.	
b. 1. *Note:* Order is important here. Recall the rule for ^nP_r.	b. $^nP_r = \dfrac{n}{(n-r)!}$
2. Substitute the given values of n and r into the permutation formula.	$^5P_3 = \dfrac{5!}{(5-3)}$ $= \dfrac{5!}{2!}$
3. Evaluate.	$= \dfrac{5 \times 4 \times 3 \times 2}{2!}$ $= 5 \times 4 \times 3$ $= 60$
4. Write the answer.	60 three-digit numbers can be made without repetition from a group of five different numbers.
5. Verify the answer obtained by using the permutation function on a calculator.	

Jade and Kelly are two of the ten members of a basketball squad. Calculate how many ways can a team of five be chosen if:

a. both Jade and Kelly are in the five

b. neither Jade nor Kelly is in the five

c. Jade is in the five but Kelly is not.

THINK	WRITE
a. 1. *Note:* Order is not important, so ^nC_r is used. Recall the rule for ^nC_r.	**a.** $^nC_r = \dfrac{n!}{(n-r)!\ r!}$
2. *Note:* If Jade and Kelly are included then there are three positions to be filled from the remaining eight players. Substitute the given values of n and r into the combination formula.	$^8C_3 = \dfrac{8!}{(8-3)!\ 3!}$ $= \dfrac{8!}{5!\ 3!}$
3. Simplify the fraction.	$= \dfrac{8 \times 7 \times 6 \times 5!}{5! \times 3 \times 2 \times 1}$ $= \dfrac{8 \times 7 \times \cancel{6}}{\cancel{3} \times \cancel{2} \times 1}$
4. Evaluate.	$= 8 \times 7$ $= 56$
5. Write the answer.	If Jade and Kelly are included, then there are 56 ways to fill the remaining three positions.
b. 1. *Note:* Order is not important, so ^nC_r is used. Recall the rule for ^nC_r.	**b.** $^nC_r = \dfrac{n!}{(n-r)!\ r!}$
2. *Note:* If Jade and Kelly are not included then there are five positions to be filled from eight players. Substitute the given values of n and r into the combination formula.	$^8C_5 = \dfrac{8!}{(8-5)!\ 5!}$ $= \dfrac{8!}{3!\ 5!}$
3. Simplify the fraction.	$= \dfrac{8 \times 7 \times 6 \times 5!}{3 \times 2 \times 1 \times 5!}$ $= \dfrac{8 \times 7 \times \cancel{6}}{\cancel{3} \times \cancel{2} \times 1}$
4. Evaluate.	$= 8 \times 7$ $= 56$
5. Answer the question.	If Jade and Kelly are not included, then there are 56 ways to fill the five positions.
c. 1. *Note:* Order is not important, so ^nC_r is used. Recall the rule for ^nC_r.	**c.** $^nC_r = \dfrac{n!}{(n-r)!\ r!}$
2. *Note:* If Jade is included and Kelly is not then there are four positions to be filled from eight players. Substitute the given values of n and r into the combination formula.	$^8C_4 = \dfrac{8!}{(8-4)!\ 4!}$ $= \dfrac{8!}{4!\ 4!}$

3. Simplify the fraction.

$$= \frac{8 \times 7 \times 6 \times 5 \times 4!}{4 \times 3 \times 2 \times 1 \times 4!}$$

$$= \frac{\cancel{8} \times 7 \times \cancel{6}^2 \times 5}{\cancel{4} \times \cancel{3} \times \cancel{2} \times 1}$$

4. Evaluate.

$$= 7 \times 2 \times 5$$
$$= 70$$

5. Write the answer.

If Jade is included and Kelly is not, then there are 70 ways to fill the four positions.

6. Verify each of the answers obtained by using the combination function on a calculator.

1.6.2 Lotto systems

An interesting application of combinations as a technique of counting is a game that Australians spend many millions of dollars on each week — lotteries. There are many varieties of lottery games in Australia. To play Saturday Gold Lotto in Queensland, a player selects six numbers from 45 numbers. The official draw chooses six numbers and two supplementary numbers. Depending on how the player's choice of six numbers matches the official draw, prizes are awarded in different divisions.

Division 1:   Six winning numbers
Division 2:   Five winning numbers and one of the supplementary numbers
Division 3:   Five winning numbers
Division 4:   Four winning numbers
Division 5:   Three winning numbers and one of the supplementary numbers

If the official draw was:

Winning numbers						Supplementaries	
13	42	6	8	20	12	2	34

A player who chose:

8   34   13   12   20   45

would win a Division 4 prize and a player who chose:

8   34   13   12   22   45

would win a Division 5 prize.

A player may have seven lucky numbers 4, 7, 12, 21, 30, 38 and 45, and may wish to include all possible combinations of these seven numbers in a six numbers lotto entry.

This can be done as follows:

4	7	12	21	30	38
4	7	12	21	30	45
4	7	12	21	38	45
4	7	12	30	38	45
4	7	21	30	38	45
4	12	21	30	38	45
7	12	21	30	38	45

The player does not have to fill out seven separate entries to enter all combinations of these seven numbers six at a time but rather can complete a 'System 7' entry by marking seven numbers on the entry form.

A System 9 consists of all entries of six numbers from the chosen nine numbers.

WORKED EXAMPLE 21 Combinations and lottery systems

Use the information on lottery systems given above.
A player uses a System 8 entry with the numbers 4, 7, 9, 12, 22, 29, 32 and 36.
The official draw for this game was 4, 8, 12, 15, 22, 36 with supplementaries 20 and 29.
a. How many single entries are equivalent to a System 8?
b. List three of the player's entries that would have won Division 4.
c. Determine how many of the player's entries would have won Division 4.

THINK

a. 1. *Note:* Order is not important, so ^nC_r is used.
Recall the rule for ^nC_r.

2. *Note:* A System 8 consists of all entries consisting of six numbers chosen from eight.
Substitute the given values of n and r into the combination formula.

3. Simplify the fraction.

4. Evaluate.

5. Write the answer.

6. Verify each of the answers obtained by using the combination function on a calculator.

b. *Note:* Division 4 requires four winning numbers.
The player's winning numbers are 4, 12, 22 and 36.
Any of the other four numbers can fill the remaining two places.
List three of the player's entries that would have won Division 4.

WRITE

a. $^nC_r = \dfrac{n!}{(n-r)!\,r!}$

$^8C_6 = \dfrac{8!}{(8-6)!\,6!}$

$= \dfrac{8!}{2!\,6!}$

$= \dfrac{8 \times 7 \times 6!}{2 \times 7 \times 6!}$

$= \dfrac{\overset{4}{8} \times 7}{\overset{}{2} \times 1}$

$= 4 \times 7$

$= 28$

A System 8 is equivalent to 28 single entries.

b. Some of the possibilities are:

4	12	22	36	7	9
4	12	22	36	7	29
4	12	22	36	7	32

c. 1. *Note:* Order is not important, so ^nC_r is used. Recall the rule for ^nC_r.

c. $^nC_r = \dfrac{n!}{(n-r)! \; r!}$

2. *Note:* To win Division 4 the numbers $4, 12, 22$ and 36 must be included in the entry. The other two spaces can be filled with any of the other four numbers in any order.
Substitute the given values of n and r into the combination formula.

$^4C_2 = \dfrac{4!}{(4-2)! \; 2!}$

$= \dfrac{4!}{2! \; 2!}$

3. Simplify the fraction.

$= \dfrac{4 \times 3 \times \cancel{1!}}{\cancel{2} \times 1 \times \cancel{2!}}$

$= \dfrac{\cancel{4}^{2} \times 3}{\cancel{2} \times 1}$

4. Evaluate.

$= 2 \times 3$

$= 6$

5. Write the answer.

Six of the player's entries would have won Division 4.

6. Verify each of the answers obtained by using the combination function on a calculator.

Exercise 1.6 Applications of permutations and combinations **learn** on

1.6 Exercise	**1.6 Exam questions** on

Simple familiar	Complex familiar	Complex unfamiliar
1, 2, 3, 4, 5, 6, 7, 8, 9	10, 11, 12, 13, 14	15, 16

These questions are even better in jacPLUS!
- Receive immediate feedback
- Access sample responses
- Track results and progress

Find all this and MORE in jacPLUS

Simple familiar

1. **WE19** Determine how many ways there are:
   a. to draw a line segment between two points on a page with ten points on it
   b. to make a four-digit number using the digits $2, 4, 6, 8$ and 1 without repetition
   c. to choose a committee of four people from ten people
   d. for a party of 15 people to shake hands with one another.

2. Determine how many ways there are:
   a. for ten horses to fill 1st, 2nd and 3rd positions
   b. to choose a team of three cyclists from a squad of five
   c. to choose 1st, 2nd and 3rd speakers for a debating team from six candidates
   d. for 20 students to seat themselves in a row of 20 desks.

3. The French flag is known as a tricolour flag because it is composed of the three bands of colour. Determine how many different tricolour flags can be made from the colours red, white, blue and green if each colour can be used only once in one of the three bands and order is important.

4. In a taste test a market research company has asked people to taste four samples of coffee and try to identify each as one of four brands. Subjects are told that no two samples are the same brand. Determine how many different ways the samples can be matched to the brands.

5. **WE20** A volleyball team of six players is to be chosen from a squad of ten players. Calculate how many ways can this be done if:
   a. there are no restrictions
   b. Stephanie is to be in the team
   c. Stephanie is not in the team
   d. two players, Stephanie and Alison, arc not both in the team together.

6. A cross-country team of four runners is to be chosen from a squad of nine runners. Determine how many ways this can be done if:
   a. there are no restrictions
   b. Cecily is to be one of the four
   c. Cecily and Michael are in the team
   d. either Cecily or Michael but not both are in the team.

7. **MC** A netball team consists of seven different positions: goal defence, goal keeper, wing defence, centre, wing attack, goal attack and goal shooter. The number of ways a squad of ten players can be allocated to these positions is:

   A. $10!$ B. $7!$ C. $\dfrac{10!}{7!}$ D. $^{10}P_7$

8. **WE21** Use the information on lotteries given in 1.6.2 Lotto systems.
   A player uses a System 8 entry with the numbers $9, 12, 14, 17, 27, 34, 37$ and 41. The official draw for this game was $9, 13, 17, 20, 27, 41$ with supplementaries 25 and 34.
   a. How many single entries are equivalent to a System 8?
   b. List three of the player's entries that would have won Division 4.
   c. Determine how many of the player's entries would have won Division 4.

9. Use the information on lotteries given in 1.6.2 Lotto systems.
   A player uses a System 9 entry with the numbers $7, 10, 12, 15, 25, 32, 35, 37$ and 41. The official draw for this game was $7, 11, 15, 18, 25, 39$ with supplementaries 23 and 32.
   a. To how many single entries is a System 9 equivalent?
   b. List three of the player's entries that would have won Division 5.
   c. How many of the player's entries would have won Division 5?

10. A soccer team of 11 players is to be chosen from a squad of 17. If one of the squad is selected as goalkeeper and any of the remaining players can be selected in any of the positions, determine how many ways can this be done if:

   a. there are no other restrictions
   b. Karl is to be chosen
   c. Karl and Andrew refuse to play in the same team
   d. Karl and Andrew are either both in or both out.

Questions **11** and **12** refer to the following information: Keno is a popular game in clubs and pubs around Australia. In each round a machine randomly generates 20 winning numbers from 1 to 80. In one entry a player can select up to 15 numbers.

11. Suppose a player selects an entry of six numbers.

   a. Determine how many ways an entry of six numbers can contain six winning numbers.
      Suppose an entry of six numbers has exactly three winning numbers in it.
   b. In how many ways can the three winning numbers be chosen?
   c. In how many ways can the three losing numbers be chosen?
   d. How many entries of six numbers contain three winning numbers and three losing numbers?

12. Suppose a player selects an entry of 20 numbers.

   a. Determine how many ways an entry of 20 numbers can contain 20 winning numbers.
   b. Suppose an entry of 20 numbers has exactly 14 winning numbers in it.

      i. In how many ways can the 14 winning numbers be chosen?
      ii. In how many ways can the six losing numbers be chosen?
      iii. How many entries of 20 numbers contain 14 winning numbers and six losing numbers?
      iv. How many entries of 20 numbers contain no winning numbers?

13. In the gambling game roulette, if a gambler puts $1 on the winning number he will win $35. Suppose a gambler wishes to place five $1 bets on five different numbers in one spin of the roulette wheel. If there are 36 numbers in all, determine how many ways the five bets can be placed.

14. **MC** A secret chemical formula requires the mixing of three chemicals. A researcher does not remember the three chemicals but has a shortlist of ten from which to choose. Each time she mixes three chemicals and tests the result she takes 15 minutes. How long does the researcher need, to be absolutely sure of getting the right combination?

   A. 120 hours
   B. 7.5 hours
   C. 15 hours
   D. 30 hours

15. During a family reunion, people left their shoes in the entryway. There are ten different pairs of shoes, five pairs of loafers, three pairs of boots and two pairs of sneakers. If someone select a pair at random, determine the probability that they have a left and a right shoe of the same type of shoe (that is a left loafer and right loafer, or a left boot and right boot, or a left sneaker and right sneaker).

16. A yellow-footed Antechinus wants to catch and eat an unmoving beetle that is 9 cm to its east and 6 cm to its north. The antechinus also notices a flower 4 cm to its east and 3 cm to its north. If the Antechinus moves 1 cm at a time, either east or north, determine in how many ways can it reach the beetle if it checks the flower for nectar on its way to the beetle.

Fully worked solutions for this chapter are available online.

LESSON
1.7 Review

1.7.1 Summary

1.7 Exercise

learnon

1.7 Exercise	**1.7 Exam questions** on

Simple familiar	Complex familiar	Complex unfamiliar
1, 2, 3, 4, 5, 6, 7, 8, 9, 10, 11, 12	13, 14, 15, 16	17, 18, 19, 20

Simple familiar

1. **a.** State the inclusion–exclusion principle for three sets, R, S and T.
   b. Use the inclusion–exclusion principle to calculate the number of cards in a deck of 52 cards that are either red, a jack or a court card (king, queen or jack).

2. **a.** State the multiplication principle.
   b. One or two letters are to be chosen from the letters A, B, C, D, E and F. In how many different ways can this be done without replacement, if order is important?

3. **a.** State the definition of ^nP_r.
   b. Without a calculator, compute the value of $^{10}P_3$.
   c. Prove that $^nP_r = \dfrac{n!}{(n-r)!}$.

4. A free-style snowboard competition has 15 entrants. In how many ways can the first, second and third places be filled?

5. **a.** Suppose five people are to be seated. Explain why there are fewer ways of seating five people at a circular table compared with seating the group on a straight bench.
   b. How many ways can five people be seated at a round table?

6. The main cricket ground in Brisbane is called the Gabba. It is short for Woolloongabba. How many different arrangements of letters can be made from the word WOOLLOONGABBA? You may wish to use technology to answer the question.

7. Apply the concept of ^nC_r to calculate the number of ways 12 different ingredients can be chosen from a box of 30 different ingredients. What can you conclude about the ingredients left behind? Do not use algebra to explain this.

8. A committee of five men and five women is to be chosen from eight men and nine women. In how many ways can this be done?

9. A netball team of seven players is to be chosen from a squad of 11 players. Suppose any squad member can play any position. In how many ways can this be done:

   a. if each player is chosen to play a particular position
   b. if players have no particular position?

10. A ward in a city hospital has 15 nurses due to work Friday. There are three shifts needed to be staffed by five nurses on each shift. How many ways can this be done assuming each nurse works only one shift? (The order of the nurses on each shift and the order of each shift are not important.)

11. 12 pots of aromatic herbs are to be arranged in a row on a windowsill. There are two pots of basil, one of thyme, two of rosemary, three of mint and four of oregano. Determine the number of different arrangements of pots if the four oregano pots must be next to each other.

12. a. How many four-digit numbers can be made from the numbers 0–9 if the numbers can be repeated?
    b. What is the probability that one of these four-digit numbers is greater than 7000 (included) and divisible by 5?

Complex familiar

13. Assume that car number plates are sequenced as follows: DLV334 → DLV335 → ... DLV999 → DLW000 and so on. Using this sequence, how many number plates are there between DLV334 and DNU211 inclusive?

14. a. A school uses identification (ID) cards that consist of two letters from A to D followed by three digits chosen from 0 to 9. Each digit may be repeated but letters cannot be repeated. If the school receives about 800 new students each year, after how many years will they run out of unique ID numbers?
    b. For a scene in a movie, five boy–girl couples are needed. If they are to be selected from ten boys and 12 girls, in how many ways can this be done? (Assume the order of the couples does not matter.)

15. Determine the number of ways nine people can be arranged around two distinct circular tables, one with four seats, the other with five seats.

16. Juan has a bike lock with a four-digit key but has forgotten the code. He knows that he used either the year he bought the bike, 2016, or the current year, 2019. Also, he jumbled the digits, so 2016 and 2019 are not possible keys. How many different possible keys are there to unlock the bike?

17. Jane, Laura and Steff are in a netball squad of 13 players. The team only takes nine players to games. Jane, Laura and Steff's parents share the driving if they can. If two or more of the girls are playing, one parent drives them to the game and another parent picks them up. The parents do not share the driving if only one of the girls is playing. For what proportion of games would you predict the girls' parents are able to share the driving?

18. Considering that a number is divisible by 9 if the sum of its digits is also divisible by 9, what is the probability that a random four-digit number is a multiple of 9 if no digit is repeated?

19. The inclusion–exclusion principle for three sets R, S and T states:
$$|R \cup S \cup T| = |R| + |S| + |T| - |R \cap S| - |R \cap T| - |S \cap T| + |R \cap S \cap T|$$
Generalise this principle to four sets, Q, R, S and T.

20. A plane is covered in points at 1-unit spacing. Each point on the plane is coloured red, blue or white. Show there are three points of the same colour at a maximum distance $2\sqrt{2}$ from each other.

Fully worked solutions for this chapter are available online.

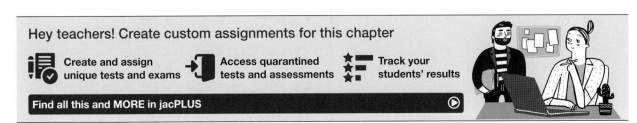

Answers

Chapter 1 Combinatorics

1.2 Counting techniques

1.2 Exercise

1. a.

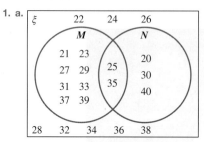

b. Sample responses can be found in the worked solutions in the online resources.

2. a.

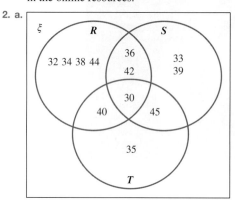

b. $|R \cup S \cup T| = 12$

c. $|R \cup S \cup T| = 12$

3. n (red, even or 4) $= 36$

4. 55 students

5. a. AB BA CA
    AC BC CB

b. 6

c. $\dfrac{1}{3}$

6. a. BG GB YB RB
    BY GY YG RG
    BR GR YR RY

b.

1st	2nd
4	3

Number of choices $= 4 \times 3$
                    $= 12$

7. a. ACB BAC CAB
    ABC BCA CBA

b.

1st	2nd	3rd
3	2	1

Number of choices $= 3 \times 2 \times 1$
                    $= 6$

8. a. 42   **b.** 210   **c.** 840   **d.** 2520   **e.** $\dfrac{1}{7}$

9. a. 24   **b.** 6   **c.** 12   **d.** 24

10. a. 49   **b.** 252

11. 126

12. a. 200   **b.** 40   **c.** 50   **d.** 290   **e.** $\dfrac{1}{40}$

13. a. 13 230

b. 17 640

Jake may wear 13 230 outfits with a jacket or 4410 outfits without a jacket. Therefore he has a total of 17 640 outfits to choose from. The assumption made with this problem is that no item of clothing is exactly the same; that is, none of the seven shirts are exactly the same.

14. C

15. D

16. 256

17. 1080

18. 6

19. 100

20. 48

21. a. 1000

b. 27

c.

271	371	471
272	372	472
273	373	473
281	381	481
282	382	482
283	383	483
291	391	491
292	392	492
293	393	493

22. There are 171 integers, between 1 and 315 included, that are divisible by 3 or by 5 or by 7.

1.3 Factorials and permutations

1.3 Exercise

1. a. $4 \times 3 \times 2 \times 1$

b. $5 \times 4 \times 3 \times 2 \times 1$

c. $6 \times 5 \times 4 \times 3 \times 2 \times 1$

d. $7 \times 6 \times 5 \times 4 \times 3 \times 2 \times 1$

2. a. 24   **b.** 120

c. 720   **d.** 3 628 800

e. $8.717 829 12 \times 10^{10}$ **f.** 362 880

g. 5040   **h.** 6

3. a. 3024   **b.** 151 200   **c.** 840   **d.** 720

4. a. $n(n-1)(n-2)(n-3)(n-4)$

b. $(n+3)(n+2)$

c. $\dfrac{1}{n(n-1)(n-2)}$

d. $\dfrac{1}{(n+2)(n+1)n(n-1)}$

5. a. $8 \times 7 = 56$

b. $7 \times 6 \times 5 \times 4 \times 3 = 2520$

c. $8 \times 7 \times 6 \times 5 \times 4 \times 3 \times 2 = 40 320$

6. a. $\dfrac{9!}{3!} = 60\,480$ **b.** $\dfrac{5!}{3!} = 20$ **c.** $\dfrac{18!}{13!} = 1\,028\,160$

7. a. $27\,907\,200$ **b.** $639\,200$ **c.** $1\,028\,160$

8. 3024

9. 2184

10. a. $\dfrac{^{15}P_4}{4}$ **b.** 8190 **c.** $\dfrac{1}{16}$

11. 3360

12. $362\,880$

13. $479\,001\,600$

14. a. $^5P_3 = 60$ **b.** $^5P_4 = 120$ **c.** $^5P_5 = 120$

15. a. 56 **b.** $\dfrac{1}{8}$

16. $358\,800$

17. 120

18. D

19. D

20. The teacher's estimations are reasonable.
Sample responses can be found in the worked solutions in the online resources.

1.4 Permutations with restrictions

1.4 Exercise

1. 10

2. 1260

3. a. $^6P_6 = 720$ **b.** $\dfrac{^6P_6}{2} = 360$

4. $83\,160$

5. $27\,720$

6. $1\,307\,504$

7. a. 5.42×10^{10} **b.** 3.63×10^9
   c. 4.00×10^{10} **d.** 1.45×10^{10}

8. a. 120 **b.** 20 **c.** 60 **d.** 60

9. a. $30\,240$ **b.** 3024 **c.** 6720 **d.** $15\,120$

10. a. 120 **b.** 48 **c.** 72 **d.** $\dfrac{3}{5}$

11. a. 1680 **b.** 180 **c.** 360 **d.** 840

12. a. 1320 **b.** 110

13. a. $80\,640$ **b.** $282\,240$
   c. $119\,750\,440$ **d.** $\dfrac{1}{6}$

14. B

15. C

16. a. 720 **b.** 24, OYSTER

17. 5.9%

18. $\dfrac{1}{6}$

1.5 Combinations

1.5 Exercise

1. a. $\dfrac{^8P_3}{3!}$ **b.** $\dfrac{^{19}P_2}{2!}$ **c.** $\dfrac{^1P_1}{1!}$ **d.** $\dfrac{^5P_0}{0!}$

2. a. $^8C_2$ **b.** $^9C_3$ **c.** $^8C_0$ **d.** $^{10}C_4$

3. a. 1 **b.** 20 **c.** 120 **d.** 220

4. 100

5. a. 28 **b.** 120

6. D

7. C

8. 1365

9. 252

10. a. 495 **b.** 11 **c.** 1
   d. $54\,264$ **e.** 100 **f.** 680

11. a. $2\,598\,960$ **b.** $65\,780$
   c. $65\,780$ **d.** $2\,467\,400$

12. a. 2002 **b.** 840 **c.** 126 **d.** $\dfrac{36}{91}$

13. $59\,400$

14. $201\,376$

15. a. i. $220, 220$ **ii.** $6435, 6435$
   iii. $10, 10$ **iv.** $56, 56$
   b. The value of ^nC_r is the same as $^nC_{n-r}$.

16. a. 120 **b.** 10 days

17. a. 57 **b.** 4 days 6 hours

18. a. 15 **b.** 1 day 4 hours

19. a. $8\,145\,060$ **b.** $11\,480$

20. $\dfrac{1}{2}(k-1)\,^nC_k$

1.6 Applications of permutations and combinations

1.6 Exercise

1. a. 45 **b.** 120 **c.** 210 **d.** 105

2. a. 720 **b.** 10 **c.** 120 **d.** 2.4×10^{18}

3. 24

4. 24

5. a. 210 **b.** 126 **c.** 84 **d.** 140

6. a. 126 **b.** 56 **c.** 21 **d.** 70

7. D

8. a. 28
   b. Sample responses include: 9 17 27 41 12 14
      9 17 27 41 12 37   9 17 27 41 12 34
   c. 6

9. a. 84
   b. Sample responses include: 7 15 25 32 10 12
      7 15 25 32 10 35   7 15 25 32 10 37
   c. 10

10. a. 8008   b. 5005   c. 5005   d. 4004

11. a. 38 760   b. 1140
   c. 34 220   d. 39 010 800

12. a. 1
   b. i. 38 760
     ii. 50 063 860
     iii. 1 940 475 213 600
     iv. 4 191 844 505 805 495

13. 376 992

14. D

15. $\dfrac{14}{95}$

16. 2450

1.7 Review

1.7 Exercise

1. a. $|R \cup S \cup T| = |R| + |S| + |T| - |R \cap S| - |R \cap T| - |S \cap T| + |R \cap S \cap T|$
   b. $n(\text{red, court card or jack}) = 32$

2. a. If there are n ways of performing operation A and m ways of performing operation B, then there are $n \times m$ ways of performing A *and* B in the order AB.
   b. 64 ways

3. a. $^{n}P_r$ is the number of ways of choosing r objects from n distinct things, when order is important. $^{n}P_r = \dfrac{n!}{(n-r)!}$
   b. $^{10}P_3 = 720$
   c. $^{n}P_r = n \times (n-1) \times (n-2) \times \ldots \times (n-r+1)$
$$= \dfrac{n \times (n-1) \times (n-2) \times \ldots \times (n-r+1) \times (n-r)!}{(n-r)!}$$
$$= \dfrac{n!}{(n-r)!}$$

4. $^{15}P_3 = 2730$

5. a. When people are sitting in a circle, we cannot tell the difference between arrangements such as between $\{A, B, C, D, E\}$ and $\{C, D, E, A, B\}$. In a circle, these represent the same arrangement. Therefore, there are fewer ways to arrange people in a circle than in a straight line.
   b. 24

6. 32 432 400

7. $^{30}C_{12} = {}^{30}C_{30-12} = {}^{30}C_{18} = 86 493 225$
Each separate time we choose r things from n distinct things, we also leave $n - r$ objects. (We can interchange taking and leaving.) Hence, $^{n}C_r = {}^{n}C_{n-r}$

8. 7056

9. a. 1 663 200   b. 330

10. 756 756

11. 15 120

12. a. 9000   b. $\dfrac{1}{15}$

13. 50 878

14. a. 15 years   b. 199 584

15. 156

16. 46 possible combinations

17. 540 out of 715

18. $\dfrac{2}{27}$

19. $|Q \cup R \cup S \cup T|$
$= |Q| + |R| + |S| + |T| - |Q \cap R| - |Q \cap s|$
$- |Q \cap T| - |R \cap S| - |R \cap T| - |S \cap T|$
$+ |Q \cap R \cap S| + |Q \cap R \cap T| + |Q \cap S \cap T|$
$+ |R \cap S \cap T| - |Q \cap R \cap S \cap T|$

20. Sample responses can be found in the worked solutions in the online resources.

2 Introduction to proof

LESSON SEQUENCE

Fully worked solutions for this chapter are available online.

EXAM PREPARATION

Access exam-style questions in every lesson, available online.

on Resources

Solutions	Solutions — Chapter 2 (sol-0394)
Exam questions	Exam question booklet — Chapter 2 (eqb-0280)
Digital documents	Learning matrix — Chapter 2 (doc-41922)
	Chapter summary — Chapter 2 (doc-41923)

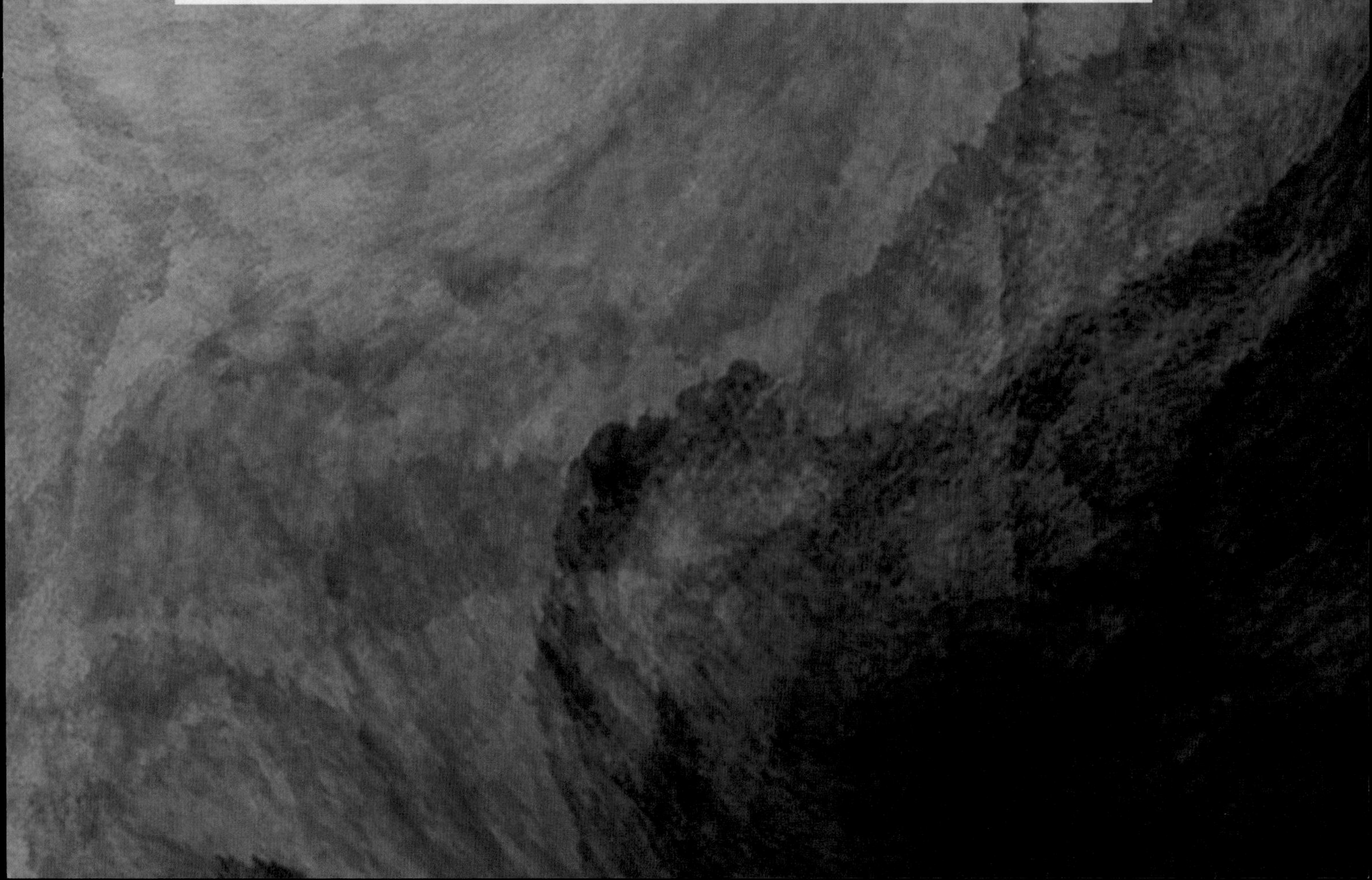

LESSON
2.1 Overview

2.1.1 Introduction

Mathematics is based on absolute truths. Theories are the best current explanation for observations, and may be later replaced by better theories, using new information, that better predict or explain how things work. Theories are useful in many fields of Science. Mathematics, however, is based on formulas and rules that are provable and always true. Provable rules include geometric relationships for calculating area and volume, Pythagoras' theorem and trigonometry formulas. Proofs make Mathematics different to the Sciences.

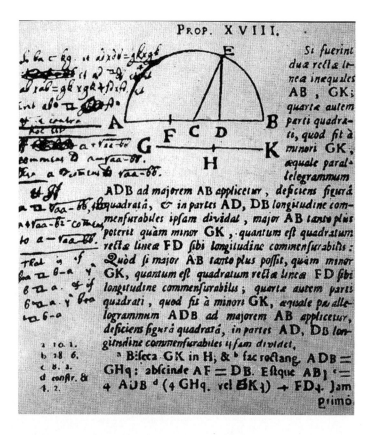

In Greece during the fifth century BCE, philosophers developed the idea of proving that a mathematical statement or proposition was true. To do this they needed to agree on the definitions of some basic terms. Also, to have starting points for their arguments, they needed to agree that some basic statements, called axioms, were true. Once a statement was proven true using a rigorous logical argument based on the definitions and axioms, it was called a theorem. Once a theorem was proved, it could be used to prove other more complicated theorems. Euclid (325–263 BCE) collected these definitions, theorems and proofs into a series of 13 books called *The Elements* — a book so influential it has been used for over 2000 years, not only in the study of mathematics, but also the development of logic. The image shows Sir Isaac Newton's copy of Euclid's work with his own notes in the margin.

However, in 1931, Kurt Gödel (1906–1978) proved that in any complex mathematical system with a certain number of axioms, there will be some statements that can be neither proved nor disproved using the axioms.

If all theorems have already been proved, why is it important to learn how to construct a proof? The first reason for a proof is to convince someone that a statement is true. The second reason is that it helps us to understand why the statement is true.

2.1.2 Syllabus links

Lesson	Lesson title	Syllabus links
2.2	**Number systems and writing propositions**	◯ Use implication, converse, equivalence, negation, contrapositive.
		◯ Use the symbols for implication ($\Rightarrow$), equivalence ($\Leftrightarrow$), and equality ($=$).
		◯ Use the quantifiers 'for all' ($\forall$) and 'there exists' ($\exists$).
		◯ Define and use set notation of number systems, including integers ($\mathbb{Z}$), positive integers ($\mathbb{Z}^+$), negative integers ($\mathbb{Z}^-$), rational numbers ($\mathbb{Q}$), irrational numbers ($\mathbb{Q}'$), and real numbers ($\mathbb{R}$).
		◯ Use the set notation symbol 'is an element of' ($\in$).
2.3	**Direct proofs using Euclidean geometry**	◯ Use implication, converse, equivalence, negation, contrapositive.
2.4	**Indirect methods of proof**	◯ Use proof by contradiction.
		◯ Prove irrationality by contradiction.
		◯ Use examples and counterexamples.
2.5	**Proofs with rational and irrational numbers**	◯ Prove results involving integers, e.g. proving that the product of two consecutive odd numbers is an odd number and $5n^2 + 3n + 6 \, \forall n \in \mathbb{Z}$ is an even number.
		◯ Express rational numbers as terminating or eventually recurring decimals and vice versa.
		◯ Prove irrationality by contradiction.

Source: Specialist Mathematics Senior Syllabus 2024 © State of Queensland (QCAA) 2024; licensed under CC BY 4.0

LESSON
2.2 Number systems and writing propositions

SYLLABUS LINKS

- Use implication, converse, equivalence, negation, contrapositive.
- Use the symbols for implication ($\Rightarrow$), equivalence ($\Leftrightarrow$), and equality ($=$).
- Use the quantifiers 'for all' ($\forall$) and 'there exists' ($\exists$).
- Define and use set notation of number systems, including integers ($\mathbb{Z}$), positive integers $\left(\mathbb{Z}^+\right)$, negative integers ($\mathbb{Z}^-$), rational numbers ($\mathbb{Q}$), irrational numbers $\left(\mathbb{Q}'\right)$, and real numbers ($\mathbb{R}$).
- Use the set notation symbol 'is an element of' ($\in$).

Source: Specialist Mathematics Senior Syllabus 2024 © State of Queensland (QCAA) 2024; licensed under CC BY 4.0

2.2.1 The real number system

The set of **real numbers** consists of all the numbers that can be thought of as points on a number line. The set of real numbers is represented by the symbol $\mathbb{R}$.

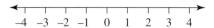

The set of real numbers has a number of subsets.

The set of **natural numbers** (or counting numbers) is represented by the symbol $\mathbb{N}$, where $\mathbb{N} = \{1, 2, 3, ...\}$.

The set of **integers** consists of all the positive and negative whole numbers and 0 (which is neither positive nor negative). Represented by the symbol $\mathbb{Z}$, the set of integers can be divided into the subsets of positive integers (or natural numbers), represented by $\mathbb{Z}^+$ (or $\mathbb{N}$), and negative integers, represented by $\mathbb{Z}^-$. That is:

$$\mathbb{Z} = \{..., -3, -2, -1, 0, 1, 2, 3, ...\}$$
$$\mathbb{Z}^+ = \{1, 2, 3, 4, 5, 6, ...\}$$
$$\mathbb{Z}^- = \{-1, -2, -3, -4, -5, -6, ...\}$$

Using set notation, $\mathbb{Z} = \mathbb{Z}^+ \cup \mathbb{Z}^- \cup \{0\}$.

Integers and natural numbers may be represented on the number line as illustrated below.

The set of integers	The set of positive integers or natural numbers	The set of negative integers

Note: Integers on the number line are marked with a solid dot to indicate that they are the only points we are interested in.

When an integer is divided by another integer, the result is a **rational number**. More formally, a rational number can be written as the ratio of two integers, a and b, in the form $\dfrac{a}{b}$ where $b \neq 0$ and a and b do not have any common factors (except 1). The set of rational numbers is represented by the symbol $\mathbb{Q}$.

In set notation,

$$\mathbb{Q} = \left\{ \frac{a}{b}, \ a, b \in \mathbb{Z}, \ b \neq 0, \ \gcd(a, b) = 1 \right\}$$

where $\gcd(a, b) = 1$ means that the greatest common divisor of a and b is 1.

If $b = 1$, then $\dfrac{a}{b} = a$. Therefore, integers are a subset of the set of rational numbers.

Some rational numbers may be expressed as terminating decimals (that is, they contain a specific number of digits). For example:

$$\frac{1}{2} = 0.5$$

$$\frac{1}{8} = 0.125$$

$$-\frac{9}{5} = -1.8$$

Other rational numbers may be expressed as recurring decimals (non-terminating or periodic decimals). For example:

$$\frac{1}{3} = 0.333\,333\,3\ldots = 0.\dot{3}$$

$$\frac{7}{11} = 0.636\,363\,63\ldots = 0.\dot{6}\dot{3} = 0.\overline{63}$$

$$\frac{53}{90} = 0.588\,888\,88\ldots = 0.58\dot{8}$$

$$\frac{15}{7} = 2.142\,857\,142\,857\ldots = 2.\dot{1}42\,85\dot{7} = 2.\overline{142\,857}$$

Note: Recurring decimals with more than one digit can be shown by either a dot over the first digit and the last digit in the recurring sequence, or an overscore over the recurring sequence.

Rational numbers can be represented on a number line as illustrated below.

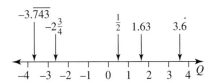

Any real numbers that are not rational numbers are called **irrational numbers** . These include **surds** (for example $\sqrt{2}$ and $\sqrt{10}$) and all other decimals that neither terminate nor recur, such as π and e. There is no common symbol for irrational numbers, but because they are the set of real numbers excluding the set of rational numbers, they can be represented by $\mathbb{R} \setminus \mathbb{Q}$.

Irrational numbers may be represented by decimals. For example,

$$\sqrt{0.03} = 0.173\,205\,080\ldots$$
$$\sqrt{18} = 4.242\,640\,68\ldots$$
$$-\sqrt{5} = -2.236.6797\ldots$$
$$-2\sqrt{7} = -5.291\,502\,62\ldots$$
$$\pi = 3.141\,592\,653\ldots$$
$$e = 2.718\,281\,828\ldots$$

These decimal approximations can be shown on a number line as illustrated below.

Using geometry, Pythagoras' **theorem** and a compass, irrational numbers in surd form can be represented on the number line exactly, as shown.

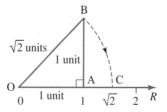

In summary, the set of real numbers can be divided into two main sets: rational and irrational numbers. These may be divided into further subsets as illustrated below.

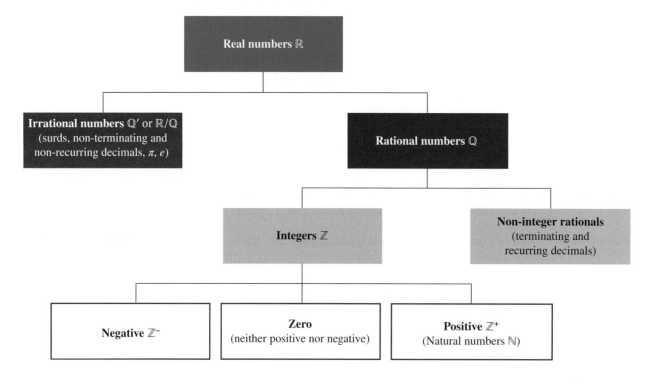

Note that 0 is an integer, but is not, in this resource, included in the natural numbers $\mathbb{N}$.

Definitions in mathematics are precise descriptions of mathematical terms, such as those given above for real numbers, integers, rational numbers and irrational numbers. Definitions make it easier for us to discuss and apply mathematical concepts.

WORKED EXAMPLE 1 Classifying numbers as rational or irrational

Identify the following numbers as either rational or irrational.

a. 7

b. $\dfrac{3}{11}$

c. 0.25

d. 0.010 110 111 …

e. $3.\dot{7}$

f. $\sqrt{5}$

THINK

a. 7 is an integer. All integers are rational.

b. $\dfrac{3}{11}$ is the quotient of two integers. It is rational.

WRITE

a. Rational

b. Rational

---

c. 0.25 is a terminating decimal. It is rational.

c. Rational

d. 0.010 110 111 ... is neither terminating nor eventually repeating. It is irrational.

d. Irrational

e. $3.\dot{7}$ is a recurring decimal. It is rational.

e. Rational

f. $\sqrt{5}$ is a surd. It is irrational.

f. Irrational

2.2.2 Converting between fractions and decimals

Rational numbers can be expressed as the quotient of two integers or as numbers whose decimal expansions are either terminating or eventually recurring. It is necessary to be able to express each rational number in both formats.

Expressing common fractions as decimals

Consider the following decimal fractions and their common fraction equivalents.

$$0.1 = \frac{1}{10} \qquad 0.37 = \frac{37}{100} \qquad 0.163 = \frac{163}{100}$$

All terminating decimals can be expressed with a power of 10 as the denominator. As the only prime factors of 10 are 2 and 5, any common fraction whose denominator can be written as powers of 2 and/or 5 will be a terminating decimal.

Consider $\frac{1}{40}$. The denominator can be written as powers of 2 and 5: $40 = 2^3 \times 5$. Therefore, the number can be expressed as a terminating decimal. $\frac{1}{40} = \frac{1}{40} \times \frac{25}{25} = \frac{25}{1000} = 0.025$.

Consider $\frac{1}{3}$. As the denominator is not a multiple of either 2 or 5, the number will be a recurring decimal.

This decimal can be calculated by division: $3\overline{)1.{}^1 0{}^1 0{}^1 0}$. Therefore, $\frac{1}{3} = 0.\dot{3}$.

Likewise, the rational number $\frac{5}{6}$ is a recurring decimal as $\frac{5}{6} = 0.8\dot{3}$, and can be calculated by division:

$6\overline{)5.{}^5 0{}^2 0{}^2 0}$. Therefore, $\frac{5}{6} = 0.8\dot{3}$.

WORKED EXAMPLE 2 Converting fractions to decimals

Express the following numbers as decimals.

a. $\dfrac{1}{50}$ b. $\dfrac{7}{9}$ c. $\dfrac{5}{11}$

THINK

a. 1. The denominator can be written as powers of 2 and 5.

WRITE

a. $50 = 5^2 \times 2$

---

2. Write the denominator as a power of 10. As $50 = 5^2 \times 2$, it needs to be multiplied by 2 in order to be written as a power of 10 ($10^2 = 100$).

$$\frac{1}{50} = \frac{1}{5^2 \times 2} \times \frac{2}{2}$$
$$= \frac{1 \times 2}{(5 \times 2)^2}$$
$$= \frac{2}{10^2}$$

3. Write the result as the decimal fraction 2 hundredths.

$$\frac{1}{50} = \frac{2}{100}$$
$$= 0.02$$

b. 1. The denominator cannot be written as powers of 2 and 5. The decimal will be recurring. Use division to determine the decimal equivalent.

b.
$$\begin{array}{r} 0.\,7\,7... \\ 9{\overline{\smash{\big)}\,7.^70^70^70}} \end{array}$$

2. Write the result as a recurring decimal.

$$\frac{7}{9} = 0.\dot{7}$$

c. 1. The denominator cannot be written as powers of 2 and 5. The decimal will be recurring. Use division to determine the decimal equivalent.

c.
$$\begin{array}{r} 0.\,4\,5\,4\,5... \\ 11{\overline{\smash{\big)}\,5.^50^60^50^60^50}} \end{array}$$

2. Write the result as a recurring decimal.

$$\frac{5}{11} = 0.\dot{4}\dot{5} \text{ (or } 0.\overline{45})$$

TI | THINK	DISPLAY/WRITE	CASIO | THINK	DISPLAY/WRITE
c. On a Calculator page, complete the entry line as $\frac{5}{11}$. Press MENU then select: 2 Number 1 Convert to Decimal Then press ENTER. Alternatively, complete the entry line as $\frac{5}{11}$ and then press ctrl ENTER.		c. On a Main screen, complete the entry line as $\frac{5}{11}$ and click on the fraction/decimal conversion button (the S↔D button).	

Expressing decimals as fractions

The fraction 0.04 can be expressed as $\dfrac{4}{100}$. This simplifies to $\dfrac{1}{25}$.

For recurring decimals, the process is demonstrated in the following example.

WORKED EXAMPLE 3 Expressing recurring decimals as fractions

Express the following recurring decimals as common fractions.
a. $0.\dot{2}$ b. $3.2\dot{1}\dot{5}$

THINK

a. 1. Let variable x equal the recurring decimal.

2. $10x$ and x will have identical recurring decimals.

WRITE

a. Let $x = 0.\dot{2}$
$\phantom{\text{Let } x} = 0.2222\ldots$

$10x = 2.2222\ldots$

3. Subtracting x from $10x$ will result in a whole number.

$$10x - x = 2.\dot{2} - 0.\dot{2}$$
$$9x = 2$$

4. Solve for x.

$$x = \frac{2}{9}$$

5. Write your concluding statement.

$$0.\dot{2} = \frac{2}{9}$$

b. 1. Let variable x equal the recurring decimal.

b. Let $x = 3.2\dot{1}\dot{5}$.

2. Calculating $10x$ will result in a number with only the recurring part after the decimal.

$$10x = 32.\dot{1}\dot{5}$$

3. $10x$ and $1000x$ will have identical recurring decimals.

$$1000x = 3215.\dot{1}\dot{5}$$

4. Subtracting $10x$ from $1000x$ will result in a whole number.

$$1000x - 10x = 3215.\dot{1}\dot{5} - 32.\dot{1}\dot{5}$$
$$990x = 3183$$

5. Solve for x and simplify.

$$x = \frac{3183}{990}$$
$$= \frac{1061}{330}$$

6. Write your concluding statement.

$$3.2\dot{1}\dot{5} = \frac{1061}{330} \text{ or}$$
$$3.2\dot{1}\dot{5} = 3\frac{71}{330}$$

TI | THINK

b. On a Calculator page, complete the entry line as 3.2151515151515. Press MENU then select:
2 Number
2 Approximate to Fraction
Then press ENTER.

DISPLAY/WRITE

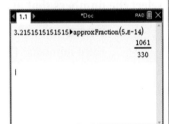

CASIO | THINK

b. On a Main screen, complete the entry line as 3.2151515151515 and click on the fraction/decimal conversion button (the S↔D button).

DISPLAY/WRITE

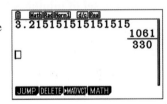

2.2.3 Propositions

A **proposition** or **mathematical statement** is a sentence that is either true or false. It may contain words and symbols. For example, '3 is a prime number' is a mathematical statement that is true. However, '$\sqrt{7} = 3$' is a false mathematical statement. Propositions may be combined to make larger, more complex propositions (that must be either true or false, but not both). In propositions, some words have very precise meanings that may differ slightly from their everyday use.

Either/or

In everyday language, 'either A or B' means that one of the options is true, but not both. For example, if you were asked if you would like salad or vegetables with your meal, the wait staff are not expecting you to choose both. However, in mathematics, saying that A or B is true means that at least one of them is true (which also means that they both might be).

A truth table can be used to display all possible true/false combinations. A truth table contains one column for each input variable, and a column showing all possible logically valid results. Each row contains the unique possible configurations of the input values.

A	B	A or B
T	T	T
T	F	T
F	T	T
F	F	F

A or B can also be presented in a Venn diagram, 'A or B' = $A \cup B$.

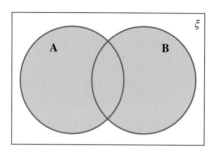

WORKED EXAMPLE 4 Classifying integers as even and/or prime 1

From the set of numbers {1, 2, 3, 4, 5, 6, 7, 8, 9}, select the numbers that are either even or prime.

THINK	WRITE
1. Select the even numbers.	Even numbers: {2, 4, 6, 8}
2. Select the prime numbers.	Prime numbers: {2, 3, 5, 7}
3. Numbers that are either even or prime will be in at least one of the sets.	Even or prime: {2, 3, 4, 5, 6, 7, 8}

And

The use of the word 'and' in mathematics and everyday language is similar. In mathematics, saying that A and B is true means both of them are true.

A truth table for A and B displays all possible true/false combinations, $A \cap B$.

A	B	A and B
T	T	T
T	F	F
F	T	F
F	F	F

A and B can also be presented in a Venn diagram, 'A and B' = A ∩ B.

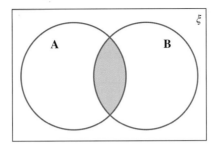

WORKED EXAMPLE 5 Classifying integers as even and prime 2

From the set of numbers {1, 2, 3, 4, 5, 6, 7, 8, 9}, select the numbers that are even and prime.

THINK	WRITE
1. Select the even numbers.	Even numbers: {2, 4, 6, 8}
2. Select the prime numbers.	Prime numbers: {2, 3, 5, 7}
3. Numbers that are even and prime will be in both sets.	Even and prime: {2}

WORKED EXAMPLE 6 Determining the truth of a statement

Consider the following statement:
'A triangle has two congruent sides, and its interior angles add up to 180 degrees.'
a. Is this statement true or false? Explain why.
b. Can you think of any specific types of triangles that fit this description?

THINK

a. Regardless of the triangle's specific type, the sum of its interior angles is always 180 degrees. This is a fundamental property of triangles.

b. A triangle that has two sides of equal length is called isosceles triangle.
For the "AND" statement to be true, both parts of the statement need to be true.

WRITE

a. True. This statement can be true; it describes an isosceles triangle, in which two sides are of equal length. Since the second part of the statement is true for all triangles, the entire statement is true for all isosceles triangles.

b. Examples of isosceles triangles:
  • Equilateral triangle: All three sides and angles are congruent.
  • Right isosceles triangle: Two sides are congruent, and one angle is a right angle (90 degrees).
  • Any isosceles triangle with two sides and two angles equal, with the third side and angle being any value, such that the angles add to 180 degrees.

Negation

Sometimes it is necessary to know what the opposite of a statement is. This is called 'negating' a statement.

The **negation** of a true statement is false, and the negation of a false statement is true.

Using symbols, the negation of statement A is written as ¬A.

The negation of the earlier true statement '3 is a prime number' is the false statement '3 is not a prime number'.

The negation of A can be shown in a truth table.

A	¬A
T	F
F	T

¬A can be presented in a Venn diagram as A′.

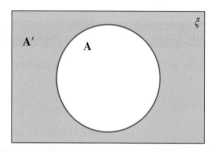

WORKED EXAMPLE 7 Writing negations of statements

Write the negation of the statement 5 > 7.

THINK	WRITE
1. The negation of 'larger than' is 'not larger than'.	$5 \not> 7$
2. If a number is not larger than a number, it must be smaller than or equal to the number. This is a better way to write the statement.	$5 \leq 7$

Negation of A or B

In Worked example 4, we looked at numbers that were even numbers or prime numbers. The numbers that weren't included were the ones that were not even and were not prime.

In general, the negation of the statement 'A or B' becomes 'not A and not B'.

Compare ¬(A or B) with ¬A and ¬B in the truth table below:

A	B	(A or B)	¬(A or B)
T	T	T	F
T	F	T	F
F	T	T	F
F	F	F	T

A	B	¬A	¬B	(¬A and ¬B)
T	T	F	F	F
T	F	F	T	F
F	T	T	F	F
F	F	T	T	T

Let us examine the second row of both tables, where A is true and B is false. In the upper table (¬(A or B)), (A or B) is true, because in an 'or' statement, at least one option must be true (but both options do not need to be true). If (A or B) is true, then ¬(A or B) must be false.

Comparing this to the same line in the lower table (¬A and ¬B), ¬A is false and ¬B is true, so (¬A and ¬B) must be false, because in an 'and' statement, both options must be true.

Hence, ¬(A or B) is the same as ¬A and ¬B. It can also be presented on a Venn diagram, $(A \cup B)'$.

The following Worked example is an extension of Worked example 4.

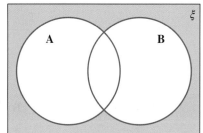

WORKED EXAMPLE 8 Using negation of A or B

From the set of numbers {1, 2, 3, 4, 5, 6, 7, 8, 9}, the set of numbers that are either even or prime is {2, 3, 4, 5, 6, 7, 8}.
a. Select the numbers that are neither even nor prime.
b. Identify another way this set of numbers could be described.

THINK	WRITE
a. From {1, 2, 3, 4, 5, 6, 7, 8, 9} remove the elements that are 'not even' and also remove the elements that are 'not prime'. Recall that 'or' means it is in one of those categories or the other, not necessarily both: The numbers that are neither even nor prime are the numbers that are not in the set {2, 3, 4, 5, 6, 7, 8}.	a. Neither even nor prime: {1, 9}
b. 1. Identify the set of numbers that are not even (i.e. the odd numbers). Identify the set of numbers that are not prime.	b. Not even (odd) numbers: {1, 3, 5, 7, 9} Not prime numbers: {1, 4, 6, 8, 9}
2. The members of the set {1, 9} are members of the set of odd numbers and the set of numbers that are not prime.	The set {1, 9} could also be described as the set of numbers that are odd **and** not prime, that is the intersection of numbers that are not even with the numbers that are not prime.

Negation of A and B

In Worked example 5, we looked at numbers from the set $\{1, 2, 3, 4, 5, 6, 7, 8, 9\}$ that were even numbers and prime numbers, and the only result was 2. The numbers that were not included were the ones that weren't even or weren't prime. Numbers that are (even **and** prime) are numbers that are not ('not even' **or** 'not prime'). The 'or' is used here because we are combining the 'not even' with the 'not prime' when deciding which elements to remove from the set.

In general, the negation of the statement 'A and B' becomes 'not A or not B'.

By comparing ¬(A and B) with ¬A or ¬B, this can be confirmed.

A	B	(A and B)	¬(A and B)
T	T	T	F
T	F	F	T
F	T	F	T
F	F	F	T

A	B	¬A	¬B	(¬A or ¬B)
T	T	F	F	F
T	F	F	T	T
F	T	T	F	T
F	F	T	T	T

As above, by examining the second row of both tables, we see that when A is true and B is false, (A and B) is false, so ¬(A and B) is true. (In 'and' statements, both options must be true.) Comparing this to the same line in the lower table (¬A or ¬B), ¬A is false and ¬B is true, so (¬A or ¬B) is true, as only one option must be true in an 'or' statement.

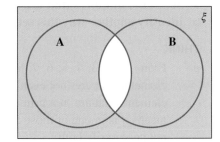

Hence, the negation of A and B, ¬(A and B), is the same as ¬A or ¬B. This can also be presented on a Venn diagram, $(A \cap B)'$ as the negation of A and B is $A' \cup B'$, or $(A \cap B)'$.

The following Worked example is an extension of Worked example 5.

WORKED EXAMPLE 9 Using negation of A and B

From the set of numbers {1, 2, 3, 4, 5, 6, 7, 8, 9}, the set of numbers that are even and prime is {2}.
a. Select the numbers that are not (even and prime).
b. Identify another way this set of numbers could be described.

THINK	WRITE
a. The numbers that are not (even and prime) are the numbers that are not in the set {2}.	Not (even and prime): {1, 3, 4, 5, 6, 7, 8, 9}
b. 1. Identify the set of numbers that are not even (that is, odd numbers). Identify the set of numbers that are not prime.	Not even (odd) numbers: {1, 3, 5, 7, 9} Not prime numbers: {1, 4, 6, 8, 9}
2. The members of the set {1, 3, 4, 5, 6, 7, 8, 9} are members of the set of odd numbers or the set of non-prime numbers.	This could also be described as the set of numbers that are not even or not prime.

If, then statements (implication)

Statements using **implication** (also called implicative statements) are normally written in the form 'if P, then Q' or 'P implies Q'.

In mathematical proof, an implication is a logical statement that connects two statements, often referred to as the hypothesis (P) and the conclusion (Q).

Here's a breakdown of its key aspects:

Conditional Statement: An implication doesn't assert that either P or Q is true on its own. It merely states that if P is true, then Q must also be true. It's a conditional relationship.

Truth Value: An implication is considered true as long as it doesn't lead to a false conclusion from a true hypothesis. In other words, the only way for an implication to be false is if P is true and Q is false.

Some examples are:
- If $x = 5$, then $x^2 = 25$.
- If $x \neq 0$, then $\dfrac{1}{x}$ is real.
- If a quadrilateral is a rectangle, then its diagonals are congruent and bisect each other.

'If P, then Q' can be written using symbols as $P \Rightarrow Q$. The above statements can be rewritten as:
- $x = 5 \Rightarrow x^2 = 25$
- $x \neq 0 \Rightarrow \dfrac{1}{x} \in \mathbb{R}$
- A quadrilateral is a rectangle $\Rightarrow$ its diagonals are congruent and bisect each other.

Implicative statements only tell you about what happens if P is true. They do not tell you what happens if Q is true.

An implicative statement can be shown on a Venn diagram. Inside the circle labelled 'P is true', the statement P is true. Outside the circle labelled 'P is true', the statement P is false.

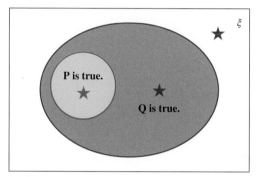

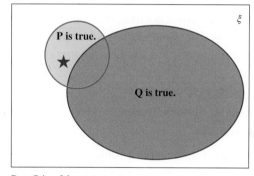

★ Whenever P is true, Q is also true.

★ Whenever P is not true, Q might or might not be true.

$P \Rightarrow Q$ is a false statement.

★ P is true but Q is not true.

To disprove the implicative statement $P \Rightarrow Q$, you need to find a case where P is true but Q is not true. As seen in the truth table for $P \Rightarrow Q$, there is only one combination of P and Q that will make the statement false.

P	Q	$P \Rightarrow Q$
T	T	T
T	F	F
F	T	T
F	F	T

Note: If P is false, the statement $P \Rightarrow Q$ is said to be 'vacuously true'. That means it is technically true but doesn't really tell us anything. 'If pigs could fly, then …' (You can put what you like there and it is vacuously true, but pigs can't fly.)

The negation of the statement $P \Rightarrow Q$ can be considered a false statement, that is, $P \Rightarrow Q$ is false if and only if (**iff**) P is true and Q is false. Hence, the negation of $P \Rightarrow Q$ is logically P and not Q (P and ¬Q). This is demonstrated in the truth table.

P	Q	¬Q	P and ¬Q
T	T	F	F
T	F	T	T
F	T	F	F
F	F	T	F

WORKED EXAMPLE 10 Propositions using implication

Determine if the following are true statements.
a. **If a shape has four sides, then it is a rectangle.**
b. **If $x^3 > 8$, then $x > 0$.**

THINK

a. 1. This is an 'if, then' statement. Identify the propositions.

   2. Consider if the statement is true.
If P is true (the shape has four sides), that means Q is also true (the shape is a rectangle).
If the statement is not true, then we should be able to determine when P is true (the shape has four sides) and Q is not true (the shape is not a rectangle).

   3. Write the concluding statement.

b. 1. This is an 'if, then' statement. Identify the propositions.

   2. Consider if the statement is true.
If P is true ($x^3 > 8$), that means Q is also true ($x > 0$).
If the statement is not true, then we should be able to determine when P is true and Q is not true.

   3. Whenever P is true, Q is also true. Write the concluding statement.

WRITE

a. P: A shape has four sides.
Q: It is a rectangle.

A parallelogram is an example of a shape that has four sides but is not a rectangle.

The statement is false.

b. P: $x^3 > 8$
Q: $x > 0$

P: $x^3 > 8$
  $x > 2$

Q is true.
P is true.

The statement is true.

Equivalence

The **converse** of the statement 'if P, then Q' is 'if Q, then P' or $Q \Rightarrow P$. The converse of a statement may or may not be true. The converse of the previous examples can be written as follows.

- If $x^2 = 25$, then $x = 5$. In this case the converse is false, as $x = -5$ is also possible. This means that it is possible for $x^2 = 25$ but $x \neq 5$.
- If $x^3 = 27$, then $x = 3$. In this case the converse is true.
- If the diagonals of a quadrilateral are congruent and bisect each other, then it is a rectangle. In this case the converse is true, remembering a square is also rectangle.

If $P \Rightarrow Q$ and $Q \Rightarrow P$, then P and Q are said to be **equivalent statements**.

Equivalent statements can also be expressed as 'P if and only if Q'. This is written symbolically as $P \Leftrightarrow Q$. Rewriting the first two examples above using symbols gives us:

- $x = 5 \Rightarrow x^2 = 25$
- $x = 3 \Leftrightarrow x^3 = 27$

$P \Leftrightarrow Q$ is true if both P and Q are true. It is also true if P and Q are both false, as illustrated in the truth table for $P \Leftrightarrow Q$.

P	Q	$P \Leftrightarrow Q$
T	T	T
T	F	F
F	T	F
F	F	T

$P \Leftrightarrow Q$ is a true statement.
★ P is true and Q is true.
★ P is false and Q is false.

WORKED EXAMPLE 11 Propositions using equivalence

For each of the following, write the converse statement and identify if implication ($\Rightarrow$) or equivalence ($\Leftrightarrow$) is the more appropriate symbol.

a. If an animal is a kangaroo, then it is a marsupial.
b. If a number, n, is odd, then n^2 is odd.
c. Events A and B are mutually exclusive if and only if the probability of both events occurring (Pr(A and B)) is equal to 0.

THINK	WRITE
a. 1. Identify the two propositions.	a. P: An animal is a kangaroo. Q: The animal is a marsupial.
2. Write the converse statement, $Q \Rightarrow P$.	The converse is 'If an animal is a marsupial, then it is a kangaroo.'
3. Determine if the converse is true.	The converse is not true.
4. The converse is not true, so implication is more appropriate.	The symbol $\Rightarrow$ should be used. $P \Rightarrow Q$
b. 1. Identify the two propositions.	b. P: n is odd. Q: n^2 is odd.
2. Write the converse statement, $Q \Rightarrow P$.	The converse is 'If n^2 is odd, then n is odd.'
3. Determine if the converse is true.	The converse is true.
4. The converse is true, so equivalence is more appropriate.	The symbol $\Leftrightarrow$ should be used. $P \Leftrightarrow Q$

c. 1. Identify the two propositions.	**c.** P: Events A and B are mutually exclusive Q: The probability of both events occurring (Pr(A and B)) is equal to 0.
2. Write the converse statement, Q ⇒ P.	Q ⇒ P: If the probability of both events occurring (Pr(A and B)) is equal to 0, then events A and B are mutually exclusive.
3. Determine if the converse is true.	If the probability of both events occurring (Pr(A and B)) is 0, it implies disjointness between them. That means, both events cannot occur simultaneously, so the converse is true.
4. Conclude. The converse is true, so equivalence is more appropriate.	P ⇔ Q

In determining if P ⇔ Q is true, it is necessary to confirm that P ⇒ Q is true and that Q ⇒ P is also true.

2.2.4 Quantifiers

The proposition $9 > 7$ is a true statement, and the proposition $4 > 7$ is a false statement. However, $x > 7$ might be true or false depending on the value of x. A proposition that includes variables is known as a **propositional function**. Because it is a function, we can use function notation to name the function. In this case, we might call it $P(x)$.

Quantifiers are used to give information about the values of the variables in propositional functions so that we can determine if the function is true or false

The universal quantifier

The **universal quantifier**, *for all*, is written with the symbol ∀. This means that all possible values for the variable are considered.

Consider the propositional function $x > 7$. Using the universal quantifier, the statement becomes 'For all real numbers x, $x > 7$.' Written with symbols, the proposition is $\forall x \in \mathbb{R}$, $x > 7$. As the proposition is not true for all possible values of x, it is a false statement.

WORKED EXAMPLE 12 Propositions using the universal quantifier

Consider the proposition $\forall x \in \mathbb{R}, x^2 \geq 0$. Write the proposition in words and determine if it is true or false.

THINK	WRITE
1. The symbol ∀ means 'for all'. $x \in \mathbb{R}$ means x is a member of the set of real numbers.	For all real numbers x, $x^2 \geq 0$.
2. If $x \geq 0$, $x^2 \geq 0$, and if $x < 0$, $x^2 > 0$. Therefore, the proposition is always true.	The proposition is true.

The existential quantifier

The **existential quantifier**, *there exists*, is written with the symbol $\exists$. This means that there is a value for the variable that would make the propositional function true.

Let us again consider the propositional function $x > 7$, but this time using the existential quantifier. The statement becomes 'There exists a real number, x, where $x > 7$.' Written in symbols, the proposition is $\exists\, x \in \mathbb{R}$, $x > 7$. In this case, because there are values of x that make the proposition true, it is true.

WORKED EXAMPLE 13 Propositions using the existential quantifier

Consider the proposition $\exists x \in \mathbb{N}, 2x + 1$ is a multiple of 3. Write the proposition in words and determine if it is true or false.

THINK	WRITE
1. The symbol $\exists$ means 'there exists'. $x \in \mathbb{N}$ means x is a member of the set of natural numbers.	There exists a natural number x where $2x + 1$ is a multiple of 3.
2. If $x = 1$, then $2x + 1$ is a multiple of 3.	The proposition is true for $x = 1$ and is therefore true.

Negating quantifiers

Consider the statement 'For all even numbers, x, x^2 is even.' This is a true statement. However, it would not be true if you could find one even number, x, where x^2 was not even.

Now consider the false statement 'All prime numbers are odd.' This can be rewritten as 'For all prime numbers, p, p is odd.' The negation of this false statement is the true statement 'There exists a prime number, p, where p is not odd.'

The negation of the statement $\forall x \in X$, $P(x)$ is the statement $\exists x \in X$, $\neg P(x)$

This also means that the negation of the statement $\exists x \in X$, $P(x)$ is the statement $\forall x \in X$, $\neg P(x)$.

Combining quantifiers

If a propositional function has more than one variable, then quantifiers are needed for all of the variables. If the quantifiers are the same type, the order that they are written in is immaterial. Also, if a quantifier is shown before a group of variables separated by commas, the quantifier applies to all of the variables in the group. For example, the following statements are equivalent.

$$\forall x \in \mathbb{R}, \forall y \in \mathbb{R}$$

$$\forall y \in \mathbb{R}, \forall x \in \mathbb{R}$$

$$\forall x, y \in \mathbb{R}$$

If the quantifiers are different, then the order becomes important. The statement $\forall x, \exists y$ means that for each possible value of x, there exists a value for y that makes a proposition true. The statement $\exists x, \forall y$ means that there is a value for x that makes the statement true for any possible value of y.

For each of the propositions below, write the proposition in words and determine if it is true or false.

a. $\forall x \in \mathbb{R}, \exists y \in \mathbb{R}, y > x^2$

b. $\exists x \in \mathbb{R}, \forall y \in \mathbb{R}, y > x^2$

THINK	WRITE
a. 1. $\forall x$, $\exists y$ means that for each possible value of x, there exists a value for y that makes a proposition true.	a. For all real numbers x, there exists a real number y where $y > x^2$.
2. For every real number, there is another real number greater than the square of the original. Therefore, the proposition is true.	True
b. 1. $\exists x$, $\forall y$ means that there is a value for x that makes the statement true for any possible value of y.	b. There exists a real number x where for all real numbers y, $y > x^2$.
2. If y is a negative number, the inequality is never true. Therefore, the proposition is false.	False

Exercise 2.2 Number systems and writing propositions

learn on

2.2 Exercise	2.2 Exam questions on

Simple familiar	Complex familiar	Complex unfamiliar
1, 2, 3, 4, 5, 6, 7, 8, 9, 10, 11, 12, 13	14, 15	16, 17

These questions are even better in jacPLUS!
- Receive immediate feedback
- Access sample responses
- Track results and progress

Find all this and MORE in jacPLUS ▶

Simple familiar

1. **WE1** Identify the following numbers as either rational or irrational.
   a. 6.2
   b. 0.321
   c. 0.25
   d. 3.121 221 222...
   e. 156
   f. $\sqrt{13}$

2. Identify the following numbers as either rational or irrational.
   a. $\sqrt{7}$
   b. $\sqrt[3]{9}$
   c. $\sqrt{81}$
   d. π

3. **WE2** Express the following numbers as decimal fractions.
   a. $\dfrac{17}{20}$
   b. $\dfrac{107}{125}$
   c. $\dfrac{2}{3}$
   d. $\dfrac{8}{15}$

4. Express the following numbers as decimal fractions.
   a. $1\dfrac{3}{5}$
   b. $\dfrac{9}{16}$
   c. $\dfrac{22}{7}$
   d. $\dfrac{13}{12}$

5. **WE3** Express the following recurring decimals as common fractions.
   a. $0.\dot{5}$
   b. $0.\dot{2}\dot{3}$
   c. $3.0\dot{1}$
   d. $6.\dot{1}\dot{6}$
   e. $7.1\dot{1}\dot{2}$
   f. $0.\dot{7}0\dot{2}$

6. From the set of numbers {0, 3, 6, 9, 12, 15, 18, 21, 24}:
   a. **WE4** select the numbers that are either odd or less than 12
   b. **WE5** select the numbers that are odd and less than 12
   c. **WE7** select the numbers that are the negation of odd
   d. select the numbers that are not less than 12
   e. **WE8** select the numbers that are neither odd nor less than 12. Identify another way that this set of numbers could be described.
   f. **WE9** select the numbers that are not (odd and less than 12). Identify another way that this set of numbers could be described.

7. **WE10** Determine if the following are true statements.
   a. If a polygon has three sides, then it is a triangle.
   b. If a polygon is a square, then it is a rectangle.
   c. If the sides of a square are doubled in length, then the area of the square is also doubled.
   d. If $x^3 > 27$, $x \in \mathbb{N}$, then $x > 2$, $x \in \mathbb{N}$
   e. If $x^2 > 16$, $x \in \mathbb{N}$, then $x > 4$, $x \in \mathbb{N}$
   f. If $x^2 < 4$, $x \in \mathbb{N}$, then $x < 2$, $x \in \mathbb{N}$

8. **WE6** Decide if the following statements are true and justify your decision.
   a. A quadrilateral is a square when it has four congruent sides and four right angles.
   b. A quadrilateral has both two right angles and two obtuse angles.
   c. In a right triangle, the longest side must be opposite the right angle and all other angles must be acute.

9. **WE11** For each of the following, write the converse statement and identify if implication ($\Rightarrow$) or equivalence ($\Leftrightarrow$) is the more appropriate symbol.
   a. If it is a fish, then it lives in water.
   b. If the quadrilateral is a rhombus, then the diagonals bisect each other.
   c. If $n - m > 0$, then $n > m$.
   d. If $n + 1$ is odd, then n is even.

10. **WE12** For each of the propositions below, write the proposition in words and determine if it is true or false.
    a. $\forall x \in \mathbb{N}$, $2x$ is even.
    b. $\forall x \in \mathbb{N}$, $2x + 1$ is a multiple of 3.
    c. $\forall x \in \mathbb{N}$, $x > 0$
    d. $\forall x \in \mathbb{R}$, $x > 0$ or $x \leq 0$.
    e. $\forall x \in \mathbb{N}$, x is even.
    f. $\forall x \in \mathbb{R}$, $x^2 + 1 \geq 0$

11. **WE13** For each of the propositions below, write the proposition in words and determine if it is true or false.
    a. $\exists x \in \mathbb{Z}$, $x + 5 = 7$
    b. $\exists x \in \mathbb{R}$, $x^2 < 0$
    c. $\exists x \in \mathbb{N}$, x is even.
    d. $\exists x \in \mathbb{R}$, $x^2 + 1 = 0$
    e. $\exists x \in \mathbb{R}$, $x^2 - 1 = 0$
    f. $\exists x \in \mathbb{Q}$, $\sqrt{x} \in \mathbb{Q}$

12. **WE14** For each of the propositions below, write the proposition in words and determine if it is true or false.
    a. $\forall x \in \mathbb{R}$, $\exists y \in \mathbb{R}$, $y = x$
    b. $\exists x \in \mathbb{R}$, $\forall y \in \mathbb{R}$, $y = x$
    c. $\forall x \in \mathbb{R}$, $\exists y \in \mathbb{R}$, $xy = 0$
    d. $\exists x \in \mathbb{R}$, $\forall y \in \mathbb{R}$, $xy = 0$

13. For each of the propositions below, determine if it is true or false.
    a. $\forall x \in \mathbb{R}$, $\exists y \in \mathbb{R}$, $x > y$
    b. $\exists x \in \mathbb{R}$, $\forall y \in \mathbb{R}$, $x > y$
    c. $\forall x \in \mathbb{R}$, $\forall y \in \mathbb{R}$, $x > y$
    d. $\forall x \in \mathbb{R}$, $\forall y \in (x, \infty)$, $x < y$

Complex familiar

14. Consider the statement '$\forall x \in \mathbb{N}$, $x(x + 1)$ is even.' Decide if this is a true statement and justify your decision.

15. The negation of A and B can be written as 'not A or not B'. Determine another way to write the negation of A and B or C in terms of 'not A', 'not B' and 'not C'. You must provide evidence to support your claim. *Note*: 'A and B or C' can be thought of as '(A and B) or C'.

16. Use the graph of $y = \sin(x) + \cos(x)$ to complete the true proposition $\forall x \in (...)$, $\sin(x) + \cos(x) > 1$

17. Use the graph of $x^2 + y^2 = 9$ to answer the following.
    a. Use quantifiers for x and y to create a true proposition.
    b. Use quantifiers for x and y to create a false proposition.

Fully worked solutions for this chapter are available online.

LESSON
2.3 Direct proofs using Euclidean geometry

SYLLABUS LINKS

- Use implication, converse, equivalence, negation, contrapositive.

Source: Specialist Mathematics Senior Syllabus 2024 © State of Queensland (QCAA) 2024; licensed under CC BY 4.0

2.3.1 Direct proofs

When trying to prove a statement directly you might think that we can simply show that the statement is true for a few randomly selected values; however, regardless of how many cases a statement is shown to be true for, this does not prove that the statement is true in general. A direct proof is a common method used to prove statements of the form 'if P then Q' or 'P implies Q', which is written mathematically as $P \Rightarrow Q$.

The method of a direct proof is to take a statement P, which we assume to be true, and use it directly to show that Q is true.

Direct proof method

1. **Identify the statements P and Q and assume that P is true.**
2. **Use the fact that P is true to directly show that Q is true.**
3. **Therefore $P \Rightarrow Q$ is true. This completes the proof.**

Before we can start using direct proof to prove mathematical statements, we must know the exact definitions of the properties that we are wishing to prove.

2.3.2 Odd and even numbers

Definitions of odd and even numbers

An even number is an integer n such that $n = 2k$, where k is an integer.

An odd number is an integer n such that $n = 2k + 1$, where k is an integer.

Note that zero in an integer.

For example, consider the number 14. Since $14 = 2 \times 7$ (and 7 is an integer), 14 is even. Alternatively, consider the number 15. Since $15 = 2 \times 7 + 1$ (and 7 is an integer), 15 is odd. Note that 15 is not even since there is no integer k such that $2k = 15$.

All integers are either odd or even, so we could have defined odd numbers as integers which are not even.

WORKED EXAMPLE 15 Direct proof

Prove that the sum of two even integers is an even integer.

THINK	WRITE
1. Write an equivalent statement to be proved. p: a and b are even integers. q: their sum $a + b$ is even.	Proof: If a and b are even integers, then their sum $a + b$ is even.
2. Assume that p is true, that is a and b are two even integers. (Note that it would be incorrect to state $b = 2j$ as that would imply $a = b$.)	Let $a = 2j$ and $b = 2k$ where $j, k \in \mathbb{Z}$.
3. Express the sum of a and b using the expressions formed in step 2. Recall that the sum of any two integers is an integer.	$a + b = 2j + 2k$ $\quad = 2(j + k)$ $\quad = 2i$ Where $i = j + k \in \mathbb{Z}$
4. Use the definition of an even number to show that q is true.	$a + b = 2i$, $i \in \mathbb{Z}$ Therefore $a + b$ is even.
5. We have shown that $p \rightarrow q$. State the conclusion.	The sum of two even integers is an even integer.

2.3.3 Axioms and postulates

As mentioned in Section 2.1.1, in mathematics there are some basic statements that are accepted as true without proving them. In general, when these statements refer to working with numbers, the statements are called **axioms**. When the statements are about geometry, they are called **postulates**.

In *The Elements*, Euclid tried to use only four postulates in his geometric **proofs**. The first four postulates state that:
- a straight line can be drawn from one point to another (that is, any two points will form a straight line)
- it is possible to produce a finite straight line
- a circle can be described with any centre and radius
- all right angles are equal to each other.

However, he found that it was necessary to introduce a fifth postulate:
- if there is a line and a point not on the line, then only one line can be drawn through the point that is parallel to the original line.

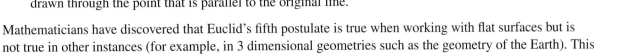

Mathematicians have discovered that Euclid's fifth postulate is true when working with flat surfaces but is not true in other instances (for example, in 3 dimensional geometries such as the geometry of the Earth). This means that geometry can be either Euclidean (a flat surface) or non-Euclidean. In this section, we are going to be constructing proofs using Euclidean geometry.

2.3.4 The fundamentals

Consider the following set of statements.
- Penguins are birds.
- Birds can fly.
- Therefore, penguins can fly.
  Or can they? Where does the logic break down?

There are a number of methods of combining relationships to produce new relationships. These techniques assist in the logical construction of proofs. We must also state the postulates that will be the basis for our proofs.

Transitive property of equality

Consider the following:

$$\left.\begin{array}{l} a=b \\ b=c \end{array}\right\} \Rightarrow a=c$$

Because both a and c are equal to b, we can conclude that $a=c$. This technique is called the transitive property of equality.

Substitution

Consider the following:

$$\left.\begin{array}{l} a+b=c \\ b=d \end{array}\right\} \Rightarrow a+d=c$$

Because $b=d$, the statement $a+b=c$ is equivalent to the statement $a+d=c$. This technique is called substitution.

Matching

Consider the following:

$$\left.\begin{array}{l} a+b=c \\ a+d=c \end{array}\right\} \Rightarrow b=d$$

Because the two statements can be rewritten as $b=c-a$ and $d=c-a$, it is clear that $b=d$. This technique is called matching.

Adding equals

Consider the following:

$$\left.\begin{array}{l} a=b \\ c=d \end{array}\right\} \Rightarrow a+c=b+d$$

As $a=b$, $a+c=b+c$.

As $c=d$, $a+c=b+d$.

This technique is called adding equals.

These techniques can be used to reach a logical conclusion that is necessarily true. This is known as deduction.

Use techniques of deduction to determine the following.

a. If $x = 20$ and $x = y$, determine y.

b. If $a + b = 16$ and $b = c$, determine $a + c$.

c. If $m + n = 45$ and $p + n = 45$, determine the relationship between m and p.

d. If $a = 6$ and $b = 10$, determine $a + b$.

e. If $x + y = 56$, $y = a$ and $a = c$, determine the relationship between x and c.

THINK

a. As x is equal to both y and 20, the transitive property of equals can be used.

b. As $b = c$, substitution can be used to replace b with c.

c. As both statements equal 45, matching can be used to show that m must equal p.

d. $a + b$ can be found by adding equals.

e. 1. Using substitution, y can be replaced by a.

   2. Using substitution again, a can be replaced by c.

WRITE

a. $x = 20$
        $x = y$
        $\therefore y = 20$

b. $a + b = 16$
         $b = c$
      $\therefore a + c = 16$

c. $m + n = 45$
     $p + n = 45$
      $\therefore m = p$

d. $a = 6$
        $b = 10$
   $\therefore a + b = 6 + 10$
          $= 16$

e. $x + y = 56$
        $y = a$
    $\therefore x + a = 56$

         $a = c$
    $\therefore x + c = 56$

2.3.5 Two additional postulates

- Adjacent angles on a straight line are supplementary; conversely, supplementary adjacent angles are on a straight line.

$\overleftrightarrow{PQR} \Leftrightarrow a + b = 180°$

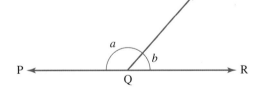

- If two lines are parallel, the corresponding angles are congruent; conversely, if corresponding angles are congruent, they are on parallel lines.

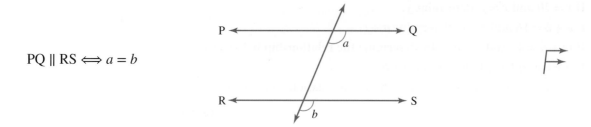

$PQ \parallel RS \iff a = b$

2.3.6 Theorems of Euclidean geometry

The direct proof is the most common form of proof in geometry. In a direct proof, the statements are made in a logical order, beginning with a statement that we know is true and ending with the statement that we wish to prove. The reason that each statement can be made is given beside the statement.

The postulates can be used to prove new theorems. Consider the following proof for vertically opposite angles:

$a + c = 180°$
$c + b = 180°$

$\therefore a = b$ (matching)

This proves that vertically opposite angles are congruent.

In a similar fashion, the following theorems can be proved in the order given using the given postulates and the previously proven theorems.
- Vertically opposite angles are congruent.

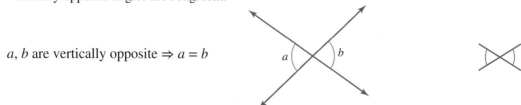

a, b are vertically opposite $\Rightarrow a = b$

- If two lines are parallel, the alternate angles are congruent; conversely, if alternate angles are congruent, the two lines are parallel.

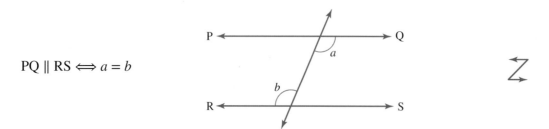

$PQ \parallel RS \iff a = b$

- If two lines are parallel, the co-interior angles are supplementary; conversely, if co-interior angles are supplementary, the two lines are parallel.

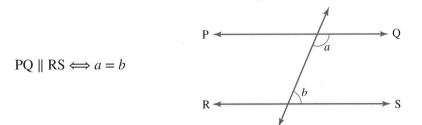

$PQ \parallel RS \Longleftrightarrow a = b$

- The exterior angle of a triangle is equal to the sum of the two interior opposite angles.

$m = a + b$

- The sum of the angles of a triangle is $180°$.

$a + b + c = 180°$

- If two sides of a triangle are congruent (i.e. the triangle is an isosceles triangle), the angles opposite those sides are also congruent. Conversely, if two angles of a triangle are congruent, the sides opposite those angles are congruent.

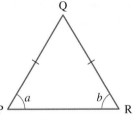

$\overline{PQ} \cong \overline{PR} \Leftrightarrow a = b$

- If three sides of a triangle are congruent (i.e. the triangle is equilateral), the angles of the triangle are also congruent and equal to $60°$. Conversely, if three angles of a triangle are congruent, the sides are congruent.

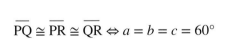 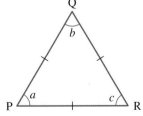

$\overline{PQ} \cong \overline{PR} \cong \overline{QR} \Leftrightarrow a = b = c = 60°$

- The sum of the angles of a quadrilateral is 360°.

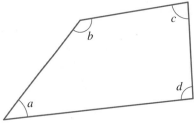

$$a + b + c + d = 360°$$

- The opposite angles of a parallelogram are congruent; conversely, if the opposite angles of a quadrilateral are congruent, it is a parallelogram.

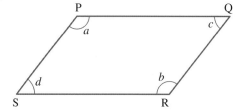

PQRS is a parallelogram
$\Longleftrightarrow a = b$ and $c = d$

- Angles at a point add to 360°.

$a + b = 180°$
$c + d = 180°$
$a + b + c + d = 360°$

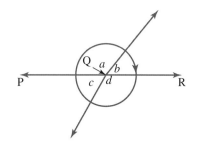

WORKED EXAMPLE 17 Direct proof and geometry

Use the diagram shown, the postulates given above and the vertically opposite angles theorem to prove that:
a. if two lines are parallel, the alternate angles are congruent
b. if alternate angles are congruent, the lines are parallel.

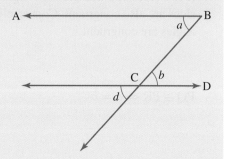

THINK

a. 1. Assume that $\overline{AB}$ is parallel to $\overline{CD}$ to prove the first half of the proof.

    2. Angles a and d are corresponding and therefore congruent.

    3. Angles b and d are vertically opposite and therefore congruent.

    4. Using the transitive property of equals, $a = b$.

    5. Write the concluding statement.

WRITE

a. Assume $\overline{AB} \parallel \overline{CD}$.

$a = d$

$b = d$

$\therefore a = b$

If two lines are parallel, the alternate angles are congruent.

b. 1. Assume that a and b are congruent.

2. Angles b and d are vertically opposite and therefore congruent.

3. Using the transitive property of equals, $a = d$

4. Angles a and d are corresponding and congruent. Therefore, $\overline{AB}$ is parallel to $\overline{CD}$.

5. Write the concluding statement.

b. Assume that $a = b$.

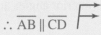

$b = d$

$\therefore a = d$

$\therefore \overline{AB} \parallel \overline{CD}$

If alternate angles are congruent, the lines are parallel.

Exercise 2.3 Direct proofs using Euclidean geometry

learn on

2.3 Exercise	2.3 Exam questions on

Simple familiar	Complex familiar	Complex unfamiliar
1, 2, 3, 4, 5, 6, 7, 8, 9, 10	11, 12, 13, 14	15, 16, 17

These questions are even better in jacPLUS!
- Receive immediate feedback
- Access sample responses
- Track results and progress

Find all this and MORE in jacPLUS ▶

Simple familiar

1. **WE15** Prove that the sum of two odd integers is an even integer.

2. **WE16** Use the transitive property of equals to determine the value of a in each of the following.

   a. $a = x$; $x = 17$

   b. $a = y$; $y = 15$

   c. $z = -15$; $a = z$

   d. $p = \dfrac{2}{3}$; $a = p$

3. Use substitution to determine a relationship between a and b in each of the following.

   a. $a + d = 7$; $d = b$

   b. $3ax = 17$; $x = b$

   c. $2ac = 5$; $c = b + 1$

   d. $b^2 - 2bc = 17$; $a = \dfrac{c}{b}$

4. Use matching to determine a third relationship in each of the following.

   a. $x + y = 15$; $x + z = 15$

   b. $a + b = 25$; $a + c = 25$

   c. $a - b = 32$; $x - b = 32$

   d. $a + b = 7$; $b - c = 7$

5. Add equals to determine a value of $x + y$ in each of the following.

   a. $x = 5$; $y = -10$

   b. $x = 27$; $y = 15$

   c. $x = 2$; $y = 0$

   d. $x = -7$; $y = 7$

6. Use techniques of deduction to determine the following.

   a. If $x = 4$ and $x = y$, determine y.

   b. If $a + b = 21$ and $b = c$, determine $a + c$.

   c. If $m + n = 19$ and $p + n = 19$, determine the relationship between m and p.

   d. If $a = 8$ and $b = 9$, determine $a + b$.

7. Use techniques of deduction to determine the following.

   a. If $2ab = 50$ and $b = c$, determine the relationship between a and c.
   b. If $x + y = 34$, $x = a$ and $y = b$, determine the relationship between a and b.

8. **WE17** Use the diagram shown and the given postulates to prove the following.

   a. If two lines are parallel, the co-interior angles are supplementary (that is, $a + b = 180$).
   b. If co-interior angles are supplementary, the lines are parallel.

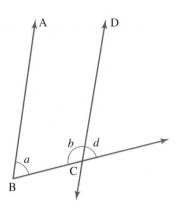

9. Use the diagram, the alternate angle theorem and the given postulates to prove that $c = a + b$.

10. Use the external angle of a triangle theorem and the given postulates to prove that the sum of angles in a triangle is 180°.

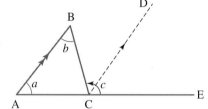

Complex familiar

11. Prove that the sum of the angles in a quadrilateral is 360°. *Hint*: Divide the quadrilateral into two triangles.

12. Prove that the opposite angles of a parallelogram are congruent, and also prove the converse.

13. Use any of the postulates and theorems given above to prove that $a = c$.

   a.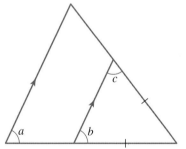

   b.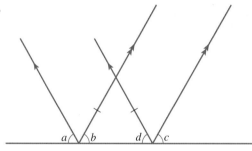

14. Use any of the postulates and theorems given above to prove that $x + y = 180°$.

   a.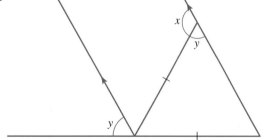

   b.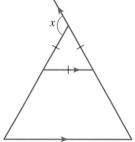

15. Prove that $a = b$ in each of the following.

a.

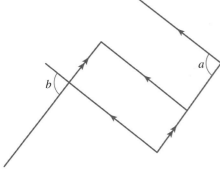

b.

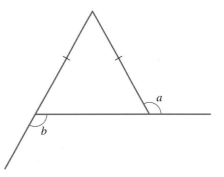

16. If $\overline{AD} \cong \overline{BD}$, prove that $a = b$.

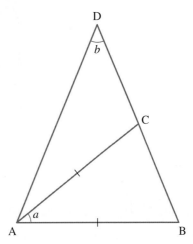

17. If $\angle BDG \cong \angle EDG$, identify a relationship between x and y. You must justify this relationship.

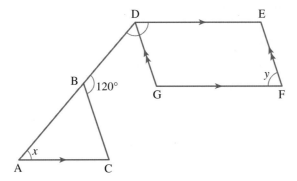

Fully worked solutions for this chapter are available online.

LESSON
2.4 Indirect methods of proof

SYLLABUS LINKS

- Use proof by contradiction.
- Prove irrationality by contradiction.
- Use examples and counterexamples.

Source: Specialist Mathematics Senior Syllabus 2024 © State of Queensland (QCAA) 2024; licensed under CC BY 4.0

2.4.1 Disproof by example/proof by counter example

When we attempt to construct a proof, sometimes the direct path is not the most efficient.

Proofs in mathematics require more than just lots of examples to demonstrate that a conjecture is correct. Examples can help to convince us that a claim is likely to be true, but they cannot prove that a statement is true.

A single example is all that is needed to show that a conjecture is false. This is called a **counter example**.

To prove that $P \Rightarrow Q$ is false, it is necessary to find an example where P is true and Q is not true.

WORKED EXAMPLE 18 Using a counter example to disprove a conjecture

Identify a counter example for each of the following conjectures:
a. **All primes are odd numbers.**
b. **All quadrilaterals with four congruent sides are squares.**
c. **If two events are independent, then their intersection must have a probability of 0.**
d. **All infinite sets only contain countable subsets.**

THINK	WRITE
a. We need to find an even number that is also prime.	a. The number 2 is both prime and even.
b. A rhombus has four congruent sides, but its angles are not all right angles, making it not a square.	b. A quadrilateral that has four congruent sides can be a rhombus
c. The independence of two events doesn't imply a zero probability for the intersection of events. Two events can be independent (the occurrence of one doesn't affect the other's probability) and still have a chance of happening together.	c. Rolling a 6 on a standard die and flipping a coin to get heads are independent events, but their intersection (rolling a 6 and getting heads) has probability $\frac{1}{12}$, not 0.
d. Infinite sets can have subsets that are not countable.	d. Consider the infinite set of all real numbers $\mathbb{R}$. A subset can have any two distinct real numbers. There are infinitely many real numbers between these two numbers you pick. Therefore, the infinite set $\mathbb{R}$ can have many uncountable subsets.

2.4.2 Contrapositive

The **contrapositive** of the conjecture 'If P, then Q' is 'If not Q, then not P.' Notice that both are P and Q are negated and the order is reversed. The contrapositive of a true statement is also true.

Consider the true proposition 'If x is even, then $x + 1$ is odd.'

The contrapositive would be 'If $x + 1$ is not odd, then x is not even' or 'If $x + 1$ is even, then x is odd'. Notice that these propositions are also true.

Previously, we used a Venn diagram to demonstrate what $P \Rightarrow Q$ looks like when it is true. You can see on the diagram that if not Q is true, then not P will also be true. Using symbols, if $P \Rightarrow Q$ is true, then $\neg Q \Rightarrow \neg P$ is also true.

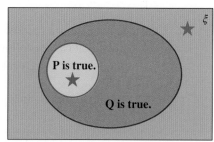

If $P \Rightarrow Q$ is true, then $\neg Q \Rightarrow \neg P$ is also true.
★ $P \Rightarrow Q$ is true.
★ Q is not true $\Rightarrow P$ is not true.

This can also be seen by comparing truth tables. $P \Rightarrow Q$ has the same truth table as $\neg Q \Rightarrow \neg P$.

We know that if a proposition is true, the contrapositive is also true. Therefore, if we prove that the contrapositive is true, the proposition is also true.

P	Q	$P \Rightarrow Q$
T	T	T
T	F	F
F	T	T
F	F	T

P	Q	$\neg Q$	$\neg P$	$\neg Q \Rightarrow \neg P$
T	T	F	F	T
T	F	T	F	F
F	T	F	T	T
F	F	T	T	T

WORKED EXAMPLE 19 Proofs using the contrapositive

Let $n \in \mathbb{Z}$ and prove the following statements by proving the contrapositive.
a. If $n + m$ is odd, then $n \neq m$.
b. If n^2 is odd, then n is odd.
c. If a shape is a square, then it is a rectangle.

THINK

a. 1. Identify the propositions and their negatives. Write the contrapositive by negating both statements and reversing the order.

 2. If $n = m$, then substitution can be used.

 3. Write concluding statements.

b. 1. Write the contrapositive.

 2. Prove the contrapositive

WRITE

a. P : $n + m$ is odd.
 $\neg$P : $n + m$ is even.
 Q : $n \neq m$.
 $\neg$Q : $n = m$.
 Contrapositive: If $n = m$ then $n + m$ is even.

 If $n = m$, then $n + m = 2n$.
 If $n = m$, then $n + m =$ is even.

 Therefore, if $n + m$ is odd, then $n \neq m$.

b. If n is even, then n^2 is even.

 Let $n = 2k (k \in \mathbb{Z})$, then
 $n^2 = (2k)^2 = 4k^2 = 2 \times (2k^2)$
 $\therefore n^2$ is even.
 Therefore, if n^2 is odd, then n is odd.

c. 1. Write the contrapositive.

c. If a shape is not a rectangle, then it is not a square.

2. Consider a square is a specific type of quadrilateral that has the following properties:
- All four sides are congruent (equal in length).
- All four angles are right angles (90°).

Also, consider a rectangle is a quadrilateral that has the following properties:
- Opposite sides are parallel and congruent (equal in length).
- All four angles are right angles (90°).

Notice how the definition of a square includes all the properties of a rectangle. Additionally, squares have the extra property of all sides being congruent.

Because every square is also a rectangle (by fulfilling all rectangle properties), if a shape is not a rectangle, it automatically cannot be a square (since squares are a specific type of rectangle).

2.4.3 Proof by contradiction

To prove something by **contradiction**, we assume that what we want to prove is not true. The proof continues until the initial assumption is contradicted. As the initial assumption must be false, the alternative is true.

If you were trying to prove that $P \Rightarrow Q$ is true by contradiction, you would try to show the following:
1. Assume the proposition to be proved is false, that is, $P \Rightarrow Q$ is not true (if P is true and Q is not true).
2. Show that the proposition contradicts the initial assumptions, that is, show that if P is true, Q is true.
3. Conclude that as the solution does not meet the initial assumptions, the original assumption must be false, and hence the proposition is true: $P \Rightarrow Q$.

WORKED EXAMPLE 20 Using proof by contradiction

Use proof by contradiction to prove the proposition 'There are no positive integer solutions to the Diophantine equation $x^2 - y^2 = 1$.'

Note: This could also have been written as 'If x and y are positive integers, then $x^2 - y^2 \neq 1$.'

THINK	WRITE
1. Assume that the proposition is not true. This means that it is possible a solution exists where x and y are both positive integers.	Assume that $x^2 - y^2 = 1$, $x, y \in \mathbb{Z}^+$.
2. Factorise the equation.	$(x + y)(x - y) = 1$

3. As $x, y \in \mathbb{Z}^+$, $x + y$ is a positive integer and $x - y$ is an integer.	Two integers, $x + y$ and $x - y$, multiply to give 1. Both integers must equal 1 or -1. But, if $x, y \in \mathbb{Z}^+$, then $x + y \in \mathbb{Z}^+$. Therefore, $x + y \neq -1$. Thus, $x + y = 1$ and $x - y = 1$.
4. Use simultaneous equations solve the equations.	$x + y = 1 \quad [1]$ $x - y = 1 \quad [2]$ $[1] + [2]: 2x = 2$ $\qquad\qquad x = 1$ Substituting $x = 1$ into [1] results in $y = 0$.
5. The initial assumption was that $x, y \in \mathbb{Z}^+$; $y = 0$ does not fit with this assumption.	$y = 0$ contradicts the initial assumption that $y \in \mathbb{Z}^+$.
6. As the only solution does not meet the original assumption, the original assumption must be false.	There are no positive integer solutions to the equation $x^2 - y^2 = 1$.

Exercise 2.4 Indirect methods of proof

learn on

2.4 Exercise	2.4 Exam questions on

Simple familiar	Complex familiar	Complex unfamiliar
1, 2, 3, 4, 5, 6, 7, 8	9, 10, 11, 12	13, 14

These questions are even better in jacPLUS!
- Receive immediate feedback
- Access sample responses
- Track results and progress

Find all this and MORE in jacPLUS ▶

Simple familiar

1. In the following list, prime numbers are marked in red.

1	2	3	4	5	6	7	8	9	10
11	12	13	14	15	16	17	18	19	20
21	22	23	24	25	26	27	28	29	30

From the list, select counter examples to disprove each of the following statements.

a. All numbers of the form $6n + 1$ are prime, $n \in \mathbb{N}$.
b. All numbers of the form $2^n + 1$ are prime, $n \in \mathbb{N}$.

2. **WE18** Identify counter examples to disprove each of the following conjectures.

a. If $x^2 = 100$, then $x = 10$.
b. If a quadrilateral has four congruent sides, then it is a square.
c. If n is a multiple of 3, then it is odd.
d. If n is a multiple of 5, then the last digit is 5.

3. Write the contrapositive statement to each of the following propositions:
   a. If today is Monday, then tomorrow is Tuesday.
   b. If A bisects a line segment, then A is the midpoint of the line segment.

4. If proof by contradiction was being used to justify that $\sqrt{ab} \le \dfrac{a+b}{2}$, what would the initial assumption be?

5. **WE19** Consider the statement 'If $n > m$, then $n - m > 0$.'
   a. Write the contrapositive statement.
   b. Prove the statement by proving the contrapositive.

6. At the beginning of a proof by contradiction 'that x is a positive number', Ailsa wrote 'Assume x is a negative number.' What has Ailsa overlooked?

7. **WE20** Use proof by contradiction to prove the proposition 'There are no integer solutions to the Diophantine equation $4x^2 - y^2 = 1$.'

8. Use proof by contradiction to prove the proposition 'If $x, y \in \mathbb{Z}$, then $2x + 4y \ne 7$.'

Complex familiar

9. Analyse the following proof and identify the flaw in the logic.
   Proposition: All integers are the same.
   Assume that there are two integers a and b that are not the same.
   Therefore, $\exists c, a = b + c$.
   Multiplying both sides by $a - b$: $a(a - b) = (b + c)(a - b)$
   $$a^2 - ab = ab - b^2 + ac - bc$$
   Expanding the brackets: $a^2 - ab - ac = ab - b^2 - bc$
   $$a(a - b - c) = b(a - b - c)$$
   Therefore, $a = b$.
   This contradicts the initial assumption.
   Therefore, all integers are the same.

10. Is the proposition 'There are no integer solutions to the equation $25 - y^2 = x^2$' true or false? Provide appropriate evidence to justify your claim.

11. Consider the statement 'If a and b are integers, there is no solution to the equation $a + b = \dfrac{1}{2}$.'

    a. Write the contrapositive statement.
    b. Prove the statement by proving the contrapositive.

12. Consider the statement 'If $x + y > 5$, then either $x > 2$ or $y > 3$, $x, y \in \mathbb{R}$.' This statement can be justified by proving the contrapositive.

    a. Write the contrapositive statement.
    b. Justify the statement by proving the contrapositive.

Complex unfamiliar

13. Prove the statement 'There are no integer solutions to $x^2 - y^2 = 2$.'

14. Prove that statement 'For every $x \in \mathbb{Q}^+$, there exists $y \in \mathbb{Q}^+$ for which $y < x$.'

Fully worked solutions for this chapter are available online.

LESSON
2.5 Proofs with rational and irrational numbers

SYLLABUS LINKS

- Prove results involving integers, e.g. proving that the product of two consecutive odd numbers is an odd number and $5n^2 + 3n + 6 \ \forall n \in \mathbb{Z}$ is an even number.
- Express rational numbers as terminating or eventually recurring decimals and vice versa.
- Prove irrationality by contradiction.

Source: Specialist Mathematics Senior Syllabus 2024 © State of Queensland (QCAA) 2024; licensed under CC BY 4.0

Earlier in this chapter, you explored ways to construct proofs. In this section, you will use the skills that you developed to write proofs with rational and irrational numbers.

2.5.1 Proofs with consecutive numbers

If a series of numbers are consecutive, they can be written as n, $n + 1$, $n + 2$, ...

Sometimes, it is easier to call the middle number n. so the series would be ... $n - 1$, n, $n + 1$.

WORKED EXAMPLE 21 Proofs with consecutive numbers

Demonstrate that the sum of three consecutive natural numbers is equal to 3 times the middle number.

THINK	WRITE
1. As the example talks about the middle number, it might be easier to let the middle number be n, where n is a natural number.	Let the middle number be n, $n \in \mathbb{N}$. The series is $n - 1$, n, $n + 1$.
2. Determine the sum of the three numbers.	$\begin{aligned} \text{Sum} &= (n - 1) + n + (n + 1) \\ &= 3n \end{aligned}$
3. Write the concluding statement.	The sum of three consecutive natural numbers is equal to 3 times the middle number.

2.5.2 Prove that a number is irrational by contradiction

A proof by contradiction can also be used to prove that a number is irrational. Begin by assuming that the number is rational and work to prove that the assumption is false.

WORKED EXAMPLE 22 Proving that a number is irrational by contradiction

Prove that the following numbers are irrational.

a. $\sqrt{2}$

b. $\log_2(5)$

THINK	WRITE
a. 1. Use a proof by contradiction and assume that $\sqrt{2}$ is rational. This means that $\sqrt{2}$ can be written as the ratio of two integers a and b, where $b \neq 0$ and a and b have no common factors, that is, the greatest common divisor of a and b is 1, written $\gcd(a, b) = 1$.	a. Assume $\sqrt{2}$ is rational. $\therefore \sqrt{2} = \dfrac{a}{b}$, $a, b \in \mathbb{Z}$, $\gcd(a, b) = 1$, $b \neq 0$

2. Multiply by b and square both sides.

$$\sqrt{2}b = a \qquad [1]$$
$$2b^2 = a^2$$

3. This means that 2 divides a^2 evenly and therefore 2 divides a evenly. *Note*: $a \mid b$ means that a divides into b evenly, or a is a factor of b.

$$\therefore 2 \mid a^2$$
$$\therefore 2 \mid a$$

4. If 2 divides a evenly, then a is a multiple of 2.

Let $a = 2m$, $m \in \mathbb{Z}$. [2]

5. Substitute [2] in [1] and repeat the process to show that 2 will divide b evenly.

Substitute [2] in [1]:
$$2b^2 = (2m)^2$$
$$2b^2 = 4m^2$$
$$b^2 = 2m^2$$
$$\therefore 2 \mid b^2$$
$$\therefore 2 \mid b$$

6. 2 is a divisor of both a and b.

2 divides both a and b.
But $\gcd(a, b) = 1$.
The assumption is false.

7. Write your concluding statement.

$\sqrt{2}$ is irrational.

b. 1. Use proof by contradiction and assume that $\log_2(5)$ is rational.

b. Assume $\log_2(5) = \dfrac{a}{b}$,
$a, b \in \mathbb{Z}$, $\gcd(a, b) = 1$, $b \neq 0$.

2. Rearrange the equation.

$$2^{\frac{a}{b}} = 5$$
$$(2^a)^{\frac{1}{b}} = 5$$
$$2^a = 5^b$$

3. The only case in which powers of 2 and 5 are equal is $2^0 = 5^0$.

As $\gcd(2, 5) = 1$, the only solution is $a = b = 0$.
But $b \neq 0$.
The assumption is false.

4. Write your concluding statement.

$\log_2(5)$ is irrational.

2.5.3 Proofs with odd and even numbers

If a proof involves odd or even numbers, you can use the fact that for any integer n, $2n$ will be even and $2n - 1$ will be odd.

WORKED EXAMPLE 23 Proofs with odd numbers

Prove that n^2 is odd if and only if n is odd.

THINK

1. 'If and only if' means you need to prove that n^2 is odd if n is odd and also that n is odd if n^2 is odd. Begin by assuming that n is odd. This means that it is 1 more than an even number.

WRITE

If n is odd, $n = 2a + 1$, $a \in \mathbb{Z}$.

2. Determine an expression for n^2.	$n^2 = (2a + 1)^2$ $\qquad = 4a^2 + 4a + 1$
3. Odd numbers can be expressed as 1 more than an even number.	$\qquad = 2(2a^2 + 2a) + 1$ This is an odd number.
4. Write your concluding statement.	Therefore, if n is odd, then n^2 is odd.
5. Now assume that n^2 is odd, meaning that it can be written as 1 more than an even number.	If n^2 is odd, then $n^2 = 2b + 1$, $b \in \mathbb{Z}$.
6. Consider $n^2 - 1$. This expression can be factorised as the difference of two squares.	$n^2 - 1 = (n + 1)(n - 1)$
7. Substitute $n^2 = 2b + 1$.	$(2b + 1) - 1 = (n + 1)(n - 1)$ $2b = (n + 1)(n - 1)$
8. If the product of two numbers is even, then one of the numbers must be even.	Either $n + 1$ is even, which means that n is odd, or $n - 1$ is even, which also means that n is odd.
9. Write your concluding statement.	Therefore, if n^2 is odd, then n is odd. Therefore, n^2 is odd if and only if n is odd.

2.5.4 Prove that a set of numbers is infinite

A proof by contradiction can be used to prove that a set of numbers is infinite (that is, it has an infinite number of members). Begin by assuming that the set is finite and work to prove that this assumption is false.

WORKED EXAMPLE 24 Proofs that a set of numbers is infinite

Prove that there are infinitely many prime numbers.

THINK	WRITE
1. Use a proof by contradiction. Begin by assuming that there are a finite number of primes.	Assume that $p_1 \dots p_n$ are the only prime numbers.
2. There must be a number that is 1 more than the product of all of the primes. Call this number q. *Note:* The notation $\displaystyle\prod_{i=1}^{n} p_i$ indicates the product of all prime numbers between 1 and n.	Let $q = \displaystyle\prod_{i=1}^{n} p_i + 1$.
3. If we divide q by any of the primes, the remainder is 1. This means that q is not a product of any of the primes p_i.	If we divide q by any of the primes, the remainder is 1. Therefore, q is prime, or there is a prime number greater than p_n that is a factor of q.
4. There are more prime numbers than initially assumed.	This means that the assumption is false, as there must be a prime number greater than p_n.
5. Write your concluding statement.	There are infinitely many prime numbers.

2.5.5 Other proofs with real numbers

It can be possible to prove that an expression is larger or smaller than another expression. It can also be possible to prove that an expression is not equal to another expression.

WORKED EXAMPLE 25 Proofs involving equalities and inequalities

a. **Prove that for $a, b \in \mathbb{R}$, $a^2 + b^2 \geq 2ab$.**

b. **Prove that $\dfrac{1}{x} + \dfrac{1}{y} \neq \dfrac{1}{x+y}$ for $x, y \neq 0$.**

THINK	WRITE
a. 1. The elements a^2, b^2 and $2ab$ are part of a perfect square expansion. A perfect square is always greater than or equal to 0.	a. $\quad (a - b)^2 \geq 0$
2. Expand the brackets.	$a^2 - 2ab + b^2 \geq 0$
3. Rearrange the terms to form the proof.	$\therefore a^2 + b^2 \geq 2ab$
b. 1. Use a proof by contradiction and begin by assuming that the equation is true. Rewrite $\dfrac{1}{x} + \dfrac{1}{y}$ as a single fraction.	b. Assume that $\dfrac{1}{x} + \dfrac{1}{y} = \dfrac{1}{x+y}$ $\dfrac{y}{xy} + \dfrac{x}{xy} = \dfrac{1}{x+y}$ $\dfrac{y+x}{xy} = \dfrac{1}{x+y}$
2. Rearrange the equation.	$(x+y)^2 = xy$
3. $(x+y)^2$ will be positive. (It cannot equal zero as $x, y \neq 0$.)	$(x+y)^2 > 0$ $\therefore xy > 0$
4. Expanding $(x+y)^2$ to show an xy term.	$(x+y)^2 = xy$ $x^2 + 2xy + y^2 = xy$ $x^2 + y^2 = -xy$
5. $x^2 + y^2$ will be positive.	$x^2 + y^2 > 0$; therefore, $-xy > 0$ $xy < 0$
6. Identify the contradiction.	But earlier, we found that $xy > 0$.
7. State your conclusions.	Therefore, the assumption is false. $\dfrac{1}{x} + \dfrac{1}{y} \neq \dfrac{1}{x+y}$

Exercise 2.5 Proofs with rational and irrational numbers

2.5 Exercise	2.5 Exam questions **on**

Simple familiar	Complex familiar	Complex unfamiliar
1, 2, 3, 4, 5, 6, 7, 8, 9, 10, 11, 12, 13, 14	15, 16, 17, 18	19, 20

These questions are even better in jacPLUS!
- Receive immediate feedback
- Access sample responses
- Track results and progress

Find all this and MORE in jacPLUS ▶

Simple familiar

1. **WE21** Demonstrate that the sum of five consecutive natural numbers is equal to 5 times the middle number.

2. Demonstrate that if n and m are multiples of 3, then $n + m$ is a multiple of 3.

3. Demonstrate that if m, n and p are consecutive natural numbers, then $n^2 - mp = 1$.

4. An even number, n, can be written in the form $n = 2a$, $a \in \mathbb{Z}$. Demonstrate that if m and n are even numbers, then $m^2 + n^2$ and $m^2 - n^2$ are both divisible by 4.

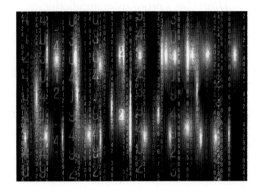

5. An odd number, n, can be written in the form $n = 2a + 1$, $a \in \mathbb{Z}$. Demonstrate that the sum of two consecutive odd numbers is divisible by 4.

6. Consider four consecutive numbers, n_1, n_2, n_3, n_4. Demonstrate that $n_1 + n_2 + n_3 + n_4 = n_3 n_4 - n_1 n_2$.

7. **WE22** a. Prove that $\sqrt{5}$ is irrational.

   b. Prove that $\sqrt{3}$ is irrational.

8. a. Prove that $\log_3(7)$ is irrational.
   b. Prove that $\log_5(11)$ is irrational.

9. Prove that for any natural number n, $\dfrac{n(n+1)}{2}$ is a natural number.

10. Use your understanding of divisibility to show the following. *Note*: $a \mid b$ means that a divides into b evenly, or a is a factor of b.

    a. If $a \mid b$ and $b \mid c$, then $a \mid c$.
    b. If $a \mid b$ and $a \mid c$, then $a \mid (b + c)$.

11. **WE23** Prove that n^2 is even if and only if n is even. *Hint:* When assuming that n^2 is even, consider factorising $n^2 - 1$.

12. Prove that all numbers of the form $n^3 - n$, $n \in \mathbb{Z}$ are multiples of 6.

13. Prove that if x is rational and y is irrational, then $x + y$ is irrational.

14. **WE24** a. Prove that there are infinitely many integers. (*Hint:* Assume that n is the largest integer.)

    b. Prove that there are infinitely many even numbers.

15. **WE25** Prove that for $a > 0$, $a + \dfrac{1}{a} \geq 2$. *Hint:* Begin by noting that $(a-1)^2 \geq 0$.

16. Prove that for $a, b \in \mathbb{R}^+ \cup \{0\}$, $\dfrac{1}{2}(a+b) \geq \sqrt{ab}$.

17. Prove that for $a, b \in \mathbb{Z}$, if $4 \mid (a^2 + b^2)$, then a and b are not both odd.

18. Two resistors have resistances of a ohms and b ohms. If the resistors are placed in series, the combined resistance is $R_S = a + b$. If the resistors are placed in parallel, the combined resistance is found using $\dfrac{1}{R_P} = \dfrac{1}{a} + \dfrac{1}{b}$. Justify that $R_S \geq R_P$.
Hint: Use a proof by contradiction.

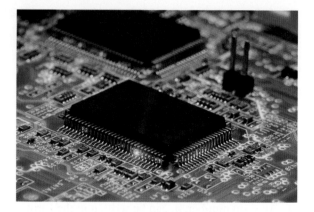

19. Consider $\sqrt{n}$ where n is an integer. There are two irrational numbers, $\sqrt{2}$ and $\sqrt{3}$, between the rational numbers $\sqrt{1} = 1$ and $\sqrt{4} = 2$. There are four irrational numbers between the next rational pair, $\sqrt{4} = 2$ and $\sqrt{9} = 3$. Determine a rule for the number of irrational numbers of the form $\sqrt{n}$ between consecutive integers m and $m + 1$. Prove that your rule works.

20. a. Arrange the integers 1 to 200 in a table similar to the one shown, marking all the primes and multiples of 6. You may like to use a spreadsheet or similar to assist with this.

1	2	3	4	5	6	7
	8	9	10	11	12	13
	14	15	16	17	18	19
	20	21	22	23	24	25
	26	27	28	29	30	31
	32	33	34	35	36	37
				Primes are in red.	Multiples of 6	Primes are in red.

Note that with the exception of the primes 2 and 3, the other primes are all 1 less than or 1 more than a multiple of 6. This means that primes can be expressed in the form $6n \pm 1$.

b. Sophie Germain primes, p, are prime numbers where $2p + 1$ is also prime. This means that 2 is a Sophie Germain prime, because 5 is also prime. Determine the other Sophie Germain primes between 1 and 200. Note that, with the exception of 2 and 3, they are all 1 less than a multiple of 6. Justify why this is so.

Fully worked solutions for this chapter are available online.

LESSON
2.6 Review

2.6.1 Summary

2.6 Exercise

learn on

2.6 Exercise	**2.6 Exam questions** on

Simple familiar	Complex familiar	Complex unfamiliar
1, 2, 3, 4, 5, 6, 7, 8, 9, 10, 11, 12	13, 14, 15, 16	17, 18, 19, 20

Simple familiar

1. Identify the following numbers as either rational or irrational.

   a. $\dfrac{23}{7}$

   b. 4.32

   c. $3.454\,545\,454\,5\ldots$

   d. $1.010\,110\,111\,011\,11\ldots$

   e. $\sqrt{7}$

   f. $\sqrt{\dfrac{25}{9}}$

2. Determine if the following are true statements.

   a. $\exists x \in \mathbb{N}, x^2 < 1$

   b. $\forall x \in \mathbb{R}, x > 3$

   c. $x > 4 \Rightarrow x > 3$

   d. $|x| > 5 \Rightarrow x > 5$

   e. $x = 7 \Leftrightarrow x^2 = 49$

   f. $x^2 < 16 \Rightarrow x < 4$

3. For each of the following, write the converse and determine if implication ($\Rightarrow$) or equivalence ($\Leftrightarrow$) is the more appropriate symbol.

   a. If an animal is a monotreme, then it is a mammal.
   b. If a number is prime, then it has exactly two factors.
   c. If $x > 2$, then $x^3 > 8$.
   d. If $x = 9$, then $x^2 = 81$.

4. Write the negation of each of the following statements.

   a. $x < 4, x \in \mathbb{R}$
   b. The number is odd or a multiple of 5.
   c. The number is an even perfect square.
   d. If x is an interesting number, then x has exactly 3 factors.

5. Express the following common fractions as decimal fractions.

   a. $\dfrac{3}{50}$

   b. $\dfrac{7}{25}$

   c. $\dfrac{13}{11}$

   d. $\dfrac{13}{6}$

   e. $\dfrac{1}{24}$

   f. $\dfrac{7}{15}$

6. Express the following decimals as simplest common fractions.

   a. 1.252

   b. $0.\dot{6}$

   c. $0.\dot{1}\dot{7}$

   d. $0.08\overline{7}$

   e. $0.7\dot{9}\dot{0}$

   f. $3.\overline{296}$

7. Using a technology of your choice, identify a counter example to disprove the conjecture 'All numbers of the form $2^{2^{n}} + 1$ are prime, $n \in \mathbb{Z}^{+} \cup \{0\}$.'

8. If Pravdeep was using proof by contradiction to prove that the diagonals of a trapezium do not bisect each other, what would her initial assumption be?

9. Prove that the following numbers are irrational.

   a. $\sqrt{7}$

   b. $\log_5(7)$

10. Prove each of the following propositions by proving the contrapositive.

    a. If n^2 is even, then n is even, $n \in \mathbb{N}$.

    b. If n^3 is odd, then n is odd, $n \in \mathbb{N}$.

11. Prove that the sum of four consecutive whole numbers is 2 less than a multiple of 4.

12. Prove that when the product of three consecutive numbers is added to the middle number, the result is the cube of the middle number.

Complex familiar

13. Identify what is wrong with each of the following proofs.

    a. Proposition: $2 = 1$

    Let x and y be two non-zero integers where $x = y$.

    $$x = y \qquad [1]$$
    $$x^2 = xy$$
    $$x^2 - y^2 = xy - y^2$$
    $$(x + y)(x - y) = y(x - y)$$
    $$x + y = y \qquad [2]$$

    Substitute [1] into [2] (for x):

    $$y + y = y$$
    $$2y = y$$
    $$2 = 1$$

    b. Proposition: $-6 = -6$

    $$9 - 15 = 4 - 10$$
    $$9 - 15 + \dfrac{25}{4} = 4 - 10 + \dfrac{25}{4}$$
    $$\left(3 - \dfrac{5}{2}\right)^2 = \left(2 - \dfrac{5}{2}\right)^2$$
    $$3 - \dfrac{5}{2} = 2 - \dfrac{5}{2}$$
    $$3 = 2$$

14. Determine if the following propositions are true.

a. $\forall x, y \in \mathbb{R}, x - y \in \mathbb{Z}$

b. $\exists y \in \mathbb{Q}, \forall x \in \mathbb{Q}, \dfrac{x}{y} \in \mathbb{Z}$

c. $\forall y \in \mathbb{Q}, \exists x \in \mathbb{Q}, \dfrac{x}{y} \in \mathbb{Z}$

d. $\exists x, y \in \mathbb{Q}, \dfrac{x}{y} \in \mathbb{Z}$

15. Prove that in a set of any three different natural numbers, there are always two whose sum is divisible by 2.

16. The numbers 1 to 100 are arranged in a 10 by 10 grid. Part of the grid is shown below, with a 2 by 2 square and its 4 corner elements highlighted.

1	2	3	4	5	6	7	8	9	10
11	12	13	14	15	16	17	18	19	20
21	22	23	24	25	26	27	28	29	20
31	32	33	34	35	36	37	38	39	40
41	42	43	44	45	46	47	48	49	50
51	52	53	54	55	56	57	58	59	60

Show that for any square in the grid:

a. the sum of the numbers inside the square is equal to the sum of the 4 corner numbers.
(In the example, this means that $25 + 26 + 35 + 36 = 14 + 17 + 44 + 47$.)

b. the sum of each diagonal is equal to the sum of the numbers inside the square.
(In the example, this means that $14 + 25 + 36 + 47 = 17 + 26 + 35 + 44 = 25 + 26 + 35 + 36$.)

Complex unfamiliar

17. If n_1 is a 2-digit number where the digit in the tens place is greater than the digit in the units place, the number n_2 is found by interchanging the two digits. Prove that $n_1 - n_2$ is a multiple of 9.

18. Consider the following conjecture: 'The sum of the squares of any pair of consecutive natural numbers is always 1 more than a multiple of 4.' Use your calculator to explore examples and demonstrate that the conjecture is likely to be true. Use algebra to prove the conjecture.

19. If $a = b$ and $c = d$, then $ac = bd$. If it is true that $a > b$ and $c > d$, then show that $ac > bd$ is false. Justify your conclusion.

20. Prove that for $a, b \in \mathbb{Z}, a^2 - 4b \neq 2$.

Fully worked solutions for this chapter are available online.

Hey teachers! Create custom assignments for this chapter

Create and assign unique tests and exams

Access quarantined tests and assessments

Track your students' results

Find all this and MORE in jacPLUS

Answers

Chapter 2 Introduction to proof

2.2 Number systems and writing propositions

2.2 Exercise

1. a. Rational   b. Rational   c. Rational
   d. Irrational   e. Rational   f. Irrational

2. a. Irrational   b. Irrational
   c. Rational   d. Irrational

3. a. 0.85 b. 0.856 c. $0.\dot{6}$ d. $0.5\dot{3}$

4. a. 1.6 b. 0.5625
   c. $3.\dot{1}42\,85\dot{7}$ or $3.\overline{142\,857}$ d. $1.08\dot{3}$

5. a. $\dfrac{5}{9}$ b. $\dfrac{23}{99}$ c. $\dfrac{271}{90}$ or $3\dfrac{1}{90}$
   d. $\dfrac{610}{99}$ or $6\dfrac{16}{99}$ e. $\dfrac{2347}{330}$ or $7\dfrac{37}{330}$ f. $\dfrac{26}{37}$

6. a. $\{0, 3, 6, 9, 15, 21\}$
   b. $\{3, 9\}$
   c. $\{0, 6, 12, 18, 24\}$
   d. $\{12, 15, 18, 21, 24\}$
   e. $\{12, 18, 24\}$; even and greater than or equal to 12
   f. $\{0, 6, 12, 15, 18, 21, 24\}$; even or greater than or equal to 12

7. a. True   b. True   c. False
   d. True   e. False   f. True

8. a. True   b. False   c. True

9. a. If it lives in water, then it is a fish. ($\Rightarrow$)
   b. If the diagonals of a quadrilateral bisect each other, it is a rhombus. ($\Rightarrow$)
   c. If $n > m$, then $n - m > 0$. ($\Leftrightarrow$)
   d. If n is even, then $n + 1$ is odd. ($\Leftrightarrow$)

10. a. For all natural numbers, x, $2x$ is even. True
    b. For all natural numbers, x, $2x + 1$ is a multiple of 3. False
    c. For all natural numbers, x, $x > 0$. True
    d. For all real numbers, x, either $x > 0$ or $x \le 0$. True
    e. For all natural numbers, x, x is even. False
    f. For all real numbers, x, $x^2 + 1 \ge 0$. True

11. a. There is an integer, x, so that $x + 5 = 7$. True
    b. There is a real number, x, so that $x^2 < 0$. False
    c. There is a natural number, x, that is even. True
    d. There is a real number, x, so that $x^2 + 1 = 0$. False
    e. There is a real number, x, so that $x^2 - 1 = 0$. True
    f. There is a rational number, x, so that $\sqrt{x}$ is also rational. True

12. a. For all real numbers, x, there exists a real number, y, so that $x = y$. True
    b. There exists a real number, x, so that for all real numbers, y, $x = y$. False
    c. For all real numbers, x, there exists a real number, y, so that $xy = 0$. True
    d. There exists a real number, x, so that for all real numbers, y, $xy = 0$. True

13. a. True   b. False   c. False   d. True

14. True

15. (Not A or not B) and not C; with symbols, ($\neg$A or $\neg$B) and $\neg$C

16. One of several solutions: $\left(0, \dfrac{\pi}{2}\right)$

17. Answers may vary. Sample answers:
    a. $\forall x, y \in (-3, 3)$, $x^2 + y^2 < 9$
    b. $\forall x \in \mathbb{R}, \exists y \in \mathbb{R}$, $x^2 + y^2 = 9$

2.3 Direct proofs using Euclidean geometry

2.3 Exercise

1. Sample responses can be found in the worked solutions in the online resources.

2. a. $a = 17$ b. $a = 15$
   c. $a = -15$ d. $a = \dfrac{2}{3}$

3. a. $a + b = 7$ b. $3ab = 17$
   c. $2a(b + 1) = 5$ d. $b^2 - 2ab^2 = 17$

4. a. $y = z$ b. $b = c$ c. $a = x$ d. $a = -c$

5. a. -5 b. 42 c. 2 d. 0

6. a. $y = 4$ b. $a + c = 21$
   c. $m = p$ d. $a + b = 17$

7. a. $ac = 25$ b. $a + b = 34$

8–16. Students will need to write their own proofs. Sample responses can be found in the worked solutions in the online resources.

17. $x + 2y = 180°$

2.4 Indirect methods of proof

2.4 Exercise

1. a. 25   b. 9

2. a. $x = -10$
   b. Counter examples include: a rhombus
   c. Counter examples include: the numbers 6, 12, 18, ...
   d. Counter examples include: the numbers 10, 20, 30, ...

3. a. If tomorrow is not Tuesday, then today is not Monday.
   b. If A is not the midpoint of a line segment, then A does not bisect the line segment.

4. $\sqrt{ab} > \dfrac{a + b}{2}$

5. a. If $n - m \le 0$, then $n \le m$.
   b. Students will need to write their own proofs. Sample responses can be found in the worked solutions in the online resources.

6. Ailsa forgot that x could equal 0.

7, 8. Students will need to write their own proofs. Sample responses can be found in the worked solutions in the online resources

9. At the 'Expanding the brackets' step, $a(a - b - c) = b(a - b - c)$ only if $a - b - c = 0$. Therefore, the proof breaks down.

10. False. One solution is $x^2 = 9$ and $y^2 = 16$ (disproof by counter example).

11. a. There is a solution to $a + b = \dfrac{1}{2}$ if a or b is not an integer.

  b. Students will need to write their own proofs. A sample response can be found in the worked solutions in the online resources.

12. a. If $x \leq 2$ and $y \leq 3$, then $x + y \leq 5$.

  b. Students will need to write their own proofs.
  A sample response can be found in the worked solutions in the online resources.

13, 14. Students will need to write their own proofs. Sample responses can be found in the worked solutions in the online resources.

2.5 Proofs with rational and irrational numbers

2.5 Exercise

1–19. Students will need to write their own proofs. Sample responses can be found in the worked solutions in the online resources.

20. a.

1	2	3	4	5	6	7
	8	9	10	**11**	12	**13**
	14	15	16	**17**	18	**19**
	20	21	22	**23**	24	25
	26	27	28	**29**	30	**31**
	32	33	34	35	36	**37**
	38	39	40	**41**	42	**43**
	44	45	46	**47**	48	49
	50	51	52	**53**	54	55
	56	57	58	**59**	60	**61**
	62	63	64	65	66	**67**
	68	69	70	**71**	72	**73**
	74	75	76	77	78	**79**
	80	81	82	**83**	84	85
	86	87	88	**89**	90	91
	92	93	94	95	96	**97**
	98	99	100			
				Primes are in **bold**.	Multiples of 6	Primes are in **bold**.

  b. The Sophie Germain primes less than 200 are 2, 3, 5, 11, 23, 29, 41, 53, 83, 89, 113, 131, 173, 179 and 191. A sample response for the proof can be found in the worked solutions in the online resources.

2.6 Review

2.6 Exercise

1. a. Rational b. Rational c. Rational
   d. Irrational e. Irrational f. Rational

2. a. False b. False c. True
   d. False e. False f. True

3. a. $\Rightarrow$ b. $\Leftrightarrow$ c. $\Leftrightarrow$ d. $\Rightarrow$

4. a. $x \geq 4, x \in \mathbb{R}$

  b. The number is even and not a multiple of 5.

  c. The number is odd or not a perfect square.

  d. x is an interesting number and it does not have exactly 3 factors.

5. a. $0.0\dot{6}$ b. $0.2\dot{8}$ c. $1.\dot{1}\dot{8}$
   d. $2.1\dot{6}$ e. $0.041\dot{6}$ f. $0.4\dot{6}$

6. a. $\dfrac{313}{250}$ or $1\dfrac{63}{250}$ b. $\dfrac{2}{3}$ c. $\dfrac{17}{99}$

   d. $\dfrac{29}{330}$ e. $\dfrac{87}{110}$ f. $\dfrac{89}{27}$ or $3\dfrac{8}{27}$

7. Examples include $4\,294\,967\,297$.

8. The diagonals of the trapezium bisect each other.

9–12. Students will need to write their own responses. Sample responses can be found in the worked solutions in the online resources.

13. a. The line $(x+y)(x-y) = y(x-y)$ does not simplify to $x+y=y$, because dividing by $x-y$ would mean dividing by 0 as $x=y$.

   b. $\left(3-\dfrac{5}{2}\right)^2 = \left(2-\dfrac{5}{2}\right)^2$ means that $\pm\left(3-\dfrac{5}{2}\right) = \pm\left(2-\dfrac{5}{2}\right)$. In this instance, $-\left(3-\dfrac{5}{2}\right) = \left(2-\dfrac{5}{2}\right)$. That is, the line $\left(3-\dfrac{5}{2}\right) = \left(2-\dfrac{5}{2}\right)$ is incorrect.

14. a. False   b. False   c. True   d. True

15–18. Students will need to write their own responses.
Sample responses can be found in the worked solutions in the online resources.

19. $ac > bd$ is true only if $a, b, c > 0$. The statement is therefore false.

20. Students will need to write their own proofs. Sample responses can be found in the worked solutions in the online resources.

3 Vectors in the plane

LESSON SEQUENCE

Fully worked solutions for this chapter are available online.

EXAM PREPARATION

Access exam-style questions in every lesson, available online.

on Resources

Solutions	Solutions — Chapter 3 (sol-0395)
Exam questions	Exam question booklet — Chapter 3 (eqb-0281)
Digital documents	Learning matrix — Chapter 3 (doc-41553)
	Chapter summary — Chapter 3 (doc-41559)

LESSON
3.1 Overview

3.1.1 Introduction

Scalars and vectors are quantities that are commonly used in mathematics and physics. The main difference between a scalar and a vector is that vectors have direction, whereas scalars do not. For instance, the speed of an object is a scalar quantity whereas its velocity is a vector quantity.

Consider the tawny owl trying to catch its prey. Knowing the speed of the owl and the mouse would not be enough to determine whether the owl will catch the mouse, you would have to know the direction of each too, that is their velocity.

Vectors are used to represent 2-dimesional or 3-dimensional space (in unit 3). In this chapter, you will learn about vectors in two dimensions. Vectors are useful tools to represent such things as position, displacement, velocity, acceleration, momentum etc.

Vectors can be used by coaches in many sporting areas to illustrate the importance of the angle of contact with the ball or the optimum position to kick a goal. By watching film clips with vectors superimposed over their golf swings, professional golfers can determine the perfect angle to hold their club to achieve the best outcome.

3.1.2 Syllabus links

Lesson	Lesson title	Syllabus links
3.2	**Vectors and scalars**	○ Examine examples of vectors including displacement, velocity and force.
		○ Understand the difference between a scalar and a vector including distance and displacement, speed and velocity, and magnitude of force and force.
		○ Define and use the magnitude and direction of a vector.
		○ Understand and use vector notation: $\overrightarrow{AB}, \underline{c}, \boldsymbol{d}$, unit vector notation $\hat{\boldsymbol{n}}$.
		○ Understand and use vector equality.
		○ Represent and use a scalar multiple of a vector.
		○ Use the triangle rule to represent the resultant vector from the sum and difference of two vectors.
		○ Represent a vector in the plane using a combination of the sum, difference and scalar multiple of other vectors.
3.3	**Vectors in two dimensions**	○ Use ordered pair notation (x, y) and column vector notation $\begin{pmatrix} x \\ y \end{pmatrix}$ to represent a position vector in two dimensions.
		○ Calculate the magnitude and direction of a vector. • $\|\boldsymbol{a}\| = \left\| \begin{pmatrix} a_1 \\ a_2 \end{pmatrix} \right\| = \sqrt{a_1^2 + a_2^2}$ • $\tan(\theta) = \dfrac{y}{x}, x \neq 0$
		○ Calculate and use a unit vector, $\hat{\boldsymbol{n}}$, in the plane. • $\hat{\boldsymbol{n}} = \dfrac{\boldsymbol{n}}{\|\boldsymbol{n}\|}$
		○ Define and use unit vectors and the perpendicular unit vectors $\hat{\boldsymbol{\imath}}$ and $\hat{\boldsymbol{\jmath}}$.
		○ Express a vector in Cartesian (component) form using the unit vectors $\hat{\boldsymbol{\imath}}$ and $\hat{\boldsymbol{\jmath}}$.
		○ Understand and express a vector in the plane in polar form using the notation (r, θ).
		○ Convert between Cartesian form and polar form, with and without technology.
		○ Understand and use the Cartesian form and polar form of a vector.

Source: Specialist Mathematics Senior Syllabus 2024 © State of Queensland (QCAA) 2024; licensed under CC BY 4.0

LESSON
3.2 Vectors and scalars

SYLLABUS LINKS

- Examine examples of vectors including displacement, velocity and force.
- Understand the difference between a scalar and a vector including distance and displacement, speed and velocity, and magnitude of force and force.
- Define and use the magnitude and direction of a vector.
- Understand and use vector notation: $\overrightarrow{AB}$, c, d, unit vector notation $\hat{n}$.
- Understand and use vector equality.
- Represent and use a scalar multiple of a vector.
- Use the triangle rule to represent the resultant vector from the sum and difference of two vectors.
- Represent a vector in the plane using a combination of the sum, difference and scalar multiple of other vectors.

Source: Specialist Mathematics Senior Syllabus 2024 © State of Queensland (QCAA) 2024; licensed under CC BY 4.0

3.2.1 Introduction

In mathematics, there is an important distinction between **scalar** quantities and **vector** quantities. Scalar quantities have **magnitude** only; vector quantities have *direction as well as magnitude*. Most of the quantities that we use are scalar, and include such measurements as temperature (for example 30 °C, 750 K), time, mass, length, surface area, volume, energy, density, speed (e.g. 30 m/s, 300 km/h) etc. For some quantities it is very important that we know both magnitude and direction.

Consider the forces involved in Linh and Kim fighting over a stuffed bear. The direction they are pulling is important, as well as how hard they pull the bear.

Linh exerts a force of 40 N and Kim exerts a force of 50 N and they apply these forces as shown in the diagram on the right, where the angle between the two forces is 150°.

In what direction will the toy move and what is the force in that direction? That is, what is the resultant force, which is the vector sum of these two forces?

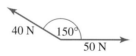

40 N   150°   50 N

The resultant force depends not only on the size of each force but the direction in which the forces are applied.

Consider the two yachts shown in the photograph. Could they be in danger of a collision? One yacht is travelling at a speed of 15 km/h and the other at a speed of 12 km/h. Their speeds give no indication of their direction. Their respective directions are as important as their speed for an analysis their path.

Examples of vectors include force, displacement, velocity, acceleration, momentum etc.

3.2.2 Vector notation

A vector is shown graphically as an arrow, or directed line, with a *tail* (start) and *head* (end). The length of the line indicates the magnitude and the orientation of the arrow indicates its direction.

In the figure, the head of the vector is at point B (indicated with an arrow head), and the tail is at point A.

When writing this vector, we can use point A to indicate start point, and point B to indicate end point. A special arrow can be used to indicate that it is a vector going from point A to point B: $\overrightarrow{AB}$. Some textbooks use a single letter, in **bold**, such as w, to denote a vector, but this is difficult to write using pen and paper, so $\underset{\sim}{w}$ can also be used. The symbol (~) is called a *tilde*.

Magnitude and direction of a vector

The magnitude of a vector is determined by the length of the line segment. The direction is determined by the start point and the end point, or the angle with respect to a fixed line.

3.2.3 Equality of vectors

Vectors are defined by *both* magnitude and direction.

Equality of vectors

Two vectors are equal if and only if *both* their magnitude and direction are equal.

In the figure, the following statements can be made:

$$\underset{\sim}{u} = \underset{\sim}{v}$$
$$\underset{\sim}{u} \neq \underset{\sim}{w} \text{ (directions are not equal)}$$
$$\underset{\sim}{u} \neq \underset{\sim}{z} \text{ (magnitudes are not equal).}$$

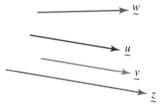

Note: equal vectors do not necessarily need to have the same start points or end points.

3.2.4 Addition of vectors — The triangle rule

Consider a vector, $\underset{\sim}{u}$, that measures the travel from A to B and another vector, $\underset{\sim}{v}$, that measures the subsequent travel from B to C. The net result is as if the person travelled directly from A to C (vector $\underset{\sim}{w}$). Therefore, we can say that $\underset{\sim}{w} = \underset{\sim}{u} + \underset{\sim}{v}$.

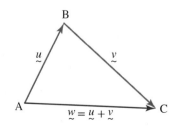

Addition of vectors

To add two vectors, take the tail of one vector and join it to the head of another. The result of this addition is the vector from the *tail of the first vector* to the *head of the second vector*.

Returning to Linh and Kim fighting over a stuffed bear, we see that the forces they apply to the toy can be represented as two vectors. The resultant force is the sum of those two vectors.

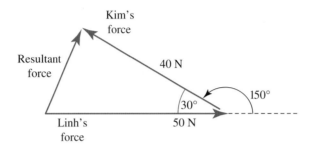

From this figure we are able to get a rough idea of the magnitude and direction of the resultant force. In the following sections, we will learn techniques for calculating the resultant magnitude and direction accurately.

3.2.5 The negative of a vector

If we walk from A to B, and then back from B to A, we have a $\underset{\sim}{0}$ net movement. Thus, the vector A to B is the opposite of the vector B to A. If we label vector A to B as $\underset{\sim}{u}$ then the vector B to A must be $-\underset{\sim}{u}$.

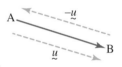

Subtraction of vectors

We can subtract vectors by adding the *negative* of the second vector to the first vector.

WORKED EXAMPLE 1 Representing the addition and subtraction of vectors using the triangle rule

Using the vectors shown, sketch the results of:

a. $\underset{\sim}{u} + \underset{\sim}{v}$ b. $-\underset{\sim}{u}$ c. $\underset{\sim}{u} - \underset{\sim}{v}$ d. $\underset{\sim}{v} - \underset{\sim}{u}$.

THINK

a. 1. Sketch $\underset{\sim}{u}$ and move $\underset{\sim}{v}$ so that its tail is at the head of $\underset{\sim}{u}$.

    2. Join the tail of $\underset{\sim}{u}$ to the head of $\underset{\sim}{v}$ to determine $\underset{\sim}{u} + \underset{\sim}{v}$.

WRITE

a.

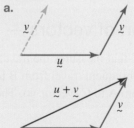

b. Reverse the arrow on $\underset{\sim}{u}$ to obtain $-\underset{\sim}{u}$.

b.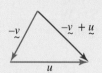

c. 1. Reverse $\underset{\sim}{v}$ to get $-\underset{\sim}{v}$.

c.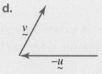

    2. Sketch $\underset{\sim}{u}$ and join the tail of $-\underset{\sim}{v}$ to the head of $\underset{\sim}{u}$ to get $-\underset{\sim}{v} + \underset{\sim}{u}$, which is the same as $\underset{\sim}{u} - \underset{\sim}{v}$ or $\underset{\sim}{u} + (-\underset{\sim}{v})$.

d. 1. Reverse $\underset{\sim}{u}$ to get $-\underset{\sim}{u}$. The vectors are now 'aligned properly' with the head of $-\underset{\sim}{u}$ joining the tail of $\underset{\sim}{v}$.

d.

    2. Join the tail of $-\underset{\sim}{u}$ to the head of $\underset{\sim}{v}$ to get $\underset{\sim}{v} - \underset{\sim}{u}$. Note that this is the same as $(-\underset{\sim}{u} + \underset{\sim}{v})$

WORKED EXAMPLE 2 Expressing a vector as a combination of the sums and differences of other vectors

The parallelogram ABCD can be defined by the two vectors $\underset{\sim}{b}$ and $\underset{\sim}{c}$.
In terms of these vectors, determine:
a. the vector from A to D
b. the vector from C to D
c. the vector from D to B.

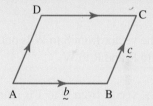

THINK	WRITE
a. The vector from A to D is equal to the vector from B to C since ABCD is a parallelogram.	**a.** $\overrightarrow{AD} = \underset{\sim}{c}$
b. The vector from C to D is equal to the vector from B to A and is the reverse of A to B (it is the same as the vector from B to A) which is $\underset{\sim}{b}$.	**b.** $\overrightarrow{CD} = -\underset{\sim}{b}$
c. The vector from D to B is obtained by adding the vector from D to A to the vector from A to B.	**c.** $\overrightarrow{DB} = -\underset{\sim}{c} + \underset{\sim}{b}$ $\quad\quad = \underset{\sim}{b} - \underset{\sim}{c}$

A cube PQRSTUVW can be defined by the three vectors $\underset{\sim}{a}$, $\underset{\sim}{b}$ and $\underset{\sim}{c}$ as shown.
Express in terms of $\underset{\sim}{a}$, $\underset{\sim}{b}$ and $\underset{\sim}{c}$:

a. the vector joining P to V
b. the vector joining P to W
c. the vector joining U to Q
d. the vector joining S to W
e. the vector joining Q to T.

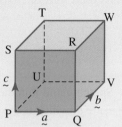

THINK

All of the opposite sides in a cube are equal in length and parallel. Therefore all opposite sides can be expressed as the same vector.

a. The vector from P to V is obtained by adding the vector from P to Q to the vector from Q to V.

b. The vector from P to W is obtained by adding the vectors P to V and V to W.

c. The vector from U to Q is obtained by adding the vectors U to P and P to Q.

d. The vector from S to W is obtained by adding the vectors S to R and R to W.

e. The vector from Q to T is obtained by adding the vectors Q to P, P to S and S to T.

WRITE

a. $\overrightarrow{PV} = \overrightarrow{PQ} + \overrightarrow{QV}$
   $\overrightarrow{PV} = \underset{\sim}{a} + \underset{\sim}{b}$

b. $\overrightarrow{PW} = \overrightarrow{PV} + \overrightarrow{VW}$
   $\overrightarrow{PW} = \underset{\sim}{a} + \underset{\sim}{b} + \underset{\sim}{c}$

c. $\overrightarrow{UQ} = \overrightarrow{UP} + \overrightarrow{PQ}$
   $\overrightarrow{UQ} = -\underset{\sim}{b} + \underset{\sim}{a}$
   $= \underset{\sim}{a} - \underset{\sim}{b}$

d. $\overrightarrow{SW} = \overrightarrow{SR} + \overrightarrow{RW}$
   $\overrightarrow{SW} = \underset{\sim}{a} + \underset{\sim}{b}$

e. $\overrightarrow{QT} = \overrightarrow{QP} + \overrightarrow{PS} + \overrightarrow{ST}$
   $\overrightarrow{QT} = -\underset{\sim}{a} + \underset{\sim}{c} + \underset{\sim}{b}$
   $= \underset{\sim}{b} + \underset{\sim}{c} - \underset{\sim}{a}$

3.2.6 Multiplying a vector by a scalar

Multiplication of a vector by a *positive* number (scalar) affects only the *magnitude* of the vector, not the direction. For example, if a vector $\underset{\sim}{u}$ has a direction of north and a magnitude of 10, then the vector $3\underset{\sim}{u}$ is still in the direction of north and magnitude of 30.

If the scalar is *negative*, then the direction is reversed. Therefore, $-2\underset{\sim}{u}$ has a direction of south and a magnitude of 20.

Scalar multiples of a vector are all parallel as multiplication by a positive scalar only affects the magnitude of a vector, and multiplication by a negative scalar affects the vector's magnitude and reverses its direction.

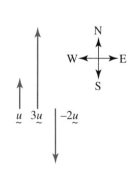

WORKED EXAMPLE 4 Representing a scalar multiple of a vector

Use the vectors shown at right to sketch the result of:
a. $2\underset{\sim}{r} + 3\underset{\sim}{s}$
b. $2\underset{\sim}{s} - 4\underset{\sim}{r}.$

THINK	WRITE
a. 1. Increase the magnitude of $\underset{\sim}{r}$ by a factor of 2 and $\underset{\sim}{s}$ by a factor of 3.	**a.** 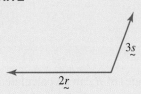
2. Move the tail of $3\underset{\sim}{s}$ to the head of $2\underset{\sim}{r}$. Then join the tail of $2\underset{\sim}{r}$ to the head of $3\underset{\sim}{s}$ to get $2\underset{\sim}{r} + 3\underset{\sim}{s}$.	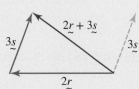
b. 1. Increase the magnitude of $\underset{\sim}{s}$ by a factor of 2 and $\underset{\sim}{r}$ by a factor of 4.	**b.** 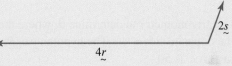
2. Reverse the arrow on $4\underset{\sim}{r}$ to get $-4\underset{\sim}{r}$.	
3. Join the tail of $-4\underset{\sim}{r}$ to the head of $2\underset{\sim}{s}$.	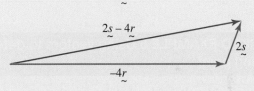

Vectors can be used to solve real-world problems involving movement and direction. Direction is typically referenced using a compass and expressed as the degrees clockwise from north. This is known as a bearing. We can then use Pythagoras' theorem and trigonometry to solve displacement and bearing problems through vector addition.

WORKED EXAMPLE 5 Using vectors to determine the displacement of an object

A boat travels 30 km north and then 40 km west.
a. **Sketch a vector drawing of the path of the boat.**
b. **Sketch the vector that represents the net displacement of the boat.**
c. **Determine the magnitude of the net displacement.**
d. **Given that vectors require a direction, calculate the bearing (clockwise from true north) of this net displacement vector.**

THINK	WRITE
a. 1. Set up vectors (tail to head), one pointing north, the other west.	**a.** 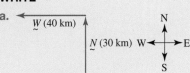
2. Indicate the distances as 30 km and 40 km respectively.	

b. Join the tail of the $\underset{\sim}{N}$ vector with the head of the $\underset{\sim}{W}$ vector.

b.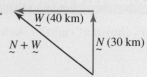

c. 1. Let R km = length of $\underset{\sim}{N} + \underset{\sim}{W}$.

c.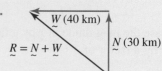

2. The length (magnitude) of $\underset{\sim}{R}$ can be calculated using Pythagoras' theorem.

$$R = \sqrt{30^2 + 40^2}$$
$$= \sqrt{900 + 1600}$$
$$= 50 \text{ km}$$

d. 1. Indicate the angle between $\underset{\sim}{N}$ and $\underset{\sim}{N} + \underset{\sim}{W}$ as θ.

d.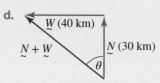

2. Use trigonometry to determine θ, where the magnitude of the opposite side is 40. The hypotenuse, $\underset{\sim}{R}$, was determined in part **c** as 50.

$$\sin(\theta) = \frac{40}{50}$$
$$= 0.8$$
$$\theta = 53.13°$$

3. The true bearing is 360° minus 53.13°.

Therefore the true bearing is:
$$360° - 53.13° = 306.87°$$

WORKED EXAMPLE 6 Using vectors to determine the distance of an object from its starting point

A student rides their bicycle 8 km south-east and then 15 km north-east.
a. Determine how far the student is from their starting point.
b. Determine how far east the student is from their starting point.
c. Determine how far north the student is from their starting point.
d. Calculate the bearing (clockwise from true north) of the net displacement vector.
Give your answers to two decimal places where appropriate.

THINK	WRITE

a. 1. Sketch a vector drawing of the bicycle ride.

a.

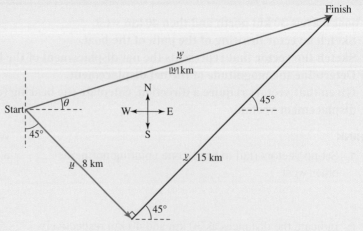

2. The magnitude of the net displacement vector w can be calculated using Pythagoras' theorem.

$$|w| = \sqrt{8^2 + 15^2}$$
$$= 17$$

The total distance from the starting point is 17 km.

b. 1. Complete your sketch with the displacement east and north.

b.

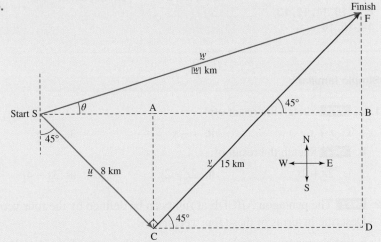

2. The total distance east from the starting point is the distance SB, which is equal to the sum of the distances SA and CD.

$$SB = SA + CD$$
$$= 8\cos(45°) + 15\cos(45°)$$
$$= 16.26$$

The student is 16.26 km east from the starting point.

c. The total distance north from the starting point is the distance BF, which is equal to the difference of the distances DF and DB.

c. $BF = DF - BF$
$$= 15\sin(45°) - 8\sin(45°)$$
$$= 4.95$$

The student is 4.95 km north from the starting point.

d. 1. The direction θ of the displacement vector can be determined using trigonometric ratios.

d. $\sin(\theta) = \dfrac{BF}{SF}$
$$= \dfrac{4.95}{17}$$
$$\theta = \sin^{-1}\left(\dfrac{4.95}{17}\right)$$
$$= 16.93°$$

2. The bearing from true north can be calculated as $90° - \theta$.

The resultant bearing is $90° - 16.93° = 73.07°$ clockwise from north.

---

on Resources

Digital document SkillSHEET Bearings (doc-26829)

Exercise 3.2 Vectors and scalars

3.2 Exercise	**3.2 Exam questions** on

Simple familiar	Complex familiar	Complex unfamiliar
1, 2, 3, 4, 5, 6, 7, 8, 9, 10, 11, 12, 13, 14, 15, 16	17, 18, 19, 20, 21	22, 23, 24, 25

These questions are even better in jacPLUS!
- Receive immediate feedback
- Access sample responses
- Track results and progress

Find all this and MORE in jacPLUS ▶

Simple familiar

1. a. **WE1** Sketch the result of:

   i. $\underset{\sim}{r} + \underset{\sim}{s}$ ii. $\underset{\sim}{r} - \underset{\sim}{s}$ iii. $\underset{\sim}{s} - \underset{\sim}{r}$

   b. **WE4** Sketch the result of:

   i. $2\underset{\sim}{r} + 2\underset{\sim}{s}$ ii. $2\underset{\sim}{r} - 2\underset{\sim}{s}$ iii. $3\underset{\sim}{s} - 4\underset{\sim}{r}$

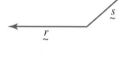

2. **WE2** The pentagon ABCDE at right can be defined by the four vectors, $\underset{\sim}{s}$, $\underset{\sim}{t}$, $\underset{\sim}{u}$ and $\underset{\sim}{v}$. Describe in terms of these four vectors:

   a. the vector from A to D
   b. the vector from A to B
   c. the vector from D to A
   d. the vector from B to E
   e. the vector from C to A.

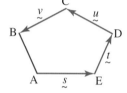

3. **MC** A girl travels 4 km north and then 2 km south. What is the net displacement vector?

   A. 6 km north   **B.** 6 km south
   C. 2 km north   **D.** −2 km north

4. In the rectangle ABCD, the vector joining A to B is denoted by $\underset{\sim}{u}$ and the vector joining B to C is $\underset{\sim}{v}$. Identify which pairs of points are joined by:

   a. $\underset{\sim}{u} + \underset{\sim}{v}$
   b. $\underset{\sim}{u} - \underset{\sim}{v}$
   c. $\underset{\sim}{v} - \underset{\sim}{u}$
   d. $3\underset{\sim}{u} + 2\underset{\sim}{v} - 2\underset{\sim}{u} - \underset{\sim}{v}$

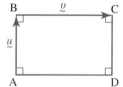

5. **MC** Consider the following relationships between vectors $\underset{\sim}{u}$, $\underset{\sim}{v}$ and $\underset{\sim}{w}$:

   $\underset{\sim}{u} = 2\underset{\sim}{v} + \underset{\sim}{w}$
   $\underset{\sim}{w} = \underset{\sim}{v} - \underset{\sim}{u}$

   Select the correct statement.

   A. $\underset{\sim}{u} = \underset{\sim}{w}$ **B.** $\underset{\sim}{u} = \underset{\sim}{v}$ **C.** $\underset{\sim}{u} = \dfrac{2}{3}\underset{\sim}{v}$ **D.** $\underset{\sim}{u} = \dfrac{3}{2}\underset{\sim}{v}$

6. **WE3** A rectangular prism (box) CDEFGHIJ can be defined by three vectors $\underset{\sim}{r}$, $\underset{\sim}{s}$ and $\underset{\sim}{t}$ as shown at right. Express in terms of $\underset{\sim}{r}$, $\underset{\sim}{s}$ and $\underset{\sim}{t}$:

   a. the vector joining C to H
   b. the vector joining C to J
   c. the vector joining G to D
   d. the vector joining F to I
   e. the vector joining H to E
   f. the vector joining D to J
   g. the vector joining C to I
   h. the vector joining J to C.

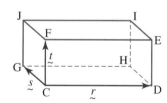

7. In terms of vectors a and b in the figure, define the vector joining O to D.

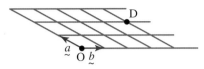

8. In terms of vectors a and b, define the vector joining E to O.

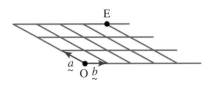

9. Identify which of the following are vector quantities. Justify your answer.
   speed velocity displacement force volume angle

10. Identify which of the following are scalar quantities. Justify your answer.
    speed time acceleration velocity length displacement

11. A 2-dimensional vector can be determined by its length and its angle with respect to (say) true north. What quantities would be needed to represent a 3-dimensional vector?

12. **WE5** A pilot plans to fly 300 km north then 400 km east.
    a. Sketch a vector drawing of her flight plan.
    b. Indicate the resulting net displacement vector.
    c. Calculate the length (magnitude) of this net displacement vector.
    d. Calculate the bearing (from true north) of this net displacement vector.

13. Another pilot plans to travel 300 km east, then 300 km north-east.
    a. Determine how far east of its starting point the plane has travelled, in km to one decimal place.
    b. Demonstrate that the resultant bearing is 67.5 degrees.

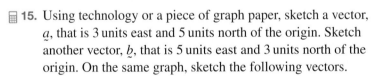

14. An aeroplane travels 400 km west, then 600 km north. How far is the aeroplane from its starting point? Determine the bearing of the resultant displacement. Give your answers to one decimal place.

15. Using technology or a piece of graph paper, sketch a vector, a, that is 3 units east and 5 units north of the origin. Sketch another vector, b, that is 5 units east and 3 units north of the origin. On the same graph, sketch the following vectors.

    a. $a + b$
    b. $a + 3b$
    c. $a - b$
    d. $b - a$
    e. $3b - 4a$
    f. $0.5a + 2.5b$
    g. $a - 2.5b$
    h. $4a$
    i. $2.5a - 1.5b$
    j. $b - 2.5a$

16. Determine the direction and magnitude of a vector joining point A to point B, where B is 10 m east and 4 m north of A.

17. Consider a parallelogram defined by the vectors a and b, and its associated diagonals, as shown.

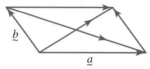

Demonstrate that the vector sum of the diagonal vectors is $2a$.

18. Demonstrate, *by construction* (accurately drawing), that for any vectors u and v:

$$3(u + v) = 3u + 3v$$

(This is called the *Distributive Law*.)

19. Demonstrate, *by construction*, that for any three vectors a, b and c:

$$(a + b) + c = a + (b + c)$$

(This is called the *Associative Law*.)

20. Demonstrate, *by construction*, that for any two vectors r and s:

$$3r - s = -(s - 3r)$$

21. A girl walks the following route: 400 m north — 300 m east — 200 m north — 500 m west — 600 m south — 200 m east.
Sketch a vector drawing of these six paths. Determine the net displacement vector.

Complex unfamiliar

22. As you will learn shortly, vectors can be represented by two values: the horizontal (or x) component and the vertical (or y) component.
Consider the vector w, defined by joining the origin to the point $(4, 5)$, and the vector v, defined by joining $(-3, -2)$ to $(-1, 1)$. Determine the horizontal and vertical components of each vector.
Demonstrate *graphically* that the sum $w + v$ has an x-component of 6, and a y-component of 8.

23. Using the same vectors, w and v, as in question **22**, demonstrate *graphically* that the difference vector, $w - v$, has an x-component of 2 and a y-component of 2.

24. Using the same vectors, w and v, as in question **22**, demonstrate *graphically* that:
    a. the vector $4w$ has an x-component of 16 and a y-component of 20
    b. the vector $-2v$ has an x-component of -4 and a y-component of -6.

25. Using the results from questions **22**, **23** and **24**, what can you deduce about an *algebraic* method (as opposed to a graphical method) of addition, subtraction and scalar multiplication of vectors?

Fully worked solutions for this chapter are available online.

LESSON
3.3 Vectors in two dimensions

SYLLABUS LINKS

- Use ordered pair notation (x, y) and column vector notation $\begin{pmatrix} x \\ y \end{pmatrix}$ to represent a position vector in two dimensions.
- Calculate the magnitude and direction of a vector.
  - $|a| = \left| \begin{pmatrix} a_1 \\ a_2 \end{pmatrix} \right| = \sqrt{a_1^2 + a_2^2}$
  - $\tan(\theta) = \dfrac{y}{x}, x \neq 0$
- Calculate and use a unit vector, $\hat{n}$, in the plane.
  - $\hat{n} = \dfrac{n}{|n|}$
- Define and use unit vectors and the perpendicular unit vectors $\hat{i}$ and $\hat{j}$.
- Express a vector in Cartesian (component) form using the unit vectors $\hat{i}$ and $\hat{j}$.
- Understand and express a vector in the plane in polar form using the notation (r, θ).
- Convert between Cartesian form and polar form, with and without technology.
- Understand and use the Cartesian form and polar form of a vector.

Source: Specialist Mathematics Senior Syllabus 2024 © State of Queensland (QCAA) 2024; licensed under CC BY 4.0

3.3.1 Cartesian form of a vector

As a vector has both magnitude and direction, it can be represented in 2-dimensional planes.

In the figure, the vector u joins the point A to point B. There are two main ways of representing this vector: component form and polar form.

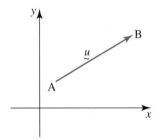

For the component form, using the Cartesian plane, an **identical vector** can be considered to join the origin with the point C.

It is easy to see that u is made up of two components: one along the x-axis and one parallel to the y-axis. Let $\hat{i}$ be a vector along the x-axis with magnitude 1. Similarly, let $\hat{j}$ be a vector along the y-axis with magnitude 1. Vectors $\hat{i}$ and $\hat{j}$ are known as **unit vectors**, and are discussed in section 2.3.5.

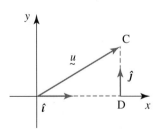

We can say the vector u is the **position vector** of point C relative to the origin.

Note: Unit vectors $\hat{i}$ and $\hat{j}$ may also be written as $\underset{\sim}{\hat{i}}$ and $\underset{\sim}{\hat{j}}$ respectively.

With vectors, it is equivalent to travel along u from the origin directly to C, or to travel first along the x-axis to D and then parallel to the y-axis to C. In either case we started at the origin and ended up at C. Clearly, then, u is made up of some multiple of $\hat{i}$ in the x-direction and some multiple of $\hat{j}$ in the y-direction.

For example, if the point C has coordinates $(6, 3)$ then $u = 6\hat{i} + 3\hat{j}$. This is the **Cartesian form of a vector**.

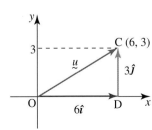

Cartesian form of a vector

The Cartesian form of a vector from the origin to the point (x, y) is given by:

$$\underset{\sim}{u} = x\hat{\imath} + y\hat{\jmath}$$

In general, the vector $\underset{\sim}{u} = x\hat{\imath} + y\hat{\jmath}$ can be expressed as an ordered pair (x, y) or in column vector notation as $\begin{pmatrix} x \\ y \end{pmatrix}$.

For instance, the vector $\underset{\sim}{u} = 6\hat{\imath} + 3\hat{\jmath}$ can be expressed as an ordered pair $(6, 3)$ or in column vector notation as $\begin{pmatrix} 6 \\ 3 \end{pmatrix}$.

3.3.2 The magnitude of a vector

By using Pythagoras' theorem on a position vector, we can determine its length, or *magnitude*.

Consider the vector $\underset{\sim}{u}$ shown.

The magnitude of $\underset{\sim}{u}$, denoted as $|\underset{\sim}{u}|$, also labelled as r, in polar form, is given by:

$$|\underset{\sim}{u}| = \sqrt{6^2 + 3^2}$$
$$= \sqrt{45}$$
$$= 3\sqrt{5}$$

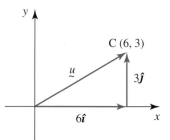

Magnitude of a vector

The magnitude of a vector, $\underset{\sim}{u} = x\hat{\imath} + y\hat{\jmath}$, is given by:

$$|\underset{\sim}{u}| = \left| \begin{pmatrix} x \\ y \end{pmatrix} \right| = \sqrt{x^2 + y^2}$$

Note that the calculation does not include $\hat{\imath}$ or $\hat{\jmath}$ under the square root.

3.3.3 The polar form (r, θ) of a vector

From what we already know about trigonometry, we can work out the angle (θ) that $\underset{\sim}{u}$ makes with the positive x-axis (that is, anticlockwise from the positive x-axis). This gives us the direction of $\underset{\sim}{u}$.

The angle θ can be calculated as:

$$\tan(\theta) = \frac{\text{opposite}}{\text{adjacent}}$$
$$= \frac{3}{6}$$

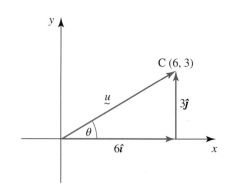

$$\theta = \tan^{-1}\left(\frac{3}{6}\right)$$
$$= \tan^{-1}(0.5)$$
$$= 0.464 \text{ radians}$$
$$= 26.6°$$

The result obtained by this method needs to be adjusted if the angle is in the 2nd, 3rd, or 4th quadrants.

This allows us to express the vector in **polar form** as $\left(3\sqrt{5},\ 26.6°\right)$.

Direction of a vector

The *direction* of a vector, $\underset{\sim}{u} = x\hat{i} + y\hat{j}$, **anticlockwise from the positive direction of the x-axis is given by** $\theta = \tan^{-1}\left(\frac{y}{x}\right)$, $x \neq 0$,

---

Note that when $x = 0$, the direction of the vector is $+90°$ if $y \geq 0$ (direction of the vector is parallel to the positive direction of the y-axis), and $-90°$ if $y \leq 0$ (direction of the vector is parallel to the negative direction of the y-axis).

To determine the direction of a vector, it might be more convenient to consider $|\theta|$ as illustrated in the diagram below:

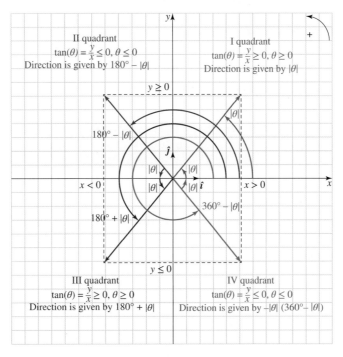

The direction of a vector in the 1st quadrant is given by $|\theta|$, the direction of a vector in the 2nd quadrant is given by $180° - |\theta|$ or $\pi - |\theta|$, the direction of a vector in the 3rd quadrant is given by $180° + |\theta|$ or $\pi + |\theta|$ and the direction of a vector in the 4th quadrant is given $-|\theta|$.

Remember that when $x = 0$, $\tan(\theta) = \frac{y}{x}$ is not defined. If $x = 0$ and $y \geq 0$, then $\theta = +90°$ (direction is parallel to the positive direction of the y-axis) and if $x = 0$ and $y \leq 0$, then $\theta = -90°$ (direction is parallel to the negative direction of the y-axis).

Note that to convert between Cartesian and polar forms of a vector, it can be useful to remember the sine and cosine exact values of common angles.

θ	0	$30°\left(\dfrac{\pi}{6}\right)$	$45°\left(\dfrac{\pi}{4}\right)$	$60°\left(\dfrac{\pi}{3}\right)$	$90°\left(\dfrac{\pi}{3}\right)$	$180°\,(\pi)$
$\cos(\theta)$	1	$\dfrac{\sqrt{3}}{2}$	$\dfrac{\sqrt{2}}{2}$	$\dfrac{1}{2}$	0	-1
$\sin(\theta)$	0	$\dfrac{1}{2}$	$\dfrac{\sqrt{2}}{2}$	$\dfrac{\sqrt{3}}{2}$	1	0

Polar form of a vector

The polar form of a vector $\underset{\sim}{u}$ is (r, θ), where $\underset{\sim}{u}$ is a vector of magnitude r, in the direction of θ anticlockwise from the positive direction of the x-axis.

Note that when adding or subtracting vectors, the component (Cartesian) form is more convenient to use.

When multiplying a vector by a scalar, both the polar and Cartesian forms are convenient to use.

We have seen already that multiplying a vector by a positive scalar affects only its magnitude, but not its direction, whereas multiplying a vector by a negative scalar will reverse its direction (that is, it will add 180° to its direction), in addition to affecting its magnitude.

Scalar multiplication of a vector in polar form

If $\underset{\sim}{u}$ is a vector of magnitude r and direction θ, and k is a scalar, then

$$k\underset{\sim}{u} = (kr, \theta) \quad \text{if } k > 0$$
$$k\underset{\sim}{u} = (|k|r, \theta + 180°) \quad \text{if } k < 0$$

WORKED EXAMPLE 7 Calculating magnitude and direction of a vector and expressing it in polar form

Using the vector shown, determine:
a. the magnitude of $\underset{\sim}{u}$
b. the direction of $\underset{\sim}{u}$ (express the angle with respect to the positive x-axis)
c. the expression of $\underset{\sim}{u}$ in polar form.
d. the true bearing of $\underset{\sim}{u}$.

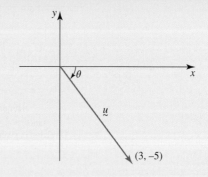

THINK

a. 1. Use Pythagoras' theorem or the rule for magnitude of a vector with the x- and y-components 3 and -5 respectively.

2. Simplify the surd.

WRITE

a. $|\underset{\sim}{u}| = \sqrt{3^2 + (-5)^2}$

$|\underset{\sim}{u}| = \sqrt{9 + 25}$
$\quad = \sqrt{34}\ (= 5.831 \text{ to three decimal places})$

b.
1. The angle is in the 4th quadrant since $x = 3$ and $y = -5$. The direction is given by $-|\theta|$. Use trigonometry to determine the angle θ from the x- and y-component values, recalling $\tan(\theta) = \dfrac{\text{opp}}{\text{adj}}$.

2. Use a calculator to simplify.

c.
1. Recall the polar form of a vector is (r, θ).

2. Write $\underset{\sim}{u}$ in polar form.

d. The negative sign implies that the direction is $59°$ clockwise from the x-axis. The true bearing from north is the angle measurement from the positive y-axis to the vector $\underset{\sim}{u}$.

b. $\theta = \tan^{-1}\left(\dfrac{-5}{3}\right)$

$\theta = -59°$

c. (r, θ)

$\underset{\sim}{u}$ is a vector of magnitude $\sqrt{34}$ in a direction of $-59°$ from the positive direction of the x-axis. In polar form, $\underset{\sim}{u} = \left(\sqrt{34}, -59°\right)$.

d. True bearing $= 90° + 59°$
$= 149°$

TI | THINK | **WRITE**

d. 1. Put the calculator in Degree mode (on the home page, select Settings, then documents settings, and for Angle, select degree). On a Calculator page, complete the entry line as: $\begin{bmatrix} 3 & -5 \end{bmatrix}$ Press MENU, then select:
7: Matrix & Vector
C: Vector
4: Convert to Polar then press ENTER.
Note: The matrix template can be found by pressing the Templates button.

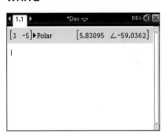

2. The answer appears on the screen.
Note: The calculator will give decimal values, not exact values.

$\underset{\sim}{u} = (5.831, -59°)$

CASIO | THINK | **WRITE**

d. 1. Put the calculator in Degree mode (set-up screen, scroll down to Angle, select deg). On the Run-Matrix screen, press OPTN, then press F6 to scroll across to more menu options. Select ANGLE by pressing F5, press F6 to scroll across to more menu options, then select Pol(by pressing F1. Complete the entry line as: Pol$(3, -5)$ then press EXE.

2. The answer appears on the screen.

$\underset{\sim}{u} = (\sqrt{34}, -59°)$

3.3.4 Components of a vector

As we have seen, any vector u is composed of x- and y- components denoted by $x\hat{i}, y\hat{j}$. The vectors, $\hat{i}$ and $\hat{j}$ are called unit vectors, as they each have a magnitude of 1. This allows us to resolve a vector into its components.

> ### Converting from polar to component form
>
> **If a 2-dimensional vector u makes an angle of θ with the positive x-axis and it has a magnitude of r, then we can determine its x- and y-components using the formulas:**
>
> $$x = r\cos(\theta)$$
> $$y = r\sin(\theta)$$

WORKED EXAMPLE 8 Determining the components of a vector

Consider the vector u, whose magnitude is 30 and whose bearing (from N) is 310°. Determine its x- and y-components and write u in terms of $\hat{i}$ and $\hat{j}$, that is, in Cartesian form.

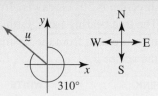

THINK	WRITE
1. Change the bearing into an angle with respect to the positive x-axis (θ).	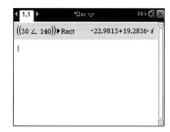
2. The angle between u and the positive y-axis is $360° - 310° = 50°$.	
3. Calculate θ.	$\theta = 90° + 50°$ $= 140°$
4. Determine the x- and y-components using trigonometry.	$x = \|r\|\cos(\theta)$ $y = \|r\|\sin(\theta)$ $= 30\cos(140°)$ $= 30\sin(140°)$ $= -22.98$ $= 19.28$
5. Express u as a vector in Cartesian form.	$u = -22.98\hat{i} + 19.28\hat{j}$

TI | THINK	WRITE	CASIO | THINK	WRITE
1. Determine the angle that u makes with the positive x-axis.	$\theta = 50° + 90° = 140°$	1. Determine the angle that u makes with the positive x-axis.	$\theta = 50° + 90° = 140°$
2. Put the calculator in Degree mode. On a Calculator page, complete the entry line as:(30∠140) Press MENU, then select: 7: Matrix & Vector C: Vector 5: Convert to Rectangular then press ENTER. *Note:* The ∠ symbol can be found by pressing CTRL and the Catalogue button.		2. Put the calculator in Degree mode. On the Run-Matrix screen, press OPTN, then press F6 to scroll across to more menu options. Select ANGLE by pressing F5, press F6 to scroll across to more menu options, then select Rec(by pressing F2. Complete the entry line as: Rec (30, 140) then press EXE.	

3. The answer appears on the screen. The first value is the x-component and the second value is the y-component.

$u = -22.98\hat{\imath} + 19.28\hat{\jmath}$

3. The answer appears on the screen. The first value is the x-component and the second value is the y-component.

$u = -22.98\hat{\imath} + 19.28\hat{\jmath}$

3.3.5 Vector between two points

You will see in chapter 4 that if A and B are points defined by position vectors $a = \begin{pmatrix} x_1 \\ y_1 \end{pmatrix}$ and $b = \begin{pmatrix} x_2 \\ y_2 \end{pmatrix}$

respectively, then $\overrightarrow{AB} = \begin{pmatrix} x_2 - x_1 \\ y_2 - y_1 \end{pmatrix} = b - a$

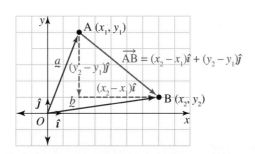

WORKED EXAMPLE 9

For the following pair of points, A $(-3, 4)$ and B $(2, -5)$:
a. express the vector $\overrightarrow{AB}$ in component (Cartesian) form
b. express the vector $\overrightarrow{AB}$ in polar form. Give your answer to one decimal place
c. express the vector $\overrightarrow{BA}$ in component (Cartesian) form
d. express the vector $\overrightarrow{BA}$ in polar form. Give your answer to one decimal place.

THINK

a. 1. Remember that the vector $\overrightarrow{AB}$ between two points A (x_1, y_1) and B (x_2, y_2) is given by $\overrightarrow{AB} = \begin{pmatrix} x_2 - x_1 \\ y_2 - y_1 \end{pmatrix}$.

   2. Express the vector $\overrightarrow{AB}$ in component.

WRITE

a. $\overrightarrow{AB} = \begin{pmatrix} x_2 - x_1 \\ y_2 - y_1 \end{pmatrix}$

$= \begin{pmatrix} 2 - (-3) \\ -5 - 4 \end{pmatrix}$

$= \begin{pmatrix} 5 \\ -9 \end{pmatrix}$

$\overrightarrow{AB} = 5\hat{\imath} - 9\hat{\jmath}$

b. 1. Calculate the magnitude of vector $\overrightarrow{AB}$.

b. $\left|\overrightarrow{AB}\right| = \sqrt{5^2 + (-9)^2}$

$= \sqrt{106}$

$= 10.3$

2. The angle θ is in the 4th quadrant since $x = 5$ and $y = -9$. The direction is given by $-|\theta|$. Recall $\tan(\theta) = \dfrac{y}{x}$.

$\tan(\theta) = \dfrac{-9}{5}$

$\theta = -60.9°$

The direction of $\overrightarrow{AB}$ is $-60.9°$.

$\overrightarrow{AB} = (10.3, -60.9°)$

3. Express the vector $\overrightarrow{AB}$ in polar form.

c. Use $\overrightarrow{BA} = -\overrightarrow{AB}$

c. $\overrightarrow{BA} = -5\hat{\imath} + 9\hat{\jmath}$

d. Remember that multiplying a scalar by -1 does not modify its magnitude but reverse its direction, thus if the direction of a vector $\overrightarrow{AB}$ is θ, then the direction of $\overrightarrow{BA} = -\overrightarrow{AB}$ is $180° + \theta \, (\pm 360°)$.

d. $\overrightarrow{BA} = (10.3, 180° - 60.9°)$

$= (10.3, 119.1°)$

WORKED EXAMPLE 10 Determining the magnitude and direction of a sum of vectors

A bushwalker walks 16 km in a direction of bearing 050°, then walks 12 km in a direction of bearing 210°. Determine the resulting position of the hiker giving magnitude and direction from the starting point. State the resultant vector in polar form. Give your answer to two decimal places.

THINK

1. Sketch a clear diagram to represent the situation.

2. Determine the angles a and b make with the horizontal axis.

WRITE

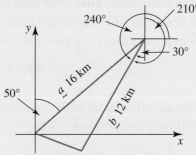

a makes an angle of 50° clockwise to the vertical, thus it makes an angle of 40° to the horizontal.
b makes an angle of 210° clockwise to the vertical, thus it makes an angle of $-120°$ (or $+240°$)

3. Resolve the position vectors into their x- and y-components by recalling the component formulas, $x = r\cos(\theta)$ and $y = r\sin(\theta)$, where θ is the angle from the x-axis.

$$\underset{\sim}{a} = 16\cos(40°)\hat{\imath} + 16\sin(40°)\hat{\jmath}$$
$$\underset{\sim}{b} = 12\cos(240°)\hat{\imath} + 12\sin(240°)\hat{\jmath}$$

4. Simplify position vectors.

$$\underset{\sim}{a} = 12.2567\hat{\imath} + 10.2846\hat{\jmath}$$
$$\underset{\sim}{b} = -6\hat{\imath} - 10.3923\hat{\jmath}$$

5. Use the triangle rule of addition of vectors.

$$\underset{\sim}{a} + \underset{\sim}{b} = (12.2567 - 6)\hat{\imath} + (10.2846 - 10.3923)\hat{\jmath}$$
$$= 6.2567\hat{\imath} - 0.1077\hat{\jmath}$$

6. Determine the angle θ by recalling the direction formula, $\theta = \tan^{-1}\left(\dfrac{y}{x}\right)$.

$$\theta = \tan^{-1}\left(\frac{-0.1077}{6.2567}\right)$$
$$= -0.986°$$

7. Determine the magnitude.

$$|\underset{\sim}{a} + \underset{\sim}{b}| = \sqrt{x^2 + y^2}$$
$$= \sqrt{6.26^2 + (-0.11)^2}$$
$$= \sqrt{39.2}$$
$$= 6.26$$

8. State the resultant vector polar form.

The bushwalker's final position is 6.26 km at an angle of $-0.99°$ from the starting point: $(6.26, -0.99°)$.

3.3.6 Unit vectors

We have seen that the Cartesian form of a vector uses its x and y components and the unit vectors $\hat{\imath}$ and $\hat{\jmath}$. Unit vectors can also be found in the direction of *any* vector. This is merely the original vector divided by its magnitude.

Unit vector

A unit vector is any vector of magnitude 1. We can scale any vector $\underset{\sim}{u}$ to be of magnitude 1 using the formula:

$$\hat{u} = \frac{\underset{\sim}{u}}{|\underset{\sim}{u}|} = \frac{\underset{\sim}{u}}{r}$$

Note that the unit vector $\hat{u}$ and vector $\underset{\sim}{u}$ have the same direction.

Determine the unit vector in the direction of $\underset{\sim}{u}$. Confirm the unit vector has a magnitude of 1.

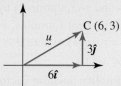

THINK

1. Express the vector in component form.

2. Calculate the magnitude of the vector $\underset{\sim}{u}$ using Pythagoras' theorem.

3. Divide each component of the original vector by the magnitude to get $\hat{u}$.

4. Confirm that $\hat{u}$ has a magnitude of 1.

WRITE

$\underset{\sim}{u} = 6\hat{i} + 3\hat{j}$

$r = \sqrt{6^2 + 3^2}$

$\quad = \sqrt{45}$

$\quad = 3\sqrt{5}$

$\hat{\underset{\sim}{u}} = \dfrac{6}{3\sqrt{5}}\hat{i} + \dfrac{3}{3\sqrt{5}}\hat{j}$

$\quad = \dfrac{2\sqrt{5}}{5}\hat{i} + \dfrac{\sqrt{5}}{5}\hat{j}$

$|\hat{\underset{\sim}{u}}| = \sqrt{x^2 + y^2}$

$\quad = \sqrt{\dfrac{20}{25} + \dfrac{5}{25}}$

$\quad = \sqrt{\dfrac{25}{25}}$

$\quad = 1$

TI	THINK	WRITE	CASIO	THINK	WRITE
1. On a Calculator page, press MENU, then select: 7: Matrix & Vector C: Vector 1: Unit Vector Complete the entry line as: unitV $\left(\begin{bmatrix}6 & 3\end{bmatrix}\right)$ then press ENTER. *Note:* The matrix template can be found by pressing the templates button.		1. On the Run-Matrix screen, press OPTN then select MAT/VCT by pressing F2. Press F6 twice to scroll across to more menu options, then select UnitV(by pressing F5. Press EXIT twice to return to the main menu, then select MATH by pressing F4. Select MAT/VCT by pressing F1, then press F6 to scroll across and select 1×2 by pressing F1. Complete the entry line as: UnitV $\left(\begin{bmatrix}6 & 3\end{bmatrix}\right)$ then press EXE.	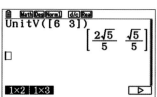
2. The answer appears on the screen. *Note:* The calculator will give decimal values, not exact values.	$\hat{\underset{\sim}{u}} = 0.894\hat{i} + 0.447\hat{j}$	2. The answer appears on the screen.	$\hat{\underset{\sim}{u}} = \dfrac{2\sqrt{5}}{5}\hat{i} + \dfrac{\sqrt{5}}{5}\hat{j}$

3.3 Exercise	**3.3 Exam questions** on

Simple familiar	Complex familiar	Complex unfamiliar
1, 2, 3, 4, 5, 6, 7, 8, 9, 10, 11, 12, 13, 14, 15, 16, 17, 18	19, 20, 21, 22, 23	24, 25, 26, 27

These questions are even better in jacPLUS!
- Receive immediate feedback
- Access sample responses
- Track results and progress

Find all this and MORE in jacPLUS ▶

Simple familiar

1. State the x, y components of the following vectors.
   a. $3\hat{\imath} + 4\hat{\jmath}$ b. $6\hat{\imath} - 3\hat{\jmath}$ c. $3.4\hat{\imath} + \sqrt{2}\hat{\jmath}$

2. **WE7a, b** For each of the following, determine:
   i. the magnitude of the vector
   ii. the direction of each vector. (Express the direction with respect to the positive x-axis.)

   a. 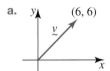   b. 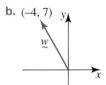   c. 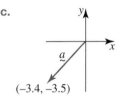   d.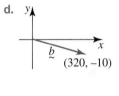

3. a. **WE7c, d** Determine the true bearing of each vector in question **2**.
   b. Express each vector in question **2** in polar form.

4. **WE8** Consider the vector w shown. Its magnitude is 100 m/s and its bearing is 210° True.

   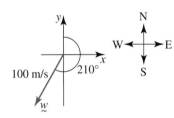

   Determine the x- and y-components of w, and express them as *exact* values (surds). State the answer in the form $w = x\hat{\imath} + y\hat{\jmath}$.

5. **MC** A vector with a bearing of 60 degrees from N and a magnitude of 10 has:

   A. x-component $= \dfrac{\sqrt{3}}{2}$, y-component $= \dfrac{1}{2}$ B. x-component $= \dfrac{1}{2}$, y-component $= \dfrac{\sqrt{3}}{2}$

   C. x-component $= 5\sqrt{3}$, y-component $= 5$ D. x-component $= 5$, y-component $= 5\sqrt{3}$

6. An aeroplane travels on a bearing of 147 degrees for 457 km. Express its position as a vector in terms of $\hat{\imath}$ and $\hat{\jmath}$ (to one decimal place).

7. A ship travels on a bearing of 331 degrees for 125 km. Express its position as a vector in terms of $\hat{i}$ and $\hat{j}$ (to one decimal place).

8. **WE11** Determine unit vectors in the direction of the given vector for the following:

a.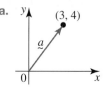

b.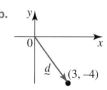

c. $\underset{\sim}{b} = 4\hat{i} + 3\hat{j}$

d. $\underset{\sim}{e} = -4\hat{i} + 3\hat{j}$

e. $\underset{\sim}{c} = \hat{i} + \sqrt{2}\hat{j}$

9. **MC** A unit vector in the direction of $3\hat{i} - 4\hat{j}$ is:

A. $\dfrac{3}{5}\hat{i} + \dfrac{4}{5}\hat{j}$
B. $\dfrac{3}{5}\hat{i} - \dfrac{4}{5}\hat{j}$
C. $\hat{i} - \hat{j}$
D. $\dfrac{3}{25}\hat{i} - \dfrac{4}{25}\hat{j}$

10. Not all unit vectors are smaller than the original vectors. Consider the vector $\underset{\sim}{v} = 0.3\hat{i} + 0.4\hat{j}$. Show that the unit vector in the direction of $\underset{\sim}{v}$ is twice as long as $\underset{\sim}{v}$.

11. Determine the unit vector in the direction of $\underset{\sim}{w} = -0.1\hat{i} - 0.02\hat{j}$. Give your answer correct to three decimal places.

12. Consider the points A $(0, 1)$ and B $(4, 5)$ in the figure. A vector joining A to B can be drawn.

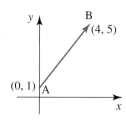

a. Show that an equivalent position vector is given by: $4\hat{i} + 4\hat{j}$.

b. Similarly, show that an equivalent position vector joining B to A is given by: $-4\hat{i} - 4\hat{j}$.

13. **WE9a, b** For each of the following pairs of points A and B:

i. express the vector $\overrightarrow{AB}$ in component (Cartesian) form

ii. express $\overrightarrow{AB}$ in polar form. Give your answers to two decimal places where appropriate.

a. A $(0, 2)$, B $(4, -5)$
b. A $(2, 3)$, B $(5, 4)$
c. A $(2, 0)$, B $(-4, 5)$
d. A $(4, 5)$, B $(3, 2)$
e. A $(3, 7)$, B $(5, 7)$
f. A $(3, 2)$, B $(3, -5)$

14. **WE9c, d** Determine the position vectors $\overrightarrow{BA}$ from question **13** in:

i. Cartesian form

ii. polar form.

15. Determine unit vectors $\overset{\frown}{AB}$ in the direction of the position vectors $\overrightarrow{AB}$ for each of the vectors of question **13**. Give your answers in exact form.

16. Let $\underset{\sim}{u} = 5\hat{i} - 2\hat{j}$ and $\underset{\sim}{e} = -2\hat{i} + 3\hat{j}$.

a. Determine:

i. $|\underset{\sim}{u}|$
ii. $|\underset{\sim}{e}|$
iii. $\hat{u}$
iv. $\hat{e}$
v. $\underset{\sim}{u} + \underset{\sim}{e}$
vi. $|\underset{\sim}{u} + \underset{\sim}{e}|$

b. Confirm or reject the statement that $|\underset{\sim}{u}| + |\underset{\sim}{e}| = |\underset{\sim}{u} + \underset{\sim}{e}|$

17. Let $u = -3\hat{\imath} + 4\hat{\jmath}$ and $e = 5\hat{\imath} - \hat{\jmath}$.

   a. Determine:

     i. $|u|$ **ii.** $|e|$ **iii.** $\hat{u}$ **iv.** $\hat{e}$ **v.** $u + e$ **vi.** $|u + e|$

   b. Confirm or reject the statement that $|u| + |e| = |u + e|$.

18. Consider the vectors represented on Cartesian plane

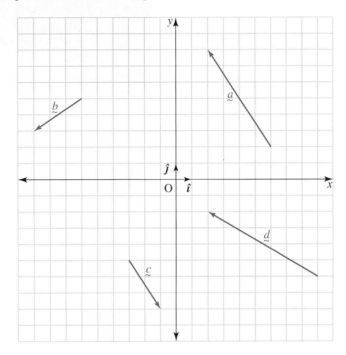

   Write the vectors a to d in component (Cartesian) form.

Complex familiar

19. **WE10** A pilot flies 420 km in a direction 45° south of east and then 200 km in a direction 60° south of east. Calculate the resultant displacement from the starting position giving both magnitude and direction. State the final vector in polar form to one decimal place.

20. The instructions to Black-eye the Pirate's hidden treasure say: 'Take 20 steps in a north-easterly direction and then 30 steps in a south-easterly direction'. However, a rockfall blocks the first part of the route in the north-easterly direction. How could you head directly to the treasure?

21. Two scouts are in contact with home base. Scout A is 15 km from home base in a direction 30° north of east. Scout B is 12 km from home base in a direction 40° west of north. Determine how far is scout B from scout A.

22. To calculate the *distance* between two vectors, a and b, simply determine $|a - b|$. Calculate the distance between these pairs of vectors:

   a. $3\hat{\imath} + 2\hat{\jmath}$ and $2\hat{\imath} + 3\hat{\jmath}$ **b.** $5\hat{\imath} - 2\hat{\jmath}$ and $2\hat{\imath} + 5\hat{\jmath}$

23. A river flows through the jungle from west to east at a speed of 3 km/h. An explorer wishes to cross the river by boat, and attempts this by travelling at 5 km/h due north. Determine:

  a. the vector representing the velocity of the river
  b. the vector representing the velocity of the boat
  c. the resultant (net) vector of the boat's journey
  d. the bearing of the boat's journey
  e. the magnitude of the net vector.

Complex unfamiliar

24. Consider the data from question **23.** At what bearing should the boat travel so that it arrives at the opposite bank of the river due north of the starting position?

25. A boat travels east at 20 km/h, while another boat travels south at 15 km/h. Determine the bearing of the difference vector.

26. Consider the vector $\underset{\sim}{u} = 3\hat{\imath} + 4\hat{\jmath}$ and the vector $\underset{\sim}{v} = 4\hat{\imath} - 3\hat{\jmath}$. Determine the angles of each of these vectors with respect to the x-axis. Show that these two vectors are perpendicular to each other and show that the products of each vector's corresponding x- and y-components add up to 0.

27. A river has a current of 4 km/h westward. A boat which is capable of travelling at 12 km/h is attempting to cross the river by travelling due north. Determine how long it takes to cross the river, if the river is 500 m wide (from north to south). (*Hint:* The maximum 'speed' of the boat is still 12 km/h.)

Fully worked solutions for this chapter are available online.

LESSON
3.4 Review

3.4.1 Summary

3.4 Exercise

learn on

3.4 Exercise	3.4 Exam questions on

Simple familiar	Complex familiar	Complex unfamiliar
1, 2, 3, 4, 5, 6, 7, 8, 9, 10, 11, 12	13, 14, 15, 16	17, 18, 19, 20

Simple familiar

1. **a.** Explain how a vector differs from a scalar.
   b. Give an example of a vector and an example of a scalar.

2. Consider the vectors $\underset{\sim}{u}$ and $\underset{\sim}{v}$ represented on the Cartesian plane.

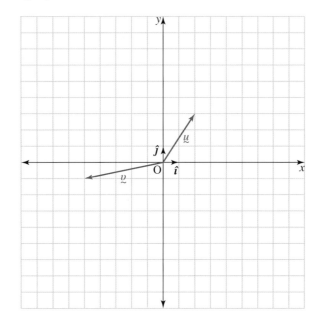

Sketch the results from the following.

    a. $\underset{\sim}{u} + \underset{\sim}{v}$
    b. $\underset{\sim}{v} - \underset{\sim}{u}$
    c. $-2\underset{\sim}{v} - \underset{\sim}{u}$
    d. $3\underset{\sim}{u} + \dfrac{1}{2}\underset{\sim}{v}$

3. Calculate the magnitude of the following vectors and express them as exact values.

   a. $u = \begin{pmatrix} -3 \\ 4 \end{pmatrix}$

   b. $v = 6\hat{\imath} + 8\hat{\jmath}$

   c. $s = 7 \begin{pmatrix} -2 \\ -1 \end{pmatrix}$

   d.

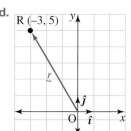

 4. Determine the direction of the following vectors and give your answers to one decimal place.

   a. $u = \begin{pmatrix} -2 \\ 5 \end{pmatrix}$

   b.

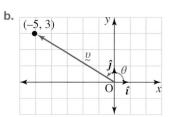

5. Determine the polar form of the following vectors. Give your answers to one decimal place.

   a. $u = \begin{pmatrix} -1 \\ \sqrt{2} \end{pmatrix}$

   b. $v = 4\hat{\imath} + 3\hat{\jmath}$

6. The hexagon ABCDEF can be defined by the three vectors $u = \overrightarrow{AB}$, $v = \overrightarrow{AF}$ and $w = \overrightarrow{BC}$.

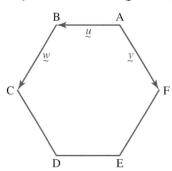

In terms of these three vectors u, v, and w, express the following:

   a. the vector from A to D
   b. the vector from F to C
   c. the vector from D to B.

7. **MC** Consider the relationship between vectors u, v, and w.
$$u = 2v - 3w$$
$$2w = u - v$$
Which of the following is true?

   A. $\dfrac{1}{3}u = v + w$
   B. $w = 5v$
   C. $7u = 5v$
   D. $u = 7w$

8. Express the following vectors in Cartesian form.

   a. $(2, 45°)$
   b. $(3, 90°)$
   c. $(4, -120°)$
   d. $(3, -30°)$

9. **MC** Which of the following is the unit vector of $(2, -60°)$?

   A. $\dfrac{\sqrt{3}}{2}\hat{\imath} + \dfrac{1}{2}\hat{\jmath}$
   B. $\dfrac{1}{2}\hat{\imath} + \dfrac{\sqrt{3}}{2}\hat{\jmath}$
   C. $\dfrac{1}{2}\hat{\imath} - \dfrac{\sqrt{3}}{2}\hat{\jmath}$
   D. $-\hat{\imath} - \sqrt{3}\hat{\jmath}$

10. **MC** Consider the vectors represented on the grid.

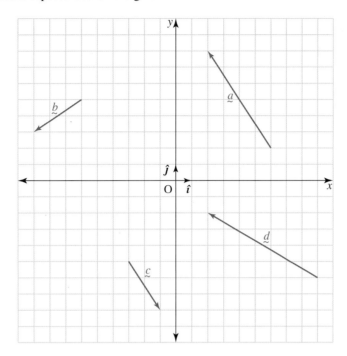

Which of the following is true?

A. $\underset{\sim}{a} = 2\underset{\sim}{c}$ **B.** $|\underset{\sim}{a}| = |\underset{\sim}{d}|$ **C.** $\underset{\sim}{a} + \underset{\sim}{b} = \underset{\sim}{d}$ **D.** $\underset{\sim}{b} + \underset{\sim}{c} = -\underset{\sim}{d}$

11. A kid travels 15 m east and then 36 m south and finally 5 m north-west.

   a. Sketch a vector drawing of the path of the kid.
   b. Determine the net displacement vector from the starting point and express it in column vector notation.
   c. Determine the net displacement in polar form. Give your answer to one decimal place.

12. **MC** Which of the following is true?

   A. $\left| \begin{pmatrix} -2 \\ 3 \end{pmatrix} \right| = \sqrt{5}$

   B. $(3, -45°) = \dfrac{3\sqrt{2}}{2}\hat{i} - \dfrac{3\sqrt{2}}{2}\hat{j}$

   C. $|-4\underset{\sim}{u}| = -4\,|\underset{\sim}{u}|$

   D. $\dfrac{1}{2}\hat{i} - \dfrac{\sqrt{3}}{2}\hat{j} = \left(\dfrac{1}{2}, 60° \right)$

Complex familiar

13. A boat travels 21 km in a direction of bearing 30° and 72 km in a direction of bearing 315°. Express its net displacement vector from the starting point in polar form. Give your answer to one decimal place.

14. Let $\underset{\sim}{a} = 2\hat{i} - 4\hat{j}$, $\underset{\sim}{b} = -3\hat{i} + \dfrac{1}{2}\hat{j}$ and $\underset{\sim}{c} = \dfrac{3}{4}\hat{i} - \hat{j}$

   Determine:

   a. $|\underset{\sim}{a} - \underset{\sim}{b} + \underset{\sim}{c}|$ to two decimal places

   b. $\hat{\underset{\sim}{b}}$ in Cartesian form

   c. $\underset{\sim}{b} + 2\underset{\sim}{c}$ in polar form to two decimal places.

15. Two bushwalkers starts their hike from the same starting point. The first one walks 6 km east and then 3 km north-east while the second one walks 9 km south-west and then 2 km west.
    Determine the final distance between the two bushwalkers. Give your answer to one decimal place.

16. A student is trying to use vectors to determine the side length of a regular pentagon inscribed in a circle of radius 1.

a. Use their diagram to complete their work and determine the side length of the pentagon. Give your answer to two decimal places.

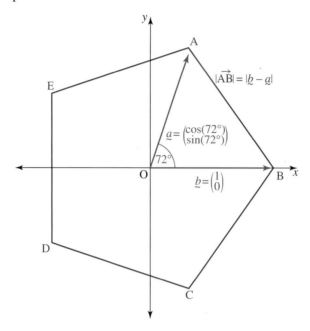

b. Use a similar method to calculate the side length of an octagon inscribed in a circle of radius 1. Give your answer to two decimal places.

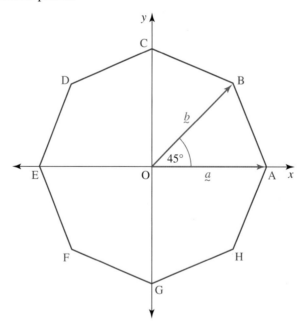

Complex unfamiliar

17. Two students are discussing which form is the most useful when trying to determine whether vectors are perpendicular. They both write a statement with an illustrative diagram.
Student A

A vector perpendicular to $\underset{\sim}{u} = \begin{pmatrix} x \\ y \end{pmatrix}$ is $k \begin{pmatrix} -y \\ x \end{pmatrix}$ where k is a non-zero scalar.

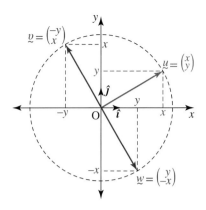

Student B

A vector perpendicular to $\underset{\sim}{u} = (r, \theta)$ is $\underset{\sim}{v} = \left(kr, \theta \pm \dfrac{\pi}{2} \right)$ where k is a strictly positive scalar.

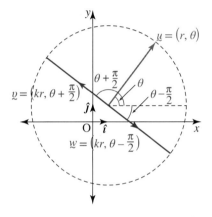

a. Demonstrate student A's statement.
b. Demonstrate student B's statement.
c. Using the notation or your choice, give an example of a vector perpendicular to
$$\underset{\sim}{u} = \begin{pmatrix} -3 \\ 4 \end{pmatrix} = (5,\ 143.1°)$$

18. Using vectors, determine the value of y so that the points $A(3, -4), B(-1, -1)$ and $C(-3, y)$ are aligned.

19. Let A, B and C be the points $A(4, -3), B(-2, 1)$ and $C(0, y)$ respectively.
Using vectors, determine the value of y so that $\overrightarrow{BA}$ and $\overrightarrow{BC}$ are perpendicular.

20. If $\underset{\sim}{a} = \begin{pmatrix} 2x \\ -y \end{pmatrix}$ and $\underset{\sim}{b} = \begin{pmatrix} -3x \\ 4y \end{pmatrix}$, $|\underset{\sim}{b} - \underset{\sim}{a}| = 15$ and $|3\underset{\sim}{a} + 2\underset{\sim}{b}| = 20$ determine the values of x and y.

Fully worked solutions for this chapter are available online.

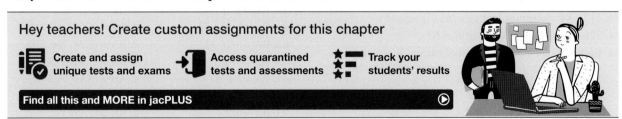

Answers

Chapter 3 Vectors in the plane

3.2 Vectors and scalars

3.2 Exercise

1. a. i. 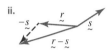   ii.

   iii.

  b. i. Same as **1 a i** except scaled by a factor of 2
    ii. Same as **1 a ii** except scaled by a factor of 2
    iii.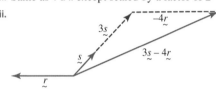

2. a. $\underset{\sim}{s} + \underset{\sim}{t}$ b. $\underset{\sim}{s} + \underset{\sim}{t} + \underset{\sim}{u} + \underset{\sim}{v}$ c. $-\underset{\sim}{s} - \underset{\sim}{t}$
   d. $-\underset{\sim}{u} - \underset{\sim}{v} - \underset{\sim}{t}$ e. $-\underset{\sim}{u} - \underset{\sim}{t} - \underset{\sim}{s}$

3. C

4. a. A to C   b. D to B
   c. B to D   d. A to C

5. D

6. a. $\underset{\sim}{r} + \underset{\sim}{s}$ b. $\underset{\sim}{s} + \underset{\sim}{t}$ c. $\underset{\sim}{r} - \underset{\sim}{s}$
   d. $\underset{\sim}{r} - \underset{\sim}{s}$ e. $\underset{\sim}{t} - \underset{\sim}{s}$ f. $\underset{\sim}{s} + \underset{\sim}{t} - \underset{\sim}{r}$
   g. $\underset{\sim}{r} + \underset{\sim}{s} + \underset{\sim}{t}$ h. $-\underset{\sim}{s} - \underset{\sim}{t}$

7. $2\underset{\sim}{a} + 4\underset{\sim}{b}$

8. $-3\underset{\sim}{a} - 4\underset{\sim}{b}$

9. Displacement, velocity, force

10. Speed, time, length

11. 1 magnitude and 2 angles

12. a, b
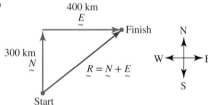
   c. 500 km
   d. 53.1° clockwise from N

13. 512.1 km; determine bearing using trigonometry

14. 721.1 km, 326.3° (clockwise from N)

15. Each part of the answer has coordinates as shown in the diagram $a, b, \dots j$. The original vectors $\underset{\sim}{a}$ and $\underset{\sim}{b}$ are also drawn.
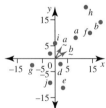

16. Magnitude = 10.77, direction 68.2° True.

17–20. Sample responses can be found in the worked solutions in the online resources.

21. $\underset{\sim}{0}$

22–24. Sample responses can be found in the worked solutions in the online resources.

25. One can deduce that x and y components can be added/subtracted/multiplied separately.

3.3 Vectors in two dimensions

3.3 Exercise

1. a. 3, 4   b. 6, −3   c. 3.4, $\sqrt{2}$

2. a.i. $6\sqrt{2}$ ii. 45°
   b.i. $\sqrt{65}$ ii. 119.7°
   c.i. 4.88   ii. 225.8°
   d.i. 320.16   ii. 358.2°

3. i. a. 045°   ii. a. $\left(6\sqrt{2}, 45°\right)$
    b. 330.3°
    c. 224.2°   b. $\left(\sqrt{65}, -60.3°\right)$
    d. 091.8°   c. (4.88, 225.8°)
                   d. (320.16, 358.2°)

4. $\underset{\sim}{w} = -50\hat{\imath} - 50\sqrt{3}\hat{\jmath}$

5. C

6. $248.9\hat{\imath} - 383.3\hat{\jmath}$

7. $60.6\hat{\imath} + 109.3\hat{\jmath}$

8. a. $\frac{3}{5}\hat{\imath} + \frac{4}{5}\hat{\jmath}$ b. $\frac{3}{5}\hat{\imath} - \frac{4}{5}\hat{\jmath}$
   c. $\frac{4}{5}\hat{\imath} + \frac{3}{5}\hat{\jmath}$ d. $-\frac{4}{5}\hat{\imath} + \frac{3}{5}\hat{\jmath}$
   e. $\frac{1}{\sqrt{3}}\hat{\imath} - \frac{\sqrt{2}}{\sqrt{3}}\hat{\jmath}$

9. B

10. Sample responses can be found in the worked solutions in the online resources.

11. $-0.98\hat{\imath} - 0.20\hat{\jmath}$

12. Sample responses can be found in the worked solutions in the online resources.

13. a.i. $\overrightarrow{AB} = 4\hat{\imath} - 7\hat{\jmath}$ ii. $\overrightarrow{AB} = (8.06, -60.26°)$
   b.i. $\overrightarrow{AB} = 2\hat{\imath} + \hat{\jmath}$ ii. $\overrightarrow{AB} = (2.24, 26.57°)$
   c.i. $\overrightarrow{AB} = 6\hat{\imath} - 5\hat{\jmath}$ ii. $\overrightarrow{AB} = (7.81, -140.19°)$
   d.i. $\overrightarrow{AB} = -\hat{\imath} - 3\hat{\jmath}$ ii. $\overrightarrow{AB} = (3.16, 251.57°)$
   e.i. $\overrightarrow{AB} = 2\hat{\imath}$ ii. $\overrightarrow{AB} = (2, 0°)$
   f.i. $\overrightarrow{AB} = -7\hat{\jmath}$ ii. $\overrightarrow{AB} = (7, -90°)$

14. a.i. $\overrightarrow{BA} = -4\hat{\imath} + 7\hat{\jmath}$ ii. $\overrightarrow{BA} = (8.06, 119.74°)$
   b.i. $\overrightarrow{BA} = -2\hat{\imath} - \hat{\jmath}$ ii. $\overrightarrow{BA} = (2.24, 206.57°)$
   c.i. $\overrightarrow{AB} = -6\hat{\imath} + 5\hat{\jmath}$ ii. $\overrightarrow{BA} = (7.81, 39.81°)$
   d.i. $\overrightarrow{BA} = \hat{\imath} + 3\hat{\jmath}$ ii. $\overrightarrow{AB} = (3.16, 71.57°)$
   e.i. $\overrightarrow{BA} = -2\hat{\imath}$ ii. $\overrightarrow{BA} = (2, 180°)$
   f.i. $\overrightarrow{BA} = 7\hat{\jmath}$ ii. $\overrightarrow{BA} = (7, 90°)$

15. a. $\widehat{AB} = \dfrac{4}{\sqrt{65}}\hat{\imath} - \dfrac{7}{\sqrt{65}}\hat{\jmath}$ b. $\widehat{AB} = \dfrac{2}{\sqrt{5}}\hat{\imath} + \dfrac{1}{\sqrt{5}}\hat{\jmath}$

   c. $\widehat{AB} = \dfrac{6}{\sqrt{61}}\hat{\imath} - \dfrac{5}{\sqrt{61}}\hat{\jmath}$ d. $\widehat{AB} = -\dfrac{1}{\sqrt{10}}\hat{\imath} - \dfrac{3}{\sqrt{10}}\hat{\jmath}$

   e. $\widehat{AB} = \hat{\imath}$ f. $\widehat{AB} = -\hat{\jmath}$

16. a.   i. $\sqrt{29}$ ii. $\sqrt{13}$

   iii. $\dfrac{5}{\sqrt{29}}\hat{\imath} - \dfrac{2}{\sqrt{29}}\hat{\jmath}$ iv. $-\dfrac{2}{\sqrt{13}}\hat{\imath} + \dfrac{3}{\sqrt{13}}\hat{\jmath}$

   v. $3\hat{\imath} + \hat{\jmath}$ vi. $\sqrt{10}$

  b. Reject, because magnitude is different.

17. a.   i. 5   ii. $\sqrt{26}$

   iii. $-\dfrac{3}{5}\hat{\imath} - \dfrac{4}{5}\hat{\jmath}$ iv. $\dfrac{5}{\sqrt{26}}\hat{\imath} - \dfrac{1}{\sqrt{26}}\hat{\jmath}$

   v. $2\hat{\imath} + 3\hat{\jmath}$ vi. $\sqrt{13}$

  b. Reject, because the magnitude is different.

18. $a = -4\hat{\imath} + 7\hat{\jmath}$
$b = -3\hat{\imath} - 2\hat{\jmath}$
$c = 2\hat{\imath} - 3\hat{\jmath}$
$d = -7\hat{\imath} + 4\hat{\jmath}$

19. $(615.4, -49.8°)$

20. 36 steps 11.3° south of east

21. 20.8 km

22. a. $\sqrt{2}$ b. $\sqrt{58}$

23. a. $3\hat{\imath}$ b. $5\hat{\jmath}$ c. $3\hat{\imath} + 5\hat{\jmath}$
   d. 031.0° T   e. $\sqrt{34}$ km/h

24. 329.0°

25. 053.1°

26. 53.1°, −36.9° Difference = 90°; sample responses can be found in the worked solutions in the online resources.

27. 0.0417 h or 2.5 minutes

3.4 Review

3.4 Exercise

1. a. Vectors are characterised by their magnitude and their direction, whereas scalars do not depend on a direction and only have a magnitude.

  b. Answers will vary.
Examples of vectors include: force, displacement, velocity, acceleration, momentum etc.
Examples of scalars include: temperature, time, mass, length, surface area, volume, energy, density, speed etc.

2. a.

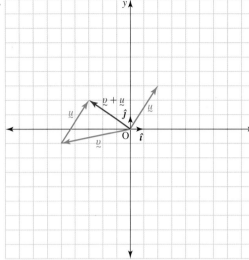

b.

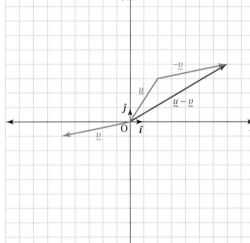

c.

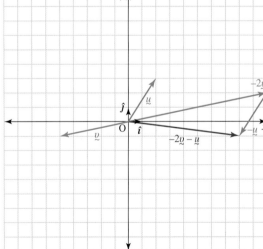

d.

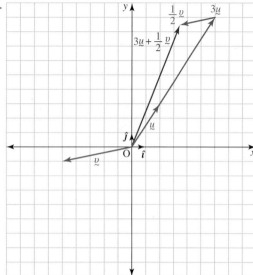

3. a. $|u| = 5$ **b.** $|v| = 10$
   c. $|s| = 7\sqrt{5}$ **d.** $|\tilde{r}| = \sqrt{34}$

4. a. $\theta = 111.8°$ **b.** $\theta = -31.0°$
   c. $\theta = 149.0°$

5. a. $u = \left(\sqrt{3}, 125.3°\right).$ **b.** $b = (5, 36.9°)$

6. a. $\overrightarrow{AD} = u + w + v$ **b.** $\overrightarrow{FC} = u + w - v$
   c. $\overrightarrow{DB} = -w - v$

7. D

8. a. $(2, 45°) = \sqrt{2}\hat{\imath} + \sqrt{2}\hat{\jmath}$
   b. $(3, 90°) = 3\hat{\jmath}$
   c. $(4, -120°) = -2\hat{\imath} + 2\sqrt{3}\hat{\jmath}$
   d. $(3, -30°) = \dfrac{3\sqrt{3}}{2}\hat{\imath} - \dfrac{3}{2}\hat{\jmath}$

9. C

10. C

11. a.

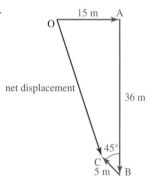

b. Net displacement vector $= \begin{pmatrix} 15 - \dfrac{5\sqrt{2}}{2} \\ \dfrac{5\sqrt{2}}{2} - 36 \end{pmatrix}$

c. Net displacement vector $= (34.4, 109.5°)$

12. B

13. Net displacement $= (80.0, 120.0°)$

14. a. $|a - b + c| == \dfrac{5\sqrt{13}}{4} \simeq 4.51$

   b. $\hat{b} = \dfrac{-6\sqrt{37}}{37}\hat{\imath} + \dfrac{\sqrt{37}}{37}\hat{\jmath}$

   c. $b + 2c = (2.30, 192.53°)$

15. The final distance between the two bushwalkers is 18.5 km.

16. a. The side length of a pentagon inscribed in a circle of radius 1 is 1.18.
   b. The side length of an octagon inscribed in a circle of radius 1 is 0.77.

17. a. Sample responses can be found in the worked solutions in the online resources.
   b. Sample responses can be found in the worked solutions in the online resources.
   c. Answers will vary.
   An example of a vector perpendicular to $v = \begin{pmatrix} -3 \\ 4 \end{pmatrix} = (5, 143.1°)$ is $\begin{pmatrix} 4 \\ 3 \end{pmatrix} = (5, 53.1°)$.

18. The points $A(3, -4)$, $B(-1, -1)$ and $C(-3, y)$ are aligned if $y = \dfrac{1}{2}$.

19. $\overrightarrow{BA}$ and $\overrightarrow{BC}$ are perpendicular if $y = 2$

20. $x = \pm\dfrac{9}{5}$ and $y = \pm 4$.

4 Algebra of vectors in two dimensions

LESSON SEQUENCE

Fully worked solutions for this chapter are available online.

EXAM PREPARATION

Access exam-style questions in every lesson, available online.

 Resources

Solutions	Solutions — Chapter 4 (sol-0396)
Exam questions	Exam question booklet — Chapter 4 (eqb-0282)
Digital documents	Learning matrix — Chapter 4 (doc-41924)
	Chapter summary — Chapter 4 (doc-41925)

LESSON
4.1 Overview

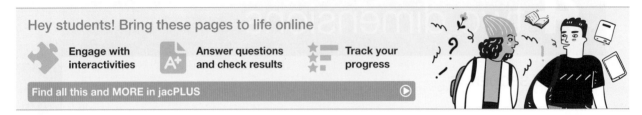
4.1.1 Introduction

In Chapter 3, we discussed the theory of vectors, which were defined as a directed line segment with both magnitude and direction. Vectors can be added, subtracted and multiplied by a scalar. Vectors are used in physics to describe and analyse any quantity that has both magnitude and direction.

Vectors are the ideal mathematical tool for dealing with the motion of objects and the forces acting on them. In this chapter, we will consider operations with vectors in more detail, and consider applications of vectors — displacement, velocity, forces in equilibrium and relative velocity.

Sir Isaac Newton was an English mathematician and physicist in the seventeenth century. He is considered one of the most influential scientists of all time. The publication of Newton's Philosophiae Naturalis Principia Mathematica in 1687 was one of the keys to unlocking the mysteries of the physical world. His three simple laws accurately describe the motion of objects, and with his mathematical descriptions of mechanics and gravity, lunar and planetary motion can be explained with the help of vectors.

4.1.2 Syllabus links

Lesson	Lesson title	Syllabus links
4.2	**Vector representations**	◯ Determine a vector between two points.
		◯ Define and use multiplication by a scalar of a vector in Cartesian form.
		◯ Define and use a vector representing a section of a line segment, including the midpoint of a line segment.
4.3	**Scalar (dot) product and the projection of vectors – scalar and vector resolutes**	◯ Define and use the scalar (dot) product. • $a \cdot b = \|a\| \|b\| \cos(\theta)$ • $\begin{pmatrix} a_1 \\ a_2 \end{pmatrix} \cdot \begin{pmatrix} b_1 \\ b_2 \end{pmatrix} = a_1 b_1 + a_2 b_2$
		◯ Examine properties of parallel and perpendicular vectors and determine if two vectors are parallel or perpendicular.
		◯ Define and use scalar and vector projections of vectors. • scalar projection of a on b: $\|a\| \cos(\theta) = a \cdot \hat{b}$ • vector projection of a on b: $\|a\| \cos(\theta)\hat{b} = (a \cdot \hat{b})\hat{b} = \left(\dfrac{a \cdot b}{b \cdot b}\right) b$
		◯ Apply the scalar product to vectors expressed in Cartesian form.
		◯ Resolve vectors into $\hat{i}$ and $\hat{j}$ components.
4.4	**Operations with vectors**	◯ Examine and use addition and subtraction of vectors in Cartesian form.
4.5	**Applications of vectors in two dimensions**	◯ Model and solve problems that involve displacement, force, velocity and relative velocity using the above concepts.
		◯ Model and solve problems that involve motion of a body in equilibrium situations, including vector applications related to smooth inclined planes (excluding situations with pulleys and connected bodies).

LESSON
4.2 Vector representations

SYLLABUS LINKS

- Determine a vector between two points.
- Define and use multiplication by a scalar of a vector in Cartesian form.
- Define and use a vector representing a section of a line segment, including the midpoint of a line segment.

Source: Specialist Mathematics Senior Syllabus 2024 © State of Queensland (QCAA) 2024; licensed under CC BY 4.0

4.2.1 A vector between two points

In the figure $\underset{\sim}{a}$ is the position vector of point A, $\overrightarrow{OA}$, and $\underset{\sim}{b}$ is the position vector of point B, $\overrightarrow{OB}$, relative to the origin.

The vector describing the location of A relative to B, $\overrightarrow{BA}$, is easily found using vector addition as $-\underset{\sim}{b} + \underset{\sim}{a}$ or $\underset{\sim}{a} - \underset{\sim}{b}$.

Similarly, the vector describing the location of B relative to A, $\overrightarrow{AB}$, is $\underset{\sim}{b} - \underset{\sim}{a}$.

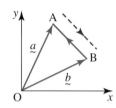

> ### Vector between two points
>
> **If A and B are points defined by position vectors $\underset{\sim}{a}$ and $\underset{\sim}{b}$ respectively, then**
>
> $$\overrightarrow{AB} = \underset{\sim}{b} - \underset{\sim}{a}$$

WORKED EXAMPLE 1 Calculating the distance between two points

a. **Determine the position vector locating point B $(3, -3)$ from point A $(2, 5)$, that is vector $\overrightarrow{AB}$.**
b. **Calculate the length of this vector.**

THINK

a. 1. Express the point A as a position vector $\underset{\sim}{a}$.

    2. Express the point B as a position vector $\underset{\sim}{b}$.

    3. The location of B relative to A, $\overrightarrow{AB}$, is $\underset{\sim}{b} - \underset{\sim}{a}$.

b. The length of $\overrightarrow{AB}$ is $|\underset{\sim}{b} - \underset{\sim}{a}|$.

WRITE

a. Let $\overrightarrow{OA} = \underset{\sim}{a} = 2\hat{\imath} + 5\hat{\jmath}$
    Let $\overrightarrow{OB} = \underset{\sim}{b} = 3\hat{\imath} - 3\hat{\jmath}$
$$\begin{aligned} \overrightarrow{AB} &= \underset{\sim}{b} - \underset{\sim}{a} \\ &= 3\hat{\imath} - 3\hat{\jmath} - (2\hat{\imath} + 5\hat{\jmath}) \\ &= \hat{\imath} - 8\hat{\jmath} \end{aligned}$$

b. $$\begin{aligned} \left|\overrightarrow{AB}\right| &= |\underset{\sim}{b} - \underset{\sim}{a}| \\ &= \sqrt{1^2 + (-8)^2} \\ &= \sqrt{65} \text{ (or 8.06)} \end{aligned}$$

Further consideration of Worked example 1 demonstrates that the direction of vector $\overrightarrow{AB}$ is $\theta = -82.87°$ (using $\tan(\theta) = \dfrac{-8}{1}$ and, $\overrightarrow{AB}$ being in the 4th quadrant, the direction θ is given by $-|\theta|$).

We are now in a position to resolve the problem of finding, accurately, the resultant force acting on the stuffed bear from chapter 3 when Kim and Linh are pulling on it.

First redraw the diagram to show the addition of vectors.

Taking the direction of Linh's force as the $\hat{\imath}$ direction:

Linh's force, $\underset{\sim}{a} = 50\hat{\imath} + 0\hat{\jmath}$

Kim's force, $\underset{\sim}{b} = 40\cos(150°)\hat{\imath} + 40\sin(150°)\hat{\jmath}$

$\qquad = -34.6410\hat{\imath} + 20\hat{\jmath}$

Resultant force, $\underset{\sim}{a} + \underset{\sim}{b} = (50 - 34.6410)\hat{\imath} + 20\hat{\jmath}$

$\qquad = 15.3590\hat{\imath} + 20\hat{\jmath}$

Magnitude $= |\underset{\sim}{a} + \underset{\sim}{b}|$

$\qquad = 25.2\text{N}$

Direction $\theta = \tan^{-1}\left(\dfrac{20}{15.3590}\right)$

$\qquad = 52.48°$

4.2.2 Equality of two vectors

When a vector in $\hat{\imath}$ and $\hat{\jmath}$ form is multiplied by a scalar, each coefficient is multiplied by the scalar.

> **Multiplication by a scalar**
>
> If $\underset{\sim}{a} = x\hat{\imath} + y\hat{\jmath}$, then $k\underset{\sim}{a} = k(x\hat{\imath} + y\hat{\imath}) = kx\hat{\imath} + ky\hat{\jmath}.$

For example, if $\underset{\sim}{a} = \hat{\imath} - 2\hat{\jmath}$, then $2\underset{\sim}{a} = 2\hat{\imath} - 4\hat{\jmath}$ and $-\underset{\sim}{a} = -\hat{\imath} + 2\hat{\jmath}$.

Given the vectors $\overrightarrow{OA} = \underset{\sim}{a} = x_1\hat{\imath} + y_1\hat{\jmath}$ and $\overrightarrow{OB} = \underset{\sim}{b} = x_2\hat{\imath} + y_2\hat{\jmath}$, the two vectors are equal, $\underset{\sim}{a} = \underset{\sim}{b}$, if and only if $x_1 = x_2$ and $y_1 = y_2$.

A vector in the form $k\hat{\imath}$, with a $\hat{\jmath}$ component equal to zero is parallel to the x-axis, and a vector in the form $k\hat{\jmath}$, with a $\hat{\imath}$ component equal to zero is parallel to the y-axis.

WORKED EXAMPLE 2 Using vector equality

Given that $\underset{\sim}{a} = x\hat{\imath} - 3\hat{\jmath}$ and $\underset{\sim}{b} = 4\hat{\imath} - 5\hat{\jmath}$, determine the value of x that places the vector $\underset{\sim}{c} = 2\underset{\sim}{a} - 3\underset{\sim}{b}$ parallel to the y-axis.

THINK	WRITE
1. Substitute the vectors $\underset{\sim}{a}$ and $\underset{\sim}{b}$ into $\underset{\sim}{c} = 2\underset{\sim}{a} - 3\underset{\sim}{b}$.	$\underset{\sim}{c} = 2\underset{\sim}{a} - 3\underset{\sim}{b}$ $= 2(x\hat{\imath} - 3\hat{\jmath}) - 3(4\hat{\imath} - 5\hat{\jmath})$
2. Multiply the vectors $\underset{\sim}{a}$ and $\underset{\sim}{b}$ by the scalars 2 and -3.	$\underset{\sim}{c} = (2x\hat{\imath} - 6\hat{\jmath}) - (12\hat{\imath} - 15\hat{\jmath})$
3. Simplify	$\underset{\sim}{c} = (2x - 12)\hat{\imath} + (-6 + 15)\hat{\jmath}$ $\underset{\sim}{c} = (2x - 12)\hat{\imath} + 9\hat{\jmath}$
4. As vector $\underset{\sim}{c}$ is parallel to the y-axis, its $\hat{\imath}$ component must be zero. Equate the $\hat{\imath}$ component to zero.	$2x - 12 = 0$
5. Solve for x.	$2x = 12$ $x = 6$

Vector addition and scalar multiplication can be used to define a vector that represents the midpoint of a line segment.

WORKED EXAMPLE 3 Representing the midpoint of a line segment with a vector

Determine the position vector of midpoint of the line segment AB if A = (1, 2) and B = (4, 3).

THINK	WRITE
1. Sketch a diagram to represent this situation.	Let M be the midpoint of AB. 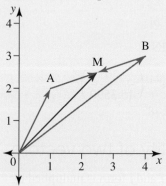
2. Express $\overrightarrow{OM}$ as a combination of $\overrightarrow{OA}$ and $\overrightarrow{AB}$.	$\overrightarrow{OM} = \overrightarrow{OA} + \dfrac{\overrightarrow{AB}}{2}$
3. Write $\overrightarrow{OA}$ and $\overrightarrow{OB}$ in component form.	$\overrightarrow{OA} = \hat{\imath} + 2\hat{\jmath}$ $\overrightarrow{OB} = 4\hat{\imath} + 3\hat{\jmath}$
4. Write $\overrightarrow{AB}$ in component form.	$\overrightarrow{AB} = \overrightarrow{AO} + \overrightarrow{OB}$ $= \overrightarrow{OB} - \overrightarrow{OA}$ $= 4\hat{\imath} + 3\hat{\jmath} - \hat{\imath} - 2\hat{\jmath}$ $= 3\hat{\imath} + \hat{\jmath}$

5. Write $\overrightarrow{OM}$ in component form.

$$\overrightarrow{OM} = \overrightarrow{OA} + \frac{1}{2}\overrightarrow{AB}$$

$$= \hat{\imath} + 2\hat{\jmath} + \frac{1}{2}(3\hat{\imath} + \hat{\jmath})$$

$$= \frac{5}{2}\hat{\imath} + \frac{5}{2}\hat{\jmath}$$

Further consideration of Worked example 3 demonstrates that the, we can also express $\overrightarrow{OM}$ as a combination of $\overrightarrow{OB}$ and $\overrightarrow{AB}$ $\left(\overrightarrow{OM} = \overrightarrow{OB} - \dfrac{\overrightarrow{AB}}{2}\right)$ or as a combination of $\overrightarrow{OA}$ and $\overrightarrow{OB}$ $\left(\overrightarrow{OM} = \dfrac{\overrightarrow{OA} + \overrightarrow{OB}}{2}\right)$.

The vector position representing the midpoint, M, of the line segment AB can be determined using the formula:

Midpoint M of a line segment AB

$$\overrightarrow{OM} = \frac{\overrightarrow{OA} + \overrightarrow{OB}}{2}$$

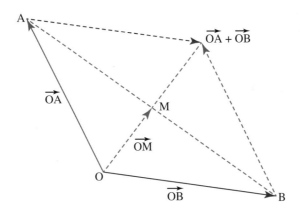

4.2.3 Solving vector problems

When solving vector problems, we may often need to equate coefficients, solve simultaneous, linear or even non-linear equations to determine the unknown variables in each case.

WORKED EXAMPLE 4 Solving vector problems

Given the vectors $\underset{\sim}{a} = 2\hat{\imath} - 4\hat{\jmath}$, $\underset{\sim}{b} = 3\hat{\imath} - 8\hat{\jmath}$ and $\underset{\sim}{c} = \hat{\imath} + 4\hat{\jmath}$, determine the values of the scalars m and n if $\underset{\sim}{c} = m\underset{\sim}{a} + n\underset{\sim}{b}$.

THINK

1. Substitute the vectors $\underset{\sim}{a}$ and $\underset{\sim}{b}$ into $\underset{\sim}{c} = m\underset{\sim}{a} + n\underset{\sim}{b}$.

2. Multiply the vectors $\underset{\sim}{a}$ and $\underset{\sim}{b}$ by the scalars m and n.

WRITE

$\underset{\sim}{c} = m\underset{\sim}{a} + n\underset{\sim}{b}$

$\hat{\imath} + 4\hat{\jmath} = m(2\hat{\imath} - 4\hat{\jmath}) + n(3\hat{\imath} - 8\hat{\jmath})$

$\hat{\imath} + 4\hat{\jmath} = (2m\hat{\imath} - 4m\hat{\jmath}) + (3n\hat{\imath} - 8n\hat{\jmath})$

▶

3. On the right hand side, rearrange and group $\hat{i}, \hat{j}$ components.	$\hat{i} + 4\hat{j} = (2m + 3n)\hat{i} - (4m + 8n)\hat{j}$
4. Equate the components to match the coefficients on the left and right hand side of the equation.	Since the $\hat{i}$ components are equal: $1 = 2m + 3n$ [1] Since the $\hat{j}$ components are equal: $4 = -4m - 8n$ [2]
5. Solve the simultaneous equations.	$2 \times [1] \quad 2 = 4m + 6n$ $\qquad\qquad 4 = -4m - 8n$ [2] Adding gives $6 = -2n \Rightarrow n = -3$
6. Substitute into [1] to solve for m.	$2m = 1 - 3n$ $2m = 1 + 9 = 10$ $\quad m = 5$
7. Write the answer.	$\underset{\sim}{c} = 5\underset{\sim}{a} - 3\underset{\sim}{b}$

4.2.4 Parallel vectors

Two vectors $\underset{\sim}{a}$ and $\underset{\sim}{b}$ are parallel if $\underset{\sim}{a} = k\underset{\sim}{b}$, where $k \in \mathbb{R}$. Conversely, if $\underset{\sim}{a} = k\underset{\sim}{b}$, then $\underset{\sim}{a}$ and $\underset{\sim}{b}$ are parallel.

WORKED EXAMPLE 5 Solving more vector problems

Given the vectors $\underset{\sim}{r} = x\hat{i} + 3\hat{j}$ and $\underset{\sim}{s} = 5\hat{i} - 6\hat{j}$, calculate the value of x in each case if:
a. the length of the vector $\underset{\sim}{r}$ is 7
b. the vector $\underset{\sim}{r}$ is parallel to the vector $\underset{\sim}{s}$.

THINK

a. 1. Determine the magnitude of the vector in terms of the unknown value.

    2. Equate the length of the vector to the given value.

    3. Square both sides and solve for the unknown value. Both answers are acceptable values.

WRITE

a. $\underset{\sim}{r} = x\hat{i} + 3\hat{j}$

      $|\underset{\sim}{r}| = \sqrt{x^2 + 3^2}$

         $= \sqrt{x^2 + 9}$

     Since $|\underset{\sim}{r}| = 7$:

     $\sqrt{x^2 + 9} = 7$

     $x^2 + 9 = 49$
        $x^2 = 40$
         $x = \pm\sqrt{40}$
           $= \pm\sqrt{4 \times 10}$
           $= \pm 2\sqrt{10}$

b. 1. If two vectors are parallel, then one is a scalar multiple of the other. Substitute for the given vectors and expand.

b.
$$\underset{\sim}{r} = k\underset{\sim}{s}$$
$$x\hat{i} + 3\hat{j} = k(5\hat{i} - 6\hat{j})$$
$$= 5k\hat{i} - 6k\hat{j}$$

2. For the two vectors to be equal, both components must both be equal.

From the $\hat{j}$ component, $3 = -6k$, so
$$k = -\frac{1}{2}.$$
From the $\hat{i}$ component,
$$x = 5k.$$

3. Solve for the unknown value in this case.

$x = 5k$ but $k = -\frac{1}{2}$, so
$$x = -\frac{5}{2}.$$

Collinear vectors are parallel vectors that lie along the same line or parallel lines. Two vectors are collinear if they are parallel to the same line irrespective of their magnitudes and direction.

Exercise 4.2 Vector representations

4.2 Exercise	4.2 Exam questions **on**

Simple familiar	Complex familiar	Complex unfamiliar
1, 2, 3, 4, 5, 6, 7, 8, 9, 10	11, 12	13, 14

Simple familiar

1. **WE1** For each of the following pairs of points, determine the position vector locating the second point from the first point, that is vector $\overrightarrow{AB}$.
   - **a.** A $(2, 6)$ and B $(5, 10)$
   - **b.** A $(3, 0)$ and B $(4, -7)$
   - **c.** A $(10, 0)$ and B $(7, 0)$
   - **d.** A $(21, 5)$ and B $(-4, -5)$

2. Consider the position vectors $\overrightarrow{OA} = \underset{\sim}{a} = 5\hat{i} + 3\hat{j}$ and $\overrightarrow{OB} = \underset{\sim}{b} = 7\hat{i} - 6\hat{j}$. Calculate the distance between points A and B by first determining the position vector that locates one point from the other. Give your answer to one decimal place.

3. **WE2** If $\underset{\sim}{a} = 4\hat{i} - 5\hat{j}$ and $\underset{\sim}{b} = 3\hat{i} + y\hat{j}$, determine the value of y if the vector $\underset{\sim}{c} = 3\underset{\sim}{a} + 2\underset{\sim}{b}$ is parallel to the x-axis.

4. The vector $\overrightarrow{CD} = 7\hat{i} - 5\hat{j}$, the coordinates of point C are $(x, -3)$ and the coordinates of point D are $(4, y)$. Determine the values of x and y.

5. **WE3** Determine the position vector of the midpoint of the line segment AB if A $= (3, \ 4)$ and B $= (7, \ 8)$.

6. **a.** Determine the value of x if the vector $x\hat{i} + 3\hat{j}$ is parallel to the vector $5\hat{i} - 6\hat{j}$.
   b. Determine the value of y if the vector $-4\hat{i} + 5\hat{j}$ is parallel to the vector $6\hat{i} + y\hat{j}$.

7. **a.** For the vectors $\underset{\sim}{a} = -2\hat{\imath} + 3\hat{\jmath}$ and $\underset{\sim}{b} = 4\hat{\imath} - 2\hat{\jmath}$, show that the vector $2\underset{\sim}{a} + 3\underset{\sim}{b}$ is parallel to the x-axis.
   b. For the vectors $\underset{\sim}{c} = 3\hat{\imath} - 5\hat{\jmath}$ and $\underset{\sim}{d} = 4\hat{\imath} - 3\hat{\jmath}$, show that the vector $4\underset{\sim}{c} - 3\underset{\sim}{d}$ is parallel to the y-axis.
   c. Given the vectors $\underset{\sim}{a} = x\hat{\imath} - 5\hat{\jmath}$ and $\underset{\sim}{b} = 5\hat{\imath} + 3\hat{\jmath}$, determine the value of x if the vector $3\underset{\sim}{a} + 4\underset{\sim}{b}$ is parallel to the y-axis.
   d. If $\underset{\sim}{c} = 5\hat{\imath} - 3\hat{\jmath}$ and $\underset{\sim}{d} = 4\hat{\imath} + y\hat{\jmath}$, determine the value of y if the vector $5\underset{\sim}{c} + 7\underset{\sim}{d}$ is parallel to the x-axis.

8. **a.** Given the points A $(5, -2)$ and B $(-1, 3)$, determine:
   - **i.** the vector $\overrightarrow{AB}$
   - **ii.** the distance between the points A and B
   - **iii.** a unit vector parallel to $\overrightarrow{AB}$.

   b. Two points C and D are given by $(-4, 3)$, $(2, -5)$ respectively. Determine:
   - **i.** the vector $\overrightarrow{CD}$
   - **ii.** the distance between the points C and D
   - **iii.** a unit vector parallel to $\overrightarrow{DC}$.

9. **WE4** Given the vectors $\underset{\sim}{a} = 5\hat{\imath} + 2\hat{\jmath}$, $\underset{\sim}{b} = -3\hat{\imath} - 4\hat{\jmath}$ and $\underset{\sim}{c} = \hat{\imath} - 8\hat{\jmath}$, determine the values of the scalars m and n if $\underset{\sim}{c} = m\underset{\sim}{a} + n\underset{\sim}{b}$.

10. Given the vectors $\underset{\sim}{r} = x\hat{\imath} + y\hat{\jmath}$ and $\underset{\sim}{s} = 4\hat{\imath} - 3\hat{\jmath}$, determine the values of x and y if $2\underset{\sim}{r} + 3\underset{\sim}{s} = 8\hat{\imath} + \hat{\jmath}$.

Complex familiar

11. **WE5** Given the vectors $\underset{\sim}{r} = x\hat{\imath} - 4\hat{\jmath}$ and $\underset{\sim}{s} = 3\hat{\imath} + 5\hat{\jmath}$, calculate the value of x if:
    a. the length of the vector $\underset{\sim}{r}$ is 6
    b. the vector $\underset{\sim}{r}$ is parallel to the vector $\underset{\sim}{s}$.

12. Calculate the value of y for the vectors $\underset{\sim}{a} = 4\hat{\imath} + y\hat{\jmath}$ and $\underset{\sim}{b} = 2\hat{\imath} - 5\hat{\jmath}$ if:
    a. the vectors $\underset{\sim}{a}$ and $\underset{\sim}{b}$ are equal in length
    b. the vector $\underset{\sim}{a}$ is parallel to the vector $\underset{\sim}{b}$.

Complex unfamiliar

13. **a.** Show that the points A $(3, -2)$, B $(4, -5)$ and C $(1, 4)$ are collinear.
    b. Show that the points $(5, -3)$, $(2, 1)$ and $(8, -7)$ are collinear.

14. **a.** Given the vectors $\underset{\sim}{a} = x\hat{\imath} + 3\hat{\jmath}$ and $\underset{\sim}{b} = -2\hat{\imath} + y\hat{\jmath}$, and also given that $3\underset{\sim}{a} + 4\underset{\sim}{b} = 4\hat{\imath} + \hat{\jmath}$, determine the values of x and y.
    b. For the vectors $\underset{\sim}{a} = 4\hat{\imath} - 5\hat{\jmath}$, $\underset{\sim}{b} = -7\hat{\imath} + 3\hat{\jmath}$ and $m\underset{\sim}{a} + n\underset{\sim}{b} = 8\hat{\imath} + 13\hat{\jmath}$, determine the values of m and n.
    c. If $\underset{\sim}{a} = 5\hat{\imath} - 6\hat{\jmath}$, $\underset{\sim}{b} = -2\hat{\imath} + 4\hat{\jmath}$ and $m\underset{\sim}{a} + n\underset{\sim}{b} = 2\hat{\jmath}$, determine the values of m and n.

Fully worked solutions for this chapter are available online.

LESSON
4.3 Scalar (dot) product and the projection of vectors – scalar and vector resolutes

4.3.1 Calculating the dot product

In a previous section we studied the result of multiplying a vector by a scalar. In this section, we study the multiplication of two vectors.

The scalar or *dot* product of two vectors, a and b, is denoted by $a \cdot b$.

There are two ways of calculating the **dot product**. The first method is based on its definition. (The second method is shown later.) Consider the two vectors a and b.

> ### Dot product
>
> **By definition, the dot product $a \cdot b$ is given by:**
>
> $$a \cdot b = |a|\,|b|\cos(\theta)$$
>
> **where θ is the angle between (the positive directions of) a and b.**
>
>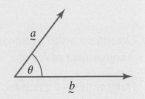

Note: The vectors are not aligned head to tail as for addition or subtraction, but rather their two tails are joined. That means, vector a, b need to have the same start point.

Properties of the dot product

- The dot product is a scalar. It is the result of multiplying three scalar quantities: the magnitudes of the two vectors and the cosine of the angle between them.
- The order of multiplication is unimportant (*commutative* property); thus,

$$a \cdot b = b \cdot a$$

- The dot product is *distributive*; thus,

$$c \cdot (a + b) = c \cdot a + c \cdot b$$

- Since the angle between a and itself is $0°$, and since $\cos(0) = 1$, then

$$a \cdot a = |a| \times |a| \cos(0)$$
$$= |a|^2$$

WORKED EXAMPLE 6 Calculating the dot product of vectors using magnitude and direction

Determine $a \cdot b$ using the information provided in the following cases.

a. Let $|a| = 4$ and $|b| = \dfrac{1}{3}$ and θ, the angle between the positive directions of a and b is $135°$.

b. $a = 3\hat{i} + 4\hat{j}$ and $b = 3\hat{i}$.

c. $a = 2\hat{i} - 3\hat{j}$ and $b = -\hat{i} + 3\hat{j}$.

THINK	WRITE
a. Determine $a \cdot b$ using the equation with the angle between two vectors.	**a.** $a \cdot b = \lvert a\rvert\,\lvert b\rvert \cos(\theta)$ $= 4 \times \dfrac{1}{3} \times \cos(135°)$ $= 4 \times \dfrac{1}{3} \times -\dfrac{\sqrt{2}}{2}$ $= -\dfrac{\sqrt{2}}{3}$
b. 1. Determine the magnitudes of a and b.	**b.** $\lvert a\rvert = \sqrt{3^2 + 4^2}$ $= 5$ $\lvert b\rvert = \sqrt{3^2}$ $= 3$
2. Sketch a right-angled triangle showing the angle that a makes with the positive x-axis since b is along the x-axis.	
3. Determine $\cos(\theta)$, knowing that $a = 5$ and the x-component of a is 3.	$\cos(\theta) = \dfrac{3}{5}$
4. Determine $a \cdot b$ using the equation with the angle between two vectors	$a \cdot b = \lvert a\rvert \times \lvert b\rvert \times \cos(\theta)$ $= 5 \times 3 \times \dfrac{3}{5}$ $= 9$
5. Simplify.	
c. 1. Determine the magnitudes of a and b.	**c.** $a = \sqrt{2^2 + (-3)^2}$ $= \sqrt{13}$ $b = \sqrt{(-1)^2 + 3^2}$ $= \sqrt{10}$

2. Determine the direction of a, the angle α and the direction of b, the angle β.

$$\tan(\alpha) = \frac{-3}{2}$$
$$\alpha = -56.31°$$

a is in the 4th quadrant, its direction is given by $-|\alpha|$. The direction of a is $-56.31°$.

$$\tan(\beta) = \frac{3}{-1}$$
$$\beta = -71.57°$$

b is in the 2nd quadrant, its direction is given by $180° - |\beta|$.
The direction of b is $180° - 71.57° = 108.43°$.

3. Sketch a diagram to represent the situation.

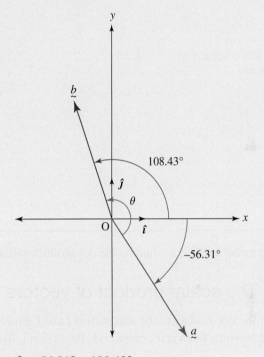

4. Evaluate θ, the angle between the positive directions of a and b.

$$\theta = 56.31° + 108.43°$$
$$= 164.74°$$

5. Determine $a \cdot b$ using the equation with the angle between two vectors.

$$a \cdot b = |a|\,|b|\cos(\theta)$$
$$= \sqrt{10} \times \sqrt{13} \times \cos(164.74°)$$
$$= 10.999\ldots$$
$$\approx -11$$

| TI | THINK | WRITE | CASIO | THINK | WRITE |

b. 1. On a Calculator page, press MENU, then select:
7: Matrix & Vector
C: Vector
3: Dot Product
Complete the entry line as:
dotP $\left(\begin{bmatrix} 3 & 4 \end{bmatrix}, \begin{bmatrix} 3 & 0 \end{bmatrix}\right)$
then press ENTER.
Note: The matrix template can be found by pressing the templates button.

2. The answer appears on the screen.

$$a \cdot b = 9$$

b. 1. On the Run-Matrix screen, select MAT/VCT by pressing F3.
Select M ⇔ V by pressing F6 to switch from the Matrix screen to the Vector screen.
To define a as Vector A, highlight VCT A and press EXE. Change the dimension to 1×2, then press EXE.
Enter the values 3 and 4 into the matrix template, then press EXIT.
Repeat these steps to define b as Vector B.

2. On the Run-Matrix screen, press OPTN, then select MAT/VCT by pressing F2. Press F6 twice to scroll across, then select DotP(by pressing F2.
Complete the entry line as:
DotP(VctA,Vct B)
then press EXE.
Note: Press F1 for 'Vct'.

3. The answer appears on the screen.

$$a \cdot b = 9$$

Note: An easier method for finding the dot product will now be shown.

4.3.2 The scalar product of vectors expressed in component form

Consider the dot product of the unit vectors $\hat{\imath}$ and $\hat{\jmath}$. Firstly, consider $\hat{\imath} \cdot \hat{\imath}$ in detail. By definition, $|\hat{\imath}| = 1$ and, since the angle between them is $0°$, $\cos(\theta) = 1$, thus $\hat{\imath} \cdot \hat{\imath} = 1$. To summarise these results:

$$\hat{\imath} \cdot \hat{\imath} = 1 \ (\text{since } \theta = 0°)$$

$$\hat{\jmath} \cdot \hat{\jmath} = 1 \ (\text{since } \theta = 0°)$$

$$\hat{\imath} \cdot \hat{\jmath} = 0 \ (\text{since } \theta = 90°)$$

Using this information, we can develop another way to calculate the dot product of any vector when in component form.

Let $a = x_1\hat{\imath} + y_1\hat{\jmath}$ and $b = x_2\hat{\imath} + y_2\hat{\jmath}$ where x_1, y_1, x_2, y_2, are constants.

We can write $a \cdot b$ as:

$$a \cdot b = (x_1\hat{\imath} + y_1\hat{\jmath}) \cdot (x_2\hat{\imath} + y_2\hat{\jmath})$$

$$= x_1x_2(\hat{\imath} \cdot \hat{\imath}) + x_1y_2(\hat{\imath} \cdot \hat{\jmath}) + y_1x_2(\hat{\jmath} \cdot \hat{\imath}) + y_1y_2(\hat{\jmath} \cdot \hat{\jmath})$$

Considering the various unit vector dot products (in brackets), the 'like' products ($\hat{i} \cdot \hat{i}$ and $\hat{j} \cdot \hat{j}$) are 1; the rest are 0. Therefore:

> ### Dot product using coordinates
>
> Let $\underset{\sim}{a} = x_1\hat{i} + y_1\hat{j}$ and $\underset{\sim}{b} = x_2\hat{i} + y_2\hat{j}$ where x_1, y_1, x_2, y_2, are constants.
>
> $$\underset{\sim}{a} \cdot \underset{\sim}{b} = x_1x_2 + y_1y_2$$

This is a very important result.

Going back to Worked example 6, part **c** for instance, the dot product between $\underset{\sim}{a} = 2\hat{i} - 3\hat{j}$ and $\underset{\sim}{b} = -\hat{i} + 3\hat{j}$ can be easily determined by:

$$\begin{aligned} \underset{\sim}{a} \cdot \underset{\sim}{b} &= 2 \times (-1) + (-3) \times 3 \\ &= -11 \end{aligned}$$

We only need to multiply the corresponding x- and y-components of two vectors to determine their dot product. This is also why the dot product is called scalar product. If you multiply two vectors using the dot product, the result will be a scalar.

WORKED EXAMPLE 7 Calculating the dot product of vectors using their coordinates

Let $\underset{\sim}{a} = 3\hat{i} + 4\hat{j}$ and $\underset{\sim}{b} = 6\hat{i} - 4\hat{j}$. Determine $\underset{\sim}{a} \cdot \underset{\sim}{b}$.

THINK	WRITE
1. Write down $\underset{\sim}{a} \cdot \underset{\sim}{b}$ using the equation $\underset{\sim}{a} \cdot \underset{\sim}{b} = x_1x_2 + y_1y_2$.	$\underset{\sim}{a} \cdot \underset{\sim}{b} = (3\hat{i} + 4\hat{j}) \cdot (6\hat{i} - 4\hat{j})$
2. Multiply the corresponding components.	$\underset{\sim}{a} \cdot \underset{\sim}{b} = 3 \times 6 + 4 \times -4$
3. Simplify.	$= 18 - 16$
	$= 2$

4.3.3 Determining the angle between two vectors

We now have two formulas for calculating the dot product:

$$\underset{\sim}{a} \cdot \underset{\sim}{b} = |\underset{\sim}{a}| \, |\underset{\sim}{b}| \cos(\theta)$$
$$\underset{\sim}{a} \cdot \underset{\sim}{b} = x_1x_2 + y_1y_2$$

Combining these two formulas allows us to determine the angle between the vectors.

Rearranging the two equations, we obtain the result:

> ### Angle between two vectors
>
> $$\cos(\theta) = \frac{x_1x_2 + y_1y_2}{|\underset{\sim}{a}||\underset{\sim}{b}|}$$

Note: The angle will always be between $0°$ and $180°$ as $180°$ is the maximum angle between two vectors.

WORKED EXAMPLE 8 Determining the angle between two vectors using the dot product

Let $\underset{\sim}{a} = 4\hat{i} + 3\hat{j}$ and $\underset{\sim}{b} = 2\hat{i} - 3\hat{j}$. Calculate the angle between them to the nearest degree.

THINK	WRITE
1. Determine the dot product using the equation $\underset{\sim}{a} \cdot \underset{\sim}{b} = x_1 x_2 + y_1 y_2$.	$\underset{\sim}{a} \cdot \underset{\sim}{b} = (4\hat{i} + 3\hat{j}) \cdot (2\hat{i} - 3\hat{j})$ $= 4 \times 2 + 3 \times -3$
2. Simplify.	$= -1$
3. Determine the magnitude of each vector.	$\|\underset{\sim}{a}\| = \sqrt{4^2 + 3^2}$ $= \sqrt{25}$ $= 5$ $\|\underset{\sim}{b}\| = \sqrt{2^2 + (-3)^2}$ $= \sqrt{13}$
4. Substitute results into the equation $\cos(\theta) = \dfrac{x_1 x_2 + y_1 y_2}{\|\underset{\sim}{a}\| \, \|\underset{\sim}{b}\|}$.	$\cos(\theta) = \dfrac{-1}{5\sqrt{13}}$
5. Simplify the result for $\cos(\theta)$.	$= -0.05547$
6. Take $\cos^{-1}$ of both sides to obtain θ and round the answer to the nearest degree.	$\theta = \cos^{-1}(-0.05547)$ $= 93°$

TI | THINK	WRITE	CASIO | THINK	WRITE
1. On a Calculator page, complete the entry line as $\cos^{-1}\left(\dfrac{\text{dotP}([4\ 3], [2\ -3])}{\text{norm}([4\ 3]) \times \text{norm}([2\ -3])} \right)$ then press ENTER. *Note:* 'dotP(' can be found by pressing MENU and then selecting 7: Matrix & Vector C: Vector 3: Dot Product. Similarly, 'norm' can be found by pressing MENU and then selecting 7: Matrix & Vector 7: Norms 1: Norm.	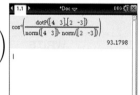	1. On the Run-Matrix screen, select MAT/VCT by pressing F3. Select M⇔V by pressing F6 to switch from the Matrix screen to the Vector screen. To define $\underset{\sim}{a}$ as Vector A, highlight VCT A and press EXE. Change the dimension to 1×2, then press EXE. Enter the values 4 and 3 into the matrix template, then press EXIT. Repeat these steps to define $\underset{\sim}{b}$ as Vector B.	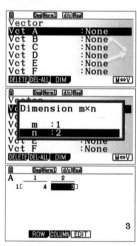
2. The answer appears on the screen.	The angle is 93°.	2. On the Run-Matrix screen, press OPTN, then select MAT/VCT by pressing F2. Press F6 twice to scroll across, then select Angle (by pressing F4). Complete the entry line as: Angle(VctA,Vct B) then press EXE. *Note:* Press F1 for 'Vct'.	
		3. The answer appears on the screen.	The angle is 93°.

4.3.4 Special results of the dot product

Perpendicular vectors

If two vectors are perpendicular then the angle between them is $90°$ and the equation $\underset{\sim}{a} \cdot \underset{\sim}{b} = |\underset{\sim}{a}| \, |\underset{\sim}{b}| \cos(\theta)$ becomes:

$$\underset{\sim}{a} \cdot \underset{\sim}{b} = |\underset{\sim}{a}| \, |\underset{\sim}{b}| \cos(90°)$$
$$= |\underset{\sim}{a}| \, |\underset{\sim}{b}| \times 0 \qquad \text{(Since } \cos(90°) = 0)$$
$$= 0$$

Dot product for perpendicular vectors

If $\underset{\sim}{a} \cdot \underset{\sim}{b} = 0$, then $\underset{\sim}{a}$ and $\underset{\sim}{b}$ are perpendicular.

If $\underset{\sim}{a}$ and $\underset{\sim}{b}$ are perpendicular, then $\underset{\sim}{a} \cdot \underset{\sim}{b} = 0$.

WORKED EXAMPLE 9 Perpendicular vectors and dot product

Determine the constant m if the vectors $\underset{\sim}{a} = 4\hat{\imath} + 3\hat{\jmath}$ and $\underset{\sim}{b} = -3\hat{\imath} + m\hat{\jmath}$ are perpendicular.

THINK	WRITE
1. Determine the dot product using the equation $\underset{\sim}{a} \cdot \underset{\sim}{b} = x_1 x_2 + y_1 y_2$.	$\underset{\sim}{a} \cdot \underset{\sim}{b} = (4\hat{\imath} + 3\hat{\jmath}) \cdot (-3\hat{\imath} + m\hat{\jmath})$
2. Simplify.	$= -12 + 3m$
3. Set $\underset{\sim}{a} \cdot \underset{\sim}{b}$ equal to zero since $\underset{\sim}{a}$ and $\underset{\sim}{b}$ are perpendicular.	$-12 + 3m = 0$
4. Solve the equation for m.	$m = 4$

Parallel vectors

If vector $\underset{\sim}{a}$ is parallel to vector $\underset{\sim}{b}$, then $\underset{\sim}{a} = k\underset{\sim}{b}$ where $k \in \mathbb{R}$.

Note: When applying the dot product to parallel vectors, θ (the angle between them) may be either $0°$ or $180°$ depending on whether the vectors are in the same or opposite directions.

WORKED EXAMPLE 10 Parallel vectors and dot product

Let $\underset{\sim}{a} = 5\hat{\imath} + 2\hat{\jmath}$. Determine a vector parallel to $\underset{\sim}{a}$ such that the dot product is 87.

THINK	WRITE
1. Let the required vector $\underset{\sim}{b} = k\underset{\sim}{a}$.	Let $\underset{\sim}{b} = k(5\hat{\imath} + 2\hat{\jmath})$ $= 5k\hat{\imath} + 2k\hat{\jmath}$
2. Determine the dot product of $\underset{\sim}{a} \cdot \underset{\sim}{b}$.	$\underset{\sim}{a} \cdot \underset{\sim}{b} = (5\hat{\imath} + 2\hat{\jmath}) \cdot (5k\hat{\imath} + 2k\hat{\jmath})$
3. Simplify.	$= 25k + 4k$ $= 29k$
4. Equate the result to the given dot product 87.	$29k = 87$
5. Solve for k.	$k = 3$
6. Substitute $k = 3$ into vector $\underset{\sim}{b}$.	$\underset{\sim}{b} = 15\hat{\imath} + 6\hat{\jmath}$

4.3.5 Projection of vectors – scalar and vector resolutes

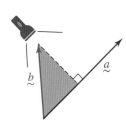

Previously, we have resolved a vector parallel and perpendicular to the *x*- and *y*-axes. In this section we consider a generalisation of this process where we resolve one vector parallel and perpendicular to another vector.

Consider the two vectors, $\underset{\sim}{a}$ and $\underset{\sim}{b}$, shown. The angle between them, as for a dot product, is given by θ. For example, a beam of light shoots from above vector $\underset{\sim}{b}$ perpendicular to vector $\underset{\sim}{a}$, and casts a shadow vector from vector $\underset{\sim}{b}$ on to the direction of vector $\underset{\sim}{a}$. How can we represent the shadow vector? It can be shown that $\underset{\sim}{b}$ is made up of a **projection** acting in the direction of $\underset{\sim}{a}$ and another projection acting perpendicular to $\underset{\sim}{a}$.

Firstly we wish to determine the projection in the direction of $\underset{\sim}{a}$.

Scalar projection and scalar resolute

To obtain the projection of $\underset{\sim}{b}$ in the direction of $\underset{\sim}{a}$, we perform the following construction:

1. Drop a perpendicular line from the head of $\underset{\sim}{b}$ to $\underset{\sim}{a}$ (this is perpendicular to $\underset{\sim}{a}$). This line joins $\underset{\sim}{a}$ at point A.
2. We wish to determine the length of the line OA.

This construction is shown.

Let the length of $\underset{\sim}{b}$ (its magnitude) be denoted by $|\underset{\sim}{b}|$. Then, from trigonometry:

$$OA = |\underset{\sim}{b}| \; \cos(\theta)$$

But from the definition of the dot product:

$$\underset{\sim}{a} \cdot \underset{\sim}{b} = |\underset{\sim}{a}| \, |\underset{\sim}{b}| \; \cos(\theta)$$

(from the equation $\underset{\sim}{a} \cdot \underset{\sim}{b} = |\underset{\sim}{a}| \, |\underset{\sim}{b}| \; \cos(\theta)$)

$$\underset{\sim}{a} \cdot \underset{\sim}{b} = |\underset{\sim}{a}| \; OA$$

Therefore, solving for OA:

$$OA = \frac{\underset{\sim}{a} \cdot \underset{\sim}{b}}{|\underset{\sim}{a}|}$$

$$= \left(\frac{\underset{\sim}{a}}{|\underset{\sim}{a}|} \right) \underset{\sim}{b}$$

But we know that $\dfrac{\underset{\sim}{a}}{|\underset{\sim}{a}|} = \hat{\underset{\sim}{a}}$, the unit vector in the direction of $\underset{\sim}{a}$, and therefore

$$OA = \hat{\underset{\sim}{a}} \cdot \underset{\sim}{b}$$

This quantity, the length OA, is called the scalar projection, or **scalar resolute**, of $\underset{\sim}{b}$ on $\underset{\sim}{a}$. It effectively indicates 'how much' of $\underset{\sim}{b}$ is in the direction of $\underset{\sim}{a}$.

Vector $\underset{\sim}{a}$ is called *base vector*. When the base vector changes, the vector resolute and scalar resolute will change.

Scalar projection

The scalar projection of $\underset{\sim}{b}$ on $\underset{\sim}{a}$, or scalar resolute of $\underset{\sim}{b}$ on $\underset{\sim}{a}$, is given by $\hat{a} \cdot \underset{\sim}{b} = \dfrac{\underset{\sim}{a} \cdot \underset{\sim}{b}}{|\underset{\sim}{a}|} = |\underset{\sim}{b}| \cos(\theta)$

where θ is the angle between vectors $\underset{\sim}{a}$ and $\underset{\sim}{b}$.

Similarly, the scalar projection of $\underset{\sim}{a}$ on $\underset{\sim}{b}$, or scalar resolute of $\underset{\sim}{a}$ on $\underset{\sim}{b}$, is given by

$\underset{\sim}{a} \cdot \hat{b} = \dfrac{\underset{\sim}{a} \cdot \underset{\sim}{b}}{|\underset{\sim}{b}|} = |\underset{\sim}{a}| \cos(\theta)$.

WORKED EXAMPLE 11 Determining scalar resolutes

Let $\underset{\sim}{a} = 3\hat{\imath} + 4\hat{\jmath}$ and $\underset{\sim}{b} = 6\hat{\imath} - 2\hat{\jmath}$. Determine:

a. the scalar resolute of $\underset{\sim}{b}$ on $\underset{\sim}{a}$

b. the scalar resolute of $\underset{\sim}{a}$ on $\underset{\sim}{b}$.

THINK	WRITE
a. 1. Determine the magnitude of $\underset{\sim}{a}$.	a. $\begin{aligned}|\underset{\sim}{a}| &= \sqrt{3^2 + 4^2} \\ &= 5\end{aligned}$
2. Determine $\hat{a}$ by dividing $\underset{\sim}{a}$ by $|\underset{\sim}{a}|$.	$\begin{aligned}\hat{a} &= \dfrac{\underset{\sim}{a}}{|\underset{\sim}{a}|} \\ &= \dfrac{\underset{\sim}{a}}{5}\end{aligned}$
3. Simplify.	$\begin{aligned} &= \dfrac{1}{5}(3\hat{\imath} + 4\hat{\jmath}) \\ &= \dfrac{3}{5}\hat{\imath} + \dfrac{4}{5}\hat{\jmath}\end{aligned}$
4. Determine the scalar resolute of $\underset{\sim}{b}$ on $\underset{\sim}{a}$ using $\hat{a} \cdot \underset{\sim}{b}$.	$\hat{a} \cdot \underset{\sim}{b} = \left(\dfrac{3}{5}\hat{\imath} + \dfrac{4}{5}\hat{\jmath}\right) \cdot (6\hat{\imath} - 2\hat{\jmath})$
5. Simplify.	$\begin{aligned} &= \dfrac{18}{5} - \dfrac{8}{5} \\ &= \dfrac{10}{5} \\ &= 2\end{aligned}$
b. 1. Determine the magnitude of $\underset{\sim}{b}$.	b. $\begin{aligned}|\underset{\sim}{b}| &= \sqrt{6^2 + (-2)^2} \\ &= \sqrt{40}\end{aligned}$
2. Determine $\hat{b}$ by dividing $\underset{\sim}{b}$ by $|\underset{\sim}{b}|$.	$\begin{aligned}\hat{b} &= \dfrac{\underset{\sim}{b}}{|\underset{\sim}{b}|} \\ &= \dfrac{1}{\sqrt{40}}(6\hat{\imath} - 2\hat{\jmath})\end{aligned}$

3. Determine the scalar resolute of $\underset{\sim}{a}$ on $\underset{\sim}{b}$ using $\hat{\underset{\sim}{b}} \cdot \underset{\sim}{a}$.

$$\hat{\underset{\sim}{b}} \cdot \underset{\sim}{a} = \frac{1}{\sqrt{40}}(6\hat{\imath} - 2\hat{\jmath}) \cdot (3\hat{\imath} + 4\hat{\jmath})$$

4. Simplify.

$$= \frac{1}{\sqrt{40}}(18 - 8)$$

$$= \frac{10}{\sqrt{40}}$$

$$= \frac{\sqrt{10}}{2}$$

Notes:
- The two scalar resolutes are not equal.
- The scalar resolute of $\underset{\sim}{b}$ on $\underset{\sim}{a}$ can easily be evaluated as $\dfrac{\underset{\sim}{b} \cdot \underset{\sim}{a}}{|\underset{\sim}{a}|}$.

Vector projection and vector resolutes

Consider, now, the vector joining O to A at right. Its magnitude is just the scalar resolute $(\hat{\underset{\sim}{a}} \cdot \underset{\sim}{b})$, while its direction is the same as $\underset{\sim}{a}$, that is $\hat{\underset{\sim}{a}}$. This quantity is called the **vector resolute**, or vector projection, of $\underset{\sim}{b}$ parallel to $\underset{\sim}{a}$ and is denoted by the symbol $\underset{\sim}{b}_{\parallel}$.

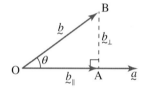

Vector projection (resolute) parallel to another vector

The vector resolute of $\underset{\sim}{b}$ parallel to $\underset{\sim}{a}$ is given by:

$$\underset{\sim}{b}_{\parallel} = (\hat{\underset{\sim}{a}} \cdot \underset{\sim}{b})\, \hat{\underset{\sim}{a}}$$

Consider the geometry of the above figure. The original vector $\underset{\sim}{b}$ can be seen to be the sum of two other vectors, namely $\underset{\sim}{b}_{\parallel}$ and $\underset{\sim}{b}_{\perp}$. This second vector $\underset{\sim}{b}_{\perp}$ is called the vector resolute of $\underset{\sim}{b}$ perpendicular to $\underset{\sim}{a}$ is and can be computed simply as follows:

$$\underset{\sim}{b} = \underset{\sim}{b}_{\parallel} + \underset{\sim}{b}_{\perp} \text{ (by addition of vectors)}$$

$$\underset{\sim}{b}_{\perp} = \underset{\sim}{b} - \underset{\sim}{b}_{\parallel} \text{ (by rearranging the vector equation)}$$

By substitution for $\underset{\sim}{b}_{\parallel}$ from the equation $\underset{\sim}{b}_{\parallel} = (\hat{\underset{\sim}{a}} \cdot \underset{\sim}{b})\, \hat{\underset{\sim}{a}}$, we can determine the vector resolute of $\underset{\sim}{b}$ perpendicular to $\underset{\sim}{a}$:

Vector resolute perpendicular to another vector

The vector resolute of $\underset{\sim}{b}$ perpendicular to $\underset{\sim}{a}$ is given by:

$$\underset{\sim}{b}_{\perp} = \underset{\sim}{b} - (\hat{\underset{\sim}{a}} \cdot \underset{\sim}{b})\, \hat{\underset{\sim}{a}}$$

In practice, once $b_{\parallel}$ has been calculated, simply subtract it from b to get $b_{\perp}$.

Note that $b_{\parallel}$ and $b_{\perp}$ are both referred to as vector resolutes. The parallel resolute is also called the vector projection while the perpendicular resolute is sometimes called the vector rejection.

Note that the vector projection parallel to another vector can be summarised with a single formula.

Vector projections

The vector projection of b on a, or vector resolute of b on a, is given by

$$b_{\parallel} = (\hat{a} \cdot b)\,\hat{a} = \left(\frac{a \cdot b}{a \cdot a} \right) a = |b| \cos(\theta)\,\hat{a} \text{ where } \theta \text{ is the angle between vectors } a \text{ and } b.$$

Similarly, the vector projection of a on b, or vector resolute of a on b, is given by

$$a_{\parallel} = \left(a \cdot \hat{b} \right) \hat{b} = \left(\frac{a \cdot b}{b \cdot b} \right) b = |a| \cos(\theta)\,\hat{b}.$$

WORKED EXAMPLE 12 Determining vector resolutes

Let $a = -2\hat{i} + 3\hat{j}$ and $b = 4\hat{i} + 2\hat{j}$. Determine:
a. the scalar resolute of b on a
b. the vector resolute of b parallel to a, namely $b_{\parallel}$
c. the vector resolute of b perpendicular to a, namely $b_{\perp}$.

THINK	WRITE
a. 1. Calculate the magnitude of a.	a. $\|a\| = \sqrt{(-2)^2 + 3^2}$
	$= \sqrt{13}$
2. Determine $\hat{a}$.	$\hat{a} = \dfrac{a}{\|a\|}$
	$= \dfrac{1}{\sqrt{13}}(-2\hat{i} + 3\hat{j})$
3. Determine the scalar resolute using $\hat{a} \cdot b$.	$\hat{a} \cdot b = \dfrac{1}{\sqrt{13}}(-2\hat{i} + 3\hat{j}) \cdot (4\hat{i} + 2\hat{j})$
4. Simplify.	$= \dfrac{1}{\sqrt{13}}(-8 + 6)$
	$= -\dfrac{2}{\sqrt{13}}$

b. 1. Determine $b_{\parallel}$ using equation $b_{\parallel} = (\hat{a} \cdot b)\, \hat{a}$.

b. $b_{\parallel} = (\hat{a} \cdot b)\, \hat{a}$

2. Simplify.

$$= \left(-\frac{2}{\sqrt{13}}\right)\left[\frac{1}{\sqrt{13}}(-2\hat{i} + 3\hat{j})\right]$$

$$= -\frac{2}{13}(-2\hat{i} + 3\hat{j})$$

$$= \frac{4}{13}\hat{i} - \frac{6}{13}\hat{j}$$

c. 1. Determine $b_{\perp}$ by subtraction of $b_{\parallel}$ from b as in equation $b_{\perp} = b - b_{\parallel}$.

c. $b_{\perp} = b - b_{\parallel}$

$$= 4\hat{i} + 2\hat{j} - \left(\frac{4}{13}\hat{i} - \frac{6}{13}\hat{j}\right)$$

2. Simplify.

$$b_{\perp} = \frac{48}{13}\hat{i} + \frac{32}{13}\hat{j}$$

Exercise 4.3 Scalar (dot) Product and the projection of vectors **learn** on

4.3 Exercise	4.3 Exam questions on

Simple familiar	Complex familiar	Complex unfamiliar
1, 2, 3, 4, 5, 6, 7, 8, 9, 10, 11, 12, 13	14, 15, 16, 17	18

These questions are even better in jacPLUS!
- Receive immediate feedback
- Access sample responses
- Track results and progress

Find all this and **MORE in jacPLUS** ▶

Simple familiar

1. **WE6** Determine the dot product of the vectors $3\hat{i} + 3\hat{j}$ and $6\hat{i} + 2\hat{j}$ using the equation $a \cdot b = |a|\, |b| \cos(\theta)$.

2. **WE7** Determine $a \cdot b$ in each of the following cases.

　　a. $a = 2\hat{i} + 3\hat{j}, \ b = 3\hat{i} + 3\hat{j}$　　　　　**b.** $a = 4\hat{i} - 2\hat{j}, \ b = 5\hat{i} + \hat{j}$
　　c. $a = -\hat{i} + 4\hat{j}, \ b = 3\hat{i} - 7\hat{j}$　　　　　**d.** $a = 5\hat{i} + 9\hat{j}, \ b = 2\hat{i} - 4\hat{j}$

3. Determine $a \cdot b$ in each of the following cases.

　　a. $a = -3\hat{i} + \hat{j}, \ b = \hat{i} + 4\hat{j}$　　　　　**b.** $a = 10\hat{i}, \ b = -2\hat{i}$
　　c. $a = 3\hat{i} + 5\hat{j}, \ b = \hat{i}$　　　　　　　　**d.** $a = 6\hat{i} - 2\hat{j}, \ b = -\hat{i} - 4\hat{j}$

4. **MC** The dot product of $a = 3\hat{i} - 3\hat{j}$ and $b = \hat{i} - 2\hat{j}$ is:

　　A. 3　　　　　　　**B.** 12　　　　　　　**C.** 21　　　　　　　**D.** 9

5. Let $a = x\hat{i} + y\hat{j}$. Show that $a \cdot a = x^2 + y^2$.

6. Let $a = 3\hat{i} + 2\hat{j}, b = \hat{i} - 2\hat{j}$ and $c = 5\hat{i} - 2\hat{j}$. Demonstrate, using these vectors, the property:

$$c \cdot (a - b) = c \cdot a - c \cdot b$$

Formally, this means that vectors are *distributive* over subtraction.

7. Repeat question **6** for the property:

$$\underset{\sim}{c} \cdot (\underset{\sim}{a} + \underset{\sim}{b}) = \underset{\sim}{c} \cdot \underset{\sim}{a} + \underset{\sim}{c} \cdot \underset{\sim}{b}$$

Formally, this means that vectors are *distributive* over addition.

8. **MC** If $\underset{\sim}{a} = 5i + 4j$ which of the following is perpendicular to $\underset{\sim}{a}$?

   A. $-5\hat{\imath} - 4\hat{\jmath}$ **B.** $3\hat{\imath} + 4\hat{\jmath}$ **C.** $-5\hat{\imath}$ **D.** $-4\hat{\imath} + 5\hat{\jmath}$

9. **MC** If $(\underset{\sim}{a} - \underset{\sim}{b}) \cdot (\underset{\sim}{a} + \underset{\sim}{b}) = 0$, then:

   A. $\underset{\sim}{a}$ is parallel to $\underset{\sim}{b}$ **B.** $\underset{\sim}{a}$ and $\underset{\sim}{b}$ have equal magnitudes

   C. $\underset{\sim}{a}$ is perpendicular to $\underset{\sim}{b}$ **D.** $\underset{\sim}{a}$ is a multiple of $\underset{\sim}{b}$

10. **MC** If $(\underset{\sim}{a} - \underset{\sim}{b}) \cdot (\underset{\sim}{a} + \underset{\sim}{b}) = |\underset{\sim}{b}|^2$, then:

    A. $\underset{\sim}{a} = \underset{\sim}{b}$ **B.** $\underset{\sim}{a}$ must be equal to the zero vector, $\underset{\sim}{0}$

    C. $\underset{\sim}{a}$ is perpendicular to $\underset{\sim}{b}$ **D.** $|\underset{\sim}{a}|$ must be equal to $\sqrt{2}\,|\underset{\sim}{b}|$

11. **WE11** For each of the following pairs of vectors, determine:

    i. the scalar resolute of $\underset{\sim}{a}$ on $\underset{\sim}{u}$.
    ii. the scalar resolute of $\underset{\sim}{u}$ on $\underset{\sim}{a}$.

    a. $\underset{\sim}{u} = 2\hat{\imath} + 3\hat{\jmath}$ and $\underset{\sim}{a} = 4\hat{\imath} + 5\hat{\jmath}$ b. $\underset{\sim}{u} = 5\hat{\imath} - 2\hat{\jmath}$ and $\underset{\sim}{a} = 3\hat{\imath} - \hat{\jmath}$ c. $\underset{\sim}{u} = -2\hat{\imath} + 6\hat{\jmath}$ and $\underset{\sim}{a} = \hat{\imath} - 4\hat{\jmath}$
    d. $\underset{\sim}{u} = 3\hat{\imath} - 2\hat{\jmath}$ and $\underset{\sim}{a} = -4\hat{\imath} - 3\hat{\jmath}$ e. $\underset{\sim}{u} = 8\hat{\imath} - 6\hat{\jmath}$ and $\underset{\sim}{a} = -5\hat{\imath} + \hat{\jmath}$

12. **WE12** For each pair of vectors $\underset{\sim}{a}$ and $\underset{\sim}{b}$, determine:

    i. the scalar resolute of $\underset{\sim}{b}$ on $\underset{\sim}{a}$
    ii. the vector resolute of $\underset{\sim}{b}$, parallel to $\underset{\sim}{a}$, namely $\underset{\sim}{b}_{\parallel}$
    iii. the vector resolute of $\underset{\sim}{b}$, perpendicular to $\underset{\sim}{a}$, namely $\underset{\sim}{b}_{\perp}$.

    a. $\underset{\sim}{a} = 3\hat{\imath} - \hat{\jmath}$; $\underset{\sim}{b} = 2\hat{\imath} + 5\hat{\jmath}$ b. $\underset{\sim}{a} = 4\hat{\imath} + 5\hat{\jmath}$; $\underset{\sim}{b} = 8\hat{\imath} + 10\hat{\jmath}$ c. $\underset{\sim}{a} = 4\hat{\imath} + 3\hat{\jmath}$; $\underset{\sim}{b} = -3\hat{\imath} + 4\hat{\jmath}$
    d. $\underset{\sim}{a} = \hat{\imath} + \hat{\jmath}$; $\underset{\sim}{b} = 2\hat{\imath} + \hat{\jmath}$ e. $\underset{\sim}{a} = 2\hat{\imath} + 3\hat{\jmath}$; $\underset{\sim}{b} = 2\hat{\imath} - 3\hat{\jmath}$ f. $\underset{\sim}{a} = 3\hat{\imath} + \hat{\jmath}$; $\underset{\sim}{b} = 2\hat{\jmath}$

13. **MC** The angle between the vectors $2\hat{\imath} - 3\hat{\jmath}$ and $-4\hat{\imath} + 6\hat{\jmath}$ is closest to:

    A. $0°$ **B.** $69°$ **C.** $111°$ **D.** $180°$

Complex familiar

14. **WE9** Determine the constant m, if the vectors $\underset{\sim}{b} = m\hat{\imath} + 3\hat{\jmath}$ and $\underset{\sim}{a} = 6\hat{\imath} - 2\hat{\jmath}$ are perpendicular.

15. Determine the constant m, such that $\underset{\sim}{b} = m\hat{\imath} - 2\hat{\jmath}$ is perpendicular to $\underset{\sim}{a} = 4\hat{\imath} - 3\hat{\jmath}$.

16. **WE10** Let $\underset{\sim}{a} = 2\hat{\imath} + 4\hat{\jmath}$. Determine a vector parallel to $\underset{\sim}{a}$ such that their dot product is 40.

17. Determine the dot product of the following pairs of vectors.

    a. $4\hat{\imath} - 3\hat{\jmath}$ and $7\hat{\imath} + 4\hat{\jmath}$ b. $\hat{\imath} + 2\hat{\jmath}$ and $-9\hat{\imath} + 4\hat{\jmath}$ c. $8\hat{\imath} + 3\hat{\jmath}$ and $2\hat{\imath} - 3\hat{\jmath}$ d. $5\hat{\imath} - 5\hat{\jmath}$ and $5\hat{\imath} + 5\hat{\jmath}$

Complex unfamiliar

18. **WE8** Calculate the angle between each pair of vectors in question **16** to the nearest degree.

Fully worked solutions for this chapter are available online.

LESSON
4.4 Operations with vectors

SYLLABUS LINKS

- Examine and use addition and subtraction of vectors in Cartesian form.

Source: Specialist Mathematics Senior Syllabus 2024 © State of Queensland (QCAA) 2024; licensed under CC BY 4.0

4.4.1 Applications of vector addition

Technology will be required throughout this chapter where trigonometric ratios for angles other than the angles 30°, 45° and 60° need to be determined.

Vectors have applications in surveying, navigation, orienteering and many other areas where problems need to be solved that involve displacement and velocity.

When considering a vector problem, sketch a diagram and use the properties of two-dimensional vectors to help solve the problem.

> **WORKED EXAMPLE 13 Calculating position vector and displacement using vector addition**
>
> A man walks 4 km due north, then 3 km due west. If $\hat{\imath}$ and $\hat{\jmath}$ represent unit vectors of 1 kilometre in the directions of east and north respectively, determine:
> a. his position vector
> b. his displacement and true bearing from his starting point.

THINK	WRITE
1. Let the man start at point O. He then walks 4 km north to point A, then 3 km west to point B.	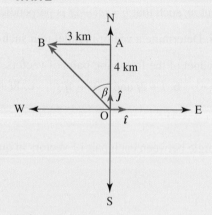
2. Determine the vectors $\overrightarrow{OA}$ and $\overrightarrow{AB}$. As $\hat{\imath}$ represents a unit vector in an easterly direction, the vector $\overrightarrow{AB}$ in a westerly direction will be negative.	$\overrightarrow{OA} = 4\hat{\jmath}$ $\overrightarrow{AB} = -3\hat{\imath}$

3. Use vector addition, $\overrightarrow{OB} = \overrightarrow{OA} + \overrightarrow{AB}$, to determine his position vector.

$\overrightarrow{OB} = \overrightarrow{OA} + \overrightarrow{AB}$
$\overrightarrow{OB} = -3\hat{\imath} + 4\hat{\jmath}$

4. The distance OB is the magnitude of this vector. It can be abbreviated to $d\,(OB)$.

$d\,(OB) = \sqrt{(-3)^2 + 4^2}$
$= \sqrt{9 + 16}$
$= \sqrt{25}$
$= 5\,\text{km}$

5. Use direction tangents to determine the bearing.

$\tan(\beta) = \dfrac{3}{4}$
$\beta = \tan^{-1}\left(\dfrac{3}{4}\right)$
$\beta = 36.87°$
$\beta = 36°52'$
Alternatively, $360 - 36°52' = 323°8'$ as a true bearing.

6. State the final answer in a sentence.

The man's displacement is 5 km on a bearing 323°8'T.

Although the next two worked examples could be solved using the sine and cosine rules, they can be now solved using the method of resolution of vectors, which is usually an easier method.

WORKED EXAMPLE 14 Calculating position vector and displacement using sine and cosine rules

A group of hikers walk 3 km due south, then turn and walk 2 km on a bearing S25°W. If $\hat{\imath}$ and $\hat{\jmath}$ represent unit vectors of magnitude 1 kilometre in the directions of east and north respectively, determine:
a. the position vector of the hikers from their initial starting point
b. the hikers' distance and bearing from their starting point.
Give your answers correct to three decimal places.

THINK

a. 1. Sketch a diagram, labelling the starting point as the origin, O, the first point as A and their final point as B.

WRITE

a.

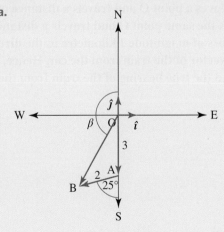

2. The hikers walk 3 km south from O to the point A. As $\hat{j}$ represents a unit vector in a northerly direction, the southerly vector $\overrightarrow{OA}$ will be negative.

$$\overrightarrow{OA} = -3\hat{j}$$

3. They walk from A to B, a distance of 2 km on a bearing S25°W. Resolve the vector $\overrightarrow{AB}$.

$$\overrightarrow{AB} = -2\sin(25°)\hat{i} - 2\cos(25°)\hat{j}$$

4. Use vector addition to determine $\overrightarrow{OB} = \overrightarrow{OA} + \overrightarrow{AB}$, which represents their position vector, and group like components.

$$\begin{aligned}\overrightarrow{OB} &= \overrightarrow{OA} + \overrightarrow{AB}\\ &= -3\hat{j} + (-2\sin(25°)\hat{i} - 2\cos(25°)\hat{j})\\ &= -2\sin(25°)\hat{i} - (3 + 2\cos(25°))\hat{j}\end{aligned}$$

5. Use a calculator to determine the position vector correct to three decimal places.

$$\overrightarrow{OB} = -0.845\hat{i} - 4.813\hat{j}$$

b. 1. The distance from O to B is the magnitude of this vector.

b. $d(OB) = \left|\overrightarrow{OB}\right|$

$$= \sqrt{(-0.845)^2 + (-4.813)^2}$$
$$= \sqrt{23.88}$$
$$= 4.886$$

2. The bearing is the angle that the vector $\overrightarrow{OB}$ makes with the y-axis or the north direction.

$$\cos(\beta) = \frac{-4.813}{4.8857}$$
$$\beta = \cos^{-1}(0.2016)$$
$$= 170.04°$$
$$= 170°2'$$

3. Alternative bearings are possible.

$360° - 170°2' = 189°58'$ as a true bearing.

4. State the final answer in a sentence.

The hikers are 4.89 km from their starting point on a bearing 189°58'T.

4.4.2 Applications using vector subtraction

Sometimes when solving vector problems, we may need to subtract two vectors.

WORKED EXAMPLE 15 Solving vector problems using vector subtraction

A train leaves a point O and travels a distance of 45 km on a bearing N25°W. At the same time a car leaves the same point O and travels a distance of 65 km on a bearing S55°W. If $\hat{i}$ and $\hat{j}$ represent unit vectors of magnitude 1 kilometre in the directions of east and north respectively, determine the position vector of the train from the car. Hence, calculate the final distance correct to two decimal places and the true bearing of the train from the car.

THINK	WRITE

THINK

1. Sketch a diagram, labelling the origin, O, the train at point T and the car at the point C. The vectors $\overrightarrow{OT}$ and $\overrightarrow{OC}$ represent the position vectors of the train and car respectively.

2. By resolving the vector $\overrightarrow{OT}$, write down the position vector of the train.

3. By resolving the vector $\overrightarrow{OC}$, write down the position vector of the car.

4. The position vector of the train from the car is given by $\overrightarrow{CT}$. This vector is found using subtraction of the two vectors $\overrightarrow{OT}$ and $\overrightarrow{OC}$.

5. Substitute for the two vectors.

6. Use the rules for subtraction of vectors.

7. Using a calculator, give the position vector correct to three decimal places.

8. The distance of the train from the car is the magnitude of this vector.

9. The bearing is the angle that the vector $\overrightarrow{CT}$ makes with the north direction.

10. State the final answers in a sentence. Note that other possible equivalent bearings are also acceptable.

WRITE

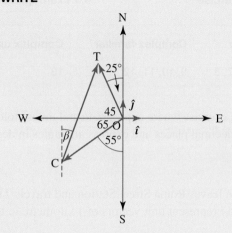

$\overrightarrow{OT} = -45 \sin(25°)\,\hat{i} + 45 \cos(25°)\,\hat{j}$

$\overrightarrow{OC} = -65 \sin(55°)\,\hat{i} - 65 \cos(55°)\,\hat{j}$

$\overrightarrow{OT} = \overrightarrow{OC} + \overrightarrow{CT}$
$\overrightarrow{CT} = \overrightarrow{OT} - \overrightarrow{OC}$

$\overrightarrow{CT} = (-45 \sin(25°)\,\hat{i} + 45 \cos(25°)\,\hat{j})$
$\qquad - (-65 \sin(55°)\,\hat{i} - 65 \cos(55°)\,\hat{j})$

$\overrightarrow{CT} = (65 \sin(55°) - 45 \sin(25°))\,\hat{i}$
$\qquad + (45 \cos(25°) + 65 \cos(55°))\,\hat{j}$

$\therefore \overrightarrow{CT} = 34.227\hat{i} + 78.066\hat{j}$

$\left|\overrightarrow{CT}\right| = \sqrt{34.227^2 + 78.066^2}$
$\qquad = \sqrt{7265.788}$
$\qquad = 85.24$

$\cos(\beta) = \dfrac{78.066}{85.24}$
$\qquad \beta = \cos^{-1}(0.9158)$
$\qquad\quad = 23.675$
$\qquad\quad = 23°40'$

The train is at a final distance of 85.24 km from the car on a bearing 23°40′T.

For the following exercise where appropriate and unless told otherwise, round all vector components and distances to two decimal places and express all angles in degrees and minutes correct to the nearest minute.

Simple familiar

1. **WE13** A train leaves Roma Street Station and travels 7 km due east, then turns a corner and travels 24 km south. If $\hat{i}$ and $\hat{j}$ represent unit vectors of 1 kilometre in the directions of east and north respectively, determine:
   a. the position vector of the train
   b. the train's displacement from its starting point.

2. A jogger runs 1200 metres due south, then 500 metres due west. If $\hat{i}$ and $\hat{j}$ represent unit vectors of 1 metre in the directions of east and north respectively, determine:
   a. the position vector of the jogger from her initial starting point
   b. her displacement from her starting point.

3. **WE14** A plane flies 35 km due south, then turns and travels 25 km on a bearing S35°W. If $\hat{i}$ and $\hat{j}$ represent unit vectors of magnitude 1 kilometre in the directions of east and north respectively, determine:
   a. the plane's position vector from its initial starting point
   b. the plane's distance (to two decimal places) and bearing from its starting point.

4. A ship moves 70 km due west, then turns and moves 240 km on a bearing N53°W. If $\hat{i}$ and $\hat{j}$ represent unit vectors of magnitude 1 kilometre in the directions of east and north respectively, determine:
   a. the ship's position vector from its initial starting point
   b. the ship's distance and bearing from its starting point.

5. **WE15** Amanda leaves a point O and runs a distance of 6 km on a bearing N37°E. At the same time Brianna leaves the same point O and walks a distance of 2 km on a bearing S48°E. If $\hat{i}$ and $\hat{j}$ represent unit vectors of magnitude 1 kilometre in the directions of east and north respectively, determine:
   a. the position vector of Amanda from Brianna
   b. the final distance (correct to two decimal places) and the true bearing of Amanda from Brianna.

6. A train leaves a point O and travels at a speed of 100 km/h on a bearing N34°W. At the same time a car leaves the same point O and travels at a speed of 60 km/h on a bearing S50°W. If $\hat{\imath}$ and $\hat{\jmath}$ represent unit vectors of magnitude 1 kilometre in the directions of east and north respectively, determine:

    a. the position vector of the train from the car after 30 minutes

    b. the distance between the car and the train and the true bearing of the train from the car at this time.

7. Petra is an avid horse rider. One day she leaves the stable and rides 3 km due east, then turns and rides 2 km 138°T. If $\hat{\imath}$ and $\hat{\jmath}$ represent unit vectors of magnitude 1 kilometre in the directions of east and north respectively, determine:

    a. Petra's position vector from the stable

    b. Petra's distance in kilometres, correct to two decimal places, and true bearing from the stable.

8. A yacht moves 1600 metres due south from a buoy, then turns and travels 800 metres on a bearing 158°T. If $\hat{\imath}$ and $\hat{\jmath}$ represent unit vectors of magnitude 1 metre in the directions of east and north respectively, determine (to two decimal places):

    a. the yacht's position vector from the buoy

    b. the yacht's distance, correct to the nearest metre, and true bearing from the buoy.

Complex familiar

9. A bus leaves a depot and travels 4 km due west, then turns and travels 6 km on a bearing N58°W. If $\hat{\imath}$ and $\hat{\jmath}$ represent unit vectors of magnitude 1 kilometre in the directions of east and north respectively, determine the bus's distance in kilometres, correct to two decimal places, and bearing from the depot.

10. A water skier leaves a point O and travels a distance of 300 metres on a bearing N27°W. At the same time a boat leaves the same point O and travels a distance of 800 metres on a bearing S39°W. If $\hat{\imath}$ and $\hat{\jmath}$ represent unit vectors of magnitude 1 metre in the directions of east and north respectively, determine the distance in metres, correct to one decimal place, and the true bearing of the water skier from the boat in their final position.

11. A plane leaves an airport and travels at 300 km/h on a bearing S48°W. At the same time a helicopter leaves the same airport and travels at 90 km/h on a bearing S26°E. If $\hat{\imath}$ and $\hat{\jmath}$ represent unit vectors of magnitude 1 kilometre in the directions of east and north respectively, determine the distance, correct to the nearest kilometre, and the true bearing of the helicopter from the plane after 30 minutes.

12. A police car leaves a station and travels a distance of d_1 km due east, then turns and travels a further distance of d_2 on a bearing $\theta°$T, where $0° < \theta < 90°$. If $\hat{\imath}$ and $\hat{\jmath}$ represent unit vectors of magnitude 1 kilometre in the directions of east and north respectively, determine a function to represent the position vector and the distance of the police car from the station.

Complex unfamiliar

13. A taxi leaves a point O and travels a distance of d_1 km on a bearing $E\theta_1°N$, where $0° < \theta_1 < 90°$. At the same time a motorbike leaves the same point O and travels a distance of d_2 km on a bearing $E\theta_2°S$, where $0° < \theta_2 < 90°$. If $\hat{\imath}$ and $\hat{\jmath}$ represent unit vectors of magnitude 1 kilometre in the directions of east and north respectively, determine the final position vector and distance of the taxi from the motorbike.

Fully worked solutions for this chapter are available online.

LESSON
4.5 Applications of vectors in two dimensions

SYLLABUS LINKS

- Model and solve problems that involve displacement, force, velocity and relative velocity using the above concepts.
- Model and solve problems that involve motion of a body in equilibrium situations, including vector applications related to smooth inclined planes (excluding situations with pulleys and connected bodies).

Source: Specialist Mathematics Senior Syllabus 2024 © State of Queensland (QCAA) 2024; licensed under CC BY 4.0

4.5.1 Forces

A **force** is a 'push' or a 'pull'. The force due to gravity acts on us all the time. A bar magnet repels a second bar magnet. The force of friction slows down the wheel of a bike when the brakes are applied. Air resistance retards the motion of athletes. In the latter two cases there is relative motion between two objects. The strings of a tennis racquet exert a force on a tennis ball while the strings and ball are in contact.

In all these examples, objects which have **unbalanced forces** acting on them tend to undergo a change in their motion; objects which have a **balanced set of forces** acting on them maintain their motion. Objects under balanced forces are said to be in **equilibrium**.

The forces that we will discuss in this chapter can be classified as one of three types: field forces, applied forces, and resistive forces.

Field forces

Field forces act without physical contact between objects. The weight force which acts on an object due to the presence of an external gravitational field or the electric and magnetic forces which influences the motion of a charged particle are examples of field forces.

Applied forces

Applied forces, such as hitting a tennis ball with a racket, are the pushes or pulls exerted on objects due to contact. They are forces with which we have daily experience. The normal reaction force acting upwards on a book resting on a table or on us as we stand on the floor or sit on a chair are examples of applied forces. Other examples of applied forces include tensile forces in taut strings and cables (as in the cable used by a crane or rescue helicopter), and compressive forces acting on weight-bearing rods.

Resistive forces

Air drag and friction are examples of **resistive forces**. This type of force occurs when two objects move or attempt to move relative to one another. Air drag has been put to good use in the design of hang glider; it is also found in the resistance between a moving body like a car and the air. An example of friction is seen in a bicycle that is slowing down on level ground, even without the brakes being applied.

What is a particle?

We are all familiar with the notion of a **particle**, but in Newtonian dynamics a particle is used to model an object and is taken to be a point. That is, the size of the object is not relevant and any internal movements such as spin and change in shape are not included in the model.

For example, if we were to describe the various forces acting on a car, we would include the upwardly directed reaction forces exerted on the car by the road at each of the four tyres in addition to frictional forces exerted on the tyres by the road. If the car was moving, we would add the propulsion force of the engine and the drag forces due to the movement of the car through the air. There would be the weight force acting on the car due to the gravitational field of the Earth. If we were to treat the car as a particle, we would describe all these forces as acting through a single point. The word *particle* serves to define the position of an object and sets it apart from the rest of the immediate environment.

What assumptions do we make in Newtonian dynamics?

In modelling motion we make the following assumptions so that problems can be solved to give reasonably accurate predictions.

Term	Meaning
Light (body or string)	Object has no mass.
Smooth	No frictional forces are exerted.
Inextensible	Strings or ropes do not stretch.
Rigid	Objects do not change shape when forces are applied to them.
Perfectly elastic	Applied forces do not permanently deform an object (for example, a spring).

In many cases we ignore the presence of forces which would be insignificant, such as air drag on slowly moving objects, to simplify the mathematics. These will be studied in more depth in Unit 4.

The resultant force, R

Crucial to a good understanding of Newton's laws of motion is the concept of a **net force** or **resultant force** acting on a particle. Force is a vector quantity because a force has not only a magnitude but also a direction. The unit of force is the **newton (N)**. (This is a *derived unit*, which is simply one defined in terms of the standard units, namely distance, time and mass. This will be discussed later.)

Forces are vectors, and thus can be described in two ways:
1. Using $\hat{\imath}$ and $\hat{\jmath}$ notation
   For example, $F = 2\hat{\imath} - 4\hat{\jmath}$ N (only coplanar forces will be considered).
2. Using size and direction notation.
   For example, $F = 200$ N, N45°E is a force of magnitude 200 N directed at an angle 45° from north towards east.

The net or resultant force is simply the vector sum of all real physical forces acting on the particle. It represents the sum or total force acting on a particle representing an object. It is not in itself a real force, only the sum of real forces.

> ### Resultant (net) force
>
> **The net or resultant force acting on a particle is the vector sum of all real forces acting on that particle.**

Force diagrams

Individual forces are one of three types — field, applied or resistive — and are drawn as vectors which indicate their direction and magnitude. Note that you will see in section 4.5.2 that when there is no acceleration, the net force is zero, and thus the forces are balanced (see Worked example 16). Similarly, when the net force is zero, then there is no acceleration.

TIP

When sketching force diagrams, it is useful to remember that when an object is not moving, or moving at a constant velocity (that is, when its acceleration is equal to zero), then the net force is 0 and the forces in the force diagram are balanced.

Sketch 'vector diagrams' to represent the forces involved in the situations shown below as a set of vectors acting on a particle. Indicate the relative size of the force by the length of the vector arrows. Further indicate the nature of each of the forces acting by labelling them $\underset{\sim}{W}$ for weight, $\underset{\sim}{N}$ for normal reaction, $\underset{\sim}{F}$ for friction, $\underset{\sim}{A}$ for applied force and $\underset{\sim}{D}$ for air drag or air resistance.

a. A stationary person

b. Constant velocity

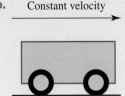

c. Accelerating

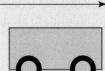

d. Constant velocity

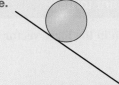

Cricket ball through the air

e.

Ball rolling down a slope

f.

Constant velocity

THINK

In the force diagrams for **a** to **f**, treat each object as a particle.

a. There are two equal, opposing forces: the weight force down, which is a field force, and the normal reaction force up.

b. 1. A cart moving at constant velocity has balanced horizontal forces.
  2. Any resistive forces ($\underset{\sim}{D}$) must be balanced by an applied force ($\underset{\sim}{A}$) to keep the cart moving at constant velocity.
  3. The cart will have balanced vertical forces arising from the weight force of the cart and the normal reaction force.

c. An accelerating cart will have balanced vertical forces as in **a** and **b** but unbalanced horizontal forces giving rise to an acceleration.

d. 1. A ball moving through the air will have the vertical force of weight and the resistive force of air drag.
  2. As the ball is moving at constant velocity, the drag force will be in the opposite direction to that of the ball and the applied force acting on it to maintain constant velocity.

WRITE

a.

b.

c.

d.

e. A ball rolling down a slope will have a weight force directed vertically down, a normal reaction force perpendicular to the slope and a resistive or frictional force.

e.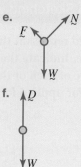

f. The parachutist has a constant velocity (i.e. is not accelerating) so will have two balanced vertical forces: one down due to the weight force and the second upwards due to air drag.

f.

WORKED EXAMPLE 17 Determining resultant forces

Three forces F_1, F_2 and F_3 act on a ball as shown in the force vector diagram. The three forces are described by the vectors:

$F_1 = 4\hat{\imath} - 5\hat{\jmath}$

$F_2 = 10\hat{\imath} + 2\hat{\jmath}$

$F_3 = -6\hat{\imath} + 7\hat{\jmath}.$

a. Determine the resultant force R, the sum of the three forces F_1, F_2 and F_3.

b. Determine the magnitude of the resultant force $|R|$.

c. Determine the angle that the vector R makes with the $\hat{\imath}$ vector.

d. The force F_1 is changed so that $R = 0$; that is, the resultant force equals zero. Determine the force F_1.

THINK	WRITE
a. 1. The resultant force R is the sum of the forces F_1, F_2 and F_3.	**a.** 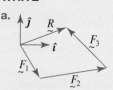
2. Evaluate R.	$R = \sum F$ $= F_1 + F_2 + F_3$ $= 4\hat{\imath} - 5\hat{\jmath} + 10\hat{\imath} - 2\hat{\jmath} - 6\hat{\imath} + 7\hat{\jmath}$ $= 8\hat{\imath} + 4\hat{\jmath}$
b. 1. The magnitude of the vector R is its length.	**b.** $|R| = \sqrt{8^2 + 4^2}$
2. The symbol $|R|$ or R is used to represent the magnitude of a vector R.	$= \sqrt{80}$ $= 4\sqrt{5}$
c. 1. Define θ.	**c.** Let the angle that R makes with $\hat{\imath}$ be θ.

2. Then use the result for the dot product of two vectors $\underset{\sim}{a} \cdot \underset{\sim}{b} = |\underset{\sim}{a}||\underset{\sim}{b}| \cos(\theta)$.

Let $\underset{\sim}{R} \cdot \hat{\imath} = |\underset{\sim}{R}||\hat{\imath}| \cos(\theta)$.

$$\cos(\theta) = \frac{(8\hat{\imath} + 4\hat{\jmath}) \cdot \hat{\imath}}{4\sqrt{5} \times 1}$$

3. Evaluate θ to the nearest tenth of a degree.

$$\cos(\theta) = \frac{8}{4\sqrt{5}}$$
$$\approx 26.6°$$

d. 1. The vector sum of all forces is now equal to zero. Set $\underset{\sim}{R} = 0$.

d. $\underset{\sim}{R} = \underset{\sim}{F}_1 + \underset{\sim}{F}_2 + \underset{\sim}{F}_3$
    $= 0$

2. Make $\underset{\sim}{F}_1$ the subject of the equation.

$\underset{\sim}{F}_1 = -(\underset{\sim}{F}_2 + \underset{\sim}{F}_3)$

3. Substitute $\underset{\sim}{F}_2$ and $\underset{\sim}{F}_3$ into the equation and simplify to determine $\underset{\sim}{F}_1$.

$\underset{\sim}{F}_1 = -(10\hat{\imath} + 2\hat{\jmath} - 6\hat{\imath} + 7\hat{\jmath})$
    $= -4\hat{\imath} - 9\hat{\jmath}$

TI | THINK **WRITE**

a. 1. On a Calculator page, complete the entry line as:
$[4 - 5] + [10\,2] + [-6\,7]$
then press ENTER.
Note: The matrix template can be found by pressing the templates button.

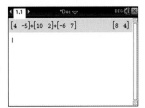

2. The answer appears on the screen.

$\underset{\sim}{R} = 8\hat{\imath} + 4\hat{\jmath}$

CASIO | THINK **WRITE**

a. 1. On the Run-Matrix screen, select MAT/VCT by pressing F3.
Select M ⇔ V by pressing F6 to switch from the Matrix screen to the Vector screen.
To define $\underset{\sim}{F}_1$ as Vector A, highlight VCT A and press EXE.
Change the dimension to 1×2, then press EXE.
Enter the values 4 and –5 into the matrix template. then press EXIT.
Repeat these steps to define $\underset{\sim}{F}_2$ as Vector B and $\underset{\sim}{F}_3$ as Vector C.

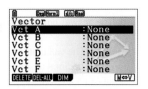

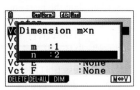

2. On the Run-Matrix screen, complete the entry line as:
Vct A+Vct B+VctC
then press EXE.
Note: To find 'Vct', press OPTN, then select MAT/VCT by pressing F2. Press F6 twice to scroll to more menu options, then select Vct (by pressing F1).

3. Press VAR and type 'r' to store the result as $\underset{\sim}{R}$.

3. The answer appears on the screen.

$\underset{\sim}{R} = 8\hat{\imath} + 4\hat{\jmath}$

4. Press the 'Store' button, then select Vct (by pressing F1) and type 'R'. This will store the result as $\underset{\sim}{R}$.

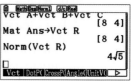

b. 1. Press MENU, then select:
7: Matrix & Vector
7: Norms
1: Norm.
Complete the entry line as:
norm(r)
then press ENTER.

b. 1. Press OPTN, then select MAT/VCT by pressing F2. Press F6 three times to scroll to more menu options, then select Norm (by pressing F1). Complete the entry line as: Norm(Vct R) then press EXE.

2. The answer appears on the screen.
Note: The calculator will give a decimal answer, not an exact answer.

$|\underset{\sim}{R}| = 8.944$

2. The answer appears on the screen.

$|\underset{\sim}{R}| = 4\sqrt{5}$

c. 1. Complete the next entry line as:
$$\cos^{-1}\left(\frac{\mathrm{dotP}(r, [1 \quad 0])}{\mathrm{norm}(r)} \right)$$
then press ENTER.
Note: 'dotP(' can be found by pressing MENU and then selecting:
7: Matrix & Vector
C: Vector
3: Dot Product.
Similarly, 'norm' can be found by pressing MENU and then selecting:
7: Matrix & Vector
7: Norms
1: Norm.

c. 1. Use the steps outlined in **a1** to define the vector $\hat{\imath} + 0\hat{\jmath}$ as Vector D.
Press OPTN, then select MAT/VCT by pressing F2. Press F6 twice to scroll to more menu options, then select Angle (by pressing F4).
Complete the entry line as:
Angle(Vct R, Vct D)
then press EXE.

2. The answer appears on the screen.

The angle is 26.6°.

2. The answer appears on the screen.

The angle is 26.6°

d. 1. Determine an expression for $\underset{\sim}{F}_1$.

$\underset{\sim}{F}_1 + \underset{\sim}{F}_2 + \underset{\sim}{F}_3 = 0$

$\therefore \underset{\sim}{F}_1 = 0 - \underset{\sim}{F}_2 - \underset{\sim}{F}_3$

d. 1. Determine an expression for $\underset{\sim}{F}_1$.

$\underset{\sim}{F}_1 + \underset{\sim}{F}_2 + \underset{\sim}{F}_3 = 0$

$\therefore \underset{\sim}{F}_1 = 0 - \underset{\sim}{F}_2 - \underset{\sim}{F}_3$

2. Complete the next entry line as:
$[0 \quad 0] - [10 \quad 2] - [-6 \quad 7]$
then press ENTER.

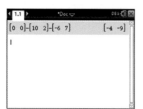

2. Use the steps outlined in **a1** to define the vector $0\hat{\imath} + 0\hat{\jmath}$ as Vector E.
Complete the next entry line as: Vct E–Vct B–Vct C then press EXE.

3. The answer appears on the screen.

$\underset{\sim}{F}_1 = -4\hat{\imath} - 9\hat{\jmath}$

3. The answer appears on the screen.

$\underset{\sim}{F}_1 = -4\hat{\imath} - 9\hat{\jmath}$

The triangle of forces

If three non-parallel forces acting on a particle have a resultant force of zero then the three forces can be represented in a triangle since the vector sum of the forces is zero.

From the figure, if $R = X + Y + Z = 0$, then the three forces X, Y and Z can be represented in a triangle representing the magnitude and direction of the three forces.

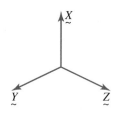

The advantage of representing three forces in a triangle is that the sine rule and/or cosine rule can be used to solve some problems involving three forces whose vector sum is zero. That is:

The triangle of forces

Sine rule: $\dfrac{a}{\sin(A)} = \dfrac{b}{\sin(B)} = \dfrac{c}{\sin(C)}$

Cosine rule: $a^2 = b^2 + c^2 - 2bc\,\cos(A)$

WORKED EXAMPLE 18 Applying the triangle of forces

Three forces A, B and C act on an object such that the resultant force is zero. The force A acts at an angle of 150° to the force B and they have the same magnitude of 20 N.
a. Determine the magnitude of C.
b. Calculate the angle that the force C makes with B to the nearest degree.

THINK

a. 1. Sketch a force vector diagram of three forces acting through a point with an angle of 150° between A and B and an angle of θ between B and C.

2. To determine the angle between A and B, draw a sketch.

WRITE

a.

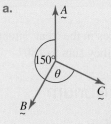

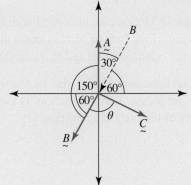

3. Since the resultant of the three forces is zero, place them in a triangle of forces.

4. Mark the angle between A and B as $180° - 150° = 30°$ and the angle between B and C as $180° - \theta$.

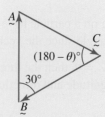

5. Recall the cosine rule and substitute $a = C$, $b = 20$, $c = 20$ and $A = 30°$.

$$a^2 = b^2 + c^2 - 2bc\cos(A)$$
$$C^2 = 20^2 + 20^2 - 2 \times 20 \times 20\cos(30°)$$
$$= 107.18$$

6. Solve for C to determine the magnitude of C.

$$C = 10.35$$

7. State the solution.

The magnitude of force C is approximately 10.35 newtons.

b. 1. Recall the sine rule and substitute $a = 20$, $A = 180° - \theta$, $b = 10.35$ and $B = 30$.

b. $$\frac{a}{\sin(A)} = \frac{b}{\sin(B)}$$

$$\frac{20}{\sin(180 - \theta)°} = \frac{10.35}{\sin(30°)}$$

2. Invert both sides of the equation.

3. Make $\sin(180 - \theta)°$ the subject of the equation.

$$\frac{\sin(180 - \theta)°}{20} = \frac{\sin(30°)}{10.35}$$

4. Determine the value of $180° - \theta$.

$$\sin(180 - \theta)° = \frac{20\sin(30°)}{10.35}$$
$$= 0.9662$$

5. Solve for θ.

$$180° - \theta = \sin^{-1}(0.9662)$$
$$= 75.06°$$

6. State the solution to the nearest degree.

$$\theta = 104.94°$$

The angle between forces B and C is approximately 105°.

Note: The angle between two forces in the 'real' situation and the angle between them in a triangle of three forces are supplementary; that is, they sum to 180°.

4.5.2 Forces and equilibrium

Newton's First Law of Motion

Newton's First Law of Motion concerns itself with the motion of a particle when the resultant force, R, acting on it is zero. Newton's First Law of Motion states that an object at rest remains at rest, and an object in motion remains in motion at constant speed and in a straight line unless acted on by an unbalanced force.

In the absence of an unbalanced force (that is, $R = 0$) acting on a body, the body would continue in a state of uniform motion. The phrase 'uniform motion' is used to describe the motion of a body with zero acceleration; that is, the velocity of the body is a constant.

Today we would write this as:

If $R = 0$ (the resultant force acting on the body is zero)

then $v = k$ (the velocity of the body is a constant)

and thus $a = 0$ (the acceleration of the body is zero).

Balanced forces and acceleration

If the resultant force acting on a body is zero, then its acceleration is zero.

It is important to recognise that the inference above is equally valid in reverse. That is,

If $a = 0$ (the acceleration of the body is zero)

then $v = k$ (the velocity of the body is a constant)

and thus $R = 0$ (the resultant force acting on the body is zero).

When the net or resultant force acting on an object is equal to zero, it is commonly said that the body is 'in equilibrium'. Such an object moves at a constant velocity. These types of situation belong to a class of problems called **statics**.

To solve statics, dynamics and **kinematics** problems there is a clear strategy:
- Read the question carefully.
- Sketch a clear diagram that contains all the information.
- Superimpose arrows depicting vectors which act on the body in question. This is called a *force vector diagram*.
- Determine the resultant force vector that is the sum of all forces acting on the body.

Mass and weight

We know from experience that some objects are harder to push than others; that is, it is more difficult to modify their motion. We call this property of matter **inertial mass**, expressed in kilograms (kg). The greater the mass of an object, the harder it is to modify its acceleration.

This is reflected in Newton's Second Law of motion, stating that the net force R acting upon the object is equal to the product of the mass m of the object and its acceleration a.

The heavier an object, the harder it is to modify its acceleration.

Weight is a vector quantity. It is a force, equal to the product of the acceleration due to gravity and the mass on which it is acting. The weight W of a mass m in a gravitational field g is given by the equation:

Weight (force due to gravity)

$$W = mg$$

Where m is the mass in kg and g is the gravitational field strength in m/s^2.

Note: The value for g of 9.8 m/s^2 down is used universally in examples and problems in this textbook. Sometimes the unit for g is quoted as N/kg, which is the same as m/s^2. This is the acceleration due to gravity.

Resolving a force into its components

In many cases the size of a force acting in a particular direction needs to be calculated. For example, when a jet aircraft is taking off from a runway the engines provide a **thrust**. Part of the thrust assists in lifting the plane up into the air (the vertical **component**) and the other component exerts a horizontal push making the plane move forward. In the diagram at right, the combined thrust of the jet's engines is represented by a single vector F acting on a particle representing the aircraft.

If F is the applied force acting on the plane due to the action of the engines then we can write the force vector as the sum of two vectors F_x and F_y that are parallel to the unit vectors $\hat{\imath}$ and $\hat{\jmath}$ respectively.

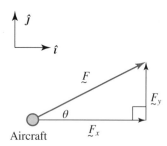

Thus
$$F = F_x + F_y$$
$$F = |F_x|\hat{\imath} + |F_y|\hat{\jmath}$$
$$F = F_x\hat{\imath} + F_y\hat{\jmath}$$

where
$$F_x = |F|\cos(\theta)$$

and
$$F_y = |F|\sin(\theta)$$

The quantities F_x and F_y are referred to as the *components* of the vector force F in the $\hat{\imath}$ and $\hat{\jmath}$ directions respectively.

Note: The $\hat{\imath}$ and $\hat{\jmath}$ components are usually horizontal and vertical components respectively, but they can be rotated to suit a particular problem.

Resolving a force into its components

A force F can be resolved into perpendicular components:

$$F = F\cos(\theta)\hat{\imath} + F\sin(\theta)\hat{\jmath}$$

where $F = |F|$ and θ is the angle between F and $\hat{\imath}$.

For example, if the thrust of the engines is 2×10^6 N and the plane during take-off had a 15° elevation angle, then the vertical component would be:

$$F_y = |F|\sin(\theta)$$
$$= 2 \times 10^6 \times \sin(15°)$$
$$\approx 5.18 \times 10^5 \text{ N}$$

while the horizontal component would be:

$$F_x = |\underset{\sim}{F}| \cos(\theta)$$
$$= 2 \times 10^6 \times \cos(15°)$$
$$= 1.93 \times 10^6 \text{ N}$$

or $\underset{\sim}{F} \approx (1.93 \times 10^6)\hat{\imath} + (5.18 \times 10^5)\hat{\jmath}$.

WORKED EXAMPLE 19 Resolving a force into its components

A water skier is being pulled along by rope attached to a speed boat across a horizontal lake. The rope makes an angle of 5° to the horizontal and exerts a force of 6000 N on the skier.

a. Calculate the horizontal component of the force exerted on the skier by the rope.

b. Calculate the vertical component of the force exerted on the skier by the rope. The skier is moving with a constant velocity.

c. Calculate the size of the horizontal resistance forces on the skier.

THINK	WRITE
1. The skier has four forces acting on him: the weight force, $\underset{\sim}{W}$, acting vertically downwards; the normal reaction, $\underset{\sim}{N}$, of the water on the skier, acting vertically upwards; the tension force, $\underset{\sim}{T}$, in the rope, acting 5° to the horizontal; and the resistance forces, $\underset{\sim}{F}$, acting horizontally against the direction of motion.	
2. Sketch a force vector diagram showing the forces acting on the skier.	
a. Evaluate the magnitude of the horizontal component of $\underset{\sim}{T}$, T_H.	a. $T_H = 6000 \cos(5°)$ ≈ 5997 N
b. Evaluate the magnitude of the vertical component of $\underset{\sim}{T}$, T_V.	b. $T_V = 6000 \sin(5°)$ ≈ 523 N
c. 1. Constant velocity means the acceleration is zero and in turn the resultant of the horizontal forces is zero.	c. Since acceleration $= 0$,
2. The magnitude of the horizontal resistance forces, F, is equal to the horizontal component of the tension force, $\underset{\sim}{T}$.	$F = T_H \approx 5997$ N

WORKED EXAMPLE 20 Modelling and solving problems that involve a body in equilibrium

In a science laboratory a 1.0-kg mass is suspended by two taut strings as shown. The tension forces in string 1 and string 2 are T_1 and T_2 respectively.

a. Sketch a force vector diagram showing all three forces which act on the 1.0 kg mass.

b. By resolving vectors into $\hat{i}$ and $\hat{j}$ components, determine the magnitudes of T_1 and T_2 respectively.

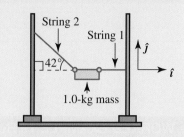

THINK	WRITE
a. 1. The mass has three forces acting on it: the weight force vertically downwards and the two tension forces. One tension force acts horizontally; the second acts at an angle of 42° to the horizontal.	a.
2. Sketch the force vector diagram.	
b. 1. The weight vector can be written as $-g\hat{j}$.	b. $\underset{\sim}{W} = -g\hat{j}$
2. The first tensile force can be written as $T_1\hat{i}$.	$\underset{\sim}{T_1} = T_1\hat{i}$
3. The second tensile force can be resolved as $\underset{\sim}{T_2} = -T_2\cos(42°)\hat{i} + T_2\sin(42°)\hat{j}$.	$\underset{\sim}{T_2} = -T_2\cos(42°)\hat{i}$ $+ T_2\sin(42°)\hat{j}$
4. Express the resultant force, $\underset{\sim}{R}$, as the sum of the three forces.	$\underset{\sim}{R} = (T_1 - T_2\cos(42°))\hat{i}$ $+ (T_2\sin(42°) - g)\hat{j}$
5. Set the sum of the three vectors to zero in accordance with Newton's First Law of Motion.	But $\underset{\sim}{R} = 0\hat{i} + 0\hat{j}$.
6. Set the $\hat{i}$ component of $\underset{\sim}{R}$ to zero and call it equation [1].	$T_1 - T_2\cos(42°) = 0 \qquad [1]$
7. Set the $\hat{j}$ component of $\underset{\sim}{R}$ to zero and call it equation [2].	$T_2\sin(42°) - g = 0 \qquad [2]$
8. Solve equation [2] for T_2.	$T_2 = \dfrac{g}{\sin(42°)}$ $\approx 14.6\,\text{N}$
9. Solve for T_1 by substituting T_2 into equation [1].	$T_1 = T_2\cos(42°)$ $\approx 10.9\,\text{N}$

Note: Part **b** could also be solved by sketching a triangle of forces and solving using trigonometry.

$$W = g = T_1 \tan(42°)$$

$$T_1 = \frac{g}{\tan(42°)}$$

$$\approx 10.9 \, \text{N}$$

and $W = g = T_2 \sin(42°)$

$$T_2 = \frac{g}{\sin(42°)}$$

$$\approx 14.6 \, \text{N (as in part } \textbf{b} \text{ above)}$$

WORKED EXAMPLE 21 Modelling and solving problems involving motion of a body in equilibrium situations

A car of mass 800 kg is parked in a street which has an angle of elevation of 15°. The $\hat{i}$ direction is parallel down the street and the $\hat{j}$ direction is perpendicular to the street. The car is subject to three forces, namely its weight, W, the normal reaction force, N, of the road acting on the car and the applied force of the brake (this is actually a static friction force) F.

a. Sketch a vector diagram indicating the three forces, W, N and F, acting on the car, taking the car as a particle.
b. Determine the magnitude of the resultant force R.
c. Resolve the weight, W, into its components and express it as a vector using $\hat{i} - \hat{j}$ notation.
d. Calculate the magnitude of N, the normal reaction force.
e. Calculate the magnitude of the applied force of the brake F.

THINK

a. 1. A stationary car parked on a street will have a vertical weight force, a normal reaction force and a static frictional force resisting its sliding or rolling down the street.

 2. Sketch the force vector diagram.

b. 1. The car is in equilibrium since it is stationary.

 2. Apply Newton's First Law of Motion: the resultant force, R, must be zero.

 3. Therefore, the magnitude of the resultant force, R, is zero.

WRITE

a.

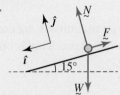

b. $R = 0$

$R = 0 \, \text{N}$

c. 1. Sketch a diagram showing the resolution of the weight, W, into components parallel to $\hat{\imath}$ and $\hat{\jmath}$.

c.

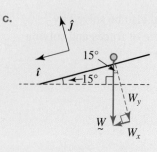

2. The component of W parallel to $\hat{\imath}$, W_x, is $W\sin(15°)$.

3. Substitute $W = 800g$ and evaluate.

$$W_x = 800g\sin(15°)$$
$$\approx 207g$$

4. The component of W parallel to $\hat{\jmath}$, W_y, is $W\cos(15°)$.

$$W_y = W\cos(15°)$$

5. Substitute $W = 800g$ and evaluate.

$$= -800g\cos(15°)$$
$$\approx -773g$$

6. Express W in vector notation.

$$\underset{\sim}{W} = W_x\hat{\imath} + W_y\hat{\jmath}$$
$$\underset{\sim}{W} = 207g\hat{\imath} - 773g\hat{\jmath}$$

d. 1. The component of the net force parallel to the $\hat{\jmath}$ vector is zero.

d. $(N - 773g)\hat{\jmath} = 0\hat{\jmath}$

2. Solve for the magnitude of the normal N.

$$N = 773g$$

e. 1. The component of the net force parallel to the $\hat{\imath}$ vector is zero.

e. $(207g - F)\hat{\imath} = 0\hat{\imath}$

2. Solve for the magnitude of the applied force of the brake, F.

$$F = 207g$$

Note: In general, when an object of mass m and hence weight mg is on an inclined plane with incline angle θ, the weight vector can be resolved into two components:

Weight on an inclined plane

Component 1 (W_x): magnitude $mg\sin(\theta)$, acting down the plane.

Component 2 (W_y): magnitude $mg\cos(\theta)$, acting perpendicular to the plane. This is opposed by the normal reaction force, N.

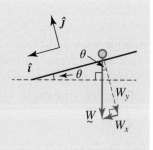

Friction

A body resting on a table is acted on by a number of forces. As we have discussed the weight W acts downwards and is given by the formula

$$\underset{\sim}{W} = \text{mass} \times \underset{\sim}{g}.$$

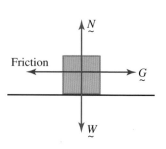

The normal reaction, N, is the force exerted by the table on the body which opposes and balances the weight.

Suppose the body is further acted on by a horizontal force G. If the table is smooth, there is no friction and the mass will move to the right. If the table is not smooth a frictional force will oppose the motion. The frictional force depends on the roughness of the surface and the normal reaction N.

Friction

Friction $= \mu \times$ normal reaction, where μ is the coefficient of friction.

$$F = \mu \times N$$

Note that this formula gives the maximum value for friction. If, in the diagram, G is less than μN, then the frictional force will just balance the force G.

WORKED EXAMPLE 22 Vector applications related to smooth horizontal planes

A body of mass 4 kg, at rest on a table, is acted on by a horizontal force, P, as shown in the figure. If the body just begins to move when P is 12 N, calculate the coefficient of friction between the body and the table.

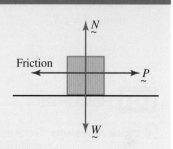

THINK	WRITE
1. The weight of the body is $m \times g$.	$W = m \times g$ $= 4 \times 9.8$ $= 39.2 \, \text{N}$
2. The normal force, N, must balance the weight.	$N = W$ $= 39.2 \, \text{N}$
3. The friction will oppose P and its maximum value is $\mu \times N$.	$F = \mu \times N$ $12 = \mu \times 39.2$ $\mu = 0.3$

WORKED EXAMPLE 23 Vector applications related to smooth inclined planes

A body of mass 4 kg rests on a plane inclined at an angle of 30° to the horizontal. It is just prevented from moving by friction. Calculate:
a. the weight, $|W|$
b. the magnitude of the normal reaction, N
c. the magnitude of the frictional force, F
d. the coefficient of friction, μ.

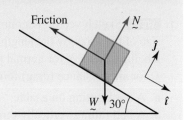

THINK	WRITE
a. Recall the formula for weight, $W = m \times g$, and substitute the known values: $m = 4 \, \text{kg}$ and $g = 9.8 \, \text{m/s}^2$. The units are newtons.	a. $|W| = m \times g$ $= 4 \times 9.8$ $= 39.2 \, \text{N}$

b. 1. First resolve each force into its components. Take the $\hat{i}$ direction to be down the plane and the $\hat{j}$ direction to be perpendicular to the plane.

b. $N = N\hat{j}$
$W = 39.2 \sin(30)\hat{i} + -39.2\cos(30)\hat{j}$
$\quad = 19.6\hat{i} + -33.9\hat{j}$
$F = -\mu N\hat{i}$

2. If a body is at rest, the vector sum of forces is $\underset{\sim}{0}$.

$F + N + W = 0$
$-F\hat{i} + N\hat{j} + 19.6\hat{i} + -33.9\hat{j} = \underset{\sim}{0}$

3. If $a\hat{i} + b\hat{j} + c\hat{i} + d\hat{j} = \underset{\sim}{0}$, then $a = -c$ and $b = -d$.

The $\hat{i}$ and $\hat{j}$ components add to 0.
$-F = -19.6$
$N = 33.9$
$\therefore N = 33.9\,\text{N}$

c. Take the value of F from part **b3**.

c. $F = 19.6\,\text{N}$

d. The coefficient of friction, μ, can be calculated using $F = \mu N$.

d. $F = \mu N$
$19.6 = \mu \times 33.9$
$\mu = 0.58$

 Resources

Digital documents SkillSHEET The sine rule (doc-26831)
SkillSHEET The cosine rule (doc-26832)
SpreadSHEET Vector addition (doc-26833)

Exercise 4.5 Applications of vectors in two dimensions   learn**on**

4.5 Exercise	4.5 Exam questions **on**

These questions are
even better in jacPLUS!
• Receive immediate feedback
• Access sample responses
• Track results and progress

Find all this and MORE in jacPLUS ▶

Simple familiar	Complex familiar	Complex unfamiliar
1, 2, 3, 4, 5, 6, 7, 8, 9, 10, 11, 12, 13, 14	15	16, 17

Simple familiar

1. **WE16** Sketch vector diagrams to represent forces which act on the following objects. Indicate the relative size of the force by the length of the vector arrows. Further indicate the nature of each of the forces acting by labelling them: N for normal reaction forces, W for gravitational forces (that is, weight), A for applied forces, D for air resistance (drag) forces, and F for friction forces.

a. A book sitting on a table.
b. A ball falling vertically through the air at constant speed.
c. A car driving on a horizontal road at a constant speed.
d. A boat drifting through the water at constant speed.
e. A body sliding across a smooth horizontal surface at constant velocity.

2. Sketch vector diagrams to represent forces which act on the following objects. Indicate the relative size of the force by the length of the vector arrows. Further indicate the nature of each of the forces acting by labelling them: N for normal reaction forces, W for gravitational forces (that is, weight), A for applied forces, D for air resistance (drag) forces, and F for friction forces.

   a. A car accelerating on a horizontal road.
   b. A body at rest on an inclined plane.
   c. A body sliding down an inclined plane at constant speed.
   d. A ball travelling vertically up (include air resistance).
   e. A ball travelling vertically down (include air resistance).

3. Refer to the diagram to answer the following questions.

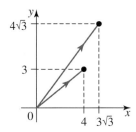

   a. Using $\hat{i}$ and $\hat{j}$ notation, determine the exact value of the resultant force.
   b. Determine the magnitude of a third force for the resultant force to be zero.

4. **WE17** Two forces F_1 and F_2 act on an object. They are described by the vectors:
   i. $F_1 = 13\hat{i} - 5\hat{j}, F_2 = -4\hat{i} + 9\hat{j}$ ii. $F_1 = 8\hat{i} + 6\hat{j}, F_2 = -14\hat{i} + 9\hat{j}$
   iii. $F_1 = 2\sqrt{2}\hat{i} - 3\hat{j}, F_2 = -3\sqrt{5}\hat{i} + 2\hat{j}$.
   For each of the above, determine:

   a. the resultant force, R, acting on the object
   b. the magnitude of the resultant force, $|R|$
   c. a third force, F_3, applied to the body so that the resultant force is equal to zero.

5. Three forces F_1, F_2 and F_3 act on an object. They are described by the vectors:
   i. $F_1 = 3\hat{i} - 5\hat{j}, F_2 = 4\hat{i} + 9\hat{j}$ and $F_3 = -6\hat{i} - 2\hat{j}$ ii. $F_1 = \hat{i} - \hat{j}, F_2 = -2\hat{i} + 3\hat{j}$ and $F_3 = \sqrt{2}\hat{i} - 2\hat{j}$.
   For each of the above, determine:

   a. the resultant force, R, acting on the object
   b. the magnitude of the resultant force, $|R|$
   c. the angle that the vector R makes to the $\hat{i}$ vector, to one decimal place
   d. a fourth force, F_4, applied to the body so that the resultant force is equal to zero.

6. **WE18** Three forces X, Y and Z act on an object such that the resultant force $R = 0$. The force Y acts at an angle of $120°$ to the force X and has the same magnitude as the force X, which is $10\,\text{N}$.

   a. Determine the magnitude of Z.
   b. Calculate the angle that the force Z makes to X.

Note: Use $g = 9.8\,\text{m/s}^2$ down for all problems involving weight.

7. **WE19** A girl is pulling along her baby brother in a cart attached to a rope. The cart is on a horizontal path. The rope makes an angle of $20°$ to the horizontal and exerts a force on the cart of magnitude $25\,\text{N}$.

   a. Calculate the horizontal component of the force exerted on the cart by the rope.
   b. Calculate the vertical component of the force exerted on the cart by the rope.
      The cart moves along at a constant velocity; that is, the acceleration of the cart is zero.
   c. Calculate the size of the horizontal friction force acting on the cart. Give you answer to one decimal place.

8. **MC** A child of mass 40 kg is held on a 'swinging rope' at an angle of 25° to the vertical by a horizontal force of 300 N. If T is the tension force of the rope acting on the child, then:

a. the force vector diagram which best represents this situation is:

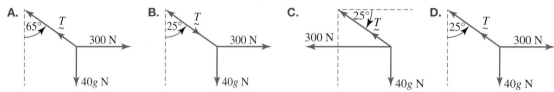

b. the horizontal component of T is:

A. $40g$ N   **B.** 300 N   **C.** 127 N   **D.** 272 N

c. the vertical component of T is:

A. 300 N   **B.** 127 N   **C.** $40g$ N   **D.** 200 N

d. the magnitude of T is nearest to:

A. 494 N   **B.** 440 N   **C.** 477 N   **D.** 92 N

9. **WE20** A swing chair of mass 8 kg is suspended by two taut ropes as shown at right. The tension forces in rope 1 and rope 2 are T_1 and T_2 respectively.

a. Sketch a force vector diagram showing all three forces which act on the swing.

b. By resolving vectors into $\hat{\imath}$–$\hat{\jmath}$ components determine the exact magnitudes of T_1 and T_2.

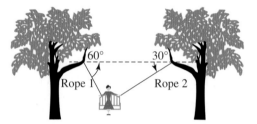

10. Sam earns some extra pocket money by mowing his neighbour's front lawn. When he pushes the lawnmower at a constant velocity he applies a force of 120 N down the shaft of the mower which is angled at 40° to the vertical. The lawnmower has a mass of 40 kg.

a. Sketch a force vector diagram illustrating all four forces acting on the lawnmower. (Treat the lawnmower as a particle.)

b. Calculate both the vertical and horizontal components of the force that Sam applies to the lawnmower, correct to one decimal place.

c. Using $\hat{\imath}$–$\hat{\jmath}$ notation, write a vector equation for the resultant force, R, acting on the lawnmower in terms of the four forces acting on the lawnmower.

d. Determine the magnitude of the force of friction acting on the lawnmower as it moves across the lawn at constant velocity.

e. Determine the magnitude of the normal reaction force.

11. **WE21** A 1.5-kg mass is placed on a smooth inclined plane angled at 30° to the horizontal. To stop it from sliding down the plane a string is attached to the upper side as shown. The unit vectors $\hat{\imath}$ and $\hat{\jmath}$ are also shown.

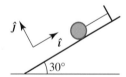

a. Sketch a force vector diagram showing the forces which act on the mass. Label the forces N for the normal reaction force, T for the tension force and W for the force arising from the effect of gravity.

b. What is the magnitude of the resultant force, R?

c. Determine the weight vector, W, using $\hat{\imath}$–$\hat{\jmath}$ notation.

d. Determine the tension force, T, using $\hat{\imath}$–$\hat{\jmath}$ notation.

e. Determine the magnitude of the normal reaction force, N.

12. A 1.5 kg mass is placed on a smooth inclined plane angled at 30° to the horizontal. To stop it from sliding down the plane a horizontal force is applied to the mass as shown. The unit vectors $\hat{\imath}$ and $\hat{\jmath}$ are also shown.

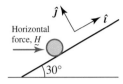

a. Sketch a force vector diagram showing the forces which act on the mass. Label the forces $\underset{\sim}{N}$ for the normal reaction force, $\underset{\sim}{H}$ for the horizontal force and $\underset{\sim}{W}$ for the force due to gravity.

b. Determine the weight vector, $\underset{\sim}{W}$, using $\hat{\imath}$–$\hat{\jmath}$ notation.

c. Determine the horizontal force, $\underset{\sim}{H}$, using $\hat{\imath}$–$\hat{\jmath}$ notation.

d. Determine the normal reaction force, $\underset{\sim}{N}$, using $\hat{\imath}$–$\hat{\jmath}$ notation.

e. Show that $\underset{\sim}{H} = \dfrac{|\underset{\sim}{W}|}{\sqrt{3}} = \dfrac{\sqrt{3}\,|\underset{\sim}{W}|}{3}$.

13. **WE22** A box containing books has a mass of 40 kg. It requires a force of 300 N to move the box across the floor. Calculate the coefficient of friction between the box and the floor.

14. **WE23** A body of mass 6 kg rests on a plane inclined at an angle of 40° to the horizontal. It is just prevented from moving by friction. Calculate the:

   a. weight   b. normal reaction   c. frictional force   d. coefficient of friction.

Complex familiar

15. Three forces $\underset{\sim}{X}$, $\underset{\sim}{Y}$ and $\underset{\sim}{Z}$ add to give a resultant force equal to zero.

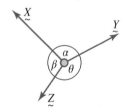

a. Calculate β (to the nearest degree) if $\alpha = 135°$, $|\underset{\sim}{Z}| = 200$ N and $|\underset{\sim}{X}| = 150$ N.

b. Determine $|\underset{\sim}{Z}|$ (to one decimal place) if $\alpha = 100°$, $\beta = 135°$ and $|\underset{\sim}{Y}| = 27$ N.

c. If $|\underset{\sim}{X}| = |\underset{\sim}{Y}|$ and $\alpha = 60°$, demonstrate that $|\underset{\sim}{Z}| = \sqrt{3}\,|\underset{\sim}{X}|$.

Complex unfamiliar

16. An object of mass 2 kg rests on a plane inclined at an angle of 40° to the horizontal. If the coefficient of friction between the object and the plane is 0.2, calculate the resultant force down the plane.

17. a. OAB is a triangle in which M is the midpoint of AB. Let $\overrightarrow{OA} = \underset{\sim}{a}$ and $\overrightarrow{OB} = \underset{\sim}{b}$. Express OM and AM in terms of $\underset{\sim}{a}$ and $\underset{\sim}{b}$ and hence show that $\left|\overrightarrow{OA}\right|^2 + \left|\overrightarrow{OB}\right|^2 = 2\left|\overrightarrow{AM}\right|^2 + 2\left|\overrightarrow{OM}\right|^2$.
   This result is known as Apollonius' theorem.

   b. If $\underset{\sim}{a} = \cos(A)\hat{\imath} + \sin(A)\hat{\jmath}$ and $\underset{\sim}{b} = \cos(B)\hat{\imath} + \sin(B)\hat{\jmath}$, determine $\underset{\sim}{a} \cdot \underset{\sim}{b}$ and hence show that $\cos(A - B) = \cos(A)\cos(B) + \sin(A)\sin(B)$.

Fully worked solutions for this chapter are available online.

LESSON
4.6 Review

4.6.1 Summary

4.6 Exercise

learnon

4.6 Exercise	4.6 Exam questions on

Simple familiar	Complex familiar	Complex unfamiliar
1, 2, 3, 4, 5, 6, 7, 8, 9, 10, 11, 12	13, 14, 15, 16	17, 18, 19, 20

Simple familiar

1. If $a = 3\hat{\imath} + a\hat{\jmath}$, $b = 2a\hat{\imath} - a\hat{\jmath}$, and it is known that a is perpendicular to b, calculate a.

2. Let $a = 4\hat{\imath} + 3\hat{\jmath}$, $b = -\hat{\imath} + 2\hat{\jmath}$. Calculate the angle between the two vectors, in degrees, to two decimal places.

Using the vectors $a = \hat{\imath} - 2\hat{\jmath}$ and $b = 2\hat{\imath} + 3\hat{\jmath}$ answer questions **3** and **4**.

3. Determine the scalar resolute of a on b.

4. Determine the vector resolute of b parallel to a.

5. a. Given the points A(2, 5) and B(−4, 7), determine the vector $\overrightarrow{OM}$ where M is the midpoint of AB.
   b. Show that the points $(-2, -6)$, $(1, -3)$ and $(5, 1)$ are collinear.
   c. Given the points A(3, −5) and B(−3, 4), determine the vector $\overrightarrow{OP}$ where P is a point on AB such that $\overrightarrow{AP} = \frac{1}{3}\overrightarrow{AB}$.

6. A body is in equilibrium under the action of three forces f_1, f_2 and f_3.
   $f_1 = -3\hat{\imath} + 7\hat{\jmath}$ and $f_2 = 5\hat{\imath} + 2\hat{\jmath}$
   Determine the magnitude of the vector f_3

7. Two forces, equal in magnitude, are sufficient to keep the mass shown at right in equilibrium against the force of gravity.
   Calculate the magnitude of each force, to one decimal place.

50°

2 kg

8. A body of mass 6.0 kg is suspended from a ceiling by two ropes. The angle between the two ropes is 90° and they are connected to the ceiling at points A and B respectively. The tension, T_2, in rope 2 is twice the tension, T_1, in rope 1. Give you answers to one decimal place where appropriate.

   a. Sketch a force vector diagram for the 6.0 kg mass.
   b. Calculate the magnitudes of the tensions T_1 and T_2.
   c. Determine the angles that rope 1 and rope 2 make with the horizontal.
   d. If rope 1 has a length x, show that the distance AB is $\dfrac{\sqrt{5}x}{2}$.

9. A mass of 4 kg rests on a plane which has a coefficient of friction of 0.25.

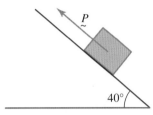

   Calculate the minimum force P, acting parallel to the plane, needed to prevent the mass slipping down the plane. Give your answer to one decimal place.

10. The vector $\dfrac{1}{2}\hat{i} + y\hat{j}$ makes an angle of 150° with the y-axis. Determine the value of y.

11. **MC** Given the vectors $a = 2\hat{i} - 3\hat{j}$ and $b = -3\hat{i} + \hat{j}$:

   A. a is parallel to b
   B. the vector $2a + 3b$ is parallel to the x-axis
   C. the vector $2a + 3b$ is parallel to the y-axis
   D. None of the above.

12. **MC** If $v_a = 2\hat{i} + 3\hat{j}$ and $v_b = -2\hat{i} + 2\hat{j}$, the value of $v_a - v_b$ is

   A. $5\hat{j}$ B. $4\hat{i} + \hat{j}$ C. $-4\hat{i} - \hat{j}$ D. $\hat{j}$

Complex familiar

13. Given the two vectors $a = 4\hat{i} - 5\hat{j}$ and $b = 2\hat{i} - \hat{j}$, determine:

   a. a unit vector parallel to b
   b. the scalar resolute of a in the direction of b
   c. the angle between the vectors a and b
   d. the vector resolute of a in the direction of b
   e. the vector resolute of a perpendicular b .

14. A ship travels a distance d km on a bearing of Nθ°W where $0° < \theta < 90°$ from an origin O. If $\hat{i}$ and $\hat{j}$ represent unit vectors in the directions of east and north respectively, determine the position vector of the ship from the origin.

15. A river flows from north to south at 5 km/h. A boat heads directly across the river from the river bank to the other side at 11 km/h. Determine the true velocity of the boat.

16. While on holidays, Lilly hired a Segway as a great way to move around and explore the city. One day she travelled 500 metres on a bearing S28°E then turned and travelled 800 metres on a bearing S67°W. If $\hat{i}$ and $\hat{j}$ represent unit vectors of magnitude 1 metre in the directions of east and north respectively, determine her position vector from the starting point. Hence, determine her displacement, correct to one decimal place, and her true bearing from her starting point.

Complex unfamiliar

17. Given the vectors $\underset{\sim}{a} = x\hat{i} + y\hat{j}$ and $\underset{\sim}{b} = 2\hat{i} + 3\hat{j}$, determine the values of x and y if the length of the vector $\underset{\sim}{a}$ is $\sqrt{34}$ and $\underset{\sim}{a} \cdot \underset{\sim}{b} = 9$.

18. Arnie is pushing against a trailer, preventing it from rolling down a hill. The trailer has a mass of 200 kg and the hill is on an incline of 15° to the horizontal. At the moment there is no problem because Arnie is capable of pushing with a force of 1000 N parallel to the plane.

However, it is raining and the trailer is filling with water at a rate of 25 litres per minute. How long will Arnie be able to hold the trailer and stop it from running down the hill? Note that the mass of 1 L of water is 1 kg.

19. A ball of mass 0.20 kg is shot vertically in the air. It decelerates under the action of two forces: the weight force and the force of air resistance. When the ball moves with a speed of 40 m/s, it has an air resistance of 1.0 N. When the ball is stationary, the air resistance force is equal to zero. Determine the magnitude and direction of the resultant force acting on the ball in each of the following situations.

   a. When the ball is moving upwards at a speed of 40 m/s.

   b. When the ball is at its maximum height above the ground and its velocity is zero.
     Later, the ball is travelling toward the ground at 40 m/s.

   c. Determine the magnitude and direction of the resultant force acting on the ball when it is moving downwards at a speed of 40 m/s.

20. When the surf gets big at Kirra there is always a sweep running from south to north.

Jodie is heading out at 8 km/h. If she heads directly out to sea and she wants to get to the take-off area, how far up the beach should she walk before paddling out?

Fully worked solutions for this chapter are available online.

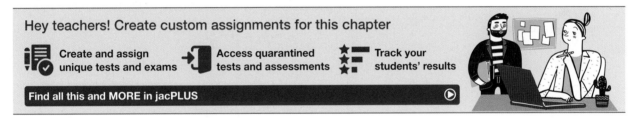

Answers

Chapter 4 Algebra of vectors in two dimensions

4.2 Vector representations

4.2 Exercise

1. a. $\overrightarrow{AB} = 3\hat{\imath} + 4\hat{\jmath}$ b. $\overrightarrow{AB} = \hat{\imath} - 7\hat{\jmath}$
   c. $\overrightarrow{AB} = -3\hat{\imath}$ d. $\overrightarrow{AB} = -25\hat{\imath} - 10\hat{\jmath}$

2. $|\overrightarrow{AB}| \approx 9.2$

3. $\dfrac{15}{2}$

4. $x = -3, y = -8$

5. $\overrightarrow{OM} = 5\hat{\imath} + 6\hat{\jmath}$

6. a. $-\dfrac{5}{2}$ b. $-\dfrac{15}{2}$

7. a, b. Sample responses can be found in the worked solutions in the online resources.
   c. $-\dfrac{20}{3}$ d. $\dfrac{15}{7}$

8. a. i. $-6\hat{\imath} + 5\hat{\jmath}$ ii. $\sqrt{61}$
      iii. $\dfrac{1}{\sqrt{61}}(-6\hat{\imath} + 5\hat{\jmath})$
   b. i. $6\hat{\imath} - 8\hat{\jmath}$ ii. 10
      iii. $\dfrac{1}{5}(-3\hat{\imath} + 4\hat{\jmath})$

9. $m = 2, n = 3$

10. $x = -2, y = 5$

11. a. $\pm 2\sqrt{5}$ b. $-\dfrac{12}{5}$

12. a. $\pm\sqrt{13}$ b. $-\dfrac{5}{2}$

13. Sample responses can be found in the worked solutions in the online resources.

14. a. $x = 4, y = -2$ b. $m = -5, n = -4$
    c. $m = \dfrac{1}{2}, n = \dfrac{5}{4}$

4.3 Scalar (dot) product and the projection of vectors – scalar and vector resolutes

4.3 Exercise

1. 23.99

2. a. 15 b. 18 c. -31 d. -26

3. a. 1 b. -20 c. 3 d. 2

4. D.

5, 6, 7. Sample responses can be found in the worked solutions in the online resources.

8. D

9. B

10. D

11. a. i. $\dfrac{23\sqrt{13}}{13}$ ii. $\dfrac{23\sqrt{41}}{41}$
    b. i. $\dfrac{17\sqrt{29}}{29}$ ii. $\dfrac{17\sqrt{10}}{10}$
    c. i. $-\dfrac{13\sqrt{10}}{10}$ ii. $-\dfrac{26\sqrt{17}}{17}$
    d. i. $-\dfrac{6\sqrt{13}}{13}$ ii. $-\dfrac{6}{5}$
    e. i. $-\dfrac{23}{5}$ ii. $\dfrac{23\sqrt{26}}{13}$

12. a. i. $\dfrac{1}{\sqrt{10}}$ ii. $\underset{\sim}{b}_{\parallel} = \dfrac{3}{10}\hat{\imath} - \dfrac{1}{10}\hat{\jmath}$
       iii. $\underset{\sim}{b}_{\perp} = \dfrac{17}{10}\hat{\imath} + \dfrac{51}{10}\hat{\jmath}$
    b. i. $2\sqrt{41}$ ii. $\underset{\sim}{b}_{\parallel} = 8\hat{\imath} + 10\hat{\jmath}$
       iii. $\underset{\sim}{b}_{\perp} = 0$
    c. i. 0 ii. $\underset{\sim}{b}_{\perp} = \underset{\sim}{0}$
       iii. $\underset{\sim}{b}_{\perp} = -3\hat{\imath} + 4\hat{\jmath}$
    d. i. $\dfrac{3}{\sqrt{2}}$ ii. $\underset{\sim}{b}_{\parallel} = \dfrac{3}{2}\hat{\imath} + \dfrac{3}{2}\hat{\jmath}$
       iii. $\underset{\sim}{b}_{\perp} = \dfrac{1}{2}\hat{\imath} - \dfrac{1}{2}\hat{\jmath}$
    e. i. $-\dfrac{5}{\sqrt{13}}$ ii. $\underset{\sim}{b}_{\parallel} = \dfrac{-10}{13}\hat{\imath} - \dfrac{15}{13}\hat{\jmath}$
       iii. $\underset{\sim}{b}_{\perp} = \dfrac{36}{13}\hat{\imath} - \dfrac{24}{13}\hat{\jmath}$
    f. i. $\dfrac{2}{\sqrt{10}}$ ii. $\underset{\sim}{b}_{\parallel} = \dfrac{3}{5}\hat{\imath} + \dfrac{1}{5}\hat{\jmath}$
       iii. $\underset{\sim}{b}_{\perp} = -\dfrac{3}{5}\hat{\imath} + \dfrac{9}{5}\hat{\jmath}$

13. D

14. $m = 1$

15. $m = \dfrac{3}{2}$

16. $4\hat{\imath} + 8\hat{\jmath}$

17. a. 16 b. -1 c. -25 d. 0

18. a. $67°$ b. $93°$ c. $144°$ d. $90°$

4.4 Operations with vectors

4.4 Exercise

1. a. $7\hat{\imath} - 24\hat{\jmath}$ b. $25\,\text{km}\ 163°44'\text{T}$

2. a. $-500\hat{\imath} - 1200\hat{\jmath}$ b. $1300\,\text{m}\ 202°37'\text{T}$

3. a. $-14.34\hat{\imath} - 55.48\hat{\jmath}$ b. $57.26\,\text{km}\ \text{S}14°30'\text{W}$

4. a. $-261.67\hat{\imath} + 144.44\hat{\jmath}$ b. $298.89\,\text{km}\ \text{N}61°6'\text{W}$

5. a. $-2.13\hat{\imath} - 6.13\hat{\jmath}$ b. $6.49\,\text{km}\ 19°5'\text{T}$

6. a. $-4.98\hat{\imath} + 60.74\hat{\jmath}$ b. $60.94\,\text{km}\ \text{N}4°39'\text{W}$

7. a. $4.34\hat{\imath} - 1.49\hat{\jmath}$ b. $4.59\,\text{km}\ 109°4'\text{T}$

8. a. $299.69\hat{\imath} - 2341.75\hat{\jmath}$ **b.** 2360.85 m $172°42'$ T

9. 9.63 km N $70°43'$ W

10. 961.9 m $22°27'$ T

11. 172 km $62°2'$ T

12. $\left(d_1 + d_2\sin(\theta)\right)\hat{\imath} + d_2\cos(\theta)\hat{\jmath}$,
$\sqrt{d_1^2 + d_2^2 + 2d_1d_2\sin(\theta)}$

13. $\left(d_1\cos(\theta_1) - d_2\cos(\theta_2)\right)\hat{\imath} + \left(d_1\sin(\theta_1) + d_2\sin(\theta_2)\right)\hat{\jmath}$,
$\sqrt{d_1^2 + d_2^2 - 2d_1d_2\cos(\theta_1 + \theta_2)}$

4.5 Applications of vectors in two dimensions

4.5 Exercise

1. a.

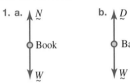

b.

Ball

c.

$F \leftarrow$ Car $\rightarrow A$

d.

$F \leftarrow$ Boat $\rightarrow A$

e.

Sliding object

2. a.

$F \leftarrow$ Accelerating car $\rightarrow A$

b.

Body at rest

c.

Sliding body

d.

Ball moving down

e.

Ball moving down

3. a. $(4 + 3\sqrt{3})\hat{\imath} + (3 + 4\sqrt{3})\hat{\jmath}$

   b. $\sqrt{100 + 48\sqrt{3}} = 13.53$ N

4. a. i. $9\hat{\imath} + 4\hat{\jmath}$ **ii.** $-6\hat{\imath} - 3\hat{\jmath}$
   iii. $(2\sqrt{2} - 3\sqrt{5})\hat{\imath} - \hat{\jmath}$

   b i. $\sqrt{97}$ **ii.** $3\sqrt{5}$
   iii. $\sqrt{54 - 12\sqrt{10}}$

   c i. $-9\hat{\imath} - 4\hat{\jmath}$ **ii.** $6\hat{\imath} + 3\hat{\jmath}$
   iii. $(3\sqrt{5} - 2\sqrt{2})\hat{\imath} + \hat{\jmath}$

5. a. i. $\hat{\imath} + 2\hat{\jmath}$ **ii.** $(\sqrt{2} - 1)\hat{\imath}$

   b. i. $\sqrt{5}$ **ii.** $\sqrt{2} - 1$

   c. i. $63.4°$ **ii.** $0°$

   d. i. $-\hat{\imath} - 2\hat{\jmath}$ **ii.** $(1 - \sqrt{2})\hat{\imath}$

6. a. 10 N   **b.** $120°$

7. a. 23.5 N   **b.** 8.6 N   **c.** 23.5 N

8. a. D   **b.** B   **c.** C   **d.** A

9. a.

   b. $T_1 = 4g$ N;
   $T_2 = 4\sqrt{3}g$ N

10. a.

$A = 120$ N at $40°$ to vertical

   b. $A_V = 91.9$ N down
   $A_H = 77.1$ N left

   c. $R = \left(F_{\text{friction}} - 77.1\right)\hat{\imath} + (N - 483.9)\hat{\jmath} = 0$

   d. 77.1 N

   e. 483.9 N

11. a.

Mass

   b. 0 N

   c. $-7.4\hat{\imath} - 12.7\hat{\jmath}$ N

   d. 7.4 N

   e. 12.7 N

12. a.

Mass H

   b. $-7.4\hat{\imath} - 12.7\hat{\jmath}$ N

   c. $7.4\hat{\imath} - 4.3\hat{\jmath}$ N

   d. $17\hat{\jmath}$ N

   e. Sample responses can be found in the worked solutions in the online resources.

13. 0.77

14. a. 58.8 N   **b.** 45.04 N   **c.** 37.8 N   **d.** 0.84

15. a. $77°$

   b. 37.6 N

   c. Sample responses can be found in the worked solutions in the online resources.

16. 9.6 N

17. Sample responses can be found in the worked solutions in the online resources.

4.6 Review

4.6 Exercise

1. 6

2. $79.70°$

3. $\dfrac{-4}{\left(\sqrt{3}\right)}$

4. $-\dfrac{4}{5}(\hat{\imath} - 2\hat{\jmath})$

5. a. $-\hat{i} + 6\hat{j}$
   b. Sample responses can be found in the worked solutions in the online resources.
   c. $\hat{i} - 2\hat{j}$
6. $\sqrt{85}$
7. 10.8 N
8. a. Sample responses can be found in the worked solutions in the online resources.
   b. 26.3 N, 52.6 N
   c. 26.6°, 63.4°
   d. Sample responses can be found in the worked solutions in the online resources.
9. 17.7 N
10. $-\dfrac{\sqrt{3}}{2}$
11. D
12. $-d\sin(\theta)\hat{i} + d\cos(\theta)\hat{j}$
13. a. $\dfrac{1}{\sqrt{5}}(2\hat{i} - \hat{j})$ b. $\dfrac{13}{\sqrt{5}}$ c. 24.775°
    d. $\dfrac{13}{5}(2\hat{i} - \hat{j})$ e. $-\dfrac{6}{5}(\hat{i} + 2\hat{j})$
14. B
15. 12.1 km/h at 65.6° to the bank
16. $-501.7\hat{i} - 754.1\hat{j}$; distance 905.7 km, bearing 213°38'T
17. $x = -3, y = 5$ or $x = \dfrac{75}{13}, y = -\dfrac{11}{13}$
18. 7 minutes 46 seconds
19. a. 2.96 N down   b. 1.96 N down   c. 0.96 N down
20. 60 m

5 Matrices

LESSON SEQUENCE

Fully worked solutions for this chapter are available online.

EXAM PREPARATION

Access exam-style questions in every lesson, available online.

on Resources

Solutions	Solutions — Chapter 5 (sol-0397)
Exam question	Exam question booklet — Chapter 5 (eqb-0283)
Digital documents	Learning matrix — Chapter 5 (doc-41926)
	Chapter summary — Chapter 5 (doc-41927)

LESSON
5.1 Overview

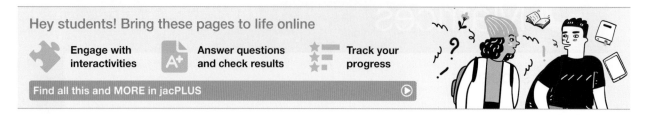
5.1.1 Introduction

Matrices were first used to solve systems of linear equations. Today they are used in many fields, including engineering, physics, economics and statistics. They are used primarily in the encoding and decoding of information. You have probably used matrix technology without realising it in encrypted messaging and internet banking.

The British mathematician Alan Turing led a team of mathematicians who were responsible for breaking the Germans' Enigma code during World War II. Their work helped the Allies to win the war in Europe and also led to the creation of the first computer. Turing built a machine to help with the decoding process. This was the start of the computer age, and now every computer-generated image is the result of matrix mathematics.

5.1.2 Syllabus links

Lesson	Lesson title	Syllabus links
5.2	**Matrix definition and notation**	○ Understand the matrix definition and notation.
5.3	**Addition, subtraction and scalar multiplication of matrices**	○ Define and use addition and subtraction of matrices and scalar multiplication.
5.4	**Matrix multiplication**	○ Define and use matrix multiplication, multiplicative identity and multiplicative inverse.
5.5	**Determinants and inverses**	○ Calculate the determinant and multiplicative inverse of 2×2 matrices, with and without technology. • If $A = \begin{bmatrix} a & b \\ c & d \end{bmatrix}$ then $\det(A) = ad - bc$ • $A^{-1} = \begin{bmatrix} a & b \\ c & d \end{bmatrix}^{-1} = \dfrac{1}{\det(A)} \begin{bmatrix} d & -b \\ -c & a \end{bmatrix}$, $\det(A) \neq 0$
5.6	**Matrix algebra**	○ Use matrix algebra properties, including • $A + B = B + A$ (commutative law for addition) • $A + 0 = A$ (additive identity) • $A + (-A) = 0$ (additive inverse) • $AI = A = IA$ (multiplicative identity) • $AA^{-1} = I = A^{-1}A$ (multiplicative inverse) • $A(B + C) = AB + AC$ (left distributive law) • $(B + C)A = BA + CA$ (right distributive law) ○ Recognise that matrix multiplication in general is not commutative.
5.7	**Matrix equations**	○ Use matrix algebra to solve matrix equations that involve matrices of up to dimension 2×2, including those of the form $AX = B$, $XA = B$ and $AX + BX = C$, with and without technology. ○ Model and solve problems that involve matrices of up to dimension 2×2, including the solution of systems of linear equations, with and without technology.

Source: Specialist Mathematics Senior Syllabus 2024 © State of Queensland (QCAA) 2024; licensed under CC BY 4.0

LESSON
5.2 Matrix definition and notation

SYLLABUS LINKS

• Understand the matrix definition and notation.

Source: Specialist Mathematics Senior Syllabus 2024 © State of Queensland (QCAA) 2024; licensed under CC BY 4.0

5.2.1 Introduction to matrices

The table shows the final medal tally for the top four countries at the 2020 Tokyo Olympic Games.

Country	Gold	Silver	Bronze
United States of America	39	41	33
People's Republic of China	38	32	18
Japan	27	14	17
Great Britain	22	21	22

This information can be presented in a matrix, without the country names, and without the headings for gold, silver and bronze:

$$\begin{bmatrix} 39 & 41 & 33 \\ 38 & 32 & 18 \\ 27 & 14 & 17 \\ 22 & 21 & 22 \end{bmatrix}$$

The data is presented in a rectangular array arranged in rows and columns, and is called a **matrix**. It conveys information such as that the second country won 38 gold, 32 silver and 18 bronze medals. This matrix has four rows and three columns. A matrix is usually displayed in square brackets with no borders between the rows and columns.

5.2.2 Defining matrices

Matrices come in different dimensions (sizes). We define the dimensions of a matrix by the number of rows and columns. Rows run horizontally and columns run vertically. This is known as the **order** of the matrix and is written as rows × columns.

In the following matrix A, there are 2 rows and 3 columns, which means it has an order of 2×3 (two by three).

$$\begin{array}{c} \text{Column} 1 \quad 2 \quad 3 \\ A = \begin{array}{c} \text{Row 1} \\ \text{Row 2} \end{array} \begin{bmatrix} 5 & 8 & 2 \\ 1 & 3 & 0 \end{bmatrix} \end{array}$$

rows × columns ⟶ **2 × 3** ⟶ This is a **2 × 3** matrix.

Matrices are represented by capital letters. In this example, A represents a matrix.

Instead of only characterising matrices by the number of rows and columns, we also use names for specific types of matrix dimensions.

Types of matrices		
Type of matrix	**Description**	**Example**
Column matrix	A matrix that has a single column is known as a **column matrix**. It is of size $n \times 1$.	$B = \begin{bmatrix} 1 \\ 4 \\ 2 \end{bmatrix}$ $C = \begin{bmatrix} 2 \\ 8 \\ 5 \\ 9 \end{bmatrix}$ B and C are both column matrices.
Row matrix	A matrix that has a single row is known as a **row matrix**. It is of size $1 \times n$.	$D = [2 \quad 3]$ $E = [9 \quad 4 \quad 3]$ D and E are both row matrices.
Square matrix	A matrix that has the same number of columns as rows is known as a **square matrix**. It is of size $n \times n$, e.g. 2×2 or 5×5.	$F = [8]$ $G = \begin{bmatrix} 5 & 9 \\ -4 & 1 \end{bmatrix}$ $H = \begin{bmatrix} 7 & 6 & 4 \\ -3 & 1 & 7 \\ 4 & 9 & 8 \end{bmatrix}$ F, G and H are all square matrices.
Identity matrix	A square matrix in which all of the elements on the diagonal line from the top left to bottom right (the **leading diagonal**) are 1s and all of the other elements are 0s is known as the **identity matrix**.	$A = \begin{bmatrix} 1 & 0 \\ 0 & 1 \end{bmatrix}$ and $B = \begin{bmatrix} 1 & 0 & 0 \\ 0 & 1 & 0 \\ 0 & 0 & 1 \end{bmatrix}$ A and B are both identity matrices.
Zero matrix	A square matrix that consists entirely of 0 elements is known as the **zero matrix** or null matrix.	$A = \begin{bmatrix} 0 & 0 \\ 0 & 0 \end{bmatrix}$ A is a zero matrix.

You will learn more about the identity matrix and the zero matrix in the lesson 5.3.

WORKED EXAMPLE 1 Constructing a column matrix

At High Vale College, 150 students are studying General Mathematics and 85 students are studying Mathematical Methods. Construct a column matrix to represent the number of students studying General Mathematics and Mathematical Methods, and state the order of the matrix.

THINK	**WRITE**
1. Read the question and highlight the key information.	150 students study General Mathematics. 85 students study Mathematical Methods.
2. Display this information in a column matrix.	$\begin{bmatrix} 150 \\ 85 \end{bmatrix}$
3. How many rows and columns are there in this matrix?	The order of the matrix is 2×1.

5.2.3 Elements of matrices

The numbers in a matrix are called **elements**. They are represented by a lowercase letter with the row and column in subscript. For matrix B, b_{ij} is the element in row i, column j.

$$\text{row 1}\ B = \begin{bmatrix} 1 & 2 & \textcircled{3} \\ 4 & 5 & 6 \\ 7 & 8 & 9 \end{bmatrix} \xrightarrow{\hspace{1cm}} \begin{array}{l} b_{13} \text{ is an element of} \\ \text{matrix B, in row 1 and} \\ \text{column 3.} \end{array}$$

$$\text{column 3}$$

The respective elements in matrix B are:

$$B = \begin{bmatrix} 1 & 2 & 3 \\ 4 & 5 & 6 \\ 7 & 8 & 9 \end{bmatrix} \rightarrow \begin{bmatrix} b_{11} & b_{12} & b_{13} \\ b_{21} & b_{22} & b_{23} \\ b_{31} & b_{32} & b_{33} \end{bmatrix}$$

WORKED EXAMPLE 2 Identifying elements in a matrix

For matrix C below, determine the following elements.

$$C = \begin{bmatrix} 3 & 8 & 2 & 4 \\ -5 & 1 & -7 & 0 \\ 4 & 9 & 8 & -4 \end{bmatrix}$$

a. c_{12} **b.** c_{31} **c.** c_{13}

THINK	WRITE
a. c_{12} is the element of matrix C that is in row 1 and column 2.	**a.** $c_{12} = 8$
b. c_{31} is the element of matrix C that is in row 3 and column 1.	**b.** $c_{31} = 4$
c. c_{13} is the element of matrix C that is in row 1 and column 3.	**c.** $c_{13} = 2$

5.2.4 Equality of matrices

Two matrices are equal, if and only if, they have the same order and each of the corresponding elements are equal.

For example, if $\begin{bmatrix} x \\ y \\ z \end{bmatrix}_{3 \times 1} = \begin{bmatrix} 1 \\ -2 \\ 3 \end{bmatrix}_{3 \times 1}$, then $x = 1$, $y = -2$ and $z = 3$; if $\begin{bmatrix} a & b \\ c & d \end{bmatrix}_{2 \times 2} = \begin{bmatrix} 3 & 5 \\ 4 & 7 \end{bmatrix}_{2 \times 2}$, then $a = 3, b = 5, c = 4$

and $d = 7$.

5.2.5 Describing matrices

The following table shows the number of participants attending three different dance classes (hip-hop, salsa and bachata) over the two days of a weekend. The matrix below displays the information presented in the table.

Number of participants attending the dance classes:

	Saturday	Sunday
Hip-hop	9	13
Salsa	12	8
Bachata	16	14

The number of participants attending the dance classes can be represented in a matrix:

$$\begin{bmatrix} 9 & 13 \\ 12 & 8 \\ 16 & 14 \end{bmatrix}$$

WORKED EXAMPLE 3 Creating a matrix from a table

The table shows the number of adults and children who attended three different events over the school holidays. Construct a matrix to represent this information.

	Circus	Zoo	Show
Adults	140	58	85
Children	200	125	150

THINK

1. A matrix is like a table that stores information. Determine what information needs to be displayed.

2. Write down how many adults and children attend each of the three events.

3. Write this information in a matrix. Remember to use square brackets.

WRITE

The information to be displayed is the number of adults and children attending the three events: circus, zoo and show.

	Circus	Zoo	Show
Adults	140	58	85
Children	200	125	150

$$\begin{bmatrix} 140 & 58 & 85 \\ 200 & 125 & 150 \end{bmatrix}$$

5.2.6 Interpreting matrices

At the Moomba festival, attendees (parents and children) visit three stages (A, B and C). The information is presented in the following matrix, E.

$$E = \begin{array}{c} \\ \text{Stage A} \\ \text{Stage B} \\ \text{Stage C} \end{array} \begin{array}{cc} \text{Parents} & \text{Children} \\ \begin{bmatrix} 654 & 620 \\ 543 & 600 \\ 320 & 300 \end{bmatrix} \end{array}$$

Columns	Rows
• Column 1 gives the number of parents at each stage. • Column 2 gives the number of children at each stage.	• Row 1 gives the number of attendees at stage A. • Row 2 gives the number of attendees at stage B. • Row 3 gives the number of attendees at stage C.

From the matrix above, we can also get the following information:
- There are a total of $320 + 300 = 620$ attendees at stage C.
- At Stage A there are 620 children.
- The total number of parents is $654 + 543 + 320 = 1517$.

WORKED EXAMPLE 4 Interpreting information from a matrix

A friend runs a cupcake business from home and sells cupcakes throughout the week. Their sales figures over two days were recorded in the following matrix.

$$S = \begin{array}{c} \\ \\ \end{array} \begin{array}{ccc} \text{Red} & & \\ \text{velvet} & \text{Chocolate} & \text{Vanilla} \\ \begin{bmatrix} 7 & 12 & 5 \\ 4 & 8 & 13 \end{bmatrix} \end{array} \begin{array}{c} \text{Saturday} \\ \text{Sunday} \end{array}$$

a. **State the order of matrix S.**
b. **State what information the element s_{12} shows.**
c. **State which element gives the number of red velvet cupcakes sold on Sunday.**
d. **Determine the total number of vanilla cupcakes sold over the weekend.**
e. **Determine the total number of cupcakes sold on Sunday.**

THINK	WRITE
a. The order is rows × columns.	a. The order of matrix S is 2×3.
b. The element s_{12} is in the first row (Saturday) and second column (chocolate). So, it represents the amount of chocolate cupcakes sold on Saturday.	b. There were 12 chocolate cupcakes sold on Saturday.
c. Sunday is row 2 and red velvet is column 1, so the element that gives the number of red velvet cupcakes sold on Sunday must be in row 2, column 1.	c. The number of red velvet cupcakes sold on Sunday is given by s_{21}.
d. To calculate the total number of vanilla cupcakes, add the numbers in column 3.	d. $5 + 13 = 18$
e. To calculate the total number of cupcakes sold on Sunday, add the numbers in row 2.	e. $4 + 8 + 13 = 25$

These questions are even better in jacPLUS!
- Receive immediate feedback
- Access sample responses
- Track results and progress

Find all this and MORE in jacPLUS ▶

Simple familiar	Complex familiar	Complex unfamiliar
1, 2, 3, 4, 5, 6, 7, 8	9, 10, 11, 12	N/A

Simple familiar

1. **WE1** An energy-saving store stocks shower water savers and energy-saving light globes. In one month they sold 45 shower water savers and 30 energy-saving light globes. Construct a column matrix to represent the number of shower water savers and energy-saving light globes sold during this month, and state the order of the matrix.

2. State the order of the following matrices.

   a. $\begin{bmatrix} 7 & 3 \\ 8 & 5 \end{bmatrix}$

   b. $\begin{bmatrix} 3 \\ 1 \end{bmatrix}$

   c. $[2 \quad -5 \quad 8 \quad 7]$

   d. $\begin{bmatrix} 7 & 4 & 9 \\ -5 & 2 & 4 \end{bmatrix}$

   e. $\begin{bmatrix} 9 & 2 & 3 \\ 4 & 5 & 6 \\ 8 & 1 & 9 \end{bmatrix}$

   f. $\begin{bmatrix} 4 & 1 \\ 6 & 4 \\ 8 & 3 \\ 7 & 7 \end{bmatrix}$

3. **WE2** For matrix Y, determine the following elements.

$$Y = \begin{bmatrix} 8 & 7 & 5 & 4 \\ 1 & 3 & 9 & 8 \\ 1 & 6 & 2 & 4 \\ 2 & 3 & 0 & 7 \end{bmatrix}$$

   a. y_{13} b. y_{24} c. y_{31} d. y_{43} e. y_{41} f. y_{12}

4. **WE3** Cheap Auto sells three types of vehicles: cars, vans and motorbikes. They have two outlets at Redwood and Newtown. The number of vehicles in stock at each of the two outlets is shown in the table.

	Cars	Vans	Motorbikes
Redwood	18	12	8
Newtown	13	10	11

Construct a matrix to represent this information.

5. Newton and Isaac played a match of tennis. Newton won the match in five sets with a final score of 6–2, 4–6, 7–6, 3–6, 6–4. Construct a matrix of order 2×5 to represent this information.

6. Determine the following elements in the matrices shown.

   $A = \begin{bmatrix} 8 & 1 \\ 6 & 9 \end{bmatrix}$ $B = \begin{bmatrix} 1 & 5 \\ 1 & 6 \end{bmatrix}$ $C = \begin{bmatrix} 6 & 3 \\ 7 & 5 \end{bmatrix}$

   a. b_{12} b. c_{11} c. a_{21} d. a_{11} e. c_{12} f. b_{22}

7. Consider the matrix $E = \begin{bmatrix} \dfrac{2}{3} & 0 & \dfrac{1}{4} \\ -1 & -\dfrac{1}{2} & -3 \end{bmatrix}$.

   a. Explain why the element e_{24} does not exist.
   b. State which element has a value of -3.
   c. Nadia was asked to write down the value of element e_{12} and wrote -1. Explain Nadia's mistake and state the correct value of element e_{12}.

8. The elements of matrix H are shown.

   $h_{12} = 3 \qquad h_{11} = 4 \qquad h_{21} = -1$
   $h_{31} = -4 \qquad h_{32} = 6 \qquad h_{22} = 7$

   a. State the order of matrix H.
   b. Construct matrix H.

Complex familiar

9. WE4 Tickets were sold for a series of concerts by popular band The Nightingales. The tickets were separated into Section A and Section B tickets, sold at different prices. The concerts would be playing over three days: Thursday, Friday and Saturday. The ticket sales are given in matrix K.

$$K = \begin{array}{c} \\ \text{Thursday} \\ \text{Friday} \\ \text{Saturday} \end{array} \begin{array}{c} \begin{array}{cc} A & B \end{array} \\ \begin{bmatrix} 300 & 600 \\ 350 & 800 \\ 400 & 700 \end{bmatrix} \end{array}$$

   a. State the order of matrix K.
   b. State the information shown by the element k_{22}.
   c. Determine which element gives the number of Section A tickets sold on Saturday.
   d. Calculate the total number of Section B tickets sold over Thursday, Friday and Saturday.

10. Happy Greens Golf Club held a three-day competition from Friday to Sunday. Participants were grouped into three different categories: experienced, beginner and club member. The table shows the total entries for each type of participant on each of the days of the competition.

Category	Friday	Saturday	Sunday
Experienced	19	23	30
Beginner	12	17	18
Club member	25	33	36

   a. State the number of participants in the competition on Friday.
   b. Calculate the total number of entries for the three-day competition.
   c. Construct a row matrix to represent the number of beginners participating in the competition on each of the three days.

11. The land area and population of each Australian state and territory were recorded in June 2023 and summarised in the table.

State/territory	Land area (km^2)	Population (millions)
Australian Capital Territory	2 358	0.5
Queensland	1 727 200	5.4
New South Wales	801 428	8.3
Northern Territory	1 346 200	0.3
South Australia	984 000	1.9
Western Australia	2 529 875	2.9
Tasmania	68 330	0.6
Victoria	227 600	6.8

a. Construct an 8 × 1 matrix that displays the population, in millions, of each state and territory in the order shown in the table.

b. Construct a row matrix that represents the land area of each of the states in ascending order.

c. Town planners place the information on land area, in km^2, and population, in millions, for New South Wales, Victoria and Queensland respectively in a matrix.

    i. State the order of this matrix.   ii. Construct this matrix.

12. The estimated number of Aboriginal and Torres Strait Islander persons living in each state and territory in Australia, recorded during the 2021 census, is shown in the table.

State and territory	Number of Aboriginal and Torres Strait Islander persons	% of population that are Aboriginal or Torres Strait Islander persons
New South Wales	339 710	4.2
Victoria	78 696	1.2
Queensland	273 119	5.2
South Australia	52 039	2.9
Western Australia	120 006	4.4
Tasmania	33 857	6.0
Northern Territory	76 487	30.8
Australian Capital Territory	9 525	2.1

a. Construct an 8 × 2 matrix to represent this information.

b. Determine the total number of Aboriginal and Torres Strait Islander persons living in the following states and territories in 2021.

    i. Northern Territory

   ii. Tasmania

  iii. Queensland, New South Wales and Victoria (combined)

c. Determine the total number of Aboriginal and Torres Strait Islander persons who were estimated to be living in Australia in 2021.

Fully worked solutions for this chapter are available online.

LESSON
5.3 Addition, subtraction and scalar multiplication of matrices

SYLLABUS LINKS

- Define and use addition and subtraction of matrices and scalar multiplication.

Source: Adapted from Specialist Mathematics Senior Syllabus 2024 © State of Queensland (QCAA) 2024; licensed under CC BY 4.0

5.3.1 Addition and subtraction of matrices

Only two matrices of the same order can be added or subtracted. To add or subtract two matrices, we add or subtract the elements in the corresponding positions. For example, if $P = \begin{bmatrix} 2 & 3 \\ -1 & 5 \end{bmatrix}$ and $Q = \begin{bmatrix} 4 & -2 \\ 6 & 3 \end{bmatrix}$, then:

$$P + Q = \begin{bmatrix} 2 & 3 \\ -1 & 5 \end{bmatrix} + \begin{bmatrix} 4 & -2 \\ 6 & 3 \end{bmatrix} = \begin{bmatrix} 2+4 & 3-2 \\ -1+6 & 5+3 \end{bmatrix} = \begin{bmatrix} 6 & 1 \\ 5 & 8 \end{bmatrix}$$

$$P - Q = P + (-Q) = \begin{bmatrix} 2 & 3 \\ -1 & 5 \end{bmatrix} - \begin{bmatrix} 4 & -2 \\ 6 & 3 \end{bmatrix} = \begin{bmatrix} 2-4 & 3+2 \\ -1-6 & 5-3 \end{bmatrix} = \begin{bmatrix} -2 & 5 \\ -7 & 2 \end{bmatrix}$$

Matrices can be added or subtracted only if they are of the same order.

To add matrices, simply add the elements in the corresponding positions. To subtract matrices, simply subtract the elements in the corresponding positions.

Addition and subtraction of matrices

For example, if $A = \begin{bmatrix} a_{11} & a_{12} & \cdots & a_{1n} \\ a_{21} & a_{22} & \cdots & a_{2n} \\ \vdots & \vdots & \ddots & \vdots \\ a_{m1} & a_{m2} & \cdots & a_{mn} \end{bmatrix}$ and $B = \begin{bmatrix} b_{11} & b_{12} & \cdots & b_{1n} \\ b_{21} & b_{22} & \cdots & b_{2n} \\ \vdots & \vdots & \ddots & \vdots \\ b_{m1} & b_{m2} & \cdots & b_{mn} \end{bmatrix}$, then:

$$A + B = \begin{bmatrix} a_{11}+b_{11} & a_{12}+b_{12} & \cdots & a_{1n}+b_{1n} \\ a_{21}+b_{21} & a_{22}+b_{22} & \cdots & a_{2n}+b_{2n} \\ \vdots & \vdots & \ddots & \vdots \\ a_{m1}+b_{m1} & a_{m2}+b_{m2} & \cdots & a_{mn}+b_{mn} \end{bmatrix}$$

and

$$A - B = \begin{bmatrix} a_{11}-b_{11} & a_{12}-b_{12} & \cdots & a_{1n}-b_{1n} \\ a_{21}-b_{21} & a_{22}-b_{22} & \cdots & a_{2n}-b_{2n} \\ \vdots & \vdots & \ddots & \vdots \\ a_{m1}-b_{m1} & a_{m2}-b_{m2} & \cdots & a_{mn}-b_{mn} \end{bmatrix}$$

Note that none of the matrices defined by

$$A = \begin{bmatrix} 3 & 5 \\ 4 & 7 \end{bmatrix}, B = \begin{bmatrix} 2 & -5 & -3 \\ -4 & 2 & -5 \\ 1 & 3 & 4 \end{bmatrix}, C = \begin{bmatrix} 1 \\ -2 \\ 3 \end{bmatrix}, D = \begin{bmatrix} 3 & -2 \end{bmatrix}, E = \begin{bmatrix} 3 & 5 \\ -4 & 2 \\ -1 & 3 \end{bmatrix}, F = \begin{bmatrix} 2 & 3 & 4 \\ 4 & -5 & -2 \end{bmatrix},$$
$$ 2 \times 2 \qquad\qquad 3 \times 3 \qquad\qquad 3 \times 1 \qquad 1 \times 2 \qquad\qquad 3 \times 2 \qquad\qquad 2 \times 3$$

can be added or subtracted from one another, as they are all of different orders.

WORKED EXAMPLE 5 Adding row matrices

At a football match one food outlet sold 280 pies, 210 hotdogs and 310 boxes of chips. Another food outlet sold 300 pies, 220 hotdogs and 290 boxes of chips.
Represent each data set as a 1×3 matrix, and determine the total number of pies, hotdogs and chips sold by these two outlets.

THINK	WRITE
1. Use a 1×3 matrix to represent the number of pies, hotdogs and chips sold.	$\begin{bmatrix} \text{pies} & \text{hotdogs} & \text{chips} \end{bmatrix}$
2. Write the matrix for the sales from the first outlet.	$S_1 = \begin{bmatrix} 280 & 210 & 310 \end{bmatrix}$
3. Write the matrix for the sales from the second outlet.	$S_2 = \begin{bmatrix} 300 & 220 & 290 \end{bmatrix}$
4. Use the rules of addition of matrices to determine the sum of these two matrices.	$S_1 + S_2 = \begin{bmatrix} 280 + 300 & 210 + 220 & 310 + 290 \end{bmatrix}$ $ = \begin{bmatrix} 580 & 430 & 600 \end{bmatrix}$

5.3.2 Scalar multiplication of matrices

A **scalar** is an entity with magnitude only; that is, it is a real number. To multiply any matrix by a scalar, we multiply every element in the matrix by the scalar.

$$\text{If } P = \begin{bmatrix} 2 & 3 \\ -1 & 5 \end{bmatrix}$$

then:

$$2P = 2 \begin{bmatrix} 2 & 3 \\ -1 & 5 \end{bmatrix}$$
$$= \begin{bmatrix} 2 \times 2 & 2 \times 3 \\ 2 \times (-1) & 2 \times 5 \end{bmatrix}$$
$$= \begin{bmatrix} 4 & 6 \\ -2 & 10 \end{bmatrix}$$

WORKED EXAMPLE 6 Scalar multiplication of matrices

Given the matrices $A = \begin{bmatrix} -2 \\ 3 \end{bmatrix}$, $B = \begin{bmatrix} 1 & 0 & -4 \end{bmatrix}$ and $C = \begin{bmatrix} 12 & -8 \\ -16 & 24 \end{bmatrix}$, calculate the following:

a. $3A$ b. $-5B$ c. $-\dfrac{1}{4}C$

THINK

WRITE

a. Multiply every element in the matrix A by the scalar 3.

a. $3A = 3 \times \begin{bmatrix} -2 \\ 3 \end{bmatrix}$

$= \begin{bmatrix} 3 \times (-2) \\ 3 \times 3 \end{bmatrix}$

$= \begin{bmatrix} -6 \\ 9 \end{bmatrix}$

b. Multiply every element in the matrix B by the scalar -5.

b. $-5B = -5 \times \begin{bmatrix} 1 & 0 & -4 \end{bmatrix}$

$= \begin{bmatrix} -5 \times 1 & -5 \times 0 & -5 \times (-4) \end{bmatrix}$

$= \begin{bmatrix} -5 & 0 & 20 \end{bmatrix}$

c. Multiply every element in the matrix C by the scalar $-\dfrac{1}{4}$.

c. $-\dfrac{1}{4}C = -\dfrac{1}{4} \times \begin{bmatrix} 12 & -8 \\ -16 & 24 \end{bmatrix}$

$= \begin{bmatrix} -\dfrac{1}{4} \times 12 & -\dfrac{1}{4} \times (-8) \\ -\dfrac{1}{4} \times (-16) & -\dfrac{1}{4} \times 24 \end{bmatrix}$

$= \begin{bmatrix} -3 & 2 \\ 4 & -6 \end{bmatrix}$

WORKED EXAMPLE 7 Addition and scalar multiplication of matrices

Given the matrices $A = \begin{bmatrix} 3 \\ -5 \end{bmatrix}$, $B = \begin{bmatrix} -5 \\ 4 \end{bmatrix}$ and $C = \begin{bmatrix} 2 \\ y \end{bmatrix}$ determine the values of x and y if $xA + 2B = C$.

THINK

WRITE

1. Substitute for the given matrices.

$xA + 2B = C$

$x \begin{bmatrix} 3 \\ -5 \end{bmatrix} + 2 \begin{bmatrix} -5 \\ 4 \end{bmatrix} = \begin{bmatrix} 2 \\ y \end{bmatrix}$

2. Apply the rules for scalar multiplication.

$\begin{bmatrix} 3x \\ -5x \end{bmatrix} + \begin{bmatrix} -10 \\ 8 \end{bmatrix} = \begin{bmatrix} 2 \\ y \end{bmatrix}$

3. Apply the rules for addition of matrices.

$\begin{bmatrix} 3x - 10 \\ -5x + 8 \end{bmatrix} = \begin{bmatrix} 2 \\ y \end{bmatrix}$

4. Apply the rules for equality of matrices.

$3x - 10 = 2$

$-5x + 8 = y$

5. Solve the first equation for x.

$3x = 12$

$x = 4$

6. Substitute for x into the second equation and solve this equation for y.	$-5x + 8 = y$ $y = 8 - 20$ $y = -12$
7. Write the answer.	$\therefore x = 4,\ y = -12$

5.3.3 Special matrices

The zero matrix

The 2×2 **null matrix** or **zero matrix** O, with all elements equal to zero, is given by $O = \begin{bmatrix} 0 & 0 \\ 0 & 0 \end{bmatrix}$.

If $A = \begin{bmatrix} 2 & 4 \\ 5 & -3 \end{bmatrix}$ then,

$$A + O = \begin{bmatrix} 2 & 4 \\ 5 & -3 \end{bmatrix} + \begin{bmatrix} 0 & 0 \\ 0 & 0 \end{bmatrix} = A$$

Similarly, $O + A = A$.

The identity matrix

The 2×2 **identity matrix** I is defined by $I = \begin{bmatrix} 1 & 0 \\ 0 & 1 \end{bmatrix}$. This square matrix has ones on the leading diagonal and zeros on the other diagonal. An example of an identity matrix and of the null matrix is shown in Worked example 8.

WORKED EXAMPLE 8 Various operations on matrices

Given the matrices $A = \begin{bmatrix} 3 & 5 \\ 4 & 7 \end{bmatrix}$ and $B = \begin{bmatrix} -5 & -3 \\ 3 & 4 \end{bmatrix}$, determine the matrix X if:

a. $X = 2A - 3B$ b. $A + X = O$ c. $X = B + 2A - 3I$

THINK	WRITE
a. 1. Substitute for the given matrices.	a. $X = 2A - 3B$ $= 2 \begin{bmatrix} 3 & 5 \\ 4 & 7 \end{bmatrix} - 3 \begin{bmatrix} -5 & -3 \\ 3 & 4 \end{bmatrix}$
2. Apply the rules for scalar multiplication.	$= \begin{bmatrix} 6 & 10 \\ 8 & 14 \end{bmatrix} - \begin{bmatrix} -15 & -9 \\ 9 & 12 \end{bmatrix}$
3. Apply the rules for subtraction of matrices.	$= \begin{bmatrix} 21 & 19 \\ -1 & 2 \end{bmatrix}$

▶

b. 1. Transpose the equation to make X the subject.   **b.** $A + X = O$

$$X = O - A$$

2. State the final answer.

$$= \begin{bmatrix} 0 & 0 \\ 0 & 0 \end{bmatrix} - \begin{bmatrix} 3 & 5 \\ 4 & 7 \end{bmatrix}$$

$$= \begin{bmatrix} -3 & -5 \\ -4 & -7 \end{bmatrix}$$

c. 1. Substitute for the given matrices.   **c.** $X = B + 2A - 3I$

$$= \begin{bmatrix} -5 & -3 \\ 3 & 4 \end{bmatrix} + 2 \begin{bmatrix} 3 & 5 \\ 4 & 7 \end{bmatrix} - 3 \begin{bmatrix} 1 & 0 \\ 0 & 1 \end{bmatrix}$$

2. Apply the rules for scalar multiplication.

$$= \begin{bmatrix} -5 & -3 \\ 3 & 4 \end{bmatrix} + \begin{bmatrix} 6 & 10 \\ 8 & 14 \end{bmatrix} - \begin{bmatrix} 3 & 0 \\ 0 & 3 \end{bmatrix}$$

3. Apply the rules for addition and subtraction of matrices.

$$= \begin{bmatrix} -5+6-3 & -3+10-0 \\ 3+8-0 & 4+14-3 \end{bmatrix}$$

4. State the final answer.

$$= \begin{bmatrix} -2 & 7 \\ 11 & 15 \end{bmatrix}$$

TI | THINK	DISPLAY/WRITE	CASIO | THINK	DISPLAY/WRITE
a. 1. On a Calculator page, press the template button and complete the entry line as: $\begin{bmatrix} 3 & 5 \\ 4 & 7 \end{bmatrix}$ Press CTRL, then press VAR, then type a and press ENTER to store matrix A. Repeat this step to store matrix B.	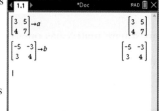	**a. 1.** On a Run-Matrix screen, select MAT/VCT by pressing F3. Select Matrix A by pressing EXE and change its dimensions to 2 by 2. Press EXE. Enter the elements of Matrix A, then press EXIT. Repeat this step to store Matrix B, then press EXIT to return to the Run-Matrix screen.	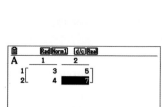
2. Complete the next entry line as $2a - 3b$ then press ENTER.	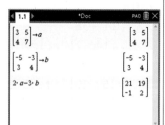	**2.** Complete the entry line as: $2 \times$ Mat A $- 3 \times$ Mat B then press EXE. *Note:* To find 'Mat', press OPTN, select MAT/VCT by pressing F2, then select Mat by pressing F1.	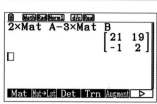
		3. The answer appears on the screen.	$X = \begin{bmatrix} 21 & 19 \\ -1 & 2 \end{bmatrix}$

Exercise 5.3 Addition, subtraction and scalar multiplication of matrices

5.3 Exercise	5.3 Exam questions

Simple familiar	Complex familiar	Complex unfamiliar
1, 2, 3, 4, 5, 6, 7, 8, 9, 10	11, 12	13, 14, 15

Simple familiar

1. **WE5** At football matches, commentators often quote player statistics. In one particular game, the top ranked player on the ground had 25 kicks, eight marks and ten handballs. The second ranked player on the same team on the ground had 20 kicks, six marks and eight handballs, while the third ranked player on the same team on the ground had 18 kicks, five marks and seven handballs.
Represent each data set as a 1×3 matrix and determine the total number of kicks, marks and handballs by these three players.

2. At the end of a doubles tennis match, one player had two aces, three double faults, 25 forehand winners and ten backhand winners, while his partner had four aces, five double faults, 28 forehand winners and seven backhand winners.
Represent this data as 2×4 matrices and determine the total number of aces, double faults, forehand and backhand winners for these players.

3. **WE6** Given the matrices $A = \begin{bmatrix} -2 & 0 & 1 & 3 \end{bmatrix}$, $B = \begin{bmatrix} 2 \\ -3 \\ 4 \\ -1 \end{bmatrix}$ and $C = \begin{bmatrix} -6 & 9 & -\frac{3}{2} \\ 0 & -1 & \frac{1}{2} \end{bmatrix}$, calculate the following:

   a. $6A$ b. $-5B$ c. $\frac{2}{3}C$

4. **WE7** Given the matrices $A = \begin{bmatrix} -3 \\ 4 \end{bmatrix}$, $B = \begin{bmatrix} 4 \\ 5 \end{bmatrix}$ and $C = \begin{bmatrix} 2 \\ y \end{bmatrix}$, determine the values of x and y if $xA + 2B = C$.

5. Given the matrices $A = \begin{bmatrix} 4 \\ -2 \\ 3 \end{bmatrix}$, $B = \begin{bmatrix} 3 \\ 5 \\ -1 \end{bmatrix}$ and $C = \begin{bmatrix} 6 \\ y \\ z \end{bmatrix}$, determine the values of x, y and z if $xA - 2B = C$.

6. **WE8** If $A = \begin{bmatrix} -2 & 4 \\ 3 & 5 \end{bmatrix}$ and $B = \begin{bmatrix} 2 & 4 \\ -1 & -3 \end{bmatrix}$, determine the matrix X given the following.

   a. $X = 3A - 2B$ b. $2A + X = O$

7. If $A = \begin{bmatrix} -1 \\ 2 \end{bmatrix}$ and $B = \begin{bmatrix} 3 \\ 5 \end{bmatrix}$, determine the matrix C given the following.

    a. $C = A + B$ **b.** $3A + 2C = 4B$

8. If $A = \begin{bmatrix} 1 & 4 \\ -3 & 2 \end{bmatrix}$, $B = \begin{bmatrix} 4 & -2 \\ 3 & 5 \end{bmatrix}$ and $O = \begin{bmatrix} 0 & 0 \\ 0 & 0 \end{bmatrix}$, determine the matrix C given the following.

    a. $3A = C - 2B$ **b.** $C + 3A - 2B = O$

9. If $A = \begin{bmatrix} x & -3 \\ 2 & x \end{bmatrix}$ and $B = \begin{bmatrix} 2 & y \\ y & -3 \end{bmatrix}$, determine the values of x and y given the following.

    a. $A + B = \begin{bmatrix} 7 & 4 \\ 9 & 2 \end{bmatrix}$ **b.** $B - A = \begin{bmatrix} -1 & 1 \\ -4 & -6 \end{bmatrix}$

10. If $D = \begin{bmatrix} 1 & 4 & 5 \\ -3 & 2 & -2 \end{bmatrix}$ and $E = \begin{bmatrix} 2 & -2 & 4 \\ 1 & 4 & -3 \end{bmatrix}$, determine the matrix C given the following.

    a. $C = D + E$ **b.** $3D + 2C = 4E$

Complex familiar

11. **a.** Given $A = \begin{bmatrix} 2 & 3 \\ -1 & 4 \end{bmatrix}$ write down the values of a_{11}, a_{12}, a_{21} and a_{22}.
    b. State the 2×2 matrix B if $b_{11} = 3$, $b_{12} = -2$, $b_{21} = -3$ and $b_{22} = 5$.

12. Given the matrices $A = \begin{bmatrix} 1 & 4 \\ -3 & 2 \end{bmatrix}$, $B = \begin{bmatrix} 4 & -2 \\ 3 & 5 \end{bmatrix}$, $O = \begin{bmatrix} 0 & 0 \\ 0 & 0 \end{bmatrix}$ and $I = \begin{bmatrix} 1 & 0 \\ 0 & 1 \end{bmatrix}$, determine the matrix C if the following apply.

    a. $3A + C - 2B + 4I = O$ **b.** $4A - C + 3B - 2I = O$

Complex unfamiliar

13. If $A = \begin{bmatrix} a & b \\ c & d \end{bmatrix}$ and $B = \begin{bmatrix} 2 & 4 \\ -1 & -3 \end{bmatrix}$, determine the values of a, b, c and d given the following.

    a. $A + 2I - 2B = O$ **b.** $3I + 4B - 2A = O$

14. **a.** Determine the 2×2 matrix A whose elements are $a_{ij} = 2i - j$ for $j \neq i$ and $a_{ij} = ij$ for $j = i$.
    b. Determine the 2×2 matrix A whose elements are $a_{ij} = i + j$ for $i < j$, $a_{ij} = i - j + 1$ for $i > j$ and $a_{ij} = i + j + 1$ for $i = j$.

15. The **trace of a matrix** A denoted by $\mathrm{tr}(A)$ is equal to the sum of leading diagonal elements. For 2×2 matrices, if $A = \begin{bmatrix} a_{11} & a_{12} \\ a_{21} & a_{22} \end{bmatrix}$ then $\mathrm{tr}(A) = a_{11} + a_{22}$. Consider the following matrices: $A = \begin{bmatrix} 2 & 3 \\ -1 & 4 \end{bmatrix}$, $B = \begin{bmatrix} 4 & -2 \\ 3 & 5 \end{bmatrix}$ and $C = \begin{bmatrix} 1 & -2 \\ 5 & 4 \end{bmatrix}$.

    a. Determine if $\mathrm{tr}(A + B + C) = \mathrm{tr}(A) + \mathrm{tr}(B) + \mathrm{tr}(C)$.
    b. Determine if $\mathrm{tr}(2A + 3B - 4C) = 2\mathrm{tr}(A) + 3\mathrm{tr}(B) - 4\mathrm{tr}(C)$.

Fully worked solutions for this chapter are available online.

LESSON
5.4 Matrix multiplication

SYLLABUS LINKS

• Define and use matrix multiplication, multiplicative identity and multiplicative inverse.

Source: Adapted from Specialist Mathematics Senior Syllabus 2024 © State of Queensland (QCAA) 2024; licensed under CC BY 4.0

5.4.1 Multiplication of matrices

At the end of an AFL football match between Sydney and Melbourne the scores were as shown.

This information is represented in a matrix as:

$$
\begin{array}{c}
\\
\text{Sydney} \\
\text{Melbourne}
\end{array}
\begin{array}{cc}
\text{Goals} & \text{Behinds} \\
\begin{bmatrix} 12 & 15 \\ 9 & 10 \end{bmatrix}
\end{array}
$$

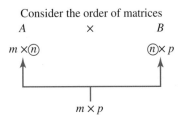

One goal in AFL football is worth 6 points and one behind is worth 1 point.

This information is represented in a matrix as:

$$
\begin{array}{c}
\text{Goals} \\
\text{Behinds}
\end{array}
\begin{bmatrix} 6 \\ 1 \end{bmatrix}
$$

It is possible for a team's total score to be calculated by matrix multiplication if matrices are selected carefully.

$$
\begin{array}{c}
\\
\text{Sydney} \\
\text{Melbourne}
\end{array}
\begin{array}{cc}
\text{Goals} & \text{Behinds} \\
\begin{bmatrix} 12 & 15 \\ 9 & 10 \end{bmatrix}
\end{array}
\times
\begin{array}{c}
\\
\text{Goals} \\
\text{Behinds}
\end{array}
\begin{bmatrix} 6 \\ 1 \end{bmatrix}
=
\begin{bmatrix} 12 \times 6 + 15 \times 1 \\ 9 \times 6 + 10 \times 1 \end{bmatrix}
=
\begin{array}{c}
\\
\text{Sydney} \\
\text{Melbourne}
\end{array}
\begin{array}{c}
\text{Points} \\
\begin{bmatrix} 87 \\ 64 \end{bmatrix}
\end{array}
$$

The matrix multiplication is written without the rows and columns headings:

Multiplying matrices in general

Two matrices A and B may be multiplied together to form the product AB if the number of columns in the first matrix A, is equal to the number of rows in the second matrix B. If matrix A is of order $m \times n$ and matrix B is of order $n \times p$, then the product AB will be of order $m \times p$. The number of columns in the first matrix must be equal to the number of rows in the second matrix otherwise matrix multiplication is not defined.

Consider the order of matrices

$$
\begin{array}{ccc}
A & \times & B \\
m \times \text{\textcircled{n}} & & \text{\textcircled{n}} \times p
\end{array}
$$

$$ m \times p $$

Columns in matrix A must equal number of rows in matrix B for matrix multiplication to be possible.

Therefore, if the two inside numbers are the same, matrix multiplication is defined and the order of the product AB will be $m \times p$ (the outside numbers).

The elements of the product matrix is obtained by adding the product of each element in each row of the first matrix and the corresponding elements of each column in the second matrix.

For an example of how this works in practice we will calculate the following matrix product:

$$\begin{bmatrix} 1 & 2 & -1 \\ 0 & 1 & 3 \end{bmatrix} \times \begin{bmatrix} 3 & 4 \\ 2 & 0 \\ 0 & -1 \end{bmatrix}.$$

The first matrix is a 2×3 matrix, and the second matrix is a 3×2 matrix, thus the product is a 2×2 matrix. To help you multiply these two matrices, start by writing an empty 2×2 matrix.

Place your finger on element 1, 1 and think row 1, column 1. Multiply the first row by the first column. $(1 \times 3) + (2 \times 2) + (-1 \times 0) = 7$ The value in row 1, column 1 is therefore 7.	$\begin{bmatrix} 1 & 2 & -1 \\ 0 & 1 & 3 \end{bmatrix} \times \begin{bmatrix} 3 & 4 \\ 2 & 0 \\ 0 & -1 \end{bmatrix} = \begin{bmatrix} 7 & \square \\ \square & \square \end{bmatrix}$
Place your finger on element 1, 2 and think row 1, column 2. Multiply the first row by the second column. $(1 \times 4) + (2 \times 0) + (-1 \times -1) = 5$ The value in row 1, column 2 is therefore 5.	$\begin{bmatrix} 1 & 2 & -1 \\ 0 & 1 & 3 \end{bmatrix} \times \begin{bmatrix} 3 & 4 \\ 2 & 0 \\ 0 & -1 \end{bmatrix} = \begin{bmatrix} 7 & 5 \\ \square & \square \end{bmatrix}$
Place your finger on element 2, 1 and think row 2, column 1. Multiply the second row by the first column. $(0 \times 3) + (1 \times 2) + (3 \times 0) = 2$ The value in row 2, column 1 is therefore 2.	$\begin{bmatrix} 1 & 2 & -1 \\ 0 & 1 & 3 \end{bmatrix} \times \begin{bmatrix} 3 & 4 \\ 2 & 0 \\ 0 & -1 \end{bmatrix} = \begin{bmatrix} 7 & 5 \\ 2 & \square \end{bmatrix}$
Place your finger on element 2, 2 and think row 2, column 2. Multiply the second row by the second column. $(0 \times 4) + (1 \times 0) + (3 \times -1) = -3$ The value in row 2, column 2 is therefore -3.	$\begin{bmatrix} 1 & 2 & -1 \\ 0 & 1 & 3 \end{bmatrix} \times \begin{bmatrix} 3 & 4 \\ 2 & 0 \\ 0 & -1 \end{bmatrix} = \begin{bmatrix} 7 & 5 \\ 2 & -3 \end{bmatrix}$

Matrix multiplication

The product AB of two matrices A and B is defined (possible) if the number of columns in A is equal to the number of rows in B.

For example, if $A = \begin{bmatrix} a_{11} & a_{12} & \cdots & a_{1n} \\ a_{21} & a_{22} & \cdots & a_{2n} \\ \vdots & \vdots & \ddots & \vdots \\ a_{m1} & a_{m2} & \cdots & a_{mn} \end{bmatrix}$ and $B = \begin{bmatrix} b_{11} & b_{12} & \cdots & b_{1p} \\ b_{21} & b_{22} & \cdots & b_{2p} \\ \vdots & \vdots & \ddots & \vdots \\ b_{n1} & b_{n2} & \cdots & b_{np} \end{bmatrix}$, then:

$$AB = \begin{bmatrix} c_{11} & c_{12} & \cdots & c_{1p} \\ c_{21} & c_{22} & \cdots & c_{2p} \\ \vdots & \vdots & \ddots & \vdots \\ c_{m1} & c_{m2} & \cdots & c_{mp} \end{bmatrix}, \text{ where } c_{ij} = a_{i1} \times b_{1j} + a_{i2} \times b_{2j} + a_{i3} \times b_{3j} + \cdots + a_{in} \times b_{nj}$$

Consider the following matrices:

$$A = \begin{bmatrix} 1 & 3 \\ -1 & 5 \\ 4 & 2 \end{bmatrix}, B = \begin{bmatrix} 2 \\ 0 \\ -2 \end{bmatrix}, C = \begin{bmatrix} 3 & 5 \\ -1 & 6 \end{bmatrix}, D = \begin{bmatrix} 1 & 7 & 4 \end{bmatrix}$$

a. Determine if matrix multiplication is defined for each of the products given below.

  i. *AB*   ii. *AC*   iii. *DB*   iv. *BD*

b. State the order of each product in part a if the matrix multiplication is defined.

THINK	WRITE
a. 1. State the order of each of the matrices.	a. *A*: 3×2 matrix *B*: 3×1 matrix *C*: 2×2 matrix *D*: 1×3 matrix

2. Determine if the product is defined by checking that the number of columns of the first matrix is the same as the number of rows of the second matrix. *Note:* Check that the inside numbers are the same.	i. *AB*: *A* *B*
	       3×2 and 3×1
	   No, *AB* is not defined.
	
	 ii. *AC*: *A* *C*
	       3×2 and 2×2
	   Yes, *AC* is defined.
	
	 iii. *DB*: *D* *B*
	       1×3 and 3×1
	   Yes, *DB* is defined.
	
	 iv. *BD*: *B* *D*
	       3×1 and 1×3
	   Yes, *BD* is defined.

b. If matrix multiplication is defined, the outside numbers give the order of the product.	b. i. Not defined
	  ii. 3×2
	  iii. 1×1
	  iv. 3×3

Given the matrices $A = \begin{bmatrix} 3 & 5 \\ 4 & 7 \end{bmatrix}$ and $B = \begin{bmatrix} -5 & -3 \\ 3 & 4 \end{bmatrix}$, determine the following matrices.

a. *AB*   b. *BA*   c. B^2

THINK	WRITE
a. 1. Substitute for the given matrices.	a. $AB = \begin{bmatrix} 3 & 5 \\ 4 & 7 \end{bmatrix} \begin{bmatrix} -5 & -3 \\ 3 & 4 \end{bmatrix}$
2. Check that matrix multiplication is defined. Since *A* and *B* are both 2×2 matrices, the product *AB* will also be a 2×2 matrix.	

3. Recall the rules for matrix multiplication and apply.

$$= \begin{bmatrix} 3 \times (-5) + 5 \times 3 & 3 \times (-3) + 5 \times 4 \\ 4 \times (-5) + 7 \times 3 & 4 \times (-3) + 7 \times 4 \end{bmatrix}$$

4. Simplify and give the final result.

$$AB = \begin{bmatrix} 0 & 11 \\ 1 & 16 \end{bmatrix}$$

b. 1. Substitute for the given matrices.

$$\textbf{b. } BA = \begin{bmatrix} -5 & -3 \\ 3 & 4 \end{bmatrix} \begin{bmatrix} 3 & 5 \\ 4 & 7 \end{bmatrix}$$

2. Check that matrix multiplication is defined. Since both A and B are 2×2 matrices, the product BA will also be a 2×2 matrix.

3. Recall the rules for matrix multiplication and apply.

$$= \begin{bmatrix} (-5) \times 3 + (-3) \times 4 & (-5) \times 5 + (-3) \times 7 \\ 3 \times 3 + 4 \times 4 & 3 \times 5 + 4 \times 7 \end{bmatrix}$$

4. Simplify and give the final result.

$$BA = \begin{bmatrix} -27 & -46 \\ 25 & 43 \end{bmatrix}$$

c. 1. $B^2 = B \times B$. Substitute for the given matrix.

$$\textbf{c. } B^2 = \begin{bmatrix} -5 & -3 \\ 3 & 4 \end{bmatrix}^2 = \begin{bmatrix} -5 & -3 \\ 3 & 4 \end{bmatrix} \begin{bmatrix} -5 & -3 \\ 3 & 4 \end{bmatrix}$$

2. Since B is a 2×2 matrix, B^2 will also be a 2×2 matrix. Apply the rules for matrix multiplication.
Note that to calculate the elements of B^2, we **do not** square the elements within B.

$$= \begin{bmatrix} (-5) \times (-5) + (-3) \times 3 & (-5) \times (-3) + -3 \times 4 \\ 3 \times (-5) + 4 \times 3 & 3 \times (-3) + 4 \times 4 \end{bmatrix}$$

3. Simplify and give the final result.

$$B^2 = \begin{bmatrix} 16 & 3 \\ -3 & 7 \end{bmatrix}$$

TI	THINK	DISPLAY/WRITE	CASIO	THINK	DISPLAY/WRITE

c. 1. On a Calculator page, press the template button and complete the entry line as:
$$\begin{bmatrix} -5 & -3 \\ 3 & 4 \end{bmatrix}$$
Press CTRL, then press VAR, then type b and press ENTER to store matrix B.

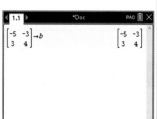

c. 1. On a Run-Matrix screen, select MAT/VCT by pressing F3.
Select Matrix B by pressing EXE and change its dimensions to 2 by 2. Press EXE.
Enter the elements of Matrix B, then press EXIT.
Press EXIT again to return to the Run-Matrix screen.

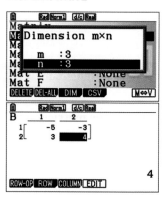

2. Complete the next entry line as b^2 then press ENTER.

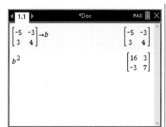

2. Complete the entry line as: Mat B^2 then press EXE.
Note: To find 'Mat', press OPTN, select MAT/VCT by pressing F2, then select Mat by pressing F1.

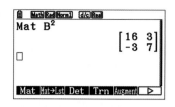

3. The answer appears on the screen.

$$B^2 = \begin{bmatrix} 16 & 3 \\ -3 & 7 \end{bmatrix}$$

Worked example 10 shows that matrix multiplication in general is not commutative: $AB \neq BA$, although there are exceptions (see section 5.6.3). It is also possible that one product is defined and the other is not defined, and that the products may have different orders.

Note that squaring a matrix (when defined) is not the square of each individual element.

WORKED EXAMPLE 11 Multiplication of 3×2 and 2×3 matrices

Given the matrices $E = \begin{bmatrix} 3 & 5 \\ -4 & 2 \\ -1 & 3 \end{bmatrix}$ and $F = \begin{bmatrix} 2 & 3 & 4 \\ 4 & -5 & -2 \end{bmatrix}$, determine the following matrices.

a. *EF*
b. *FE*

THINK	WRITE
a. 1. Substitute for the given matrices.	a. $EF = \begin{bmatrix} 3 & 5 \\ -4 & 2 \\ -1 & 3 \end{bmatrix} \begin{bmatrix} 2 & 3 & 4 \\ 4 & -5 & -2 \end{bmatrix}$
2. Check that matrix multiplication is defined. Since E is a 3×2 matrix and F is a 2×3 matrix, the product EF will be a 3×3 matrix.	
3. Recall the rules for matrix multiplication and apply.	$= \begin{bmatrix} 3 \times 2 + 5 \times 4 & 3 \times 3 + 5 \times (-5) & 3 \times 4 + 5 \times (-2) \\ (-4) \times 2 + 2 \times 4 & (-4) \times 3 + 2 \times (-5) & (-4) \times 4 + 2 \times (-2) \\ (-1) \times 2 + 3 \times 4 & (-1) \times 3 + 3 \times (-5) & (-1) \times 4 + 3 \times (-2) \end{bmatrix}$
4. Simplify and give the final result.	$EF = \begin{bmatrix} 26 & -16 & 2 \\ 0 & -22 & -20 \\ 10 & -18 & -10 \end{bmatrix}$

b. 1. Substitute for the given matrices.

b. $FE = \begin{bmatrix} 2 & 3 & 4 \\ 4 & -5 & -2 \end{bmatrix} \begin{bmatrix} 3 & 5 \\ -4 & 2 \\ -1 & 3 \end{bmatrix}$

2. Check that matrix multiplication is defined. Since F is a 2×3 matrix and E is a 3×2 matrix, the product EF will be a 2×2 matrix.

3. Recall the rules for matrix multiplication and apply.

$= \begin{bmatrix} 2 \times 3 + 3 \times (-4) + 4 \times (-1) & 2 \times 5 + 3 \times 2 + 4 \times 3 \\ 4 \times 3 + (-5) \times (-4) + (-2) \times (-1) & 4 \times 5 + (-5) \times 2 + (-2) \times 3 \end{bmatrix}$

4. Simplify and give the final result.

$FE = \begin{bmatrix} -10 & 28 \\ 34 & 4 \end{bmatrix}$

Exercise 5.4 Matrix multiplication

learn on

5.4 Exercise	5.4 Exam questions on

Simple familiar	Complex familiar	Complex unfamiliar
1, 2, 3, 4, 5, 6, 7, 8, 9, 10, 11, 12	13, 14, 15, 16, 17	18, 19, 20

Simple familiar

1. State the order of each of the following matrices.

$$A = \begin{bmatrix} 2 & 5 \\ -1 & 3 \\ 7 & -2 \end{bmatrix}, B = \begin{bmatrix} 1 \\ 5 \\ 6 \end{bmatrix}, C = \begin{bmatrix} 3 & 1 \\ -1 & 6 \end{bmatrix}, D = \begin{bmatrix} 8 & -1 & 1 \end{bmatrix}$$

2.  **WE9** Given the following matrices:

$$A = \begin{bmatrix} 2 & 5 \\ -1 & 3 \\ 7 & -2 \end{bmatrix}, B = \begin{bmatrix} 1 \\ 5 \\ 6 \end{bmatrix}, C = \begin{bmatrix} 3 & 1 \\ -1 & 6 \end{bmatrix}, D = \begin{bmatrix} 8 & -1 & 1 \end{bmatrix}$$

a. Determine if matrix multiplication is defined for each of the products given below.

   i. AB **ii.** BD **iii.** DA **iv.** CD **v.** DB **vi.** AC

b. State the order of each product in part **a** if the matrix multiplication is defined.

3. **WE10** Given the matrices $A = \begin{bmatrix} -2 & 4 \\ 3 & 5 \end{bmatrix}$ and $B = \begin{bmatrix} 2 & 4 \\ -1 & -3 \end{bmatrix}$, determine the following matrices.

   a. AB **b.** BA **c.** A^2 **d.** B^2

4. Given the matrices $A = \begin{bmatrix} a & b \\ c & d \end{bmatrix}$ and $X = \begin{bmatrix} x \\ y \end{bmatrix}$, determine the following matrices.

   a. AX

   b. XA

5. **WE11** Given the matrices $D = \begin{bmatrix} 2 & -1 \\ -3 & 5 \\ -1 & -4 \end{bmatrix}$ and $E = \begin{bmatrix} 1 & 2 & -3 \\ 2 & -4 & 5 \end{bmatrix}$, determine the following matrices.

   a. DE
   b. ED

6. Given the matrices $C = \begin{bmatrix} 1 \\ -2 \end{bmatrix}$ and $D = \begin{bmatrix} 3 & -2 \end{bmatrix}$, determine the following matrices.

   a. CD
   b. DC

7. If $A = \begin{bmatrix} x & -3 \\ 2 & x \end{bmatrix}$ and $B = \begin{bmatrix} 2 & x \\ x & -3 \end{bmatrix}$, calculate the value of x given the following.

   a. $AB = \begin{bmatrix} 3 & 18 \\ 13 & 3 \end{bmatrix}$

   b. $BA = \begin{bmatrix} -16 & 10 \\ 10 & 24 \end{bmatrix}$

   c. $A^2 = \begin{bmatrix} -2 & 12 \\ -8 & -2 \end{bmatrix}$

   d. $B^2 = \begin{bmatrix} 8 & -2 \\ -2 & 13 \end{bmatrix}$

The following information relates to questions 8 and 9.

Consider the following matrices: $A = \begin{bmatrix} -1 \\ 2 \end{bmatrix}$, $B = \begin{bmatrix} 3 & -5 \end{bmatrix}$ and $C = \begin{bmatrix} 2 & 4 \\ -3 & 5 \end{bmatrix}$.

8. Calculate the following.

   a. $A + B$
   d. AB

   b. $A + C$
   e. BA

   c. $B + C$
   f. AC

9. Calculate the following.

   a. CA
   d. ABC

   b. BC
   e. CBA

   c. CB
   f. CAB

10. Given $D = \begin{bmatrix} 1 & -4 & 2 \\ -2 & 8 & -4 \end{bmatrix}$ and $E = \begin{bmatrix} 6 & 2 \\ 3 & -1 \\ 3 & -3 \end{bmatrix}$, determine the following matrices.

   a. DE
   b. ED
   c. $E + D$
   d. D^2

11. If $A = \begin{bmatrix} 2 & 3 \\ -1 & 4 \end{bmatrix}$ and $I = \begin{bmatrix} 1 & 0 \\ 0 & 1 \end{bmatrix}$, evaluate the matrix $A^2 - 6A + 11I$.

12. The final medal tally for the top four countries at the 2020 Tokyo Olympic Games is shown in the table below.

Country	Gold	Silver	Bronze
United States of America	39	41	33
People's Republic of China	38	32	18
Japan	27	14	17
Great Britain	22	21	22

In calculating the winning country, 3 points are usually awarded for a gold medal, 2 for a silver one and 1 for a bronze one.

Determine which country won the Olympic games, and with how many points.

Complex familiar

13. We know that scalar multiplication is commutative. In particular, If k is a scalar, $k \times 0 = 0 \times k$ and $k \times 1 = 1 \times k$. Use the matrix $A = \begin{bmatrix} -2 & 4 \\ 3 & 5 \end{bmatrix}$ as an example to determine whether matrix multiplication can be commutative with special matrices.

14. a. If $P = \begin{bmatrix} -1 & 0 \\ 0 & 4 \end{bmatrix}$, determine the matrices P^2, P^3 and P^4, and deduce the matrix P^n.

b. If $Q = \begin{bmatrix} 2 & 0 \\ 0 & -3 \end{bmatrix}$, determine the matrices Q^2, Q^3 and Q^4, and deduce the matrix Q^n.

15. a. If $R = \begin{bmatrix} 1 & 0 \\ 3 & 1 \end{bmatrix}$, determine the matrices R^2, R^3 and R^4, and deduce the matrix R^n.

b. If $S = \begin{bmatrix} 0 & 3 \\ 2 & 0 \end{bmatrix}$, determine the matrices S^2, S^3 and S^4, and deduce the matrices S^8 and S^9.

16. a. If $B = \begin{bmatrix} 4 & 5 \\ -2 & -3 \end{bmatrix}$ and $I = \begin{bmatrix} 1 & 0 \\ 0 & 1 \end{bmatrix}$, evaluate the matrix $B^2 - B - 22I$.

b. If $C = \begin{bmatrix} 1 & -2 \\ 5 & 4 \end{bmatrix}$ and $I = \begin{bmatrix} 1 & 0 \\ 0 & 1 \end{bmatrix}$, evaluate the matrix $C^2 - 5C + 14I$.

17. If $D = \begin{bmatrix} d & -4 \\ -2 & 8 \end{bmatrix}$, evaluate the matrix $D^2 - 9D$.

Complex unfamiliar

18. The trace of a matrix A, denoted by $\text{tr}(A)$, is equal to the sum of leading diagonal elements. For 2×2 matrices, if $A = \begin{bmatrix} a_{11} & a_{12} \\ a_{21} & a_{22} \end{bmatrix}$, then $\text{tr}(A) = a_{11} + a_{22}$.

Consider the matrices $A = \begin{bmatrix} 2 & 3 \\ -1 & 4 \end{bmatrix}$ $B = \begin{bmatrix} 4 & -2 \\ 3 & 5 \end{bmatrix}$ and $C = \begin{bmatrix} 1 & -2 \\ 5 & 4 \end{bmatrix}$.

Determine if $\text{tr}(ABC)$ is equal to $\text{tr}(A)\,\text{tr}(B)\,\text{tr}(C)$.

19. Consider the matrices $A = \begin{bmatrix} 4 & 1 & 3 \\ 0 & -2 & k \end{bmatrix}$ and $B = \begin{bmatrix} 1 & 0 \\ 0 & -1 \\ -1 & k \end{bmatrix}$, where k is a scalar.

Determine k if $AB = \begin{bmatrix} 1 & 5 \\ -2 & 6 \end{bmatrix}$.

20. Ally loves to crochet and decides to sell some of their creations at a design and craft market.

They have 80 small keychains, 30 beanies and 25 scarves in stock. The cost of materials to make each item is as follows:

$1.50 per keychain
$7.50 per beanie
$12.25 per scarf.

Ally wants to sell their keychains for $2.00 each and thinks that they can sell the scarves for twice as much as the beanies. Determine the minimal price of a beanie for Ally to be able to turn a profit of at least $800 if they sell all their stock. Give your answer to the nearest $0.50.

Fully worked solutions for this chapter are available online.

LESSON
5.5 Determinants and inverses

SYLLABUS LINKS

- Calculate the determinant and multiplicative inverse of 2×2 matrices, with and without technology.
  - If $A = \begin{bmatrix} a & b \\ c & d \end{bmatrix}$ then $\det(A) = ad - bc$
  - $A^{-1} = \begin{bmatrix} a & b \\ c & d \end{bmatrix}^{-1} = \dfrac{1}{\det(A)} \begin{bmatrix} d & -b \\ -c & a \end{bmatrix}$, $\det(A) \neq 0$

Source: Specialist Mathematics Senior Syllabus 2024 © State of Queensland (QCAA) 2024; licensed under CC BY 4.0

5.5.1 Determinant of a 2×2 matrix

Associated with a square matrix is a single number called the **determinant of a matrix**. For a matrix A the determinant of the matrix A is denoted by $\det(A)$. Another common notation for the determinant uses straight lines around a matrix instead of square brackets:

if $A = \begin{bmatrix} a & b \\ c & d \end{bmatrix}$, $\det(A) = |A| = \begin{vmatrix} a & b \\ c & d \end{vmatrix}$.

Note: The notation using straight lines around a matrix for the determinant is not to be confused with the notation for absolute value or magnitude, for scalar and vectors. To avoid confusion, this title only uses the notation from the syllabus.

To evaluate the determinant, multiply the elements in the leading diagonal and subtract the product of the elements in the other diagonal.

Determinant of a 2×2 matrix

The determinant of a 2×2 matrix $A = \begin{bmatrix} a & b \\ c & d \end{bmatrix}$ is defined as follows:

$$\det(A) = ad - bc$$

WORKED EXAMPLE 12 Calculating the determinant of a 2×2 matrix

Calculate the determinant of the matrix $F = \begin{bmatrix} 3 & 5 \\ 4 & 7 \end{bmatrix}$.

THINK	WRITE
1. Apply the definition $\det(F) = ad - bc$ by multiplying the elements in the leading diagonal. Subtract the product of the elements in the other diagonal.	$F = \begin{bmatrix} 3 & 5 \\ 4 & 7 \end{bmatrix}$ $\det(F) = 3 \times 7 - 5 \times 4$ $= 21 - 20$
2. State the value of the determinant.	$\det(F) = 1$

5.5.2 Inverse of a 2×2 matrix

The identity matrix I, defined by $I = \begin{bmatrix} 1 & 0 \\ 0 & 1 \end{bmatrix}$, has the property that for a 2×2 non-zero matrix A, $AI = IA = A$.

When any square matrix is multiplied by its **multiplicative inverse**, the identity matrix I is obtained.

This is the same as multiplying 3 by its multiplicative inverse, $\dfrac{1}{3}$; the result is 1.

That is, $3 \times \dfrac{1}{3} = 1$ and $\dfrac{1}{3} \times 3 = 1$.

Thus, 3 is the multiplicative inverse of $\dfrac{1}{3}$ and $\dfrac{1}{3}$ is the multiplicative inverse of 3.

For a matrix, A, the multiplicative inverse is called the inverse matrix and is denoted by A^{-1}, and $AA^{-1} = A^{-1}A = I$.

Note that $A^{-1} \neq \dfrac{1}{A}$ as division of matrices is not defined.

Consider the products of $A = \begin{bmatrix} 3 & 5 \\ 4 & 7 \end{bmatrix}$ and $A^{-1} = \begin{bmatrix} 7 & -5 \\ -4 & 3 \end{bmatrix}$

$$AA^{-1} = \begin{bmatrix} 3 & 5 \\ 4 & 7 \end{bmatrix}\begin{bmatrix} 7 & -5 \\ -4 & 3 \end{bmatrix} = \begin{bmatrix} 3 \times 7 + 5 \times (-4) & 3 \times (-5) + 5 \times 3 \\ 4 \times 7 + 7 \times (-4) & 4 \times (-5) + 7 \times 3 \end{bmatrix} = \begin{bmatrix} 1 & 0 \\ 0 & 1 \end{bmatrix}$$

$$A^{-1}A = \begin{bmatrix} 7 & -5 \\ -4 & 3 \end{bmatrix}\begin{bmatrix} 3 & 5 \\ 4 & 7 \end{bmatrix} = \begin{bmatrix} 7 \times 3 + (-5) \times 4 & 7 \times 5 + (-5) \times 7 \\ (-4) \times 3 + 3 \times 4 & (-4) \times 5 + 3 \times 7 \end{bmatrix} = \begin{bmatrix} 1 & 0 \\ 0 & 1 \end{bmatrix}$$

Now for the matrix $A = \begin{bmatrix} 3 & 5 \\ 4 & 7 \end{bmatrix}$, $\det\left(\begin{bmatrix} 3 & 5 \\ 4 & 7 \end{bmatrix}\right) = 3 \times 7 - 5 \times 4 = 1$.

$A^{-1} = \begin{bmatrix} 7 & -5 \\ -4 & 3 \end{bmatrix}$ is obtained from the matrix A by swapping the elements on the leading diagonal, and placing a negative sign on the other two elements.

In general, the inverse of a 2×2 matrix $A = \begin{bmatrix} a & b \\ c & d \end{bmatrix}$ can be determined in three simple steps.

Step 1: Evaluate $\dfrac{1}{\det(A)} = \dfrac{1}{ad - bc}$.

Step 2: Swap a with d and multiply b and c by -1 to form the matrix $\begin{bmatrix} d & -b \\ -c & a \end{bmatrix}$.

Step 3: Multiply the results of the previous steps together to form the inverse $A^{-1} = \dfrac{1}{ad - bc} \begin{bmatrix} d & -b \\ -c & a \end{bmatrix}$

Inverse of a 2 × 2 matrix

The inverse of a 2×2 matrix $A = \begin{bmatrix} a & b \\ c & d \end{bmatrix}$ is defined as:

$$A^{-1} = \frac{1}{\det(A)} \begin{bmatrix} d & -b \\ -c & a \end{bmatrix} = \frac{1}{ad - bc} \begin{bmatrix} d & -b \\ -c & a \end{bmatrix}$$

Note: **A matrix A has an inverse only if $\det(A) \neq 0$.**

WORKED EXAMPLE 13 Determining the inverse of a 2 × 2 matrix

Consider the matrix $P = \begin{bmatrix} 2 & 3 \\ -1 & 5 \end{bmatrix}$

a. **Demonstrate that matrix P has an inverse.**
b. **Determine P^{-1}.**
c. **Verify that $PP^{-1} = P^{-1}P = I$.**

THINK

a. 1. Recall that a matrix has an inverse only if its determinant is not zero and calculate the determinant of P.
   If $P = \begin{bmatrix} a & b \\ c & d \end{bmatrix}$, then $\det(P) = ad - bc$

   2. Conclude.

b. To determine the inverse of matrix P, recall and apply the rule $P^{-1} = \dfrac{1}{\det(P)} \begin{bmatrix} d & -b \\ -c & a \end{bmatrix}$.

c. 1. Substitute and evaluate PP^{-1}.

   2. Apply the rules for scalar multiplication and multiplication of matrices.

WRITE

a. $\det(P) = \det\left(\begin{bmatrix} 2 & 3 \\ -1 & 5 \end{bmatrix}\right)$
   $= 2 \times 5 - 3 \times (-1)$
   $= 10 + 3$
   $= 13$

   $\det(P) \neq 0$ therefore matrix P has an inverse.

b. $P^{-1} = \dfrac{1}{13} \begin{bmatrix} 5 & -3 \\ 1 & 2 \end{bmatrix}$

c. $PP^{-1} = \begin{bmatrix} 2 & 3 \\ -1 & 5 \end{bmatrix} \times \dfrac{1}{13} \begin{bmatrix} 5 & -3 \\ 1 & 2 \end{bmatrix}$

   $= \dfrac{1}{13} \begin{bmatrix} 2 \times 5 + 3 \times 1 & 2 \times (-3) + 3 \times 2 \\ (-1) \times 5 + 5 \times 1 & (-1) \times (-3) + 5 \times 2 \end{bmatrix}$

3. Simplify the matrix product to show that $PP^{-1}=I$.

$$= \frac{1}{13}\begin{bmatrix} 13 & 0 \\ 0 & 13 \end{bmatrix} = \begin{bmatrix} 1 & 0 \\ 0 & 1 \end{bmatrix}$$

4. Substitute and evaluate $P^{-1}P$.

$$P^{-1}P = \frac{1}{13}\begin{bmatrix} 5 & -3 \\ 1 & 2 \end{bmatrix} \times \begin{bmatrix} 2 & 3 \\ -1 & 5 \end{bmatrix}$$

5. Use the rules for scalar multiplication and multiplication of matrices.

$$= \frac{1}{13}\begin{bmatrix} 5\times 2+(-3)\times(-1) & 5\times 3+(-3)\times 5 \\ 1\times 2+2\times(-1) & 1\times 3+2\times 5 \end{bmatrix}$$

6. Simplify the matrix product to show that $P^{-1}P=I$.

$$= \frac{1}{13}\begin{bmatrix} 13 & 0 \\ 0 & 13 \end{bmatrix} = \begin{bmatrix} 1 & 0 \\ 0 & 1 \end{bmatrix}$$

7. Write the answer.

$$PP^{-1} = P^{-1}P = \begin{bmatrix} 1 & 0 \\ 0 & 1 \end{bmatrix} = I$$

Singular matrices

If a matrix has a zero determinant then the inverse matrix does not exist, and the original matrix is termed a **singular matrix**.

$\dfrac{1}{\det(A)} = \dfrac{1}{0}$ is not defined as we cannot divide by zero.

Hence, if $\det(A) = 0$, then A^{-1} does not exist and matrix A is singular.

WORKED EXAMPLE 14 Verifying that a matrix is singular

Demonstrate that the matrix $\begin{bmatrix} -3 & 2 \\ 6 & -4 \end{bmatrix}$ **is singular.**

THINK	WRITE
1. Calculate the determinant.	$\det\left(\begin{bmatrix} -3 & 2 \\ 6 & -4 \end{bmatrix}\right)$ $= -3\times(-4) - 2\times 6$ $= 12 - 12$ $= 0$
2. Since the determinant is zero, the matrix $\begin{bmatrix} -3 & 2 \\ 6 & -4 \end{bmatrix}$ is singular.	$\det\left(\begin{bmatrix} -3 & 2 \\ 6 & -4 \end{bmatrix}\right) = 0$

WORKED EXAMPLE 15 Applications of matrices

If $A = \begin{bmatrix} -2 & 4 \\ 3 & 5 \end{bmatrix}$ **and** $I = \begin{bmatrix} 1 & 0 \\ 0 & 1 \end{bmatrix}$**, express the determinant of the matrix** $A - kI$ **in the form** $pk^2 + qk + r$**, stating the values of** p**,** q **and** r**. Hence evaluate the matrix** $pA^2 + qA + rI$**.**

THINK	WRITE
1. Substitute to calculate the matrix $A - kI$. Recall and apply the rules for scalar multiplication and subtraction of matrices.	$A - kI = \begin{bmatrix} -2 & 4 \\ 3 & 5 \end{bmatrix} - k\begin{bmatrix} 1 & 0 \\ 0 & 1 \end{bmatrix} = \begin{bmatrix} -2-k & 4 \\ 3 & 5-k \end{bmatrix}$

2. Evaluate the determinant of the matrix $A - kI$.

$$\det(A - kI) = \begin{vmatrix} -2 - k & 4 \\ 3 & 5 - k \end{vmatrix}$$

$$= (-2 - k)(5 - k) - 4 \times 3$$

3. Simplify the determinant of the matrix $A - kI$.

$$= -(2 + k)(5 - k) - 12$$

$$= -(10 + 3k - k^2) - 12$$

$$= k^2 - 3k - 22$$

4. Equate the matrices and state the values of p, q and r.

$$pk^2 + qk + r = k^2 - 3k - 22$$

$$\therefore p = 1; q = -3; r = -22$$

5. Determine the matrix A^2.

$$A^2 = \begin{bmatrix} -2 & 4 \\ 3 & 5 \end{bmatrix} \begin{bmatrix} -2 & 4 \\ 3 & 5 \end{bmatrix} = \begin{bmatrix} 16 & 12 \\ 9 & 37 \end{bmatrix}$$

6. Substitute for p, q and r and evaluate the matrix $A^2 - 3A - 22I$.

$$A^2 - 3A - 22I = \begin{bmatrix} 16 & 12 \\ 9 & 37 \end{bmatrix} - 3 \begin{bmatrix} -2 & 4 \\ 3 & 5 \end{bmatrix} - 22 \begin{bmatrix} 1 & 0 \\ 0 & 1 \end{bmatrix}$$

7. Simplify by applying the rules for scalar multiplication of matrices.

$$= \begin{bmatrix} 16 & 12 \\ 9 & 37 \end{bmatrix} - \begin{bmatrix} -6 & 12 \\ 9 & 15 \end{bmatrix} - \begin{bmatrix} 22 & 0 \\ 0 & 22 \end{bmatrix}$$

8. Simplify and apply the rules for addition and subtraction of matrices.

$$\therefore A^2 - 3A - 22I = \begin{bmatrix} 0 & 0 \\ 0 & 0 \end{bmatrix}$$

5.5.3 Determinant of a 3×3 matrix (Extension)

The determinant of any square matrix can be calculated. However, in Unit 1, it is only required that you can calculate the determinants of 2×2 matrices, with and without technology.

You will explore, in Unit 3, matrices beyond dimension 2×2 and learn how to calculate the determinant and the inverse of square matrices of any order, **with** technology.

This section lets you examine how the determinants of 3×3 matrices can be calculated **without** technology, to better understand matrices and further practice calculating the determinants of 2×2 matrices.

However, please note that this is beyond the scope of this course.

The determinant of a 3×3 matrix involves evaluating the determinants of three 2×2 matrices.

If $A = \begin{bmatrix} a & b & c \\ d & e & f \\ g & h & i \end{bmatrix}$, $\det(A) = a \times \det\left(\begin{bmatrix} e & f \\ h & i \end{bmatrix}\right) - b \times \det\left(\begin{bmatrix} d & f \\ g & i \end{bmatrix}\right) + c \times \det\left(\begin{bmatrix} d & e \\ g & h \end{bmatrix}\right)$

$$= a(ei - hf) - b(di - gf) + c(dh - ge)$$

The three sub-determinants are referred to as **minors**.

The coefficients of each sub-determinant are the elements of row 1 of the 3×3 matrix, that is, elements a, b and c.

The second coefficient, b, is given a negative sign.

Minors are formed by removing the row and column of each of these coefficients and using the remaining elements to give three 2×2 matrices.

For example,

$$\begin{bmatrix} a & b & c \\ d & e & f \\ g & h & i \end{bmatrix} \text{ becomes } a\begin{bmatrix} e & f \\ h & i \end{bmatrix} \qquad \begin{bmatrix} a & b & c \\ d & e & f \\ g & h & i \end{bmatrix} \text{ becomes } -b\begin{bmatrix} d & f \\ g & i \end{bmatrix} \qquad \begin{bmatrix} a & b & c \\ d & e & f \\ g & h & i \end{bmatrix} \text{ becomes } c\begin{bmatrix} d & e \\ g & h \end{bmatrix}$$

Determinant of a 3×3 matrix (Extension)

If $A = \begin{bmatrix} a & b & c \\ d & e & f \\ g & h & i \end{bmatrix}$, $\det(A) = a \times \det\left(\begin{bmatrix} e & f \\ h & i \end{bmatrix}\right) - b \times \det\left(\begin{bmatrix} d & f \\ g & i \end{bmatrix}\right) + c \times \det\left(\begin{bmatrix} d & e \\ g & h \end{bmatrix}\right)$

$= a(ei - hf) - b(di - gf) + c(dh - ge)$

Note: Although you may not be required to calculate the determinant of matrices of higher order than 2×2 without technology, the formula is included here to enhance your understanding. Worked example 17 demonstrates the calculation of higher-order determinants using technology.

WORKED EXAMPLE 16 Calculating the determinant of a 3×3 matrix without technology (Extension)

Evaluate $d = \det\left(\begin{bmatrix} 2 & 1 & 3 \\ 1 & -1 & 2 \\ -1 & 2 & 0 \end{bmatrix}\right)$.

THINK

1. Use elements of row 1 as the coefficients of the minors.

2. Evaluate the minors.

WRITE

$d = 2 \times \det\left(\begin{bmatrix} -1 & 2 \\ 2 & 0 \end{bmatrix}\right) - 1 \times \det\left(\begin{bmatrix} 1 & 2 \\ -1 & 0 \end{bmatrix}\right) + 3 \times \det\left(\begin{bmatrix} 1 & -1 \\ -1 & 2 \end{bmatrix}\right)$

$= 2(-1 \times 0 - 2 \times 2) - 1(1 \times 0 - 2 \times -1) + 3(1 \times 2 - (-1) \times -1)$

$= 2(0 - 4) - 1(0 + 2) + 3(2 - 1)$

$= -8 - 2 + 3$

$= -7$

WORKED EXAMPLE 17 Calculating the determinant of a square matrix of dimension 3×3 with technology (Extension)

If $A = \begin{bmatrix} 1 & -3 & 2 \\ -1 & 4 & 3 \\ 0 & 2 & 5 \end{bmatrix}$, use technology to calculate the determinant of A.

TI	THINK	DISPLAY/WRITE	CASIO	THINK	DISPLAY/WRITE
1. On a Calculator page, press the template button and complete the entry line as: $\begin{bmatrix} 1 & -3 & 2 \\ -1 & 4 & 3 \\ 0 & 2 & 5 \end{bmatrix}$ Press CTRL, then press VAR, then type a and press ENTER to store matrix A.	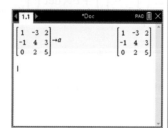	1. On a Run-Matrix screen, select MAT/VCT by pressing F3. Select Matrix A by pressing EXE and change its dimensions to 3 by 3. Press EXE. Enter the elements of Matrix A, then press EXIT. Press EXIT again to return to the Run-Matrix screen.	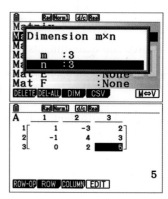

2. Press MENU then select:
   7 Matrix & Vector
   3 Determinant
   Complete the entry line as det(a) then press ENTER.

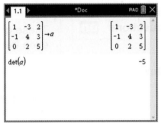

2. Complete the entry line as:
   Det Mat A then press EXE.
   Note: 'Det' and 'Mat' can be found by pressing OPTN, then selecting MAT/VCT by pressing F2.

3. The answer appears on the screen. $\det(A) = -5$

3. The answer appears on the screen. $\det(A) = -5$

5.5.4 Inverse of a 3 × 3 matrix (Extension)

Calculators can be used to determine the inverse of a 3×3 matrix. You will explore matrices of beyond 2×2 dimensions in Unit 3.

WORKED EXAMPLE 18 Determining the inverse of a 3 × 3 matrix using technology

Using technology, determine the inverse of the following matrix.

$$A = \begin{bmatrix} 3 & -1 & 3 \\ 1 & 2 & -1 \\ 5 & 3 & 4 \end{bmatrix}$$

TI | THINK

1. On a Calculator page, press the template button and complete the entry line as:

$$\begin{bmatrix} 3 & -1 & 3 \\ 1 & 2 & -1 \\ 5 & 3 & 4 \end{bmatrix}^{-1}$$

then press ENTER.

DISPLAY/WRITE

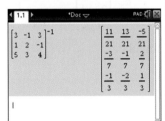

2. The answer appears on the screen.

$$A^{-1} = \begin{bmatrix} \dfrac{11}{21} & \dfrac{13}{21} & \dfrac{-5}{21} \\ \dfrac{-3}{7} & \dfrac{-1}{7} & \dfrac{3}{7} \\ \dfrac{-1}{3} & \dfrac{-2}{3} & \dfrac{1}{3} \end{bmatrix}$$

CASIO | THINK

1. On a Run-Matrix screen, select MATH by pressing F4, then select MAT/VCT by pressing F1, then select 3×3 by pressing F2. Complete the entry line as:

$$\begin{bmatrix} 3 & -1 & 3 \\ 1 & 2 & -1 \\ 5 & 3 & 4 \end{bmatrix}^{-1}$$

then press ENTER.

DISPLAY/WRITE

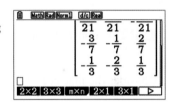

2. The answer appears on the screen.

$$A^{-1} = \begin{bmatrix} \dfrac{11}{21} & \dfrac{13}{21} & \dfrac{-5}{21} \\ \dfrac{-3}{7} & \dfrac{-1}{7} & \dfrac{3}{7} \\ \dfrac{-1}{3} & \dfrac{-2}{3} & \dfrac{1}{3} \end{bmatrix}$$

Exercise 5.5 Determinants and inverses

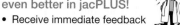

5.5 Exercise	5.5 Exam questions on

Simple familiar	Complex familiar	Complex unfamiliar
1, 2, 3, 4, 5, 6, 7, 8, 9, 10, 11	12, 13, 14, 15, 16, 17, 18, 19	20

> These questions are even better in jacPLUS!
> • Receive immediate feedback
> • Access sample responses
> • Track results and progress
>
> **Find all this and MORE in jacPLUS** ▶

Simple familiar

1. **WE12** Calculate the determinant of the matrix $G = \begin{bmatrix} -2 & 4 \\ 3 & 5 \end{bmatrix}$.

2. The matrix $\begin{bmatrix} x & 5 \\ 3 & x+2 \end{bmatrix}$ has a determinant equal to 9. Calculate the possible value(s) of x.

3. **WE16** Evaluate the determinant of following matrices.

   a. $\begin{bmatrix} 2 & 1 & 3 \\ 4 & -2 & 5 \\ -1 & 3 & 6 \end{bmatrix}$

   b. $\begin{bmatrix} -1 & -1 & 0 \\ -3 & 4 & 2 \\ 2 & 3 & 5 \end{bmatrix}$

4. **WE13** Determine the inverse of the matrix $A = \begin{bmatrix} 4 & -2 \\ 5 & 6 \end{bmatrix}$ and verify that $AA^{-1} = A^{-1}A = I$.

5. Determine the inverse matrix of each of the following matrices.

   a. $\begin{bmatrix} -1 & 0 \\ 0 & 4 \end{bmatrix}$
   b. $\begin{bmatrix} 2 & 1 \\ 0 & -3 \end{bmatrix}$
   c. $\begin{bmatrix} 2 & 0 \\ 3 & 1 \end{bmatrix}$
   d. $\begin{bmatrix} 0 & -3 \\ 2 & -1 \end{bmatrix}$

6. **WE14** Demonstrate that the matrix $\begin{bmatrix} 1 & -2 \\ -5 & 10 \end{bmatrix}$ is singular.

7. Consider the matrix $P = \begin{bmatrix} 6 & -2 \\ 4 & 2 \end{bmatrix}$.

   a. Calculate the following.
      i. $\det(P)$
      ii. P^{-1}
   b. Verify that $PP^{-1} = P^{-1}P = I$.
   c. Calculate the following.
      i. $\det\left(P^{-1}\right)$
      ii. $\det(P) \det\left(P^{-1}\right)$

8. Calculate the value(s) of x for each of the following.

   a. $\det\left(\begin{bmatrix} x & -3 \\ 4 & 2 \end{bmatrix}\right) = 6$ b. $\det\left(\begin{bmatrix} x & 3 \\ 4 & x \end{bmatrix}\right) = 4$

9. Calculate the value(s) of x if each of the following are singular matrices.

   a. $\begin{bmatrix} x & -3 \\ 4 & 2 \end{bmatrix}$
   b. $\begin{bmatrix} x & 3 \\ 4 & x \end{bmatrix}$

Questions **10** *and* **11** *refer to the following matrices*:

$$A = \begin{bmatrix} 2 & -3 \\ -1 & -4 \end{bmatrix}, B = \begin{bmatrix} 4 & 5 \\ 2 & 3 \end{bmatrix} \text{ and } C = \begin{bmatrix} 1 & -2 \\ 3 & 4 \end{bmatrix}$$

10. a. Calculate $\det(A)$, $\det(B)$ and $\det(C)$.

    b. Determine if $\det(AB) = \det(A)\ \det(B)$.

    c. Verify that $\det(ABC) = \det(A)\ \det(B)\ \det(C)$.

11. a. Calculate the matrices A^{-1}, B^{-1}, C^{-1}.

    b. Determine if $(AB)^{-1} = A^{-1}B^{-1}$.

    c. Determine if $(AB)^{-1} = B^{-1}A^{-1}$.

    d. Determine if $(ABC)^{-1} = C^{-1}B^{-1}A^{-1}$.

Complex familiar

12. Consider the matrices $B = \begin{bmatrix} -3 & 5 \\ -2 & 4 \end{bmatrix}$, $Q = \begin{bmatrix} 5 & 1 \\ 2 & 1 \end{bmatrix}$ and $I = \begin{bmatrix} 1 & 0 \\ 0 & 1 \end{bmatrix}$.

    a. Calculate the values of k for which the determinant of the matrix $B - kI = 0$.

    b. Determine the matrix $Q^{-1}BQ$.

13. The inverse of the matrix $\begin{bmatrix} 2 & 3 \\ 3 & 4 \end{bmatrix}$ is $\begin{bmatrix} p & 3 \\ 3 & q \end{bmatrix}$. Calculate the values of p and q.

14. Calculate the value(s) of x if the matrix $\begin{bmatrix} x & 4 \\ 3 & x+4 \end{bmatrix}$ is singular.

15. Given $A = \begin{bmatrix} -1 \\ 2 \end{bmatrix}$, $B = \begin{bmatrix} 3 & -5 \end{bmatrix}$ and $C = \begin{bmatrix} 2 & 4 \\ -3 & 5 \end{bmatrix}$, determine, if possible, the following matrices.

    a. $(AB)^{-1}$ **b.** A^{-1} **c.** C^{-1} **d.** $(ABC)^{-1}$

16. **WE15** If $A = \begin{bmatrix} 2 & 3 \\ -1 & 5 \end{bmatrix}$ and $I = \begin{bmatrix} 1 & 0 \\ 0 & 1 \end{bmatrix}$, express the determinant of the matrix $A - kI$ in the form $pk^2 + qk + r$, stating the values of p, q and r. Hence, evaluate the matrix $pA^2 + qA + rI$.

17. If $A = \begin{bmatrix} 2 & -3 \\ -1 & -4 \end{bmatrix}$ and $I = \begin{bmatrix} 1 & 0 \\ 0 & 1 \end{bmatrix}$, express the determinant of the matrix $A - kI$, $k \in \mathbb{R}$ in the form $pk^2 + qk + r$, stating the values of p, q and r. Hence, evaluate the matrix $pA^2 + qA + rI$.

18. **WE17** If $A = \begin{bmatrix} 4 & 6 & 8 \\ -2 & 3 & 6 \\ -3 & 2 & -1 \end{bmatrix}$, use technology to calculate the determinants of A.

19. **WE18** Use technology to determine the inverse of each of the following matrices.

    a. $A = \begin{bmatrix} 5 & -1 & 2 \\ -2 & 3 & -1 \\ 6 & 4 & -4 \end{bmatrix}$ **b.** $B = \begin{bmatrix} 12 & 8 & 0 & 4 \\ 8 & 4 & 8 & 8 \\ 4 & 12 & 8 & 12 \\ 8 & 4 & 4 & 8 \end{bmatrix}$

Complex unfamiliar

20. If $A = \begin{bmatrix} 4 & -8 \\ -3 & 2 \end{bmatrix}$ and $I = \begin{bmatrix} 1 & 0 \\ 0 & 1 \end{bmatrix}$, determine the value of k for which the determinant of the matrix $A - kI$ is equal to zero.

Fully worked solutions for this chapter are available online.

LESSON
5.6 Matrix algebra

SYLLABUS LINKS

- Use matrix algebra properties, including
  - $A + B = B + A$ (commutative law for addition)
  - $A + 0 = A$ (additive identity)
  - $A + (-A) = 0$ (additive inverse)
  - $AI = A = IA$ (multiplicative identity)
  - $AA^{-1} = I = A^{-1}A$ (multiplicative inverse)
  - $A(B + C) = AB + AC$ (left distributive law)
  - $(B + C)A = BA + CA$ (right distributive law)
- Recognise that matrix multiplication in general is not commutative.

Source: Specialist Mathematics Senior Syllabus 2024 © State of Queensland (QCAA) 2024; licensed under CC BY 4.0

We have already used some properties of matrix operations in the previous lessons. The most commonly used properties of matrix operations are summarised in this lesson and linked to known properties of operations on integers.

5.6.1 Matrix addition properties

Commutative law for matrix addition

We know that scalar addition is commutative. For example, $3 + 7 = 7 + 3$. What about matrix addition?

Let's consider two matrices of the same order: $A = \begin{bmatrix} a_{11} & a_{12} \\ a_{21} & a_{22} \end{bmatrix}, B = \begin{bmatrix} b_{11} & b_{12} \\ b_{21} & b_{22} \end{bmatrix}$, then

$$A + B = \begin{bmatrix} a_{11} & a_{12} \\ a_{21} & a_{22} \end{bmatrix} + \begin{bmatrix} b_{11} & b_{12} \\ b_{21} & b_{22} \end{bmatrix}$$

$$= \begin{bmatrix} a_{11} + b_{11} & a_{12} + b_{12} \\ a_{21} + b_{21} & a_{22} + b_{22} \end{bmatrix}$$

and

$$B + A = \begin{bmatrix} b_{11} & b_{12} \\ b_{21} & b_{22} \end{bmatrix} + \begin{bmatrix} a_{11} & a_{12} \\ a_{21} & a_{22} \end{bmatrix}$$

$$= \begin{bmatrix} b_{11} + a_{11} & b_{12} + a_{12} \\ b_{21} + a_{21} & b_{22} + a_{22} \end{bmatrix}$$

$$= \begin{bmatrix} a_{11} + b_{11} & a_{12} + b_{12} \\ a_{21} + b_{21} & a_{22} + b_{22} \end{bmatrix}$$

$$= A + B$$

Commutative law for matrix addition

If A and B are two matrices of the same order, then

$$A + B = B + A$$

Additive identity

The additive identity for a scalar is the value which, when added to a scalar, results in the original scalar. Zero is the additive identity for scalar addition. For instance, $11 + 0 = 11$.

We have seen in section 5.3.3 the null matrix O with all elements equal to zero.

For any matrix A of order $m \times n$, adding the null matrix of the same order results in the same original matrix.

For instance, with 2×2 matrices, $A = \begin{bmatrix} a_{11} & a_{12} \\ a_{21} & a_{22} \end{bmatrix}, O = \begin{bmatrix} 0 & 0 \\ 0 & 0 \end{bmatrix}$ then

$$
\begin{aligned}
A + O &= \begin{bmatrix} a_{11} & a_{12} \\ a_{21} & a_{22} \end{bmatrix} + \begin{bmatrix} 0 & 0 \\ 0 & 0 \end{bmatrix} \\
&= \begin{bmatrix} a_{11} + 0 & a_{12} + 0 \\ a_{21} + 0 & a_{22} + 0 \end{bmatrix} \\
&= \begin{bmatrix} a_{11} & a_{12} \\ a_{21} & a_{22} \end{bmatrix} \\
&= A
\end{aligned}
$$

The null matrix is the additive identity for matrix addition.

> ### Additive identity for matrix addition
>
> **For any matrix A of order $m \times n$, the null matrix of the same order is the additive identity.**
>
> $$A + 0 = A$$

Additive inverse

The **additive inverse** of a scalar k is the number that, when added to k, results in zero.

For instance, $k + (-k) = 0$, $-k$ is the additive inverse of k.

Similarly, for a matrix A of order $m \times n$, the matrix $-A$ is the additive inverse.

For example, with a 2×2 matrix:

$$
A = \begin{bmatrix} a_{11} & a_{12} \\ a_{21} & a_{22} \end{bmatrix}, \ -A = \begin{bmatrix} -a_{11} & -a_{12} \\ -a_{21} & -a_{22} \end{bmatrix}
$$

$$
\begin{aligned}
A + (-A) &= \begin{bmatrix} a_{11} & a_{12} \\ a_{21} & a_{22} \end{bmatrix} + \begin{bmatrix} -a_{11} & -a_{12} \\ -a_{21} & -a_{22} \end{bmatrix} \\
&= \begin{bmatrix} a_{11} - a_{11} & a_{12} - a_{12} \\ a_{21} - a_{21} & a_{22} - a_{22} \end{bmatrix} \\
&= \begin{bmatrix} 0 & 0 \\ 0 & 0 \end{bmatrix}
\end{aligned}
$$

> **Additive inverse for matrix addition**
>
> For any matrix A of order $m \times n$, the additive inverse of A is $-A$.
>
> $$A + (-A) = 0$$

5.6.2 Scalar multiplication properties

In section 5.3.2, we have seen that to multiply any matrix by a scalar, we multiply every element in the matrix by the scalar.

We can demonstrate that if A and B are two matrices of the same order, and k_1 is a scalar, then $k_1(A + B) = k_1 A + k_1 B$.

For instance: with 2×2 matrices,

$$A = \begin{bmatrix} a_{11} & a_{12} \\ a_{21} & a_{22} \end{bmatrix}, B = \begin{bmatrix} b_{11} & b_{12} \\ b_{21} & b_{22} \end{bmatrix}$$

$$k_1(A + B) = k_1 \begin{bmatrix} a_{11} + b_{11} & a_{12} + b_{12} \\ a_{21} + b_{21} & a_{22} + b_{22} \end{bmatrix}$$

$$= \begin{bmatrix} k_1(a_{11} + b_{11}) & k_1(a_{12} + b_{12}) \\ k_1(a_{21} + b_{21}) & k_1(a_{22} + b_{22}) \end{bmatrix}$$

and

$$k_1 A + k_1 B = k_1 \begin{bmatrix} a_{11} & a_{12} \\ a_{21} & a_{22} \end{bmatrix} + k_1 \begin{bmatrix} b_{11} & b_{12} \\ b_{21} & b_{22} \end{bmatrix}$$

$$= \begin{bmatrix} k_1 a_{11} & k_1 a_{12} \\ k_1 a_{21} & k_1 a_{22} \end{bmatrix} + \begin{bmatrix} k_1 b_{11} & k_1 b_{12} \\ k_1 b_{21} & k_1 b_{22} \end{bmatrix}$$

$$= \begin{bmatrix} k_1 a_{11} + k_1 b_{11} & k_1 a_{12} + k_1 b_{12} \\ k_1 a_{21} + k_1 b_{21} & k_1 a_{22} + k_1 b_{22} \end{bmatrix}$$

$$= \begin{bmatrix} k_1(a_{11} + b_{11}) & k_1(a_{12} + b_{12}) \\ k_1(a_{21} + b_{21}) & k_1(a_{22} + b_{22}) \end{bmatrix}$$

$$= k_1(A + B)$$

We can also demonstrate that if A and B are two matrices of the same order, and k_1 and k_2 are two scalars, then $k_1(k_2 A) = (k_1 k_2)A$ and $(k_1 + k_2)A = k_1 A + k_2 A$.

For instance,

$$A = \begin{bmatrix} a_{11} & a_{12} \\ a_{21} & a_{22} \end{bmatrix}$$

$$k_1 k_2 A = k_1 k_2 \begin{bmatrix} a_{11} & a_{12} \\ a_{21} & a_{22} \end{bmatrix}$$

$$= \begin{bmatrix} k_1 k_2 a_{11} & k_1 k_2 a_{12} \\ k_1 k_2 a_{21} & k_1 k_2 a_{22} \end{bmatrix}$$

and

$$k_1(k_2A) = k_1 \begin{bmatrix} k_2a_{11} & k_2a_{12} \\ k_2a_{21} & k_2a_{22} \end{bmatrix}$$

$$= \begin{bmatrix} k_1k_2a_{11} & k_1k_2a_{12} \\ k_1k_2a_{21} & k_1k_2a_{22} \end{bmatrix}$$

$$= k_1k_2A$$

5.6.3 Matrix multiplication properties

Multiplicative identity

The multiplicative identity for a scalar is the value which, when multiplied by the scalar, results in the original scalar. One is the multiplicative identity for scalar multiplication. For instance, $13 \times 1 = 13$.

We have seen in section 5.3.3 the square identity matrix I with ones on the leading diagonal, and all the other elements equal to zero.

For any matrix A of order $n \times n$, multiplying A by the identity matrix of the same order results in the same original matrix.

For instance, with 2×2 matrices, $A = \begin{bmatrix} a_{11} & a_{12} \\ a_{21} & a_{22} \end{bmatrix}, I = \begin{bmatrix} 1 & 0 \\ 0 & 1 \end{bmatrix}$ then

$$AI = \begin{bmatrix} a_{11} & a_{12} \\ a_{21} & a_{22} \end{bmatrix} \times \begin{bmatrix} 1 & 0 \\ 0 & 1 \end{bmatrix}$$

$$= \begin{bmatrix} 1a_{11} + 0a_{12} & 0a_{11} + 1a_{12} \\ 1a_{21} + 0a_{12} & 0a_{21} + 1a_{22} \end{bmatrix}$$

$$= \begin{bmatrix} a_{11} & a_{12} \\ a_{21} & a_{22} \end{bmatrix}$$

$$= A$$

and

$$IA = \begin{bmatrix} 1 & 0 \\ 0 & 1 \end{bmatrix} \times \begin{bmatrix} a_{11} & a_{12} \\ a_{21} & a_{22} \end{bmatrix}$$

$$= \begin{bmatrix} 1a_{11} + 0a_{21} & 1a_{12} + 0a_{22} \\ 0a_{11} + 1a_{21} & 0a_{12} + 1a_{22} \end{bmatrix}$$

$$= \begin{bmatrix} a_{11} & a_{12} \\ a_{21} & a_{22} \end{bmatrix}$$

$$= A$$

The identity matrix is the multiplicative identity for matrix multiplication.

Multiplicative identity for matrix multiplication

For any matrix A of order $n \times n$, the identity matrix of the same order is the multiplicative identity.

$$AI = A = IA$$

Multiplicative inverse

The multiplicative inverse of a non-zero scalar k is the number that, when multiplied with k, results in one.

For instance, if $k \neq 0$, $k \times \dfrac{1}{k} = 1$, $\dfrac{1}{k}$ is the multiplicative inverse of k.

Similarly, as seen in lesson 5.5, for a matrix A of order $n \times n$ with a non-zero determinant, the inverse matrix A^{-1} is the multiplicative inverse.

For example, with a 2×2 matrix:

$$A = \begin{bmatrix} a_{11} & a_{12} \\ a_{21} & a_{22} \end{bmatrix}, A^{-1} = \frac{1}{a_{11} \times a_{22} - a_{21}a_{12}} \begin{bmatrix} a_{22} & -a_{12} \\ -a_{21} & a_{11} \end{bmatrix}$$

then

$$AA^{-1} = \frac{1}{a_{11} \times a_{22} - a_{21}a_{12}} \begin{bmatrix} a_{11} \times a_{22} - a_{12} \times a_{21} & -a_{11} \times a_{12} + a_{12} \times a_{11} \\ a_{21} \times a_{22} - a_{22} \times a_{21} & -a_{21} \times a_{22} + a_{22} \times a_{11} \end{bmatrix}$$

$$= \frac{1}{a_{11} \times a_{22} - a_{21}a_{12}} \begin{bmatrix} a_{11} \times a_{22} - a_{12} \times a_{21} & 0 \\ 0 & -a_{21} \times a_{22} + a_{22} \times a_{11} \end{bmatrix}$$

$$= \begin{bmatrix} 1 & 0 \\ 0 & 1 \end{bmatrix}$$

Similarly, we can demonstrate that $A^{-1}A = I$

Multiplicative inverse for matrix multiplication

For any matrix A of order $n \times n$ with a non-zero determinant, the inverse matrix A^{-1} of the same order is the matrix such that $AA^{-1} = I = A^{-1}A$

Associative law for matrix multiplication

We know that if we multiply three numbers a, b and c, then $(a \times b) \times c = a \times (b \times c)$.

Similarly, we can show that for three matrices A, of order $m \times n$, B, of order $n \times p$, and C of order $p \times q$ that $(AB)C = A(BC)$.

For instance, if

$$A = \begin{bmatrix} -1 & 2 \end{bmatrix}, B = \begin{bmatrix} 3 & 0 & -2 \\ 1 & -1 & 4 \end{bmatrix} \text{ and } C = \begin{bmatrix} 7 & -1 \\ 0 & 4 \\ 3 & 2 \end{bmatrix}$$

$$\text{then } AB = \begin{bmatrix} -1 \times 3 + 2 \times 1 & -1 \times 0 + 2 \times -1 & -1 \times -2 + 2 \times 4 \end{bmatrix}$$

$$= \begin{bmatrix} -1 & -2 & 10 \end{bmatrix}$$

and

$$(AB)C = \begin{bmatrix} -1 \times 7 + -2 \times 0 + 10 \times 3 & -1 \times -1 + -2 \times 4 + 10 \times 2 \end{bmatrix}$$

$$= \begin{bmatrix} 23 & 13 \end{bmatrix}$$

and

$$(BC) = \begin{bmatrix} 3 \times 7 + 0 \times 0 - 2 \times 3 & 3 \times -1 + 0 \times 4 - 2 \times 2 \\ 1 \times 7 - 1 \times 0 + 4 \times 3 & 1 \times -1 - 1 \times 4 + 4 \times 2 \end{bmatrix}$$

$$= \begin{bmatrix} 15 & -7 \\ 19 & 3 \end{bmatrix}$$

and

$$A(BC) = \begin{bmatrix} -1 \times 15 + 2 \times 19 & -1 \times -7 + 2 \times 3 \end{bmatrix}$$

$$= \begin{bmatrix} 23 & 13 \end{bmatrix}$$

$$= (AB)C$$

Associative law for matrix multiplication

If A is a matrix of order $m \times n$, B a matrix of order $n \times p$, and C a matrix of order $p \times q$ then $(AB)C = A(BC)$.

Left and right distributive laws

With the distributive property of scalar multiplication, we know that for any three scalars a, b and c, $a(b + c) = ab + ac$ and $(b + c)a = ba + ca$.

We can demonstrate the distributive property of matrix multiplication.

For example, using

$$A = \begin{bmatrix} 2 \\ -1 \\ 3 \end{bmatrix}, B = \begin{bmatrix} 4 & -2 \end{bmatrix} \text{ and } C = \begin{bmatrix} -1 & -4 \end{bmatrix}$$

then

$$A(B + C) = \begin{bmatrix} 2 \\ -1 \\ 3 \end{bmatrix} \times \begin{bmatrix} 3 & -6 \end{bmatrix}$$

$$= \begin{bmatrix} 6 & -12 \\ -3 & 6 \\ 9 & -18 \end{bmatrix}$$

and

$$AB + AC = \begin{bmatrix} 8 & -4 \\ -4 & 2 \\ 12 & -6 \end{bmatrix} + \begin{bmatrix} -2 & -8 \\ 1 & 4 \\ -3 & -12 \end{bmatrix}$$

$$= \begin{bmatrix} 6 & -12 \\ -3 & 6 \\ 9 & -18 \end{bmatrix}$$

$$= A(B + C)$$

Left distributive law

If A is a matrix of order $m \times n$, B and C are matrices of order $n \times p$, then $A(B + C) = AB + AC$.

And using

$$A = \begin{bmatrix} 2 \\ -1 \\ 3 \end{bmatrix}, B = \begin{bmatrix} 0 & -1 & 3 \\ 2 & 1 & 4 \end{bmatrix} \text{ and } C = \begin{bmatrix} 2 & -2 & -3 \\ 1 & 0 & -1 \end{bmatrix}$$

then

$$(B + C)A = \begin{bmatrix} 2 & -3 & 0 \\ 3 & 1 & 3 \end{bmatrix} \times \begin{bmatrix} 2 \\ -1 \\ 3 \end{bmatrix}$$

$$= \begin{bmatrix} 7 \\ 14 \end{bmatrix}$$

and

$$BA + CA = \begin{bmatrix} 10 \\ 15 \end{bmatrix} + \begin{bmatrix} -3 \\ -1 \end{bmatrix}$$

$$= \begin{bmatrix} 7 \\ 14 \end{bmatrix}$$

$$= (B + C)A$$

Right distributive law

If B and C are matrices of order $m \times n$, and A is a matrix of order $n \times p$ then $(B + C)A = BA + CA$.

WORKED EXAMPLE 19 Calculating powers of a matrix using distributive law

Consider the matrices $A = \begin{bmatrix} 0 & 1 & 1 \\ 1 & 0 & 1 \\ 1 & 1 & 0 \end{bmatrix}$ and $I = \begin{bmatrix} 1 & 0 & 0 \\ 0 & 1 & 0 \\ 0 & 0 & 1 \end{bmatrix}$. Answer the following without using technology.

a. Show that $A^2 = A + 2I$.
b. Express A^3 as a linear combination of A and I.
c. Determine A^7.

THINK	WRITE
a. 1. Check that matrix multiplication is defined. A is a 3×3 square matrix, any power of A is defined and will be a 3×3 matrix.	a. $A^2 = \begin{bmatrix} 0 \times 0 + 1 \times 1 + 1 \times 1 & 0 \times 1 + 0 \times 0 + 1 \times 1 & 0 \times 1 + 1 \times 1 + 1 \times 0 \\ 1 \times 0 + 0 \times 1 + 1 \times 1 & 1 \times 1 + 0 \times 0 + 1 \times 1 & 1 \times 1 + 0 \times 1 + 0 \times 0 \\ 1 \times 0 + 1 \times 1 + 0 \times 1 & 1 \times 1 + 1 \times 0 + 0 \times 1 & 1 \times 1 + 1 \times 1 + 0 \times 0 \end{bmatrix}$ $= \begin{bmatrix} 2 & 1 & 1 \\ 1 & 2 & 1 \\ 1 & 1 & 2 \end{bmatrix}$

2. Recall the rules for matrix multiplication and apply.

3. Express A^2 as a linear combination of A and I

$$A^2 = \begin{bmatrix} 2 & 1 & 1 \\ 1 & 2 & 1 \\ 1 & 1 & 2 \end{bmatrix} = \begin{bmatrix} 0 & 1 & 1 \\ 1 & 0 & 1 \\ 1 & 1 & 0 \end{bmatrix} + 2\begin{bmatrix} 1 & 0 & 0 \\ 0 & 1 & 0 \\ 0 & 0 & 1 \end{bmatrix}$$

$$= A + 2I$$

b. Since $A^3 = A^2 \times A$, use the distributive law and substitute the expression of A^2 from part **a.** to express A^3 as a linear combination of A and I.

b. $A^3 = A^2 \times A$

$= (A + 2I) \times A$

$= A^2 + 2A$

$= (A + 2I) + 2A$

$= 3A + 2I$

c. 1. Repeat the process of expressing A^{n+1} as $A^n \times A$, using the distributing law and substituting the expression of the powers of A from the previous steps, until reaching A^7.

c. $A^4 = A^3 \times A$

$= (3A + 2I) \times A$

$= 3A^2 + 2A$

$= 3(A + 2I) + 2A$

$= 5A + 6I$

$A^5 = A^4 \times A$

$= (5A + 6I) \times A$

$= 5A^2 + 6A$

$= 5(A + 2I) + 6A$

$= 11A + 10I$

$A^6 = A^5 \times A$

$= (11A + 10I) \times A$

$= 11A^2 + 10A$

$= 11(A + 2I) + 10A$

$= 21A + 22I$

$A^7 = A^6 \times A$

$= (21A + 22I) \times A$

$= 21A^2 + 22A$

$= 21(A + 2I) + 22A$

$= 43A + 42I$

2. Substitute the matrices A and I and determine A^7.

$A^7 = 43A + 42I$

$$= 43\begin{bmatrix} 0 & 1 & 1 \\ 1 & 0 & 1 \\ 1 & 1 & 0 \end{bmatrix} + 42\begin{bmatrix} 1 & 0 & 0 \\ 0 & 1 & 0 \\ 0 & 0 & 1 \end{bmatrix}$$

$$= \begin{bmatrix} 42 & 43 & 43 \\ 43 & 42 & 43 \\ 43 & 43 & 42 \end{bmatrix}$$

No commutative property for matrix multiplication

With the commutative property of scalar multiplication, we know that for any two scalars a and b, $ab = ba$. However, matrix multiplication in general is not commutative.

This can easily be understood for non-square matrices by considering the order of the matrices.

If A is a matrix of order $m \times n$ and B is a matrix or order $n \times m$ so that both products AB and BA are defined, then AB is a matrix of order $m \times m$ and BA is a matrix of order $n \times n$.

These two matrices can only be equal is they have the same order, that is if $m = n$, the matrices are square matrices.

Matrix multiplication is never commutative for non-square matrices.

Let's consider, then, the case of square matrices.

Matrix multiplication can sometimes be commutative for square matrices.

For instance:

$$\begin{bmatrix} 2 & -1 \\ 3 & 4 \end{bmatrix} \times \begin{bmatrix} 5 & 1 \\ -3 & 3 \end{bmatrix} = \begin{bmatrix} 13 & -1 \\ 3 & 15 \end{bmatrix}$$

and

$$\begin{bmatrix} 5 & 1 \\ -3 & 3 \end{bmatrix} \times \begin{bmatrix} 2 & -1 \\ 3 & 4 \end{bmatrix} = \begin{bmatrix} 13 & -1 \\ 3 & 15 \end{bmatrix}$$

Matrix multiplication is generally not commutative for square matrices though.

For instance, using

$$A = \begin{bmatrix} 2 & 4 \\ -1 & 3 \end{bmatrix} \text{ and } B = \begin{bmatrix} 5 & 0 \\ -6 & 2 \end{bmatrix}$$

then

$$AB = \begin{bmatrix} -14 & 8 \\ -23 & 6 \end{bmatrix}$$

and

$$BA = \begin{bmatrix} 10 & 20 \\ -14 & -18 \end{bmatrix}$$
$$\neq AB$$

Matrix multiplication is not commutative

Matrix multiplication is never commutative for non-square matrices.

Matrix multiplication can be commutative sometimes for square matrices but generally, it is not.

Generally, $AB \neq BA$.

Simple familiar

1. Consider these matrices: $A = \begin{bmatrix} 2 & 3 \\ -1 & 4 \end{bmatrix}, B = \begin{bmatrix} 4 & 5 \\ 2 & -3 \end{bmatrix}$ and $C = \begin{bmatrix} 1 & -2 \\ 5 & 4 \end{bmatrix}$.

    a. Determine the following matrices.
       i. $B + C$
       ii. $A + B$
    b. Verify the Associative Law for matrix addition: $A + (B + C) = (A + B) + C$.

2. a. Given the matrices $A = \begin{bmatrix} 2 & 3 \\ -1 & 4 \end{bmatrix}, I = \begin{bmatrix} 1 & 0 \\ 0 & 1 \end{bmatrix}$ and $O = \begin{bmatrix} 0 & 0 \\ 0 & 0 \end{bmatrix}$, verify the following.

       i. $AI = IA = A$
       ii. $AO = OA = O$
    b. Given the matrices $A = \begin{bmatrix} a & b \\ c & d \end{bmatrix}, I = \begin{bmatrix} 1 & 0 \\ 0 & 1 \end{bmatrix}$ and $O = \begin{bmatrix} 0 & 0 \\ 0 & 0 \end{bmatrix}$, verify the following.

       i. $AI = IA = A$
       ii. $AO = OA = O$

3. a. Given the matrices $A = \begin{bmatrix} 2 & 3 \\ -1 & 4 \end{bmatrix}$ and $I = \begin{bmatrix} 1 & 0 \\ 0 & 1 \end{bmatrix}$, verify that $(I - A)(I + A) = I - A^2$.

    b. If $A = \begin{bmatrix} 3 & 0 \\ -4 & 0 \end{bmatrix}$ and $B = \begin{bmatrix} 0 & 0 \\ 2 & -1 \end{bmatrix}$, verify that $AB = O$ where $O = \begin{bmatrix} 0 & 0 \\ 0 & 0 \end{bmatrix}$. Determine if $BA = O$.

    c. If $A = \begin{bmatrix} a & 0 \\ b & 0 \end{bmatrix}$ and $B = \begin{bmatrix} 0 & 0 \\ x & y \end{bmatrix}$, verify that $AB = O$ where $O = \begin{bmatrix} 0 & 0 \\ 0 & 0 \end{bmatrix}$. Determine if $BA = O$.

Questions **4** and **5** refer to the following matrices:

$A = \begin{bmatrix} 1 & -1 \\ 3 & 2 \end{bmatrix}, B = \begin{bmatrix} -2 & 4 \\ 0 & -1 \end{bmatrix}$ and $C = \begin{bmatrix} 0 & 1 \\ -2 & 6 \end{bmatrix}$

4. Show that $A(B + C) = AB + AC$

5. Show that $(B + C)A = BA + CA$

6. Consider the matrices $A = \begin{bmatrix} 2 & 3 \\ -1 & 4 \end{bmatrix}, B = \begin{bmatrix} 4 & 5 \\ 2 & -3 \end{bmatrix}$ and $C = \begin{bmatrix} 1 & -2 \\ 5 & 4 \end{bmatrix}$.

    a. Verify the Distributive Law: $A(B + C) = AB + AC$.
    b. Verify the Associative Law for Multiplication: $A(BC) = (AB)C$.

7. Consider the matrices $A = \begin{bmatrix} 2 & -1 \\ 4 & 1 \end{bmatrix}$ and $B = \begin{bmatrix} 5 & 0 \\ -2 & -3 \end{bmatrix}$.

    a. Determine if $\det(A + B) = \det(A) + \det(B)$.
    b. Determine if $\det(AB) = \det(A) \times \det(B)$.
    c. Determine if $\det(A^3) = (\det(A))^3$.

Complex familiar

Questions **8** and **9** refer to the following matrices:

$$A = \begin{bmatrix} 3 & -1 \\ -2 & 0 \end{bmatrix} \text{ and } B = \begin{bmatrix} 0 & 1 \\ 3 & 2 \end{bmatrix}$$

8. a. Compare $(A + B)^2$ to $A^2 + 2AB + B^2$.
  b. Compare $(A + B)^2$ to $A^2 + AB + BA + B^2$.
  c. Justify whether $(A - B)^2 = A^2 - AB - BA + B^2$ is true for all square matrices A and B.

9. a. Compare $(A + B)(A - B)$ to $A^2 - B^2$.
  b. Compare $(A + B)(A - B)$ to $A^2 + BA - AB - B^2$.
  c. Give a condition on A and B for $(A + B)(A - B) = A^2 - B^2$ to be true for all square matrices A and B.

10. ▨ **WE19** Consider the matrices $A = \begin{bmatrix} 0 & 1 & -1 \\ 1 & 0 & 1 \\ -1 & 1 & 0 \end{bmatrix}$ and $I = \begin{bmatrix} 1 & 0 & 0 \\ 0 & 1 & 0 \\ 0 & 0 & 1 \end{bmatrix}$. Answer the following without using technology.
  a. Show that $A^2 = 2I - A$.
  b. Determine A^5.

11. Consider the matrices $A = \begin{bmatrix} -1 & 1 & 1 \\ 1 & -1 & 1 \\ 1 & 1 & -1 \end{bmatrix}$ and $I = \begin{bmatrix} 1 & 0 & 0 \\ 0 & 1 & 0 \\ 0 & 0 & 1 \end{bmatrix}$.
  a. Show that $A^2 = 2I - A$.
  b. Show that $\dfrac{1}{2}(A + I)$ is the inverse matrix of A.
  c. Express A^4 and A^5 as linear combinations of A and I.
  d. Show that $\left(\dfrac{5A + 11I}{16}\right)$ is the inverse of A^4.

12. Consider the matrices $A = \begin{bmatrix} 1 & 1 & 0 \\ 0 & 1 & 1 \\ 0 & 0 & 1 \end{bmatrix}, I = \begin{bmatrix} 1 & 0 & 0 \\ 0 & 1 & 0 \\ 0 & 0 & 1 \end{bmatrix}$ and $B = A - I$.
  a. Show that $B^3 = 0$.
  b. Express A^3 as a linear combination of powers of B and determine A^3.

Complex unfamiliar

13. Given $A = \begin{bmatrix} a & b \\ 0 & a \end{bmatrix}$, where a and b are non-zero integers, determine all the 2×2 matrices B such that $AB = BA$.

14. Let $A = \begin{bmatrix} 2 & -1 \\ 1 & 3 \end{bmatrix}$ and $I = \begin{bmatrix} 1 & 0 \\ 0 & 1 \end{bmatrix}$.
  Demonstrate that A^3 can be written as a linear combination of A and I.

15. Let $A = \begin{bmatrix} a & b \\ c & d \end{bmatrix}$ and $I = \begin{bmatrix} 1 & 0 \\ 0 & 1 \end{bmatrix}$.
  Demonstrate that A^2 can always be written as a linear combination of A and I.

Fully worked solutions for this chapter are available online.

LESSON
5.7 Matrix equations

SYLLABUS LINKS

- Use matrix algebra to solve matrix equations that involve matrices of up to dimension 2×2, including those of the form $AX = B$, $XA = B$ and $AX + BX = C$, with and without technology.
- Model and solve problems that involve matrices of up to dimension 2×2, including the solution of systems of linear equations, with and without technology.

Source: Specialist Mathematics Senior Syllabus 2024 © State of Queensland (QCAA) 2024; licensed under CC BY 4.0

5.7.1 Introduction to matrix equations

Inverse matrices are used to solve matrix equations as division of matrices is not possible.

Consider the matrix equations $AX = B$, $XA = B$, and $(A + B)X = AX + BX = C$ where A, B, C and X are matrices, and X needs to be found.

If $AX = B$, pre-multiply both sides by A^{-1}, the inverse of matrix A. Remember order of multiplication is important when multiplying matrices.

$$AX = B$$
$$\Rightarrow A^{-1}AX = A^{-1}B$$

Since $A^{-1}A = I$,

$$IX = A^{-1}B \qquad \text{where } I = \begin{bmatrix} 1 & 0 \\ 0 & 1 \end{bmatrix}$$
$$\therefore X = A^{-1}B.$$

If $XA = B$, multiply both sides of the equation by A^{-1}, remembering order of multiplication is important.

$$XA = B$$
$$\Rightarrow XAA^{-1} = BA^{-1}$$

Since $AA^{-1} = I$,

$$XI = BA^{-1} \qquad \text{where } I = \begin{bmatrix} 1 & 0 \\ 0 & 1 \end{bmatrix}$$
$$\therefore X = BA^{-1}.$$

If $AX + BX = C$, that is if $(A + B)X = C$, multiply both sides of the equation by $(A + B)^{-1}$, remembering order of multiplication is important.

$$AX + BX = C$$
$$(A + B)X = C$$
$$\Rightarrow (A + B)^{-1}(A + B)X = (A + B)^{-1}C$$
$$\text{since } (A + B)^{-1}(A + B) = I, \text{ where } I = \begin{bmatrix} 1 & 0 \\ 0 & 1 \end{bmatrix}$$
$$IX = (A + B)^{-1}C$$
$$\therefore X = (A + B)^{-1}C$$

Solving matrix equations

If $AX = B$, then $X = A^{-1}B$

If $XA = B$, then $X = BA^{-1}$

If $AX + BX = C$, then $X = (A + B)^{-1}C$.

WORKED EXAMPLE 20 Solving matrix equations

Given the matrices $A = \begin{bmatrix} 3 & -4 \\ 5 & -6 \end{bmatrix}$, $B = \begin{bmatrix} 0 & 10 \\ -8 & -4 \end{bmatrix}$, $C = \begin{bmatrix} -1 \\ 2 \end{bmatrix}$ and $D = \begin{bmatrix} 3 & -2 \end{bmatrix}$, determine the matrix X if:

a. $AX = C$ b. $XA = D$ c. $AX + BX = C$

THINK	WRITE
a. 1. If $AX = C$, pre-multiply both sides by the inverse matrix, A^{-1} and solve for X.	**a.** $AX = C$ $A^{-1}AX = A^{-1}C$ $IX = X = A^{-1}C$ $\therefore X = A^{-1}C$
2. Recall the determinant rule and apply to evaluate the determinant of the matrix A. $\det(A) = ad - bc$	$\det(A) = 3 \times (-6) - (-4) \times 5$ $= 2$
3. Recall the inverse matrix formula to determine the inverse matrix A^{-1}. $A^{-1} = \dfrac{1}{ad - bc}\begin{bmatrix} d & -b \\ -c & a \end{bmatrix}$	$A^{-1} = \dfrac{1}{2}\begin{bmatrix} -6 & 4 \\ -5 & 3 \end{bmatrix}$
4. Substitute for the given matrices.	$X = \dfrac{1}{2}\begin{bmatrix} -6 & 4 \\ -5 & 3 \end{bmatrix} \times \begin{bmatrix} -1 \\ 2 \end{bmatrix}$
5. X is a 2×1 column matrix. Apply the rules to multiply the matrices.	$= \dfrac{1}{2}\begin{bmatrix} (-6) \times (-1) + 4 \times 2 \\ (-5) \times (-1) + 3 \times 2 \end{bmatrix}$
6. State the answer.	$= \dfrac{1}{2}\begin{bmatrix} 14 \\ 11 \end{bmatrix}$ $= \begin{bmatrix} 7 \\ \dfrac{11}{2} \end{bmatrix}$
b. 1. If $XA = D$, multiply both sides by the inverse matrix A^{-1}, and solve for X.	**b.** $XA = D$ $XAA^{-1} = DA^{-1}$
2. Substitute for the given matrices.	$XI = X = DA^{-1}$ $\therefore X = DA^{-1}$ $X = \begin{bmatrix} 3 & -2 \end{bmatrix} \times \dfrac{1}{2}\begin{bmatrix} -6 & 4 \\ -5 & 3 \end{bmatrix}$

3. X is a 1×2 matrix. Apply the rules to multiply the matrices.

$$X = \frac{1}{2}\left[3 \times (-6) + (-2) \times (-5) \quad 3 \times 4 + (-2) \times 3\right]$$

$$= \frac{1}{2}\left[-8 \quad 6\right]$$

4. State the answer.

$$= \left[-4 \quad 3\right]$$

c. 1. If $AX + BX = C$ then $(A + B)X = C$ (left distributive law), pre-multiply both sides by the inverse matrix $(A + B)^{-1}$ and solve for X.

c.
$$(A + B)X = C$$
$$(A + B)^{-1}(A + B)X = (A + B)^{-1}C$$
$$IX = (A + B)^{-1}C$$
$$\therefore X = (A + B)^{-1}C$$

2. Calculate the matrix $(A + B)$.

$$(A + B) = \begin{bmatrix} 3 & 6 \\ -3 & -10 \end{bmatrix}$$

3. Recall the determinant rule and apply to evaluate the determinant of $(A + B)$.

$$\det(A + B) = 3 \times (-10) - (-3) \times 6$$
$$= -12$$

4. Recall the inverse matrix formula to determine the inverse matrix $(A + B)^{-1}$.

$$(A + B)^{-1} = -\frac{1}{12}\begin{bmatrix} -10 & -6 \\ 3 & 3 \end{bmatrix}$$

5. Substitute for the given matrices.

$$X = (A + B)^{-1}C$$
$$= -\frac{1}{12}\begin{bmatrix} -10 & -6 \\ 3 & 3 \end{bmatrix}\begin{bmatrix} -1 \\ 2 \end{bmatrix}$$

6. X is a 2×1 matrix. Apply the rules to multiply the matrices.

$$X = -\frac{1}{12}\begin{bmatrix} -10 \times (-1) - 6 \times 2 \\ 3 \times (-1) + 3 \times 2 \end{bmatrix}$$

$$= -\frac{1}{12}\begin{bmatrix} -2 \\ 3 \end{bmatrix}$$

7. State the answer.

$$= \begin{bmatrix} \frac{1}{6} \\ -\frac{1}{4} \end{bmatrix}$$

TI | THINK

a. 1. On a Calculator page, press the template button and complete the entry line as:
$$\begin{bmatrix} 3 & -4 \\ 5 & -6 \end{bmatrix}$$
Press CTRL, then press VAR, then type 'A' and press ENTER to store matrix A. Repeat this step to store Matrix C.

DISPLAY/WRITE

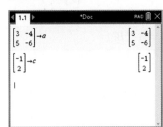

CASIO | THINK

a. 1. On a Run-Matrix screen, select MAT/VCT by pressing F3. Select Matrix A by pressing EXE and change its dimensions to 2 by 2. Press EXE. Enter the elements of Matrix A, then press EXIT. Repeat this step to store Matrix C, then press EXIT to return to the Run-Matrix screen.

DISPLAY/WRITE

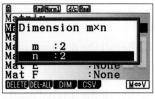

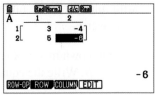

2. Complete the next entry line as: $a^{-1} \times c$ then press ENTER. 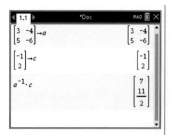	2. Complete the next entry line as: Mat $A^{-1} \times$ Mat C then press ENTER. *Note:* To find 'Mat', press OPTN, select MAT/VCT by pressing F2, then select Mat by pressing F1. 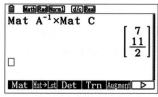
3. The answer appears on the screen. $X = \begin{bmatrix} 7 \\ 11 \\ 2 \end{bmatrix}$	3. The answer appears on the screen. $X = \begin{bmatrix} 7 \\ 11 \\ 2 \end{bmatrix}$

5.7.2 Solving 2 × 2 linear equations

Consider the two linear equations $ax + by = e$ and $cx + dy = f$.

These equations can be written in matrix form as follows:

$$\underset{\text{Coefficient matrix}}{\begin{bmatrix} a & b \\ c & d \end{bmatrix}} \quad \times \quad \underset{\text{Variable matrix}}{\begin{bmatrix} x \\ y \end{bmatrix}} \quad = \quad \underset{\text{Constant matrix}}{\begin{bmatrix} e \\ f \end{bmatrix}}$$

If we let $A = \begin{bmatrix} a & b \\ c & d \end{bmatrix}$, $X = \begin{bmatrix} x \\ y \end{bmatrix}$ and $B = \begin{bmatrix} e \\ f \end{bmatrix}$, this equation is of the form $AX = B$, which can be solved for X, as $X = A^{-1}B$.

WORKED EXAMPLE 21 Solving simultaneous equations using inverse matrices

Solve for x and y using inverse matrices.

$$4x + 5y = 6$$
$$3x + 2y = 8$$

THINK

1. First rewrite the two equations as a matrix equation.

2. Write down the matrices A, X and B.

3. Write as an equation and solve for X.

4. Recall the determinant rule and calculate for matrix A.

5. Recall the inverse matrix A^{-1} rule, and apply with rules for scalar multiplication to simplify this inverse.

6. The unknown matrix X satisfies the equation $X = A^{-1}B$. Write the equation in matrix form.

WRITE

$$\begin{bmatrix} 4 & 5 \\ 3 & 2 \end{bmatrix} \begin{bmatrix} x \\ y \end{bmatrix} = \begin{bmatrix} 6 \\ 8 \end{bmatrix}$$

$$A = \begin{bmatrix} 4 & 5 \\ 3 & 2 \end{bmatrix}, X = \begin{bmatrix} x \\ y \end{bmatrix} \text{ and } B = \begin{bmatrix} 6 \\ 8 \end{bmatrix}$$

$$AX = B$$
$$\therefore X = A^{-1}B$$

$$\det(A) = 4 \times 2 - 3 \times 5 = -7$$

$$A^{-1} = \frac{1}{-7} \begin{bmatrix} 2 & -5 \\ -3 & 4 \end{bmatrix} = \frac{1}{7} \begin{bmatrix} -2 & 5 \\ 3 & -4 \end{bmatrix}$$

$$X = \begin{bmatrix} x \\ y \end{bmatrix} = \frac{1}{7} \begin{bmatrix} -2 & 5 \\ 3 & -4 \end{bmatrix} \begin{bmatrix} 6 \\ 8 \end{bmatrix}$$

7. Apply the rules for matrix multiplication. The product is a 2×1 matrix.

$$= \begin{bmatrix} x \\ y \end{bmatrix} = \frac{1}{7} \begin{bmatrix} (-2) \times 6 + 5 \times 8 \\ 3 \times 6 + (-4) \times 8 \end{bmatrix}$$

8. Apply the rules for scalar multiplication, and the rules for equality of matrices.

$$X = \begin{bmatrix} x \\ y \end{bmatrix} = \frac{1}{7} \begin{bmatrix} 28 \\ -14 \end{bmatrix} = \begin{bmatrix} 4 \\ -2 \end{bmatrix}$$

9. State the final answer.

$$x = 4 \text{ and } y = -2$$

Note that the solution to the system of linear equations in Worked example 21 can be verified by sketching the graphs $4x + 5y = 6$ and $3x + 2y = 8$. As you will see in the next section, the solutions to the system are the points of intersections of the graphs.

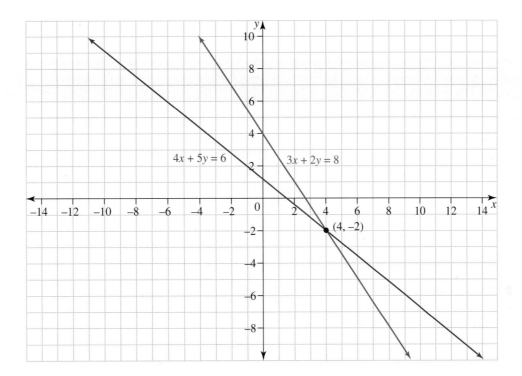

5.7.3 Geometrical interpretation of solutions (Extension)

There are three possible cases for the solutions of systems of linear equations as follows:

- A unique solution
- No solution
- Infinitely many solutions

In Unit 3, you will further study matrix algebra and systems of equations (beyond 2×2) and examine these three cases for solutions of systems of equations and the geometric interpretation of a solution of a system of equations with three variables. Here, we are only considering system of equations with two variables.

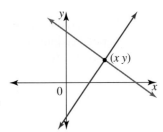

If the determinant of the coefficient matrix is non-zero, then these two equations are consistent. Graphically, the two lines have different gradients and therefore they intersect at a unique point resulting in a *unique solution*.

If the determinant of the coefficient matrix is zero, then there are two possibilities.

The lines are parallel, which indicates that there is *no solution*. Graphically the two lines have the same gradient but different *y*-intercepts.

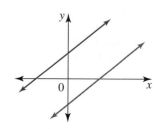

OR

The lines are multiples of one another, that is, they have the same gradient and the same *y*-intercept (they overlap).

This indicates that there is an *infinite number of solutions*.

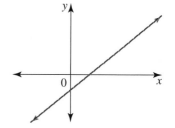

WORKED EXAMPLE 22 Interpreting the solutions of simultaneous equations with no solutions

Solve the following simultaneous linear equations using matrices and interpret the solution geometrically.

$$3x - 2y = 6$$
$$-6x + 4y = -10$$

THINK

WRITE

1. First write the two equations as a matrix equation.

$$\begin{bmatrix} 3 & -2 \\ -6 & 4 \end{bmatrix} \begin{bmatrix} x \\ y \end{bmatrix} = \begin{bmatrix} 6 \\ -10 \end{bmatrix}$$

2. Write down the matrices A, X and B.

$$A = \begin{bmatrix} 3 & -2 \\ -6 & 4 \end{bmatrix}, X = \begin{bmatrix} x \\ y \end{bmatrix} \text{ and } B = \begin{bmatrix} 6 \\ -10 \end{bmatrix}$$

3. Write as an equation and solve for X.

$$AX = B$$
$$\therefore X = A^{-1}B$$

4. Recall the determinant rule and calculate for matrix A.

$$\det(A) = 3 \times 4 - (-2) \times (-6) = 0$$

5. The inverse matrix A^{-1} does not exist. This method cannot be used to solve the simultaneous equations.

The matrix A is singular, as the determinant equals zero, which means the lines may be parallel or the exact same line.

6. Rearrange both equations into the form $y = mx + c$.

$$3x - 2y = 6 \ \rightarrow y = \frac{3x - 6}{2} \rightarrow y = \frac{3}{2}x - 3$$

$$-6x + 4y = -10 \ \rightarrow y = \frac{6x - 10}{4} \rightarrow y = \frac{3}{2}x - \frac{5}{2}$$

7. As the gradients are equal, the lines are parallel (*y*-intercepts are different). Alternatively:

The gradients of both lines are $\frac{3}{2}$, and the *y*-intercepts are -3 and $-\frac{5}{2}$. The lines are parallel.

8. Apply another method to solving simultaneous equations: the graphical method. Since both equations represent straight lines, determine the x- and y-intercepts.

Line crosses the x-axis at $(2, 0)$ and the y-axis at $(0, -3)$.
Line [2]: $-6x + 4y = -10$ crosses the x-axis at $\left(\dfrac{5}{3}, 0\right)$ and the y-axis at $\left(0, -\dfrac{5}{2}\right)$.

9. Sketch the graphs. Note that the two lines are parallel and therefore have no points of intersection.

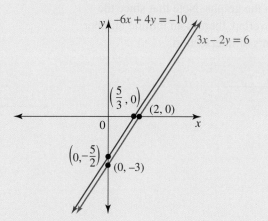

10. State the final answer.

There is no solution.

WORKED EXAMPLE 23 Interpreting the solutions of simultaneous equations with infinite solutions

Solve the following linear simultaneous equations for x and y, using matrices and interpret the solution geometrically.

$$3x - 2y = 6$$
$$-6x + 4y = -12$$

THINK

1. First write the two equations as a matrix equation.

2. Write down the matrices A, X and B.

3. Write as an equation and solve for X.

4. Recall the determinant rule and calculate for matrix A.

5. The inverse matrix A^{-1} does not exist. This method cannot be used to solve the simultaneous equations.

6. Rearrange both equations into the form $y = mx + c$.

7. The lines have the same gradients and y-intercept. They are the same line. Alternatively:

WRITE

$$\begin{bmatrix} 3 & -2 \\ -6 & 4 \end{bmatrix} \begin{bmatrix} x \\ y \end{bmatrix} = \begin{bmatrix} 6 \\ -12 \end{bmatrix}$$

$$A = \begin{bmatrix} 3 & -2 \\ -6 & 4 \end{bmatrix}, X = \begin{bmatrix} x \\ y \end{bmatrix} \text{ and } B = \begin{bmatrix} 6 \\ -12 \end{bmatrix}$$

$AX = B$
$\therefore X = A^{-1} B$

$\det(A) = 3 \times 4 - (-2) \times (-6) = 0$

The matrix A is singular, as the determinant equals zero, which means the lines may be parallel or the exact same line.

$3x - 2y = 6 \rightarrow y = \dfrac{3x - 6}{2} \rightarrow y = \dfrac{3}{2}x - 3$

$-6x + 4y = -12 \rightarrow y = \dfrac{6x - 12}{4} \rightarrow y = \dfrac{3}{2}x - 3$

Gradients $= \dfrac{3}{2}$
y-intercept $= -3$

8. Apply another method of solving simultaneous equations: the graphical method. Determine the x- and y-intercepts.

Line [1]: $3x - 2y = 6$ crosses the x-axis at $(2, 0)$ and the y-axis at $(0, -3)$.
Line [2]: $-6x + 4y = -12$ is actually the same line, since $[2] = -2 \times [1]$.

9. Sketch the graphs. Note that since the lines overlap, there is an infinite number of points of intersection.

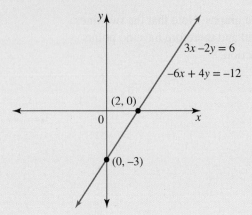

10. Let $y = t$ and then substitute for y into the equation, and solve for x.

Since $3x - 2y = 6$, $x = \dfrac{6 + 2y}{3}$.

If $y = 0$, $x = 2$: $(2, 0)$

If $y = 1$, $x = \dfrac{8}{3}$: $\left(\dfrac{8}{3}, 1\right)$

If $y = 2$, $x = \dfrac{10}{3}$: $\left(\dfrac{10}{3}, 2\right)$

If $y = 3$, $x = 4$: $(4, 3)$

In general, let $y = t$ so that $x = \dfrac{6 + 2t}{3}$.

As a coordinate: $\left(\dfrac{6 + 2t}{3}, t\right)$

11. State the final answer.

There is an infinite number of solutions of the form $\left(2 + \dfrac{2t}{3}, t\right)$ where $t \in \mathbb{R}$.

WORKED EXAMPLE 24 Solving a parameter for the three possible cases

Determine the values of k for which the equations $kx - 3y = k - 1$ and $10x - (k + 1)y = 8$ have:
a. a unique solution
b. no solution
c. an infinite number of solutions.
(You are not required to determine the solution set.)

THINK	WRITE
1. First write the two equations as matrix equations.	$\begin{bmatrix} k & -3 \\ 10 & -(k+1) \end{bmatrix} \begin{bmatrix} x \\ y \end{bmatrix} = \begin{bmatrix} k-1 \\ 8 \end{bmatrix}$
2. Write out the determinant, as it is the key to answering this question.	$\det\left(\begin{bmatrix} k & -3 \\ 10 & -(k+1) \end{bmatrix} \right) = -k(k+1) + 30$

3. Evaluate the determinant in terms of k.

$$= -k^2 - k + 30$$
$$= -\left(k^2 + k - 30\right)$$
$$= -(k+6)(k-5)$$

4. Let the determinant equal zero and solve for k.

$$k = -6, \; k = 5$$

a. If $\Delta \neq 0$, the solution is unique; that is, there is a unique solution when $k \neq -6$ and $k \neq 5$.

There is either no solution or an infinite number of solutions when $\Delta = 0$.

a. There is a unique solution when $\Delta \neq 0$, that is, when $k \neq -6$ and $k \neq 5$, or $k \in \mathbb{R} \setminus \{-6, 5\}$.

b. 1. Substitute $k = -6$ into the two equations.

b. $-6x - 3y = -7 \Rightarrow 2x + y = \dfrac{7}{3}$

$$\Rightarrow y = -2x + \dfrac{7}{3}$$

$$10x + 5y = 8 \Rightarrow 2x + y = \dfrac{8}{5}$$

$$\Rightarrow y = -2x + \dfrac{8}{5}$$

2. The gradients are the same. The two equations represent parallel lines with different y-intercepts. Interpret the answer.

When $k = -6$ there is no solution, as the lines are parallel.

c. 1. Substitute $k = 5$ into the two equations.

c. $5x - 3y = 4 \Rightarrow y = \dfrac{5x - 4}{3} \Rightarrow y = \dfrac{5}{3}x - \dfrac{4}{3}$

$$10x - 6y = 8 \Rightarrow y = \dfrac{10x - 8}{6} \Rightarrow y = \dfrac{5}{3}x - \dfrac{4}{3}$$

2. The lines have the same gradient and y-intercept. They are the same line.

Gradient $= \dfrac{5}{3}$, y-intercept $= \dfrac{-4}{3}$

3. The two equations are multiples of one another. Interpret the answer.

When $k = 5$ there are an infinite number of solutions.

Exercise 5.7 Matrix equations

5.7 Exercise	5.7 Exam questions on

Simple familiar	Complex familiar	Complex unfamiliar
1, 2, 3, 4, 5, 6, 7	8, 9, 10, 11, 12, 13, 14, 15, 16	17, 18, 19, 20

These questions are even better in jacPLUS!
- Receive immediate feedback
- Access sample responses
- Track results and progress

Find all this and MORE in jacPLUS

Simple familiar

1. **WE20** If $A = \begin{bmatrix} -2 & 4 \\ 3 & -5 \end{bmatrix}$, $C = \begin{bmatrix} -2 \\ 3 \end{bmatrix}$ and $D = \begin{bmatrix} 2 & -5 \end{bmatrix}$, determine matrix X given the following.

   a. $AX = C$
   b. $XA = D$

2. If $B = \begin{bmatrix} -5 & -3 \\ 3 & 4 \end{bmatrix}$, $C = \begin{bmatrix} -1 \\ 2 \end{bmatrix}$ and $D = \begin{bmatrix} 4 & 3 \end{bmatrix}$, determine matrix X given the following:

   a. $BX = C$

   b. $XB = D$

3. **WE21** Solve for x and y using inverse matrices.
$$3x - 4y = 23$$
$$5x + 2y = 21$$

4. Solve for x and y using inverse matrices.
$$2x + 5y = -7$$
$$3x - 2y = 18$$

5. Solve each of the following simultaneous linear equations using inverse matrices.

   a. $2x + 3y = 4$
      $-x + 4y = 9$

   b. $4x + 5y = -6$
      $2x - 3y = 8$

   c. $x - 2y = 8$
      $5x + 4y = -2$

   d. $-2x + 7y + 3 = 0$
      $3x + y + 7 = 0$

6. Consider the matrices $A = \begin{bmatrix} 1 & -2 \\ 5 & 4 \end{bmatrix}$, $B = \begin{bmatrix} 3 & 1 \\ -7 & 2 \end{bmatrix}$, $C = \begin{bmatrix} -5 \\ -19 \end{bmatrix}$ and $D = \begin{bmatrix} 7 & 14 \end{bmatrix}$.
   Calculate the matrix X in each of the following cases.

   a. $AX = C$

   b. $XA = B$

   c. $AX = B$

   d. $XA = D$

7. Consider matrices $A = \begin{bmatrix} -2 & 3 \\ 4 & 5 \end{bmatrix}$, $B = \begin{bmatrix} 2 & 19 \\ 12 & -7 \end{bmatrix}$, $C = \begin{bmatrix} 3 \\ 1 \end{bmatrix}$ and $D = \begin{bmatrix} -1 & 3 \end{bmatrix}$. Calculate the matrix X in each of the following cases.

   a. $AX = C$

   b. $XA = B$

   c. $AX = B$

   d. $XA = D$

Complex familiar

8. If $P = \begin{bmatrix} 1 & -2 \\ 3 & 4 \end{bmatrix}$, $Q = \begin{bmatrix} 2 & -1 \\ -3 & 6 \end{bmatrix}$ and $O = \begin{bmatrix} 0 & 0 \\ 0 & 0 \end{bmatrix}$, calculate the matrix X given the following.

   a. $XP - Q = O$
   b. $PX - Q = O$

9. **WE22** Solve the following simultaneous linear equations using matrices and interpret the solution geometrically.
$$4x - 3y = 12$$
$$-8x + 6y = -18$$

10. **WE23** Solve the following simultaneous linear equations using matrices and interpret the solution geometrically.
$$4x - 3y = 12$$
$$-8x + 6y = -24$$

11. Determine the value of k if the following simultaneous linear equations have no solution.
$$5x - 4y = 20$$
$$kx + 2y = -8$$

12. Calculate the value of k if the following simultaneous equations for x and y have an infinite number of solutions.
$$5x - 4y = 20$$
$$kx + 2y = -10$$

13. Determine the values of k for which the following simultaneous linear equations have:

   i. no solution
   ii. an infinite number of solutions.

   a. $x - 3y = k$
      $-2x + 6y = 6$

   b. $3x - 5y = k$
      $-6x + 10y = 10$

14. Demonstrate that each of the following does not have a unique solution. Describe the solution set and solve if possible.

    a. $x - 2y = 3$
        $-2x + 4y = -6$

    b. $2x - y = 4$
        $-4x + 2y = -7$

15. Determine the values of k for which the following systems of equations have:

    i. a unique solution
    ii. no solution
    iii. an infinite number of solutions
    (You are not required to determine the solution set.)

    a. $(k - 2)x - 2y = k - 1$
        $-4x + ky = -6$

    b. $(k + 1)x + 5y = 4$
        $6x + 5ky = k + 6$

    c. $(k - 1)x - 3y = k + 2$
        $-4x + 2ky = -10$

    d. $2x - (k - 2)y = 6$
        $(k - 5)x - 2y = k - 3$

16. **WE24** Determine the values of k for which the equations $(k + 1)x - 2y = 2k$ and $-6x + 2ky = -8$ have:

    a. a unique solution   **b.** no solution   **c.** an infinite number of solutions.

    (You are not required to determine the solution set.)

Complex unfamiliar

17. **a.** The line $\dfrac{x}{a} + \dfrac{y}{b} = 1$ passes through the points $(12, 6)$ and $(8, 3)$.

        Determine the values of a and b.

    b. The line $\dfrac{x}{a} + \dfrac{y}{b} = 1$ passes through the points $(4, 5)$ and $(-4, -15)$.
        Determine the values of a and b.

18. Determine the values of p and q for which the following systems of equations has:

    i. a unique solution
    ii. no solution
    iii. an infinite number of solutions.
    (You are not required to determine the solution set.)

    a. $-2x + 3y = p$ **b.** $4x - 2y = q$ **c.** $3x - py = 6$ **d.** $px - y = 3$
        $qx - 6y = 7$ $3x + py = 10$ $7x - 2y = q$ $-3x + 2y = q$

19. a and d are both non-zero real numbers.

    If $P = \begin{bmatrix} a & 0 \\ 0 & d \end{bmatrix}$, verify that $PP^{-1} = P^{-1}P = I$.

20. a, b, c and d are all non-zero real numbers.

    a. If $R = \begin{bmatrix} a & b \\ c & 0 \end{bmatrix}$, verify that $RR^{-1} = R^{-1}R = I$.   **b.** If $S = \begin{bmatrix} 0 & b \\ c & d \end{bmatrix}$, verify that $SS^{-1} = S^{-1}S = I$.

    c. If $A = \begin{bmatrix} a & b \\ c & d \end{bmatrix}$, verify that $AA^{-1} = A^{-1}A = I$.

Fully worked solutions for this chapter are available online.

LESSON
5.8 Review

5.8.1 Summary

doc-41927

Hey students! Now that it's time to revise this chapter, go online to:

 Access the chapter summary

Review your results

A+ Practise exam questions

Find all this and MORE in jacPLUS

5.8 Exercise

learn on

 5.8 Exercise   **5.8 Exam questions** on

These questions are even better in jacPLUS!
- Receive immediate feedback
- Access sample responses
- Track results and progress

Find all this and MORE in jacPLUS

Simple familiar	Complex familiar	Complex unfamiliar
1, 2, 3, 4, 5, 6, 7, 8, 9, 10, 11, 12	13, 14, 15, 16	17, 18, 19, 20

Simple familiar

1. Calculate the following.

   a. $\begin{bmatrix} 2 \\ -1 \end{bmatrix} + \begin{bmatrix} -3 \\ 5 \end{bmatrix}$

   b. $\begin{bmatrix} 1 & 0 & -3 \end{bmatrix} - \begin{bmatrix} 2 & -4 & 0 \end{bmatrix}$

   c. $\dfrac{3}{2} \begin{bmatrix} 0 & 6 \\ -4 & \dfrac{2}{3} \\ 8 & -2 \end{bmatrix}$

   d. $3 \begin{bmatrix} 2 \\ -1 \end{bmatrix} - 4 \begin{bmatrix} -3 \\ 5 \end{bmatrix}$

The following information relates to questions 2 and 3.

2. **MC** Consider the matrices $A = \begin{bmatrix} 1 & -2 \\ 0 & 1 \\ 2 & 3 \end{bmatrix}$, $B = \begin{bmatrix} 2 & 0 & 1 \\ -1 & 3 & 2 \end{bmatrix}$ $C = \begin{bmatrix} -1 & 2 \\ -2 & 1 \end{bmatrix}$ and $D = \begin{bmatrix} 1 & 0 & 2 \\ -1 & 2 & 1 \\ 0 & 3 & -1 \end{bmatrix}$.

   Determine which of the following statement is incorrect.

   A. The products AB, D^2 are both square matrices.
   B. The products BA and C^2 are both 2×2 matrices.
   C. Neither AC nor AD is defined.
   D. The product AC is a 2×3 matrix.

3. Calculate the following:

   a. AB
   b. $\det(BA)$
   c. C^{-1}

4. Determine the value of x in each of the following.

   a. $\det\left(\begin{bmatrix} x & x \\ 8 & 2 \end{bmatrix}\right) = 12$

   b. $\det\left(\begin{bmatrix} \dfrac{1}{x} & x \\ -2 & 3 \end{bmatrix}\right) = 7$

5. Solve each of the following simultaneous linear equations using inverse matrices.

   a. $3x + 4y = 6$
   $2x + 3y = 5$

   b. $x + 4y = 5$
   $3x - y = -11$

   c. $-4x + 3y = 13$
   $2x - y = 5$

   d. $-2x + 5y = 15$
   $3x - 2y = 16$

6. If $A = \begin{bmatrix} 2 & -3 \\ 1 & -4 \end{bmatrix}$ and $B = \begin{bmatrix} -1 & 4 \\ 3 & 5 \end{bmatrix}$, determine matrix X given the following.

   a. $AX = B$

   b. $XA = B$

7. Demonstrate that each of the following does not have a unique solution. Describe the solution set and solve if possible.

   a. $2x - 3y = 5$
   $-4x + 6y = -11$

   b. $3x - 4y = 5$
   $-6x + 8y = -10$

8. Solve each of the following using inverse matrices.

   a. $x + y = 6a$
   $4x - 3y = 3a + 14b$

   b. $3bx - 2ay = 0$
   $bx + ay = 5ab$

   c. $x - y = 6b$
   $3x - 4y = 17b - a$

9. Let $A = \begin{bmatrix} a \\ b \end{bmatrix}$ and $B = \begin{bmatrix} c & d \end{bmatrix}$.

   a. Demonstrate that AB exists but $(AB)^{-1}$ does not.
   b. Calculate BA and $(BA)^{-1}$.

10. Calculate the value(s) of x if each of the following is a singular matrix.

   a. $\begin{bmatrix} x-2 & -2 \\ 12 & 6 \end{bmatrix}$

   b. $\begin{bmatrix} x+1 & x-1 \\ 4 & 3 \end{bmatrix}$

   c. $\begin{bmatrix} x+2 & 3 \\ 5 & x \end{bmatrix}$

   d. $\begin{bmatrix} x+3 & 5 \\ 4 & x+2 \end{bmatrix}$

11. Consider the following matrices.

    $A = \begin{bmatrix} 2 & 1 \\ 3 & -4 \end{bmatrix}$, $B = \begin{bmatrix} 0 & 11 \\ -5 & -8 \end{bmatrix}$, $C = \begin{bmatrix} 2 \\ 25 \end{bmatrix}$, $D = \begin{bmatrix} 3 & -4 \end{bmatrix}$

    Determine the matrix X in each of the following cases.

    a. $AX = C$

    b. $XA = B$

    c. $AX = B$

    d. $XA = D$

12. Consider the matrices $A = \begin{bmatrix} 2 & 3 \\ -1 & 4 \end{bmatrix}$, $B = \begin{bmatrix} 4 & 5 \\ 2 & -3 \end{bmatrix}$, $C = \begin{bmatrix} 5 \\ 14 \end{bmatrix}$ and $D = [1 \ -2]$.

    Determine the matrix X in the each of the following cases.

    a. $AX = C$

    b. $XA = B$

    c. $AX = B$

    d. $XA = D$

Complex Familiar

13. Solve the following using matrices.
    $$\frac{x}{a} + \frac{y}{b} = 2a + b$$
    $$\frac{2x}{b} + \frac{3y}{a} = 2a + 6b$$

14. Determine the values of k for which the following systems of equations have:

    a. a unique solution
    b. no solution
    c. an infinite number of solutions.
    $$2x + (k+1)y = 4$$
    $$kx + 6y = k - 4$$

15. Determine the values of k for which the system of equations $2x + (k-1)y = 4$ and $kx + 6y = k + 4$ have.

    a. a unique solution

    b. no solution

    c. an infinite number of solutions.

16. If $A = \begin{bmatrix} 2 & 1 \\ 3 & -4 \end{bmatrix}$, $I = \begin{bmatrix} 1 & 0 \\ 0 & 1 \end{bmatrix}$ and $O = \begin{bmatrix} 0 & 0 \\ 0 & 0 \end{bmatrix}$, express the determinant of the matrix $A - kI$ in the form $pk^2 + qk + r$ starting the values of p, q and r. Hence show that $pA^2 + qA + rI = O$.

Complex unfamiliar

17. A Japanese food stall operates at a local market. Their four bestselling items over a few weeks are displayed in the table below. Determine the total sales in week 5.

		Items sold			
	Takoyaki	**Okonomiyaki**	**Yakisoba**	**Dango**	**Total sales**
Week 1	48	37	22	67	$927.50
Week 2	35	46	61	59	$1212
Week 3	25	38	48	63	$987.50
Week 4	62	45	50	78	$1247
Week 5	49	32	56	47	

18. Consider the matrices $A = \begin{bmatrix} 1 & 3 \\ 2 & 2 \end{bmatrix}$, $P = \begin{bmatrix} 3 & 1 \\ -2 & 1 \end{bmatrix}$ and $I = \begin{bmatrix} 1 & 0 \\ 0 & 1 \end{bmatrix}$.

Demonstrate that $P^{-1}AP$ is a diagonal matrix with a determinant of -4.

19. The matrix $M = \begin{bmatrix} -\dfrac{1}{2} & -\dfrac{\sqrt{3}}{2} \\ \dfrac{\sqrt{3}}{2} & -\dfrac{1}{2} \end{bmatrix}$ is such that $M^3 = I$, the 2×2 identity matrix. Demonstrate that $M^{-1} = M^2$.

20. Lou receives an allowance from his parents based on his willingness to do the two following chores: emptying the dishwasher (daily task) and vacuuming the whole house (once on the weekend).
In March, Lou emptied the dishwasher every two days and vacuumed once, and in April, Lou vacuumed twice but emptied the dishwasher only three times. He received $50 each time.
Lou is planning for a trip at the end of Year 11 and decides to save his allowance for the next six months (184 days, 26 weekends). Determine how much money he can save if he never forgets to do the two chores assigned.
Assume that the same amount of money Lou's parents always give him for each chore is constant throughout the year.

Fully worked solutions for this chapter are available online.

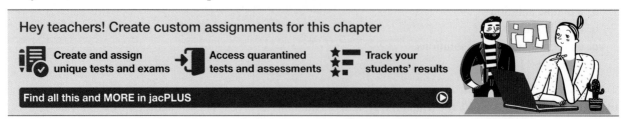

Answers

Chapter 5 Matrices

5.2 Matrix definition and notation

5.2 Exercise

1. $\begin{bmatrix} 45 \\ 30 \end{bmatrix}$, order 2×1

2. a. 2×2
   b. 2×1
   c. 1×4
   d. 2×3
   e. 3×3
   f. 4×2

3. a. $y_{13} = 5$
   b. $y_{24} = 8$
   c. $y_{31} = 1$
   d. $y_{43} = 0$
   e. $y_{41} = 2$
   f. $y_{12} = 7$

4. $\begin{bmatrix} 18 & 12 & 8 \\ 13 & 10 & 11 \end{bmatrix}$

5. $\begin{bmatrix} 6 & 4 & 7 & 3 & 6 \\ 2 & 6 & 6 & 6 & 4 \end{bmatrix}$

6. a. $b_{12} = 5$
   b. $c_{11} = 6$
   c. $a_{21} = 6$
   d. $a_{11} = 8$
   e. $c_{12} = 3$
   f. $b_{22} = 6$

7. a. There is no 4th column.
   b. e_{23}
   c. Nadia thought that e_{12} was read as 1st column, 2nd row. The correct value is 0.

8. a. 3×2
   b. $H = \begin{bmatrix} 4 & 3 \\ -1 & 7 \\ -4 & 6 \end{bmatrix}$

9. a. 3×2
   b. Element k_{22} shows that 800 Section B tickets were sold on Friday.
   c. k_{31}
   d. 2100 Section B tickets were sold in total over Thursday, Friday and Saturday.

10. a. 56 participants
    b. 213 entries
    c. $\begin{bmatrix} 12 & 17 & 18 \end{bmatrix}$

11. a. $\begin{bmatrix} 0.5 \\ 5.4 \\ 8.3 \\ 0.3 \\ 1.9 \\ 2.9 \\ 0.6 \\ 6.8 \end{bmatrix}$
    b. See matrix at the bottom of the page.*
    c. i. 3×2
    ii. $\begin{bmatrix} 801\ 428 & 8.3 \\ 227\ 600 & 6.8 \\ 1\ 727\ 200 & 5.4 \end{bmatrix}$

12. a. $\begin{bmatrix} 339\ 710 & 4.2 \\ 78\ 696 & 1.2 \\ 273\ 119 & 5.2 \\ 52\ 039 & 2.9 \\ 120\ 006 & 4.4 \\ 33\ 857 & 6.0 \\ 76\ 487 & 30.8 \\ 9\ 525 & 2.1 \end{bmatrix}$
    b. i. 76 487   ii. 33 857   iii. 691 525
    c. 983 439

5.3 Addition, subtraction and scalar multiplication of matrices

5.3 Exercise

1. $\begin{bmatrix} 63 & 19 & 25 \end{bmatrix}$; 63 kicks, 19 marks, 25 handballs

2. a. $\begin{bmatrix} 2 & 3 & 25 & 10 \\ 4 & 5 & 28 & 7 \end{bmatrix}$
   b. 6 aces, 8 double faults, 53 forehand winners and 17 backhand winners.

3. a. $\begin{bmatrix} -12 & 0 & 6 & 18 \end{bmatrix}$
   b. $\begin{bmatrix} -10 \\ 15 \\ -20 \\ 5 \end{bmatrix}$
   c. $\begin{bmatrix} -4 & 6 & -1 \\ 0 & -\dfrac{2}{3} & \dfrac{1}{3} \end{bmatrix}$

4. $x = 2;\ y = 18$

5. $x = 3;\ y = -16;\ z = 11$

6. a. $\begin{bmatrix} -10 & 4 \\ 11 & 21 \end{bmatrix}$
   b. $\begin{bmatrix} 4 & -8 \\ -6 & -10 \end{bmatrix}$

7. a. $\begin{bmatrix} 2 \\ 7 \end{bmatrix}$
   b. $\dfrac{1}{2}\begin{bmatrix} 15 \\ 14 \end{bmatrix}$

8. a. $\begin{bmatrix} 11 & 8 \\ -3 & 16 \end{bmatrix}$
   b. $\begin{bmatrix} 5 & -16 \\ 15 & 4 \end{bmatrix}$

9. a. $x = 5;\ y = 7$
   b. $x = 3;\ y = -2$

10. a. $\begin{bmatrix} 3 & 2 & 9 \\ -2 & 6 & -5 \end{bmatrix}$
    b. $\dfrac{1}{2}\begin{bmatrix} 5 & -20 & 1 \\ 13 & 10 & -6 \end{bmatrix}$

11. a. $a_{11} = 2;\ a_{12} = 3;\ a_{21} = -1;\ a_{22} = 4$
    b. $\begin{bmatrix} 3 & -2 \\ -3 & 5 \end{bmatrix}$

12. a. $\begin{bmatrix} 1 & -16 \\ 15 & 0 \end{bmatrix}$
    b. $\begin{bmatrix} 14 & 10 \\ -3 & 21 \end{bmatrix}$

13. a. $a = 2;\ b = 8;\ c = -2;\ d = -8$
    b. $a = \dfrac{11}{2};\ b = 8;\ c = -2;\ d = -\dfrac{9}{2}$

14. a. $\begin{bmatrix} 1 & 0 \\ 3 & 4 \end{bmatrix}$
    b. $\begin{bmatrix} 3 & 3 \\ 2 & 5 \end{bmatrix}$

15. a. Yes
    b. Yes

---

*11. b. $\begin{bmatrix} 2358 & 68\ 330 & 227\ 600 & 801\ 428 & 984\ 000 & 1\ 346\ 200 & 1\ 727\ 200 & 2\ 529\ 875 \end{bmatrix}$

5.4 Matrix multiplication

5.4 Exercise

1. $A = 3 \times 2$
   $B = 3 \times 1$
   $C = 2 \times 2$
   $D = 1 \times 3$

2. a. i. No   ii. Yes
      iii. Yes   iv. No
      v. Yes   vi. Yes

   b. i. –   ii. 3×3
      iii. 1×2 iv. –
      v. 1×1 vi. 3×2

3. a. $\begin{bmatrix} -8 & -20 \\ 1 & -3 \end{bmatrix}$ b. $\begin{bmatrix} 8 & 28 \\ -7 & -19 \end{bmatrix}$

   c. $\begin{bmatrix} 16 & 12 \\ 9 & 37 \end{bmatrix}$ d. $\begin{bmatrix} 0 & -4 \\ 1 & 5 \end{bmatrix}$

4. a. $\begin{bmatrix} ax + by \\ cx + dy \end{bmatrix}$ b. Does not exist.

5. a. $\begin{bmatrix} 0 & 8 & -11 \\ 7 & -26 & 34 \\ -9 & 14 & -17 \end{bmatrix}$ b. $\begin{bmatrix} -1 & 21 \\ 11 & -42 \end{bmatrix}$

6. a. $\begin{bmatrix} 3 & -2 \\ -6 & 4 \end{bmatrix}$ b. $[7]$

7. a. $x = -3$ b. $x = -4$
   c. $x = -2$ d. $x = 2$

8. a. Does not exist.   b. Does not exist.
   c. Does not exist.   d. $\begin{bmatrix} -3 & 5 \\ 6 & -10 \end{bmatrix}$
   e. $[-13]$ f. Does not exist.

9. a. $\begin{bmatrix} 6 \\ 13 \end{bmatrix}$ b. $\begin{bmatrix} 21 & -13 \end{bmatrix}$

   c. Does not exist.   d. $\begin{bmatrix} -21 & 13 \\ 42 & -26 \end{bmatrix}$

   e. Does not exist.   f. $\begin{bmatrix} 18 & -30 \\ 39 & -65 \end{bmatrix}$

10. a. $\begin{bmatrix} 0 & 0 \\ 0 & 0 \end{bmatrix}$ b. $\begin{bmatrix} 2 & -8 & 4 \\ 5 & -20 & 10 \\ 9 & -36 & 18 \end{bmatrix}$

    c. Does not exist.   d. Does not exist.

11. $\begin{bmatrix} 0 & 0 \\ 0 & 0 \end{bmatrix}$

12. United States of America, 232 points.

13. a. $\begin{bmatrix} 0 & 0 \\ 0 & 0 \end{bmatrix}$ b. $\begin{bmatrix} 0 & 0 \\ 0 & 0 \end{bmatrix}$

    c. $\begin{bmatrix} -2 & 4 \\ 3 & 5 \end{bmatrix}$ d. $\begin{bmatrix} -2 & 4 \\ 3 & 5 \end{bmatrix}$

    Observations:
    $AO = OA = O$
    $AI = IA = A$

14. a. $P^2 = \begin{bmatrix} 1 & 0 \\ 0 & 16 \end{bmatrix}$; $P^3 = \begin{bmatrix} -1 & 0 \\ 0 & 64 \end{bmatrix}$; $P^4 = \begin{bmatrix} 1 & 0 \\ 0 & 256 \end{bmatrix}$;

    $P^n = \begin{bmatrix} (-1)^n & 0 \\ 0 & 4^n \end{bmatrix}$

b. $Q^2 = \begin{bmatrix} 4 & 0 \\ 0 & 9 \end{bmatrix}$; $Q^3 = \begin{bmatrix} 8 & 0 \\ 0 & -27 \end{bmatrix}$; $Q^4 = \begin{bmatrix} 16 & 0 \\ 0 & 81 \end{bmatrix}$;

$Q^n = \begin{bmatrix} 2^n & 0 \\ 0 & (-3)^n \end{bmatrix}$

15. a. $R^2 = \begin{bmatrix} 1 & 0 \\ 6 & 1 \end{bmatrix}$; $R^3 = \begin{bmatrix} 1 & 0 \\ 9 & 1 \end{bmatrix}$; $R^4 = \begin{bmatrix} 1 & 0 \\ 12 & 1 \end{bmatrix}$;

$R^n = \begin{bmatrix} 1 & 0 \\ 3n & 1 \end{bmatrix}$

b. $S^2 = \begin{bmatrix} 6 & 0 \\ 0 & 6 \end{bmatrix}$; $S^3 = \begin{bmatrix} 0 & 18 \\ 12 & 0 \end{bmatrix}$; $S^4 = \begin{bmatrix} 36 & 0 \\ 0 & 36 \end{bmatrix}$;

$S^8 = \begin{bmatrix} 1296 & 0 \\ 0 & 1296 \end{bmatrix}$; $S^9 = \begin{bmatrix} 0 & 3888 \\ 2592 & 0 \end{bmatrix}$

16. a. $\begin{bmatrix} -20 & 0 \\ 0 & -20 \end{bmatrix}$ b. $\begin{bmatrix} 0 & 0 \\ 0 & 0 \end{bmatrix}$

17. $\begin{bmatrix} d^2 - 9d + 8 & 4 - 4d \\ 2 - 2d & 0 \end{bmatrix}$

18. No, sample responses can be found in the worked solutions in the online resources.

19. $k = 2$

20. $14.00

5.5 Determinants and inverses

5.5 Exercise

1. -22

2. $x = 4, -6$

3. a. -53 b. -33

4. $\dfrac{1}{34} \begin{bmatrix} 6 & 2 \\ -5 & 4 \end{bmatrix}$

5. a. $\dfrac{1}{4} \begin{bmatrix} -4 & 0 \\ 0 & 1 \end{bmatrix}$ b. $\dfrac{1}{6} \begin{bmatrix} 3 & 1 \\ 0 & -2 \end{bmatrix}$

   c. $\dfrac{1}{2} \begin{bmatrix} 1 & 0 \\ -3 & 2 \end{bmatrix}$ d. $\dfrac{1}{6} \begin{bmatrix} -1 & 3 \\ -2 & 0 \end{bmatrix}$

6. Determinant $= 0$

7. a. i. 20 ii. $\dfrac{1}{10} \begin{bmatrix} 1 & 1 \\ -2 & 3 \end{bmatrix}$

   b. Sample responses can be found in the worked solutions in the online resources.

   c. i. $\dfrac{1}{20}$ ii. 1

8. a. $x = -3$ b. $x = \pm 4$

9. a. $x = -6$ b. $x = \pm 2\sqrt{3}$

10. a. -11; 2; 10

    b. Yes

    c. Sample responses can be found in the worked solutions in the online resources.

11. a. $A^{-1} = -\dfrac{1}{11} \begin{bmatrix} -4 & 3 \\ 1 & 2 \end{bmatrix}$ b. No

    $B^{-1} = \dfrac{1}{2} \begin{bmatrix} 3 & -5 \\ -2 & 4 \end{bmatrix}$

    $C^{-1} = \dfrac{1}{10} \begin{bmatrix} 4 & 2 \\ -3 & 1 \end{bmatrix}$

    c. Yes   d. Yes

12. a. $k = -1, 2$ **b.** $\begin{bmatrix} -1 & 0 \\ 0 & 2 \end{bmatrix}$

13. $p = -4, \ q = -2$

14. $x = -6, 2$

15. a. Does not exist.   **b.** Does not exist.

    c. $\dfrac{1}{22} \begin{bmatrix} 5 & -4 \\ 3 & 2 \end{bmatrix}$ **d.** Does not exist.

16. $p = 1, \ q = -7, \ r = 13; \ \begin{bmatrix} 0 & 0 \\ 0 & 0 \end{bmatrix}$

17. $p = 1, \ q = 2, \ r = -11; \ \begin{bmatrix} 0 & 0 \\ 0 & 0 \end{bmatrix}$

18. $\det(A) = -140$

19. a. $\begin{bmatrix} \dfrac{4}{39} & -\dfrac{2}{39} & \dfrac{5}{78} \\[8pt] \dfrac{7}{39} & \dfrac{16}{39} & -\dfrac{1}{78} \\[8pt] \dfrac{1}{3} & \dfrac{1}{3} & -\dfrac{1}{6} \end{bmatrix}$

    b. $\begin{bmatrix} \dfrac{1}{16} & \dfrac{1}{16} & -\dfrac{1}{16} & 0 \\[8pt] \dfrac{1}{12} & 0 & \dfrac{1}{12} & -\dfrac{1}{6} \\[8pt] 0 & \dfrac{1}{4} & 0 & -\dfrac{1}{4} \\[8pt] -\dfrac{5}{48} & -\dfrac{3}{16} & \dfrac{1}{48} & \dfrac{1}{3} \end{bmatrix}$

20. $k = -2, \ 8$

5.6 Matrix algebra

5.6 Exercise

1. a. i. $\begin{bmatrix} 5 & 3 \\ 7 & 1 \end{bmatrix}$ **ii.** $\begin{bmatrix} 6 & 8 \\ 1 & 1 \end{bmatrix}$

    b. $\begin{bmatrix} 7 & 6 \\ 6 & 5 \end{bmatrix}$

    Sample responses can be found in the worked solutions in the online resources.

2. a. i. $\begin{bmatrix} 2 & 3 \\ -1 & 4 \end{bmatrix}$ **ii.** $\begin{bmatrix} 0 & 0 \\ 0 & 0 \end{bmatrix}$

    b. i. $\begin{bmatrix} a & b \\ c & d \end{bmatrix}$ **ii.** $\begin{bmatrix} 0 & 0 \\ 0 & 0 \end{bmatrix}$

3. a. $\begin{bmatrix} 0 & -18 \\ 6 & -12 \end{bmatrix}$ **b.** $\begin{bmatrix} 0 & 0 \\ 10 & 0 \end{bmatrix}$ **c.** $\begin{bmatrix} 0 & 0 \\ ax + by & 0 \end{bmatrix}$

4. Sample responses can be found in the worked solutions in the online resources.

5. Sample responses can be found in the worked solutions in the online resources.

6. a. $\begin{bmatrix} 31 & 9 \\ 23 & 1 \end{bmatrix}$ **b.** $\begin{bmatrix} 19 & -24 \\ -81 & -76 \end{bmatrix}$

7. a. $\det(A + B) \neq \det(A) + \det(B)$.

    b. $\det(AB) = \det(A) \times \det(B)$

    c. $\det(A^3) = (\det(A))^3$

8. a. $(A + B)^2 \neq A^2 + 2AB + B^2$

    b. $(A + B)^2 = A^2 + AB + BA + B^2$

    c. Sample responses can be found in the worked solutions in the online resources.

9. a. $(A + B)(A - B) \neq A^2 - B^2$

    b. $(A + B)(A - B) = A^2 + BA - AB - B^2$

    c. $(A + B)(A - B) = A^2 - B^2$ for all square matrices A and B if A and B commute.

10. a. Sample responses can be found in the worked solutions in the online resources.

    b. $A^5 = 11A - 10I$

$$= \begin{bmatrix} -10 & 11 & -11 \\ 11 & -10 & 11 \\ -11 & 11 & -10 \end{bmatrix}$$

11. a. $A^2 = \begin{bmatrix} 3 & -1 & -1 \\ -1 & 3 & -1 \\ -1 & -1 & 3 \end{bmatrix}$

        $= 2I - A$

    b. $\dfrac{1}{2}(A + I) \times A = A \times \dfrac{1}{2}(A + I) = I$ thus $\dfrac{1}{2}(A + I)$ is the inverse of A.

    c. $A^4 = 6I - 5A$

        $A^5 = 11A - 10I$

    d. $A^4 \left(\dfrac{5A + 11I}{16} \right) = \left(\dfrac{5A + 11I}{16} \right) A^4$

            $= I$

12. a. $B = \begin{bmatrix} 0 & 1 & 0 \\ 0 & 0 & 1 \\ 0 & 0 & 0 \end{bmatrix}$ **b.** Thus $A^3 = 3B^2 + 3B + I$

        $B^2 = \begin{bmatrix} 0 & 0 & 1 \\ 0 & 0 & 0 \\ 0 & 0 & 0 \end{bmatrix}$ $= \begin{bmatrix} 1 & 3 & 3 \\ 0 & 1 & 3 \\ 0 & 0 & 1 \end{bmatrix}$

        $B^3 = \begin{bmatrix} 0 & 0 & 0 \\ 0 & 0 & 0 \\ 0 & 0 & 0 \end{bmatrix}$

13. $B = \begin{bmatrix} c & d \\ 0 & c \end{bmatrix}$

14. $A^3 = 18A - 35I$

    Sample responses can be found in the worked solutions in the online resources.

15. $A^2 = (a + d)A + (bc - ad)I$

    Sample responses can be found in the worked solutions in the online resources.

5.7 Matrix equations

5.7 Exercise

1. a. $\begin{bmatrix} 1 \\ 0 \end{bmatrix}$ **b.** $\dfrac{1}{2} \begin{bmatrix} -5 & -2 \end{bmatrix}$

2. a. $\dfrac{1}{11} \begin{bmatrix} -2 \\ 7 \end{bmatrix}$ **b.** $\dfrac{1}{11} \begin{bmatrix} -7 & 3 \end{bmatrix}$

3. $x = 5, \ y = -2$

4. $x = 4, \ y = -3$

5. **a.** $x = -1, y = 2$ **b.** $x = 1, y = -2$
   c. $x = 2, y = -3$ **d.** $x = -2, y = -1$

6. **a.** $\dfrac{1}{7}\begin{bmatrix} -29 \\ 3 \end{bmatrix}$ **b.** $\dfrac{1}{14}\begin{bmatrix} 7 & 7 \\ -38 & -12 \end{bmatrix}$

   c. $\dfrac{1}{14}\begin{bmatrix} -2 & 8 \\ -22 & -3 \end{bmatrix}$ **d.** $\begin{bmatrix} -3 & 2 \end{bmatrix}$

7. **a.** $\dfrac{1}{11}\begin{bmatrix} -6 \\ 7 \end{bmatrix}$ **b.** $\begin{bmatrix} 3 & 2 \\ -4 & 1 \end{bmatrix}$

   c. $\dfrac{1}{11}\begin{bmatrix} 13 & -58 \\ 16 & 31 \end{bmatrix}$ **d.** $\dfrac{1}{22}\begin{bmatrix} 17 & 3 \end{bmatrix}$

8. **a.** $\dfrac{1}{10}\begin{bmatrix} 11 & 3 \\ -30 & 0 \end{bmatrix}$ **b.** $\dfrac{1}{10}\begin{bmatrix} 2 & 8 \\ -9 & 9 \end{bmatrix}$

9. The lines are parallel. No solution.

10. $\left(3 + \dfrac{3t}{4}, t\right), t \in \mathbb{R}$

11. $k = -\dfrac{5}{2}$

12. $k = -\dfrac{5}{2}$

13. **a. i.** $k \neq -3$ **ii.** $k = -3$
    b. i. $k \neq -5$ **ii.** $k = -5$

14. **a.** $(2t + 3, t), t \in \mathbb{R}$ **b.** No solution.

15. **a. i.** $k \in \mathbb{R}\backslash\{-2, 4\}$ **ii.** $k = -2$
    iii. $k = 4$

    b. i. $k \in \mathbb{R}\backslash\{-3, 2\}$ **ii.** $k = -3$
    iii. $k = 2$

    c. i. $k \in \mathbb{R}\backslash\{-2, 3\}$ **ii.** $k = -2$
    iii. $k = 3$

    d. i. $k \in \mathbb{R}\backslash\{1, 6\}$ **ii.** $k = 1$
    iii. $k = 6$

16. **a.** $k \in \mathbb{R}\backslash\{-3, 2\}$
    b. $k = -3$
    c. $k = 2$

17. **a. i.** $\dfrac{12}{a} + \dfrac{6}{b} = 1, \dfrac{8}{a} + \dfrac{3}{b} = 1$
    ii. $a = 4, b = -3$
    b. i. $\dfrac{4}{a} + \dfrac{5}{b} = 1, -\dfrac{4}{a} - \dfrac{15}{b} = 1$
    ii. $a = 2, b = -5$

18. **a. i.** $q \neq 4, p \in \mathbb{R}$ **ii.** $q = 4, p \neq -\dfrac{7}{2}$
    iii. $q = 4, p = -\dfrac{7}{2}$

    b. i. $p \neq -\dfrac{3}{2}, q \in \mathbb{R}$ **ii.** $p = -\dfrac{3}{2}, q \neq \dfrac{40}{3}$
    iii. $p = -\dfrac{3}{2}, q = \dfrac{40}{3}$

    c. i. $p \neq \dfrac{6}{7}, q \in \mathbb{R}$ **ii.** $p = \dfrac{6}{7}, q \neq 14$
    iii. $p = \dfrac{6}{7}, q = 14$

d. i. $p \neq \dfrac{3}{2}, q \in \mathbb{R}$ **ii.** $p = \dfrac{3}{2}, q \neq -6$
   iii. $p = \dfrac{3}{2}, q = -6$

19. $P^{-1} = \begin{bmatrix} \dfrac{1}{a} & 0 \\ 0 & \dfrac{1}{d} \end{bmatrix}$

20. **a.** $R^{-1} = \begin{bmatrix} 0 & \dfrac{1}{c} \\ \dfrac{1}{b} & -\dfrac{a}{bc} \end{bmatrix}$

    b. $S^{-1} = \begin{bmatrix} -\dfrac{d}{bc} & \dfrac{1}{c} \\ \dfrac{1}{b} & 0 \end{bmatrix}$

    c. $A^{-1} = \dfrac{1}{ad - bc}\begin{bmatrix} d & -b \\ -c & a \end{bmatrix}$

5.8 Review

5.8 Exercise

1. **a.** $\begin{bmatrix} -1 \\ 4 \end{bmatrix}$
   b. $\begin{bmatrix} -1 & 4 & -3 \end{bmatrix}$
   c. $\begin{bmatrix} 0 & 9 \\ -6 & 1 \\ 12 & -3 \end{bmatrix}$
   d. $\begin{bmatrix} 18 \\ -23 \end{bmatrix}$

2. C

3. **a.** $\begin{bmatrix} 4 & -6 & -3 \\ -1 & 3 & 2 \\ 1 & 9 & 8 \end{bmatrix}$
   b. 47
   c. $C^{-1} = \begin{bmatrix} \dfrac{1}{3} & -\dfrac{2}{3} \\ \dfrac{2}{3} & -\dfrac{1}{3} \end{bmatrix}$

4. **a.** $x = -2$ **b.** $x = \dfrac{1}{2}, 3$

5. **a.** $x = -2, y = 3$ **b.** $x = -3, y = 2$
   c. $x = 14, y = 23$ **d.** $x = 10, y = 7$

6. **a.** $\dfrac{1}{5}\begin{bmatrix} -13 & 1 \\ -7 & -6 \end{bmatrix}$ **b.** $\dfrac{1}{5}\begin{bmatrix} 0 & -5 \\ 17 & -19 \end{bmatrix}$

7. **a.** No solution. Lines are parallel.
   b. Infinite solutions, lines are identical. $\left(\dfrac{4t + 5}{3}, t\right), t \in \mathbb{R}$

8. **a.** $x = 3a + 2b, y = 3a - 2b$
   b. $x = 2a, y = 3b$
   c. $x = a + 7b, y = a + b$

9. a. $AB = \begin{bmatrix} ac & ad \\ bc & bd \end{bmatrix}$

   b. $BA = [ac + bd]$

   $(BA)^{-1} = \left[\dfrac{1}{ac + bd} \right]$

10. a. $x = -2$
    c. $x = -5,\ 3$
    b. $x = 7$
    d. $x = -7,\ 2$

11. a. $\begin{bmatrix} 3 \\ -4 \end{bmatrix}$
    b. $\begin{bmatrix} 3 & -2 \\ -4 & 1 \end{bmatrix}$

    c. $\dfrac{1}{11} \begin{bmatrix} -5 & 36 \\ 10 & 49 \end{bmatrix}$
    d. $\begin{bmatrix} 0 & 1 \end{bmatrix}$

12. a. $\begin{bmatrix} -2 \\ 3 \end{bmatrix}$
    b. $\dfrac{1}{11} \begin{bmatrix} 21 & -2 \\ 5 & -12 \end{bmatrix}$

    c. $\dfrac{1}{11} \begin{bmatrix} 10 & 29 \\ 8 & -1 \end{bmatrix}$
    d. $\dfrac{1}{11} \begin{bmatrix} 2 & -7 \end{bmatrix}$

13. $x = ab,\ y = 2ab$

14. a. $k \in \mathbb{R} \backslash \{-4, 3\}$
    b. $k = 3$
    c. $k = -4$

15. a. $k \in \mathbb{R} \backslash \{-3, 4\}$
    b. No solution when $k = -3$
    c. An infinite number of solutions when $k = 4$

16. $p = 1, q = 2, r = -11,\ A^2 + 2A - 11$
    Sample responses can be found in the worked solutions in the online resources.

17. \$1063.50

18. Sample responses can be found in the worked solutions in the online resources.

19. Sample responses can be found in the worked solutions in the online resources.

20. \$888

UNIT

2 Complex numbers, further proof, trigonometry, functions and transformations

Source: Specialist Mathematics Senior Syllabus 2024 © State of Queensland (QCAA) 2024; licensed under CC BY 4.0.

6 Complex numbers

LESSON SEQUENCE

Fully worked solutions for this chapter are available online.

EXAM PREPARATION
Access exam-style questions in every lesson, available online.

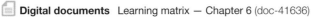 Resources

 Solutions Solutions — Chapter 6 (sol-0398)

Exam questions Exam question booklet — Chapter 6 (eqb-0284)

 Digital documents Learning matrix — Chapter 6 (doc-41636)
Chapter summary — Chapter 6 (doc-41633)

LESSON
6.1 Overview

6.1.1 Introduction

In 1545, the Italian mathematician Girolamo Cardano proposed what was then a startling mathematical expression:

$$40 = \left(5 + \sqrt{-15}\right)\left(5 - \sqrt{-15}\right)$$

This was a valid expression, yet it included the square root of a negative number, which seemed 'impossible. And in 1572, the Italian mathematician Rafael Bombelli published '*L'Algebra*', a major contribution to complex number theory, and determined that complex numbers were key in solving cubic and quartic equations.

The definition of real numbers included whole numbers, fractions, decimals, irrational and rational numbers as subsets of the real number set. Whenever the square root of a negative number was encountered, it did not fit into any of these subsets and hence could not be classified as a real number. However, the solutions to quadratic equations sometimes force us to consider the square root of a negative number. Consider the two equations $x^2 + 2x + 26 = 0$ and $x^2 - 4x + 29 = 0$. How do the solutions of these equations relate to properties of the associated parabolas?

Up to this point, you have encountered solutions of quadratic equations that were classified as having rational, irrational or no real solutions. The study of complex numbers provides solutions to these previously unsolvable equations.

Why did the square roots of negative numbers become central to the study of a new set of numbers called complex numbers? It was partly curiosity and partly because people such as the Ancient Greek mathematician Diophantus and the 17th-century German mathematician Leibniz had found that real numbers could not solve all equations. Eventually it was shown that complex numbers could solve previously unsolvable problems. They are now used extensively in the fields of physics and engineering in areas such as electrical circuits and electromagnetic waves. Combined with calculus, complex numbers form an important part of the field of mathematics known as complex analysis.

6.1.2 Syllabus links

Lesson	Lesson title	Syllabus links
6.2	**Introduction to complex numbers**	○ Define the imaginary number i as a root (solution) of the equation $x^2 = -1$.
		○ Define and use set notation of the number system for complex numbers ($\mathbb{C}$).
		○ Use complex numbers in the form $a + bi$ where a and b are the real and imaginary parts (components) $\mathrm{Re}(z)$ and $\mathrm{Im}(z)$ of a complex number z.
6.3	**Basic operations on complex numbers**	○ Perform complex-number arithmetic: addition, subtraction, multiplication and division, with and without technology.
6.4	**Complex conjugates and division of complex numbers**	○ Use complex numbers in the form $a + bi$ where a and b are the real and imaginary parts (components) $\mathrm{Re}(z)$ and $\mathrm{Im}(z)$ of a complex number z.
		○ Determine and use complex conjugates.
		○ Perform complex-number arithmetic: addition, subtraction, multiplication and division, with and without technology.
6.5	**The complex plane (the Argand plane)**	○ Sketch and use complex numbers as points in the complex plane with real and imaginary parts as Cartesian coordinates.
		○ Understand and use addition of complex numbers as vector addition in the complex plane.
6.6	**Conjugates and multiplication in the complex plane**	○ Understand and use location of complex conjugates in the complex plane.
		○ Understand and use multiplication by a complex number as a linear transformation in the complex plane.

Source: Specialist Mathematics Senior Syllabus 2024 © State of Queensland (QCAA) 2024; licensed under CC BY 4.0

LESSON
6.2 Introduction to complex numbers

SYLLABUS LINKS

- Define the imaginary number i as a root (solution) of the equation $x^2 = -1$.
- Define and use set notation of the number system for complex numbers ($\mathbb{C}$).
- Use complex numbers in the form $a + bi$ where a and b are the real and imaginary parts (components) $\text{Re}(z)$ and $\text{Im}(z)$ of a complex number z.

Source: Specialist Mathematics Senior Syllabus 2024 © State of Queensland (QCAA) 2024; licensed under CC BY 4.0

6.2.1 Square root of a negative number

The quadratic equation $x^2 + 1 = 0$ has no solutions for x in the Real Number System $\mathbb{R}$ because the equation yields $x = \pm \sqrt{-1}$ and there is no real number which, when squared, gives -1 as the result.

If, however, we define an **imaginary number** denoted by i such that:

$$i^2 = -1$$
$$i = \pm\sqrt{-1}$$

Then the answer to the quadratic $x^2 + 1 = 0$, $x = \pm \sqrt{-1}$ becomes $x = \pm \sqrt{i^2} = \pm i$.

For instance, let's solve $x^2 + 12 = 0$ using the imaginary number i:

$$x^2 = -12$$
$$x = \pm\sqrt{-12}$$
$$x = \pm\sqrt{-1 \times 12}$$
$$x = \pm\sqrt{i^2 \times 12}$$
$$x = \pm i\sqrt{12}$$
$$x = \pm 2\sqrt{3}i$$

Powers of i will produce $\pm i$ or ± 1. As $i^2 = -1$, if follows:

$i^0 = 1$	$i^1 = i$	$i^2 = -1$	$i^3 = -i$
$i^4 = 1$	$i^5 = i$	$i^6 = -1$	$i^7 = -i$
$i^8 = 1$	$i^9 = i$	$i^{10} = -1$	$i^{11} = -i$
$i^{12} = 1$	$i^{13} = i$	$i^{14} = -1$	$i^{15} = -i$

What pattern do you notice?

The pattern is, quite obviously, even powers of i result in 1 or -1 and odd powers of i result in i or $-i$.

Let's generalise, if $k \in \mathbb{Z}$:

| $i^{4k} = 1$ | $i^{4k+1} = i$ | $i^{4k+2} = -1$ | $i^{4k+3} = -i$ |

6.2.2 Definition of a complex number

The real number system, which you denoted by the symbol $\mathbb{R}$, contains the set of rational and irrational numbers. Rational numbers ($\mathbb{Q}$) and integers ($\mathbb{Z}$) are examples of subsets of the real number systems.

You will now see that $\mathbb{R}$ is itself a subset of the complex number system $\mathbb{C}$.

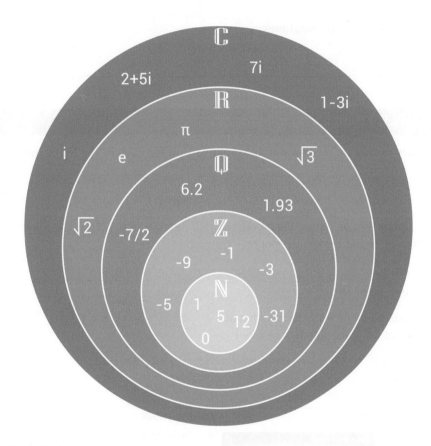

A **complex number** (generally denoted by the letter z) is defined as a quantity consisting of a real number added to a multiple of the imaginary unit i. For real numbers a and b, $a + bi$ is a complex number. This is referred to as the standard or **Cartesian form**.

$\mathbb{C} = \{z : z = a + bi$ where $a, b \in \mathbb{R}\}$ defines the set of complex numbers.

The real part of z is a and is written as $\text{Re}(z)$. That is, $\text{Re}(z) = a$.

The imaginary part of z is b and is written as $\text{Im}(z)$. That is, $\text{Im}(z) = b$.

Every real number x can be written as $a + 0i$, so the set of real numbers is a subset of the set of complex numbers. That is, $\mathbb{R} \subset \mathbb{C}$. A real number is a complex number with an imaginary part equal to zero.

Definition of a complex number

A complex number is of the form:

$$z = a + bi$$

where $a, b \in \mathbb{R}$.

$a = \text{Re}(z)$ is the real part of z, and $b = \text{Im}(z)$, is the imaginary part of z.

A number is purely real:

$$z = a + 0i$$

A number is purely imaginary:

$$z = 0 + bi$$

WORKED EXAMPLE 1 Simplifying in terms of *i*

Using the imaginary number i, write a simplified expression for each of the following.

a. $\sqrt{-16}$ **b.** $\sqrt{-5}$.

THINK	WRITE
a. 1. Express the square root of -16 as the product of the square root of 16 and the square root of -1.	**a.** $\sqrt{-16} = \sqrt{16} \times \sqrt{-1}$
2. Substitute i^2 for -1.	$= \sqrt{16} \times \sqrt{i^2}$
3. Take the square root of 16 and i^2.	$= \pm 4i$
b. 1. Express the square root of -5 as the product of the square root of 5 and the square root of -1.	**b.** $\sqrt{-5} = \sqrt{5} \times \sqrt{-1}$
2. Substitute i^2 for -1.	$= \sqrt{5} \times \sqrt{i^2}$
3. Simplify.	$= \pm \sqrt{5}i$

TI | THINK	WRITE	CASIO | THINK	WRITE
a. 1. In the Document Settings menu, change the Calculation Mode to Rectangular, then select OK.	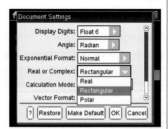	**a. 1.** In the SETCP menu, change the Complex Mode to $a + bi$, then press EXIT.	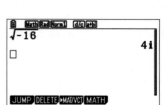
2. On a Calculator page, complete the entry line as: $\sqrt{-16}$ then press ENTER.	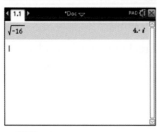	**2.** On the Run-Matrix screen, complete the entry line as: $\sqrt{-16}$ then press EXE.	
3. The answer appears on the screen.	$\sqrt{-16} = 4i$	**3.** The answer appears on the screen.	$\sqrt{-16} = 4i$

WORKED EXAMPLE 2 Decomposing complex numbers into real and imaginary parts

Write down the real and imaginary parts of the following complex numbers, z.

a $z = -3 + 2i$ **b** $z = -\dfrac{1}{2}i$ **c** $z = \dfrac{\sqrt{2}}{2}$

THINK	WRITE
a. 1. The real part is the 'non-i' term.	**a.** $\text{Re}(z) = -3$
2. The imaginary part is the coefficient of the i term.	$\text{Im}(z) = 2$
b. 1. The real part is the 'non-i' term.	**b.** $\text{Re}(z) = 0$
2. The imaginary part is the coefficient of the i term.	$\text{Im}(z) = -\dfrac{1}{2}$
c. 1. The real part is the 'non-i' term.	**c.** $\text{Re}(z) = \dfrac{\sqrt{2}}{2}$
2. The imaginary part is the coefficient of the i term.	$\text{Im}(z) = 0$

WORKED EXAMPLE 3 Evaluating powers of i

Write $i^8 + i^5$ in the form $a + bi$ where a and b are real numbers.

THINK	WRITE
1. Simplify both i^8 and i^5, remembering $i^2 = -1$ and recalling the index laws.	$i^8 = \left(i^2\right)^4 = (-1)^4 = 1$ $i^5 = i^4 \times i = \left(i^2\right)^2 \times i = (-1)^2 \times i = 1 \times i = i$
2. Add the two answers.	$i^8 + i^5 = 1 + i$

WORKED EXAMPLE 4 Simplifying complex expressions

Simplify $z = i^4 - 2i^2 + 1$ and $w = i^6 - 3i^4 + 3i^2 - 1$ and show that $z + w = -4$.

THINK	WRITE
1. Replace terms with the lowest possible powers of i (remember $i^2 = -1$).	$z = i^4 - 2i^2 + 1$ $= \left(i^2\right)^2 - 2 \times (-1) + 1$ $= (-1)^2 + 2 + 1$ $= 4$ $w = i^6 - 3i^4 + 3i^2 - 1$ $= \left(i^2\right)^3 - 3\left(i^2\right)^2 + 3 \times (-1) - 1$ $= (-1)^3 - 3(-1)^2 - 3 - 1$ $= -1 - 3 - 3 - 1$ $= -8$
2. Add the two answers.	$z + w = i^4 - 2i^2 + 1 + i^6 - 3i^4 + 3i^2 - 1$ $= 4 - 8$ $= -4$

WORKED EXAMPLE 5 Evaluating real and imaginary parts

Evaluate each of the following.

a. $\text{Re}(7 + 6i)$

b. $\text{Im}(10)$

c. $\text{Re}\left(2 + i - 3i^3\right)$

d. $\text{Im}\left(\dfrac{1 - 3i - i^2 - i^3}{2}\right)$

THINK	WRITE
a. Recall the real part of the complex number $a + bi$ is a, so the real part of $7 + 6i$ is 7.	a. $\text{Re}(7 + 6i) = 7$
b. The number 10 can be expressed in complex form as $10 + 0i$ and so the imaginary part is 0.	b. $\text{Im}(10) = \text{Im}(10 + 0i)$ $= 0$
c. 1. Simplify $2 + i - 3i^3$ to the lowest possible power of i (remember $i^2 = -1$).	c. $\text{Re}(2 + i - 3i^3) = \text{Re}(2 + i - 3i \times i^2)$ $= \text{Re}(2 + i + 3i)$ $= \text{Re}(2 + 4i)$
2. The real part is 2.	$= 2$
d. 1. Simplify the numerator of $\left(\dfrac{1 - 3i - i^2 - i^3}{2}\right)$ to the lowest possible value of i.	d. $\text{Im}\left(\dfrac{1 - 3i - i^2 - i^3}{2}\right) = \text{Im}\left(\dfrac{1 - 3i + 1 + i}{2}\right)$ $= \text{Im}\left(\dfrac{2 - 2i}{2}\right)$
2. Simplify by dividing the numerator by 2.	$= \text{Im}\dfrac{(2(1 - i))}{2}$ $= \text{Im}(1 - i)$
3. The imaginary part is -1.	$= -1$

TI | THINK	WRITE	CASIO | THINK	WRITE
c. 1. On a Calculator page, press MENU, then select: 2: Number 9: Complex Number Tools 2: Real Part. Complete the entry line as: real $(2 + i - 3i^3)$ then press ENTER. *Note:* The symbol i can be found by pressing the π button.	real$(2+i-3\cdot i^3)$ 2	c. 1. On a Run-Matrix screen, press OPTN, then select COMPLEX by pressing F3. Press F6 to scroll across to more menu options, then select ReP by pressing F1. Complete the entry line as: ReP $(2 + i - 3i^3)$ then press EXE. *Note:* The symbol i can be found by pressing SHIFT 0.	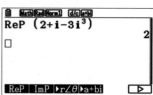
2. The answer appears on the screen.	$\text{Re}(2 + i - 3i^3) = 2$	2. The answer appears on the screen.	$\text{Re}(2 + i - 3i^3) = 2$

d. 1. On a Calculator page, press MENU, then select:
2: Number
9: Complex Number Tools
3: Imaginary Part.
Complete the entry line as:

$$\text{imag}\left(\frac{1 - 3i - i^2 - i^3}{2}\right)$$

then press ENTER.

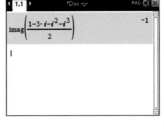

2. The answer appears on the screen.

$$\text{Im}\left(\frac{1 - 3i - i^2 - i^3}{2}\right) = -1$$

d. 1. On a Run-Matrix screen, press OPTN, then select COMPLEX by pressing F3. Press F6 to scroll across to more menu options, then select ImP by pressing F2. Complete the entry line as:

$$\text{ImP}\,\frac{1 - 3i - i^2 - i^3}{2}$$

then press EXE.

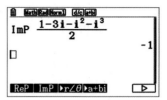

2. The answer appears on the screen.

$$\text{Im}\left(\frac{1 - 3i - i^2 - i^3}{2}\right) = -1$$

Exercise 6.2 Introduction to complex numbers

learnon

6.2 Exercise	6.2 Exam questions on

Simple familiar	Complex familiar	Complex unfamiliar
1, 2, 3, 4, 5, 6, 7, 8, 9, 10	11, 12, 13	14, 15

These questions are even better in jacPLUS!
- Receive immediate feedback
- Access sample responses
- Track results and progress

Find all this and MORE in jacPLUS ▶

Simple familiar

1. **WE1** Using the imaginary number i, write down expressions for:

   a. $\sqrt{-9}$ **b.** $\sqrt{-25}$ **c.** $\sqrt{-49}$ **d.** $\sqrt{-3}$

2. Using the imaginary number i, write down expressions for:

   a. $\sqrt{-11}$ **b.** $\sqrt{-7}$ **c.** $\sqrt{-\dfrac{4}{9}}$ **d.** $\sqrt{-\dfrac{36}{25}}$

3. **WE2** Write down the real and imaginary parts, respectively, of the following complex numbers, z.

   a. $9 + 5i$ **b.** $5 - 4i$ **c.** $-3 - 8i$ **d.** $11i - 6$

4. Write down the real and imaginary parts, respectively, of the following complex numbers, z.

   a. 27 **b.** $2i$ **c.** $-5 + i$ **d.** $-17i$

5. **WE3** Write each of the following in the form $a + bi$, where a and b are real numbers.

   a. $i^9 + i^{10}$ **b.** $i^{12} + i^{15}$ **c.** $i^7 - i^{11}$ **d.** $i^{416} - i^{263}$

6. Write each of the following in the form $a + bi$, where a and b are real numbers.

   a. $i^5 + i^6 - i^7$ **b.** $i\left(i^{13} + i^{16}\right)$ **c.** $2i - i^2 + 2i^3$ **d.** $3i + i^4 - 5i^5$

7. **WE5** Evaluate each of the following.

   a. $\text{Re}(-5 + 4i)$
   b. $\text{Re}(15 - 8i)$
   c. $\text{Re}(12i)$
   d. $\text{Im}(1 - 6i)$

8. Evaluate each of the following.

   a. $\text{Im}(3 + 2i)$

   b. $\text{Im}(8)$

   c. $\text{Re}(i^5 - 3i^4 + 6i^6)$

   d. $\text{Im}\left(\dfrac{4i^9 - 5i^{14} - 2i^7}{3}\right)$

9. Write $3 - \dfrac{i^3 - i + 2}{i^2 - i^4}$ in the form $a + bi$, where a and b are real numbers.

10. Simplify each of the following.

    a. $\dfrac{6i^3}{\sqrt{-9}}$

    b. $\dfrac{20i^4}{\sqrt{-100}}$

    c. $\dfrac{10i^5}{\sqrt{-50}}$

    d. $\dfrac{8i^6}{\sqrt{-16}}$

Complex familiar

11. **WE4** Simplify $z = i^6 + 3i^7 - 2i^{10} - 3$ and $w = 4i^8 - 3i^{11} + 3$ and show that $z + w = 5$.

12. If $f(i) = \dfrac{1 - 2i + 3i^2 - 4i^3 + 5i^4}{4}$, evaluate the following.

    a. $\text{Re}(f(i))$

    b. $\text{Im}(f(i))$

13. Evaluate the following.

    a. $\text{Re}(3(4 - 6i) + i^8)$

    b. $\text{Im}(2(2 - 5i) + i^7)$

    c. $\text{Re}(3(2 - 5i) - 4i^{10})$

    d. $\text{Im}(4(2 + 3i) - 5i^9)$

Complex unfamiliar

14. If $f(n) = 1 + i + i^2 + i^3 + i^4 + \dots + i^n$, evaluate the following without using technology.

    a. $f(2014)$

    b. $f(2015)$

15. If $g(n) = 1 - i + i^2 - i^3 + i^4 + \dots + (-i)^n$, evaluate the following without using technology.

    a. $g(2014)$

    b. $g(2015)$

Fully worked solutions for this chapter are available online.

LESSON
6.3 Basic operations on complex numbers

> **SYLLABUS LINK**
>
> - Perform complex-number arithmetic: addition, subtraction, multiplication and division, with and without technology.
>
> ***Source:*** Specialist Mathematics Senior Syllabus 2024 © State of Queensland (QCAA) 2024; licensed under CC BY 4.0

6.3.1 Complex number arithmetic

Complex numbers can be added, subtracted, multiplied and divided. In general, the solutions obtained when performing these operations are presented in the **standard form** $z = a + bi$.

Addition of complex numbers

Addition is performed by adding the real and imaginary parts separately.

You can think of the real parts of the numbers being like terms and the imaginary parts of the numbers being like terms.

> ### Addition of complex numbers
>
> If $z = m + ni$ and $w = p + qi$, then:
> $$z + w = (m + p) + (n + q)i$$

Subtraction of complex numbers

If we write $z - w$ as $z + -w$, we can use the rule for addition of complex numbers to obtain:

$$\begin{aligned} z + -w &= (m + ni) + -(p + qi) \\ &= m + ni - p - qi \\ &= (m - p) + (n - q)i \end{aligned}$$

> ### Subtraction of complex numbers
>
> If $z = m + ni$ and $w = p + qi$, then:
> $$z - w = (m - p) + (n - q)i$$

WORKED EXAMPLE 6 Addition and subtraction of complex numbers

For $z = 8 + 7i$, $w = -12 + 5i$ and $u = 1 + 2i$, calculate the following.

a. $z + w$ b. $w - z$ c. $u - w + z$

THINK

a. Use the addition rule for complex numbers.

b. Use the subtraction rule for complex numbers.

c. Use both the addition rule and the subtraction rule.

WRITE

a. $\begin{aligned} z + w &= (8 + 7i) + (-12 + 5i) \\ &= (8 - 12) + (7 + 5)i \\ &= -4 + 12i \end{aligned}$

b. $\begin{aligned} w - z &= (-12 + 5i) - (8 + 7i) \\ &= (-12 - 8) + (5 - 7)i \\ &= -20 - 2i \end{aligned}$

c. $\begin{aligned} u - w + z &= (1 + 2i) - (-12 + 5i) + (8 + 7i) \\ &= (1 + 12 + 8) + (2 - 5 + 7)i \\ &= 21 + 4i \end{aligned}$

Multiplication by a constant (or scalar)

Consider part **a** in Worked example 6, where $z + w = -4 + 12i$. The real and imaginary parts share a common factor of 4. Hence, the equation could be rewritten as:

$$\begin{aligned} z + w &= -4 + 12i \\ &= 4(-1 + 3i) \end{aligned}$$

This is the equivalent of the complex number $-1 + 3i$ multiplied by a constant (or scalar), 4.

If we consider the general form $z = a + bi$ and $k \in \mathbb{R}$, then:

Multiplying a complex number by a constant

For a complex number $z = a + bi$ multiplied by a constant $k \in \mathbb{R}$:

$$kz = k(a + bi)$$
$$= ka + kbi$$

WORKED EXAMPLE 7 Scalar multiplication of complex numbers

If $z = 3 + 5i$, $w = 4 - 2i$ and $v = 6 + 10i$, evaluate the following.

a. $4z$

b. $-3v$

c. $3z + w$

d. $2z - v$

e. $4z - 3w + 2v$

THINK	WRITE
a. Calculate $4z$ by substituting values for z.	a. $4z = 4(3 + 5i)$ $= 12 + 20i$
b. Calculate $-3v$ by substituting values for v.	b. $-3v = -3(6 + 10i)$ $= -18 - 30i$
c. 1. Calculate $3z + w$ by substituting values for z and w.	c. $3z + w = 3(3 + 5i) + (4 - 2i)$ $= (9 + 15i) + (4 - 2i)$
2. Use the rule for adding complex numbers.	$= (9 + 4) + (15 - 2)i$ $= 13 + 13i$ (or $13(1 + i)$)
d. 1. Calculate $2z - v$ by substituting values for z and v.	d. $2z - v = 2(3 + 5i) - (6 + 10i)$
2. Use the rule for subtraction of complex numbers.	$= 6 + 10i - 6 - 10i$ $= 0 + 0i$ $= 0$
e. 1. Calculate $4z - 3w + 2v$ by substituting values for z, w and v.	e. $4z - 3w + 2v = 4(3 + 5i) - 3(4 - 2i) + 2(6 + 10i)$
2. Use the addition rule and the subtraction rule to simplify.	$= 12 + 20i - 12 + 6i + 12 + 20i$ $= 12 + 46i$

TI	THINK	WRITE	CASIO	THINK	WRITE
e. 1. On a Calculator page, complete the entry line as: $3 + 5i$ Press CTRL, then press VAR, then type 'z' and press ENTER to store $z = 3 + 5i$. Repeat this step to store the complex numbers w and v. *Note:* The symbol i can be found by pressing the π button.	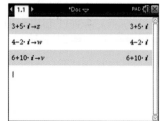	e. 1. On a Run-Matrix screen, complete the entry line as: $3 + 5i$ Press the store button, then type 'Z' and press EXE to store $z = 3 + 5i$. Repeat this step to store the complex numbers w and v. *Note:* The symbol i can be found by pressing SHIFT 0.	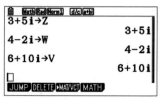

2. Complete the next
entry line as:
$$4z - 3w + 2v$$
then press ENTER.

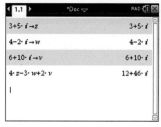

2. Complete the next entry
line as:
$$4Z - 3W + 2V$$
then press EXE.

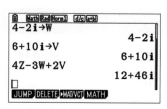

3. The answer appears on $4z - 3w + 2v = 12 + 46i$
the screen.

3. The answer appears on $4z - 3w + 2v = 12 + 46i$
the screen.

Multiplication of two complex numbers

Multiplication of two complex numbers also results in a complex number.

To multiply complex numbers, simply treat them like algebraic expressions and expand the brackets.

WORKED EXAMPLE 8 Multiplication of complex numbers (1)

If $z = 6 - 2i$ and $w = 3 + 4i$, express zw in standard form.

THINK	WRITE
1. Expand the brackets.	$zw = (6 - 2i)(3 + 4i)$ $= 18 + 24i - 6i - 8i^2$
2. Express in the form $a + bi$ by substituting -1 for i^2 and simplifying the expression using the addition and subtraction rules.	$= 18 + 24i - 6i + 8$ $= 26 + 18i$

WORKED EXAMPLE 9 Multiplication of complex numbers (2)

Simplify $(2 - 3i)(2 + 3i)$.

THINK	WRITE
1. Expand the brackets.	$(2 - 3i)(2 + 3i) = 4 + 6i - 6i - 9i^2$
2. Substitute -1 for i^2 and simplify the expression.	$= 4 - 9 \times (-1)$ $= 13$
Alternatively, you could also use the algebraic identity $(a + b)(a - b) = a^2 - b^2$	$(2 - 3i)(2 + 3i) = (2)^2 - (3i)^2$ $= 4 - (-9)$ $= 4 + 9$ $= 13$

Note that if a and b are real numbers, then $(a + bi)(a - bi) = a^2 + b^2$, which is always a real number. You will learn more about this in lesson 6.4.

b. 1. Write the left-hand side of the equation.

b. $\text{LHS} = (3 + 4i)(x + yi)$

2. Expand the left-hand side of the equation.

$= 3x + 3yi + 4xi + 4yi^2$

3. Express the left-hand side in the form $a + bi$.

$= (3x - 4y) + (4x + 3y)i$

4. Equate the real parts and imaginary parts of both sides of the equation to create a pair of simultaneous equations.

$3x - 4y = 29$ [1]
$4x + 3y = 22$ [2]

5. Simultaneously solve [1] and [2] for x and y.

$9x - 12y = 87$ [3]

Multiply equation [1] by 3 and equation [2] by 4 so that y can be eliminated.

$16x + 12y = 88$ [4]

6. Add the two new equations and solve for x.

Adding equations [3] and [4]:
$25x = 175$
$x = 7$

7. Substitute $x = 7$ into equation [1] and solve for y.

Substituting $x = 7$ into equation [1]:
$3(7) - 4y = 29$
$21 - 4y = 29$
$-4y = 8$
$y = -2$

8. State the solution.

Therefore, $x = 7$ and $y = -2$.

9. Check the solution by substituting these values into equation [2].

Check: $4 \times 7 + 3 \times -2 = 22$.

Exercise 6.3 Basic operations on complex numbers

learn on

6.3 Exercise	**6.3 Exam questions** on

Simple familiar	Complex familiar	Complex unfamiliar
1, 2, 3, 4, 5, 6, 7, 8, 9, 10, 11	12, 13, 14, 15	16

These questions are even better in jacPLUS!
- Receive immediate feedback
- Access sample responses
- Track results and progress

Find all this and MORE in jacPLUS

Simple familiar

1. **WE6** For $z = 5 + 3i$, $w = -1 - 4i$, $u = 6 - 11i$ and $v = 2i - 3$, calculate the following.

   a. $z + w$ **b.** $u - z$ **c.** $w + v$

2. For $z = 5 + 3i$, $w = -1 - 4i$, $u = 6 - 11i$ and $v = 2i - 3$, calculate the following.

   a. $u - v$ **b.** $w - z - u$ **c.** $v + w - z$

3. **WE7** If $z = -3 + 2i$, $w = -4 + i$ and $u = -8 - 5i$, evaluate the following.

   a. $3w$ **b.** $2u + z$ **c.** $4z - 3u$

4. If $z = -3 + 2i$, $w = -4 + i$ and $u = -8 - 5i$, evaluate the following.

   a. $3z + u + 2w$ **b.** $2z - 7w + 9u$ **c.** $3(z + 2u) - 4w$

5. **WE8** For $z = 5 + 3i$, $w = -1 - 4i$, $u = 6 - 11i$ and $v = 2i - 3$, calculate each of the following in the standard form $a + bi$.

    a. zw **b.** uv **c.** wu

6. For $z = 5 + 3i$, $w = -1 - 4i$, $u = 6 - 11i$ and $v = 2i - 3$, calculate each of the following in the standard form $a + bi$.

    a. zu **b.** u^2 **c.** $u(wv)$

7. **WE9** Simplify the following.

    a. $(10 + 7i)(9 - 3i)$ **b.** $(3 - 4i)(5 + 4i)$ **c.** $(8 - 2i)(4 - 5i)$

8. Simplify the following.

    a. $(5 + 6i)(5 - 6i)$ **b.** $(2i - 7)(2i + 7)$ **c.** $(9 - 7i)^2$

9. For $z = -1 - 3i$ and $w = 2 - 5i$, calculate $z^2 w$.

10. For $z = 3 + 5i$, $w = 2 - 3i$ and $u = 1 - 4i$, determine:

    a. $\text{Re}(zu) - \text{Im}\left(w^2\right)$
    b. $\text{Re}\left(z^2\right) - \text{Re}(zw) - \text{Im}(uz)$
    c. $\text{Re}\left(u^2 w\right) + \text{Im}\left(zw^2\right)$

11. For $z = 3 + 5i$, $w = 2 - 3i$ and $u = 1 - 4i$, determine:

    a. $\text{Im}\left(u^2\right)$ **b.** $\text{Re}\left(w^2\right)$ **c.** $\text{Re}(uw) + \text{Im}(zw)$

Complex familiar

12. **WE10** Determine $\text{Re}\left(z^2\right) - \text{Im}(zw)$ for $z = 1 + i$ and $w = 4 - i$.

13. **WE11** Determine the values of a and b that satisfy each of the following.

    a. $(2 + 3i)(a + bi) = 16 + 11i$ **b.** $(5 - 4i)(a + bi) = 1 - 4i$

14. Determine the values of a and b that satisfy each of the following.

    a. $(3i - 8)(a + bi) = -23 - 37i$ **b.** $(7 + 6i)(a + bi) = 4 - 33i$

15. Evaluate the values of the real numbers x and y if:

    a. $x(3 - 2i) + y(4 + 5i) = 23$
    b. $x(3 - 2i) - y(4 - 7i) = 13i$
    c. $x(4 - 3i) + y(3 - 4i) = 14 - 7i$
    d. $x(3 - 5i) + y(4 + 7i) + 7 + 43i = 0$

Complex unfamiliar

16. Let $z_1 = 2 + 5i$, $z_2 = 3 - 2i$ and $z_3 = -18 - 7i$.

    a. Verify the associative law, that is, $z_1 + (z_2 + z_3) = (z_1 + z_2) + z_3$.
    b. Evaluate $5z_1 - 2z_2 + 3z_3$.
    c. Evaluate $\text{Re}(2z_1 - 3z_2 + 4z_3) + \text{Im}(2z_1 - 3z_2 + 4z_3)$.
    d. Evaluate the real numbers x and y if $xz_1 + yz_2 = z_3$.

Fully worked solutions for this chapter are available online.

LESSON
6.4 Complex conjugates and division of complex numbers

SYLLABUS LINKS

- Use complex numbers in the form $a + bi$ where a and b are the real and imaginary parts (components) $\text{Re}(z)$ and $\text{Im}(z)$ of a complex number z.
- Determine and use complex conjugates.
- Perform complex-number arithmetic: addition, subtraction, multiplication and division, with and without technology.

Source: Specialist Mathematics Senior Syllabus 2024 © State of Queensland (QCAA) 2024; licensed under CC BY 4.0

6.4.1 The conjugate of a complex number

The **conjugate of a complex number** is obtained by changing the sign of **only** the imaginary component.

If $z = a + bi$, the conjugate $\bar{z}$ of z is defined as $\bar{z} = a - bi$.

Conjugates are useful because the multiplication (or addition) of a complex number and its conjugate results in a real number. (See Worked Example 9 again).

The conjugate of a complex number

For a complex number $z = a + bi$, the conjugate is:

$$\bar{z} = a - bi$$

Multiplication of a complex number and its conjugate results in a real number:

$$z\bar{z} = (a + bi)(a - bi)$$
$$= a^2 - abi + abi - b^2 i^2$$
$$= a^2 + b^2$$

You will use this result when dividing complex numbers.

Note: Compare this expression with the formula for the difference of two squares:

$$(m - n)(m + n) = m^2 - n^2$$

Addition of a complex number and its conjugate results in two times the real component of the complex number:

$$z + \bar{z} = a + bi + a - bi = 2a$$

WORKED EXAMPLE 12 Determining complex conjugates

Write the conjugate of each of the following complex numbers.

a. $8 + 5i$ b. $-2 - 3i$ c. $i\sqrt{5} + 4$

THINK	WRITE
a. Change the sign of the imaginary component.	a. $8 - 5i$

▶

b. Change the sign of the imaginary component.   **b.** $-2 + 3i$

c. Change the sign of the imaginary component.   **c.** $4 - i\sqrt{5}$

TI | THINK	WRITE	CASIO | THINK	WRITE
a. 1. On a Calculator page, press MENU, then select: 2: Number 9: Complex Number Tools 1: Complex Conjugate. Complete the entry line as: conj $(8 + 5i)$ then press ENTER. *Note:* The symbol i can be found by pressing the π button.	conj(8+5·i) 8−5·i	**a. 1.** On a Run-Matrix screen, press OPTN, then select COMPLEX by pressing F3. Select Conjg by pressing F4. Complete the entry line as: Conjg $(8 + 5i)$ then press EXE. *Note:* The symbol i can be found by pressing SHIFT 0.	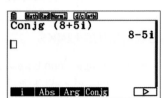
2. The answer appears on the screen. $8 - 5i$		**2.** The answer appears on the screen. $8 - 5i$	

WORKED EXAMPLE 13 Verifying the sum of conjugates is equal to the conjugate of a sum

If $z = 5 - 2i$ and $w = 7 - i$, show that $\overline{z + w} = \overline{z} + \overline{w}$.

THINK	WRITE
1. Add the conjugates $\overline{z}$ and $\overline{w}$.	$\overline{z} + \overline{w} = (5 + 2i) + (7 + i) = 12 + 3i$
2. Add z to w.	$z + w = (5 - 2i) + (7 - i) = 12 - 3i$
3. Write down the conjugate of $z + w$.	$\overline{z + w} = 12 + 3i$
4. The conjugate of $z + w$ equals the sum of the individual conjugates.	$\overline{z + w} = \overline{z} + \overline{w}$

6.4.2 Division of complex numbers

The application of conjugates to division of two complex numbers will now be investigated. When we divide two complex numbers, the solution is always written with a real denominator. In order to achieve this, we multiply both numerator and denominator by the conjugate of the denominator, as this ensures the denominator will have no imaginary part.

Consider the complex numbers $z = a + bi$ and $w = c + di$. To calculate $\dfrac{z}{w}$ in the form $x + yi$ we must multiply both the numerator and denominator by the conjugate of w to make the denominator a real number only.

$$\frac{z}{w} = \frac{a + bi}{c + di}$$

$$= \frac{a + bi}{c + di} \times \frac{c - di}{c - di} \qquad \text{Multiply by the conjugate of } c + di.$$

$$= \frac{(ac+bd)+(bc-ad)i}{c^2+d^2}$$ Simplify the expressions in the numerator and in the denominator.

$$= \frac{ac+bd}{c^2+d^2} + \frac{(bc-ad)i}{c^2+d^2}$$ Express in the form $x+yi$.

Dividing complex numbers

To express the quotient of complex numbers $\dfrac{z}{w}$ in the form $a+bi$ multiply the numerator and denominator by the conjugate of the denominator.

$$\frac{z}{w} = \frac{z}{w} \times \frac{\overline{w}}{\overline{w}}$$

WORKED EXAMPLE 14 Division of complex numbers

Express $\dfrac{2+i}{2-i}$ in the form $a+bi$.

THINK	WRITE
1. Multiply both the numerator and denominator by the conjugate of $2-i$ to make the denominator real.	$\dfrac{2+i}{2-i} = \dfrac{(2+i)}{(2-i)} \times \dfrac{(2+i)}{(2+i)}$ $= \dfrac{(2+i)^2}{(2-i)(2+i)}$
2. Expand the expressions obtained in the numerator and denominator.	$= \dfrac{4+4i+i^2}{4-i^2}$
3. Substitute -1 for i^2 and simplify the expression.	$= \dfrac{4+4i-1}{4+1}$ $= \dfrac{3+4i}{5}$ $= \dfrac{3}{5} + \dfrac{4i}{5}$

TI	THINK	WRITE	CASIO	THINK	WRITE
1. On a Calculator page, complete the entry line as: $\dfrac{2+i}{2-i}$ then press ENTER. *Note:* The symbol i can be found by pressing the π button.	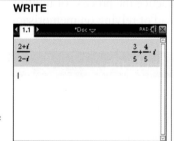	1. On a Run-Matrix screen, complete the entry line as: $\dfrac{2+i}{2-i}$ then press EXE. *Note*: The symbol i can be found by pressing SHIFT 0.	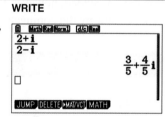
2. The answer appears on the screen.	$\dfrac{2+i}{2-i} = \dfrac{3}{5} + \dfrac{4}{5}i$	2. The answer appears on the screen.	$\dfrac{2+i}{2-i} = \dfrac{3}{5} + \dfrac{4}{5}i$

6.4.3 Multiplicative inverse of a complex number

Given a non-zero complex number z, there exists a complex number w such that $zw = 1$, with w being the **multiplicative inverse** of z denoted by $w = z^{-1} = \dfrac{1}{z}$. Hence, $zz^{-1} = 1$.

WORKED EXAMPLE 15 Determining the multiplicative inverse of a complex number

If $z = 3 + 4i$, determine z^{-1}.

THINK	WRITE
1. Write z^{-1} as a rational expression: $z^{-1} = \dfrac{1}{z}$.	$z^{-1} = \dfrac{1}{z} = \dfrac{1}{3 + 4i}$
2. Multiply both the numerator and denominator by the conjugate of $3 + 4i$.	$= \dfrac{1}{(3 + 4i)} \times \dfrac{(3 - 4i)}{(3 - 4i)}$ $= \dfrac{3 - 4i}{25}$
3. Write the expression in the form $a + bi$.	$= \dfrac{3}{25} - \dfrac{4i}{25}$

In general terms, the multiplicative inverse of a complex number can be written in standard (Cartesian) form, $a + bi$.

> **Multiplicative inverse of a complex number**
>
> If $z = a + bi$, then:
> $$z^{-1} = \frac{a - bi}{a^2 + b^2}$$

For instance, applying this formula to Worked Example 15:

$$z = 3 + 4i$$
$$z^{-1} = \frac{3 - 4i}{3^2 + 4^2}$$
$$= \frac{3}{25} - \frac{4i}{25}$$

It is recommended, however, that you understand and practice the process of multiplying numerator and denominator by the conjugates of the denominator, instead of relying on memorisation of this formula.

WORKED EXAMPLE 16 Simplifying complex quotients

If $z = 3 + i$ and $w = \dfrac{2}{4 - i}$, determine $\text{Im}(4z - w)$.

THINK	WRITE
1. Substitute for z and w in $4z - w$.	$4z - w = 4(3 + i) - \dfrac{2}{4 - i}$

2. Express $4z - w$ with a common denominator.

$$= \frac{4(3+i)(4-i) - 2}{4-i}$$

$$= \frac{4(13+i) - 2}{4-i}$$

$$= \frac{50 + 4i}{4-i}$$

3. Remove i from the denominator by multiplying the numerator and denominator by the conjugate of $4-i$.

$$= \frac{(50 + 4i)}{(4-i)} \times \frac{(4+i)}{(4+i)}$$

$$= \frac{196 + 66i}{17}$$

4. Simplify the expression so that it is in the form $a + bi$.

$$= \frac{196}{17} + \frac{66i}{17}$$

5. State the imaginary component of $4z - w$.

$$\text{Im}(4z - w) = \frac{66}{17}$$

WORKED EXAMPLE 17 Proofs with complex numbers

Prove that $\overline{z_1 z_2} = \bar{z_1} \bar{z_2}$.

THINK

1. When asked to 'prove, you should not use actual values for the pronumerals. Use general values of $z_1, z_2, \bar{z}_1$ and $\bar{z}_2$.

2. Generally, in a proof, do not work both sides of the equation at once. Calculate the LHS first.

3. Calculate the RHS and show that it equals the LHS.

WRITE

Let $z_1 = a + bi$.

$\bar{z}_1 = a - bi$.

Let $z_2 = c + di$.

$\bar{z}_2 = c - di$.

$\text{LHS} = \overline{(a + bi) \times (c + di)}$

$= \overline{ac + adi + bci + bdi^2}$

$= \overline{(ac - bd) + (ad + bc)i}$

$= (ac - bd) - (ad + bc)i$

$\text{RHS} = (a - bi)(c - di)$

$= ac - adi - bci + bdi^2$

$= (ac - bd) - (ad + bc)i$

$= \text{LHS}$

Hence, $\overline{z_1 z_2} = \bar{z}_1 \bar{z}_2$.

Exercise 6.4 Complex conjugates and division of complex numbers

6.4 Exercise	**6.4 Exam questions** on

Simple familiar	Complex familiar	Complex unfamiliar
1, 2, 3, 4, 5, 6, 7, 8, 9, 10, 11	12, 13, 14, 15, 16, 17, 18, 19	20, 21, 22

Simple familiar

1. **WE12** Write down the conjugate of each of the following complex numbers.
   a. $7 + 10i$ b. $5 - 9i$ c. $3 + 12i$ d. $\sqrt{7} - 3i$ e. $2i + 5$ f. $-6 - \sqrt{11}i$

2. **WE13** If $z = 6 + 3i$ and $w = 3 - 4i$, show that $\overline{z - w} = \overline{z} - \overline{w}$.

3. **WE14** Express $\dfrac{2 + i}{3 - i}$ in the form $a + bi$.

4. Express each of the following in the form $a + bi$.
   a. $\dfrac{1 - i}{1 + i}$ b. $\dfrac{3 - 2i}{2 + 3i}$ c. $\dfrac{2 + 5i}{4 - 3i}$

5. Express each of the following in the form $a + bi$.
   a. $\dfrac{4 - 3i}{5 + 2i}$ b. $\dfrac{4 - 5i}{2 - 7i}$ c. $\dfrac{2 + \sqrt{3}i}{\sqrt{5} - \sqrt{2}i}$

6. **WE15** Determine z^{-1} if z is equal to:
   a. $2 - i$ b. $3 + i$ c. $4 - 3i$

7. Determine z^{-1} if z is equal to:
   a. $5 + 4i$ b. $2i - 3$ c. $\sqrt{3} - i\sqrt{2}$

8. Determine the conjugate of $(5 - 6i)(3 - 8i)$.

9. If $z = -5 - 4i$ and $w = 2i$, calculate $\text{Re}(z\overline{w} + \overline{z}w)$.

10. If $z_1 = 2 + 3i$, $z_2 = -4 - i$ and $z_3 = 5 - i$, calculate:
    a. $2z_1 - z_2 - 4z_3$ b. $z_1\overline{z_2} + z_2\overline{z_3}$ c. $\overline{z_1 z_2 z_3} - \overline{z_1}\,\overline{z_2}\,\overline{z_3}$

11. If $z = 2 - 3i$ and $w = 1 - 2i$:
    a. calculate:
       i. $z\overline{z}$ ii. $w\overline{w}$
    b. show that:
       i. $\overline{z + w} = \overline{z} + \overline{w}$ ii. $\overline{zw} = \overline{z} \times \overline{w}$ iii. $\overline{\left(\dfrac{z}{w}\right)} = \dfrac{\overline{z}}{\overline{w}}$
    c. calculate:
       i. $\overline{\left(\dfrac{1}{z}\right)}$ ii. $\overline{\left(\dfrac{1}{w}\right)}$

d. calculate $\overline{z^2 + w^2}$

e. calculate $\overline{z + zw}$

f. calculate $z^{-1}w^{-1}$.

Complex familiar

12. Write $\dfrac{2+i}{1+i} + \dfrac{9-2i}{2-i} + \dfrac{7+i}{1-i}$ in the form $a + bi$.

13. Simplify $\dfrac{(2+5i)^2(5i-2)}{3(4+7i) - 2(5+8i)}$.

14. If $676z = 10 - 24i$, express z^{-1} in the form $a + bi$.

15. **WE16** If $z = 2 - i$ and $w = \dfrac{1}{3+i}$, determine each of the following.

    a. $\text{Re}(z + w)$ **b.** $\text{Im}(w - z)$ **c.** $\text{Re}(z^{-1} + w^{-1})$ **d.** $\text{Im}(3z + 2w)$ **e.** $\text{Re}(4w - 2z)$

16. Let $z = 6 + 8i$ and $w = 10 - 3i$.

Show that $\overline{zw} = \overline{z} \times \overline{w}$.

17. Let $z = 4 + i$ and $w = 1 + 3i$.

Show that $\overline{\left(\dfrac{z}{w}\right)} = \dfrac{\overline{z}}{\overline{w}}$.

18. **WE17** Let $z = a + bi$ and $w = c + di$.

    a. Show that $\overline{zw} = \overline{z} \times \overline{w}$.

    b. Show that $\overline{\left(\dfrac{z}{w}\right)} = \dfrac{\overline{z}}{\overline{w}}$.

19. If $z_1 = a + bi$ and $z_2 = c + di$, show that $(z_1 z_2)^{-1} = z_1^{-1} z_2^{-1}$.

Complex unfamiliar

20. a. If $z = 1 + i$, evaluate z^4, z^8 and z^{12}.

    b. Deduce from your results in **a** that $z^{4n} = (2i)^{2n}$, $n \in \mathbb{N}$.

21. If $z = a + bi$, determine the values of a and b such that $\dfrac{z-1}{z+1} = z + 2$.

22. Determine values for a and b so that $z = a + bi$ satisfies $\dfrac{z+i}{z+2} = i$.

Fully worked solutions for this chapter are available online.

LESSON
6.5 The complex plane (the Argand plane)

SYLLABUS LINKS

- Sketch and use complex numbers as points in the complex plane with real and imaginary parts as Cartesian coordinates.
- Understand and use addition of complex numbers as vector addition in the complex plane.

Source: Specialist Mathematics Senior Syllabus 2024 © State of Queensland (QCAA) 2024; licensed under CC BY 4.0

6.5.1 Plotting numbers in the complex plane

Thus far we have represented complex numbers in the standard form $z = a + bi$. This is referred to as the Cartesian form. We know from previous studies that an ordered pair of real numbers (x, y) can be represented on the Cartesian plane. Similarly, we can consider the complex number $z = a + bi$ as $z = x + yi$, consisting of the ordered pair of real numbers (x, y), which can be plotted as a point (x, y) on the complex number plane.

This is known as the **Argand plane** or an **Argand diagram** in recognition of the work done in this area by the Swiss mathematician Jean-Robert Argand.

The horizontal axis is referred to as the real axis and the vertical axis is referred to as the imaginary axis.

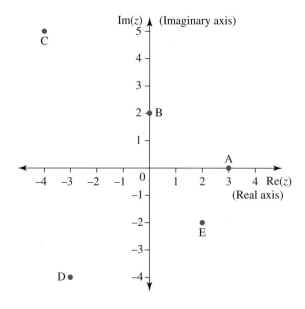

For example, the complex numbers $3 + 0i$ (A), $0 + 2i$ (B), $-4 + 5i$ (C), $-3 - 4i$ (D) and $2 - 2i$ (E) are shown on the Argand diagram.

When representing a complex number $z = a + bi$ on the Argand plane, the point corresponding to this complex number, that is the point with Cartesian coordinates (a, b) is called the **affix**. Note that, by misuse of language, the affix, which is the point on the complex plane, and the corresponding complex number are sometimes confused.

Complex numbers behave like vectors in the plane.

Note: The imaginary axis is labelled $1, 2, 3 \dots$ (not $i, 2i, 3i \dots$).

6.5.2 Geometrically multiplying a complex number by a scalar

WORKED EXAMPLE 18 Representing complex numbers on an Argand plane

Given the complex number $z = 1 + 2i$, represent the complex numbers z and $2z$ on the same Argand plane. Comment on their relative positions.

THINK	WRITE
1. Draw an Argand plane.	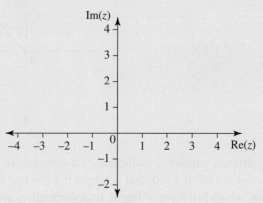
2. The complex number $z = 1 + 2i$ is of the form $z = x + yi$. It can be plotted on the Argand plane at $\text{Re}(z) = 1$ and $\text{Im}(z) = 2$.	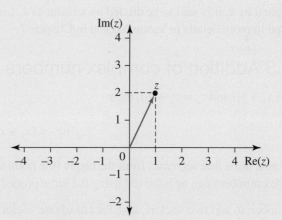
3. The complex number z has been multiplied by a scalar of 2. Calculate the complex number $2z$ from $z = 1 + 2i$.	$z = 1 + 2i$ $2z = 2 + 4i$
4. The complex number $2z = 2 + 4i$ is of the form $z = x + yi$. It can be plotted on the Argand plane at $\text{Re}(z) = 2$ and $\text{Im}(z) = 4$. *Note:* The complex numbers are represented by the points only; the arrows and lines are only drawn only for convenience.	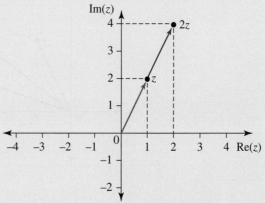 We can see that $2z$ is a complex number with twice the length of z.

If $z = x + yi$, then $kz = kx + kyi$, where $k \in \mathbb{R}$. The following diagram shows the situation for $x > 0$, $y > 0$ and $k > 1$.

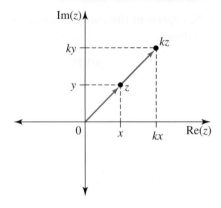

When a complex number is multiplied by a constant (or scalar), k, this produces a line segment in the same direction (or at $180°$ if $k < 0$) that is longer if $k < -1$ or $k > 1$, or shorter if $-1 < k < 0$ or $0 < k < 1$. This is called a **dilation**, which is a form of **linear transformation** using multiplication. When the complex number z is multiplied by k, it is said to be dilated by a factor of k. Linear transformations in the complex plane will be covered in more details in lesson 6.6 and in Chapter 7.

6.5.3 Addition of complex numbers in the complex plane

If $z_1 = x_1 + y_1 i$ and $z_2 = x_2 + y_2 i$, then:

$$z_1 + z_2 = (x_1 + x_2) + (y_1 + y_2) i$$

If we now draw line segments from the origin (the point $0 + 0i$) to the points z_1 and z_2, then the addition of two complex numbers can be achieved using the same procedure as adding two vectors.

Remember, to add two vectors, take the tail of one vector and join it to the head of another. The result of this addition is the vector from the tail of the first vector to the head of the second vector.

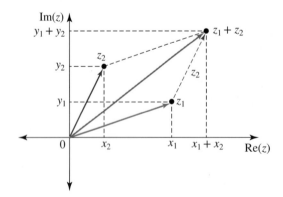

6.5.4 Subtraction of complex numbers in the complex plane

If $z_1 = x_1 + y_1 i$ and $z_2 = x_2 + y_2 i$, then:

$$z_1 - z_2 = z_1 + (-z_2)$$
$$= (x_1 + y_1 i) - (x_2 + y_2 i)$$
$$z_1 - z_2 = (x_1 - x_2) + (y_1 - y_2) i$$

The subtraction of two complex numbers can be achieved using the same procedure as subtracting two vectors.

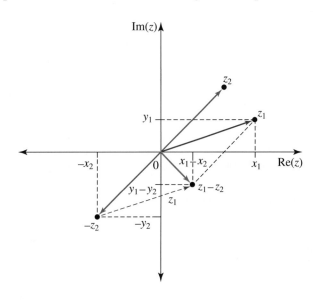

WORKED EXAMPLE 19 Representing sums and products on an Argand plane

Given the complex numbers $u = 1 + i$ and $v = 2 - 3i$, represent the following complex numbers on separate Argand planes.

a. $u + v$ **b.** $u - v$

THINK

a. 1. The complex numbers $u = 1 + i$ and $v = 2 - 3i$ are of the form $z = x + yi$. They can be plotted on an Argand plane at:
u: $\text{Re}(z) = 1$ and $\text{Im}(z) = 1$
v: $\text{Re}(z) = 2$ and $\text{Im}(z) = -3$

WRITE

a. $u = 1 + i$
$v = 2 - 3i$

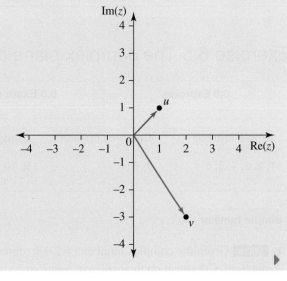

2. Evaluate the complex number $u + v$ and plot this complex number on the same Argand plane.

$u + v$: $\text{Re}(z) = -1$ and $\text{Im}(z) = -2$

$u + v = 1 + i + 2 - 3i$
$= 3 - 2i$

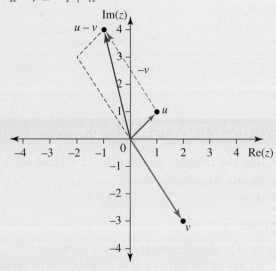

b. Plot the complex numbers u and v on an Argand plane.
Evaluate the complex number $u - v$ and plot this complex number on the same Argand plane.

$u - v$: $\text{Re}(z) = -1$ and $\text{Im}(z) = 4$

b. $u = 1 + i$
$v = 2 - 3i$
$u - v = -1 + 4i$

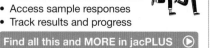

Exercise 6.5 The complex plane (the Argand plane)

learnon

6.5 Exercise	6.5 Exam questions on

These questions are even better in jacPLUS!
- Receive immediate feedback
- Access sample responses
- Track results and progress

Find all this and MORE in jacPLUS ▶

Simple familiar	Complex familiar	Complex unfamiliar
1, 2, 3, 4, 5, 6	7, 8	9, 10

Simple familiar

1. **WE18** Given the complex number $z = 2 + i$, represent the complex numbers z and $3z$ on the same Argand diagram. Comment on their relative positions.

2. Given the complex number $z = 4 - 2i$, represent the complex numbers z and $-\dfrac{z}{2}$ on the same Argand diagram. Comment on their relative positions.

3. **WE19** Given the complex numbers $u = 3 - 2i$ and $v = 1 + 2i$, represent the following complex numbers on separate Argand diagrams.

   a. $u + v$

   b. $u - v$

4. Given the complex numbers $u = 2 - 3i$ and $v = 2 + i$, evaluate and draw each of the following on separate Argand diagrams.

   a. $u + v$

   b. $u - v$

5. a. Given the complex numbers $u = -1 - 2i$ and $v = 2 + 3i$, evaluate and plot each of the following on separate Argand diagrams.

   i. $u + v$

   ii. $u - v$

   b. Given the complex numbers $u = 2 - 3i$ and $v = 1 + 4i$, evaluate and plot each of the following on separate Argand diagrams.

   i. $u + v$

   ii. $u - v$

6. Given the complex numbers $u = \dfrac{1}{2} + \dfrac{3}{4}i$ and $v = -2 + \dfrac{1}{4}i$, represent the following complex numbers on separate Argand planes.

   a. $3u + 2v$

   b. $2u - 3v$

Complex familiar

7. Consider the complex number $z = 1 + i$.

   a. Represent the complex numbers z, z^2 and $\dfrac{1}{z}$ on the Argand plane.

   b. Show that the affixes z, z^2 and $\dfrac{1}{z}$ are not aligned.

8. Consider the complex number $z = -\dfrac{1}{2} + \dfrac{\sqrt{3}}{2}i$.

   Represent the affixes z, z^2 and $\dfrac{1}{z}$ on the Argand plane.

Complex unfamiliar

9. Consider the points $A\left(z_A = 1 + \sqrt{3}i\right)$, $B\left(z_B = 2\right)$ and $C\left(z_C = \dfrac{7}{4} + \dfrac{\sqrt{3}}{4}i\right)$ on the Argand plane. Show that A, B and C are aligned.

10. Consider the complex numbers $a = 2 - i$, $b = 2 + \sqrt{3} + 2i$ and $c = i$.

   a. Represent the complex numbers on the same Argand plane.

   b. Using transformations, represent the complex ia, $\dfrac{i^2b}{2}$ and $2ic$ numbers on the same Argand plane as part **a**.

Fully worked solutions for this chapter are available online.

LESSON
6.6 Conjugates and multiplication in the complex plane

SYLLABUS LINKS

- Understand and use location of complex conjugates in the complex plane
- Understand and use multiplication by a complex number as a linear transformation in the complex plane.

Source: Specialist Mathematics Senior Syllabus 2024 © State of Queensland (QCAA) 2024; licensed under CC BY 4.0

6.6.1 Geometrical representation of a conjugate of a complex number

If $z = x + yi$, then the conjugate of z is given by $\bar{z} = x - yi$.

It can be seen that $\bar{z}$ is the **reflection** of the complex number in the real axis. Reflections are another form of linear transformation, which will be studied in more detail in Chapter 11.

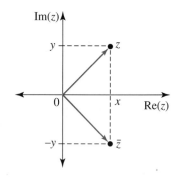

WORKED EXAMPLE 20 Representing the conjugate and multiplication by i on an Argand plane

Given the complex number $u = 1 + 2i$, represent the complex numbers $\bar{u}$ and iu on the same Argand plane and comment on their relative positions.

THINK	WRITE
1. Plot the complex number u on an Argand plane. u: $\text{Re}(z) = 1$ and $\text{Im}(z) = 2$	$u = 1 + 2i$ 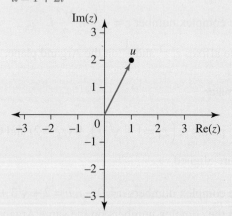

2. Evaluate the complex numbers $\bar{u}$ and iu, and plot these complex numbers on the same Argand diagram.

$\bar{u}$: $\text{Re}(z) = 1$ and $\text{Im}(z) = -2$

iu: $\text{Re}(z) = -2$ and $\text{Im}(z) = 1$

$$\bar{u} = 1 - 2i$$
$$iu = i(1 + 2i)$$
$$= i + 2i^2$$
$$= -2 + i$$

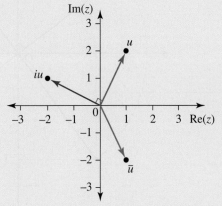

3. Comment on their relative positions.

$\bar{u}$ is the reflection of the complex number in the real axis. iu is a rotation of 90° anticlockwise from u.

6.6.2 Multiplication by i, $-i$, ki and i^2

If $z = x + yi$, then iz is given by:

$$iz = i(x + yi)$$
$$= ix + i^2 y$$
$$= -y + xi$$

It is clear from the diagram that iz is a **rotation** of 90° anticlockwise from z. Rotation is another form of transformation, using multiplication. The complex number z is said to be rotated 90° anticlockwise from z.

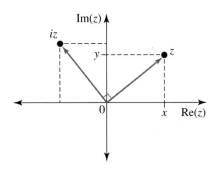

Similarly, $-iz = y - xi$ and $-iz$ is a rotation of 90° clockwise from z.

And $i^2 z = -z$, thus $i^2 z$ is a rotation of 180° anticlockwise from z.

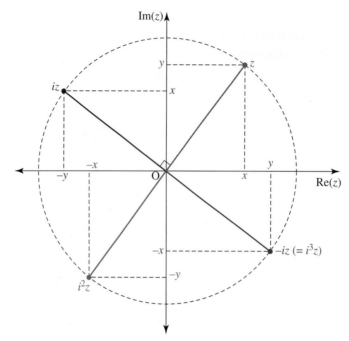

Generalising to the multiplication of z by a pure imaginary number ki, where $k \in \mathbb{R}$, we can show that kzi is a rotation of 90° anticlockwise and a dilation of k.

For example, let $z = x + yi$ and $k \in \mathbb{R}$.

$$kzi = -ky + kxi.$$

This is a rotation of 90° anticlockwise and a dilation of k.

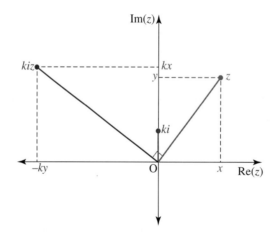

6.6.3 Scaling and rotating complex numbers using multiplication

Let $u = -3 + 4i$, represented as vector $\underset{\sim}{u} = \begin{pmatrix} -3 \\ 4 \end{pmatrix} = [5, 126.9°]$ in the Argand plane, and $v = 1 - \sqrt{3}i$, represented

as vector $\underset{\sim}{v} = \begin{pmatrix} 1 \\ -\sqrt{3} \end{pmatrix} = [2, -60°]$ in the Argand plane, then uv is given by:

$$uv = (-3 + 4i) \times \left(1 - \sqrt{3}\right)i$$
$$= \left(-3 + 4\sqrt{3}\right) + \left(3\sqrt{3} + 4\right)i$$

and is represented by the vector $\begin{pmatrix} -3 + 4\sqrt{3} \\ 3\sqrt{3} + 4 \end{pmatrix} = [10, 66.9°]$ in the Argand plane.

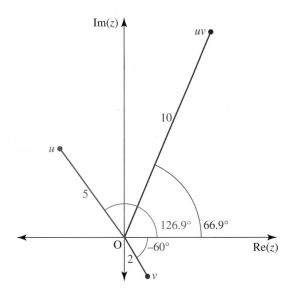

Using our knowledge of vectors in the plane, we can observe that by multiplying two complex numbers together, the resulting complex number is dilated and rotated.

In Chapter 7, using the polar form of complex numbers, you will see how to generalise the example presented here.

6.6.4 Simple linear transformations in the complex plane

In section 6.5.2, we have seen that when a complex number is multiplied by a scalar k (a complex number with an imaginary part equal to zero), this produces a dilation.

In section 6.6.1, we have seen that the conjugate of a complex number is its reflection in the real axis.

In section 6.6.2, we have seen that when a complex number is multiplied by a pure imaginary number (a complex number whose real part is equal to zero) this produces a rotation of 90° and a dilation.

In section 6.6.3, we have seen that when a complex number is multiplied by another complex number, this produces a dilation and a rotation.

Let a, b, and z be complex numbers and k be a real number.

$T(z) = az + b$ is a linear transformation of z in the complex plane.

Dilation by a factor k: $T(z) = kz$

Translation by b: $T(z) = z + b$

Rotation by 90° anticlockwise: $T(z) = iz$

Reflection in the real axis: $T(z) = \overline{z}$

Rotation and dilation: $T(z) = az$

See Chapter 7 for the generalisation of linear transformations in the complex plane using polar forms of complex numbers.

Exercise 6.6 Conjugates and multiplication in the complex plane

learnon

6.6 Exercise	6.6 Exam questions on

Simple familiar	Complex familiar	Complex unfamiliar
1, 2, 3, 4, 5, 6, 7	8, 9	10

Simple familiar

1. **WE20** Given the complex number $u = -3 - 2i$, represent the complex numbers $\bar{u}$ and iu on the same Argand diagram and comment on their relative positions.

2. Given the complex number $v = -2 + 3i$, represent the complex numbers $\bar{v}$ and i^2v on the same Argand diagram and comment on their relative positions.

3. a. Given $z = 2 - 3i$, evaluate and plot each of the following on the same Argand diagram. Comment on their relative positions.
   - i. $2z$
   - ii. $\bar{z}$
   - iii. iz

   b. Given $z = -4 - 4i$, evaluate and plot each of the following on the same Argand diagram. Comment on their relative positions.
   - i. $-\dfrac{z}{2}$
   - ii. $\bar{z}$
   - iii. iz

4. a. Given $z = 1 - i$, plot each of the following on one Argand diagram.
   - i. z
   - ii. i^2z
   - iii. i^3z

   b. Given $z = -2 + 3i$, plot each of the following on the same Argand diagram.
   - i. z
   - ii. i^2z
   - iii. i^3z

5. a. If $u = 1 + i$, evaluate and plot each of the following on one Argand diagram.
   - i. u
   - ii. $\dfrac{1}{u}$

   b. If $u = -\sqrt{3} - i$, evaluate and plot each of the following on one Argand diagram.
   - i. u
   - ii. $\dfrac{1}{u}$

6. a. Given that $u = 2 + 2i$ and $v = -1 - i$, evaluate and plot each of the following on one Argand diagram.
   - i. uv
   - ii. $\dfrac{u}{v}$

   b. Given that $u = 2i$ and $v = 1 - i$, evaluate and plot each of the following on one Argand diagram.
   - i. uv
   - ii. $\dfrac{u}{v}$

7. a. Given $z = 1 - i$, plot each of the following on one Argand diagram.
   - i. z
   - ii. z^2
   - iii. z^3

   b. Given $z = -1 + i$, plot each of the following on one Argand diagram.
   - i. z
   - ii. z^2
   - iii. z^3

8. **a.** If $u = 3i - 4j$ and $v = 4i + 3j$, show that the vectors u and v are perpendicular.

   b. Evaluate the image of the point $(3, -4)$ when it is rotated $90°$ anticlockwise.

9. Consider the vectors $u = ai + bj$ and $v = -bi + aj$ where $a, b \in \mathbb{R}$.
   Show that u and v are perpendicular.

10. Let $z = -\dfrac{3}{4} + 2i$. z is rotated by $30°$ clockwise, dilated by 2 and then reflected in the real axis, giving z'.
    Express z'.

Fully worked solutions for this chapter are available online.

LESSON
6.7 Review

6.7.1 Summary

doc-41633

Hey students! Now that it's time to revise this chapter, go online to:

 Access the chapter summary

 Review your results

 Practise exam questions

Find all this and MORE in jacPLUS

6.7 Exercise

learnon

6.7 Exercise	**6.7 Exam questions** on

Simple familiar	Complex familiar	Complex unfamiliar
1, 2, 3, 4, 5, 6, 7, 8, 9, 10, 11, 12	13, 14, 15, 16	17, 18, 19, 20

These questions are even better in jacPLUS!

- Receive immediate feedback
- Access sample responses
- Track results and progress

Find all this and MORE in jacPLUS

Simple familiar

1. Simplify $i^6 - i^3(i^2 - 1)$.

2. Let $u = 5 - i$ and $v = 4 + 3i$. Evaluate the expression $2u - v$.

3. If $z = 3 - 8i$, determine the value of:
   a. $\text{Im}\left(z^2\right)$
   b. a and b if $z^3 = a + bi$.

4. If $z = 6 - 2i$ and $w = 5 + 3i$, express $\dfrac{z}{w}$ in the form $a + bi$, $a, b \in \mathbb{R}$.

5. **MC** The value of $\text{Im}\left[i\left(2i^4 - 3i^2 + 5i\right)\right]$ is
   - **A.** 0
   - **B.** -5
   - **C.** 5
   - **D.** 10

6. **MC** If $z = 5 - 12i$, $\text{Re}\left(z^{-1}\right)$ is
   - **A.** 5
   - **B.** 12
   - **C.** $\dfrac{12}{169}$
   - **D.** $\dfrac{5}{169}$

7. **MC** If $z = 8 - 7i$ and $w = 3 + 4i$, then $\text{Im}\left(w^2\right) + \text{Re}\left(z^2\right)$ is equal to
   - **A.** 76
   - **B.** 39
   - **C.** 105
   - **D.** 56

8. **MC** The Argand diagram that correctly represents $z = 2\sqrt{5} - 4i$ is

A.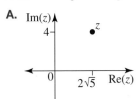

B.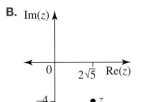

C.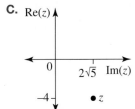

D.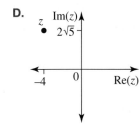

9. Let $z = \dfrac{\sqrt{2}}{2} + \dfrac{\sqrt{2}}{2}i$ and $a = z + \dfrac{1}{z}$.

   Express a in Cartesian form.

10. Let $z = -1 + i$ and n an integer.

    Show that $\operatorname{Im}\left(z^{8n}\right) = 0$.

11. Let $z = \sqrt{3} + i$ and n an integer.

    Show that $\operatorname{Re}\left(z^{3(1+2n)}\right) = 0$.

12. Let $a = 1 + \sqrt{3}i$, $b = -\sqrt{3} + i$, and $c = 2i$.

    a. Represent the affixes $a(\mathrm{A})$, $b(\mathrm{B})$ and $c(\mathrm{C})$ on the Argand plane.

    b. Show that A, B and C are on a circle of centre 0 and radius 2.

Complex familiar

13. Consider the complex number z such that $z = 3 + 2i$.

    a. Determine the value for iz, i^2z, i^3z and i^4z.

    b. Plot each number from part **a** on the same Argand diagram.

    c. Determine the transformation required to transform point z into point iz.

14. **a.** For any complex number $z = x + yi$ where both x and y are real, describe the transformation required to obtain $\dfrac{1}{z}$.

    b. For any complex number $z = x + yi$ where both x and y are real, describe the transformation required to obtain iz.

15. Let $a = 1$, $z = -2 + i$, and $z' = \dfrac{a - z}{\bar{z} - a}$.

    a. Represent the affixes $a(\mathrm{A})$, $z(\mathrm{B})$ and $z'(\mathrm{C})$ on the Argand plane.

    b. Show that A, B and C are aligned.

16. Let $z = x + iy$ and $z' = 5iz + 6i + 4$.

    Describe the transformations required to obtain z' from z.

Complex unfamiliar

17. Let $a = 3 + i$ and $b = 1 + 2i$.

    a. Represent the affixes $a(\mathrm{A})$ and $b(\mathrm{B})$ on the Argand plane.

    b. Determine the value of the angle $\theta = \left(\overrightarrow{\mathrm{OA}}, \overrightarrow{\mathrm{OB}}\right)$. Give your answer to one decimal place.

18. Let $a = -2$, $b = 1$ and $c = -\dfrac{1}{2} - \dfrac{\sqrt{3}}{2}i$ and $d = -\dfrac{1}{2} + \dfrac{\sqrt{3}}{2}i$.

    a. Represent the affixes $a(\mathrm{A})$, $b(\mathrm{B})$, $c(\mathrm{C})$ and $d(\mathrm{D})$ on the Argand plane.

    b. Determine whether ACBD is a rhombus.

19. Let $a = 3 + \sqrt{c-9}i$, and $b = 3 - \sqrt{c-9}i$ where $c \in \mathbb{R}, c > 9$.

Let A be the affix of a, B the affix of b and O the affix of 0.

Determine c so that OAB is a right-angle triangle with the right-angle at angle O.

20. Let $a = 2$, and $b = -\sqrt{2} + \sqrt{2}i$ and $c = \dfrac{a+b}{2}$.

Show that an exact value of $\cos\left(\dfrac{135°}{2}\right)$ is $\dfrac{\sqrt{2-\sqrt{2}}}{2}$.

Fully worked solutions for this chapter are available online.

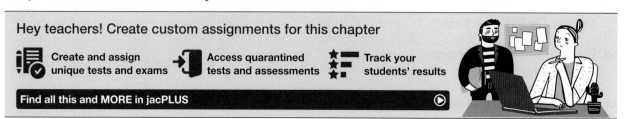

Hey teachers! Create custom assignments for this chapter

Create and assign unique tests and exams

Access quarantined tests and assessments

Track your students' results

Find all this and MORE in jacPLUS

Answers

Chapter 6 Complex numbers

6.2 Introduction to complex numbers

6.2 Exercise

1. a. $\pm 3i$ b. $\pm 5i$ c. $\pm 7i$ d. $\pm \sqrt{3}\,i$

2. a. $\pm \sqrt{11}\,i$ b. $\pm \sqrt{7}\,i$ c. $\pm \dfrac{2}{3}i$ d. $\pm \dfrac{6}{5}i$

3. a. $9, 5$ b. $5, -4$ c. $-3, -8$ d. $-6, 11$

4. a. $27, 0$ b. $0, 2$ c. $-5, 1$ d. $0, -17$

5. a. $-1 + i$ b. $1 - i$ c. $0 + 0i$ d. $1 + i$

6. a. $-1 + 2i$ b. $-1 + i$ c. $1 + 0i$ d. $1 - 2i$

7. a. -5 b. 15 c. 0 d. -6

8. a. 2 b. 0 c. -9 d. 2

9. $4 - i$

10. a. -2 b. $-2i$ c. $\sqrt{2}$ d. $2i$

11. $z = -2 - 3i$, $w = 7 + 3i$ and thus $z + w = 5$; sample responses can be found in the worked solutions in the online resources.

12. a. $\text{Re}(f(i)) = \dfrac{3}{4}$

    b. $\text{Im}(f(i)) = 2$

13. a. 13 b. -11 c. 10 d. 7

14. a. i b. 0

15. a. $-i$ b. 0

6.3 Basic operations on complex numbers

6.3 Exercise

1. a. $4 - i$ b. $1 - 14i$ c. $-4 - 2i$

2. a. $9 - 13i$ b. $-12 + 4i$ c. $-9 - 5i$

3. a. $-12 + 3i$ b. $-19 - 8i$ c. $12 + 23i$

4. a. $-25 + 3i$ b. $-50 - 48i$ c. $-41 - 28i$

5. a. $7 - 23i$ b. $4 + 45i$ c. $-50 - 13i$

6. a. $63 - 37i$ b. $-85 - 132i$ c. $176 - 61i$

7. a. $111 + 33i$ b. $31 - 8i$ c. $22 - 48i$

8. a. 61 b. -53 c. $32 - 126i$

9. $14 + 52i$

10. a. 35 b. -30 c. -115

11. a. -8 b. -5 c. -9

12. -3

13. a. $a = 5, b = -2$ b. $a = \dfrac{21}{41}, b = -\dfrac{16}{41}$

14. a. $a = 1, b = 5$ b. $a = -2, b = -3$

15. a. $x = 5, y = 2$ b. $x = 4, y = 3$
    c. $x = 5, y = -2$ d. $x = 3, y = -4$

16. a. $-13 - 4i$ b. $-50 + 8i$
    c. -89 d. $x = -3, y = -4$

6.4 Complex conjugates and division of complex numbers

6.4 Exercise

1. a. $7 - 10i$ b. $5 + 9i$ c. $3 - 12i$
   d. $\sqrt{7} + 3i$ e. $5 - 2i$ f. $-6 + \sqrt{11}\,i$

2. Sample responses can be found in the worked solutions in the online resources.

3. $\dfrac{1}{2} + \dfrac{1}{2}i$

4. a. $0 - i$
   b. $0 - i$
   c. $-\dfrac{7}{25} + \dfrac{26}{25}i$

5. a. $\dfrac{14}{29} - \dfrac{23}{29}i$

   b. $\dfrac{43}{53} + \dfrac{18}{53}i$

   c. $\dfrac{2\sqrt{5} - \sqrt{6}}{7} + \dfrac{2\sqrt{2} + \sqrt{15}}{7}i$

6. a. $\dfrac{2}{5} + \dfrac{1}{5}i$ b. $\dfrac{3}{10} - \dfrac{1}{10}i$ c. $\dfrac{4}{25} + \dfrac{3}{25}i$

7. a. $\dfrac{5}{41} - \dfrac{4}{41}i$ b. $-\dfrac{3}{13} - \dfrac{2}{13}i$ c. $\dfrac{\sqrt{3}}{5} + \dfrac{\sqrt{2}}{5}i$

8. $-33 + 58i$

9. -16

10. a. $-12 + 11i$ b. $-30 - 19i$ c. 0

11. a. i. 13 ii. 5

    b. Sample responses can be found in the worked solutions in the online resources.

    c. i. $\dfrac{2}{13} - \dfrac{3}{13}i$

       ii. $\dfrac{1}{5} - \dfrac{2}{5}i$

    d. $-8 + 16i$

    e. $-2 + 10i$

    f. $-\dfrac{4}{65} + \dfrac{7}{65}i$

12. $\dfrac{17}{2} + \dfrac{9}{2}i$

13. -29

14. $10 + 24i$

15. a. $\dfrac{23}{10}$ b. $\dfrac{9}{10}$ c. $\dfrac{17}{5}$

    d. $-\dfrac{16}{5}$ e. $-\dfrac{14}{5}$

16-19. Sample responses can be found in the worked solutions in the online resources.

20. a. $-4, 16, -64$

    b. Sample responses can be found in the worked solutions in the online resources.

21. $a = -1, b = \pm \sqrt{2}$

22. $a = -\dfrac{1}{2}, b = \dfrac{1}{2}$

6.5 The complex plane (the Argand plane)

6.5 Exercise

1. $3z = 3 + 6i$, $3z$ has a length of 3 times that of z.

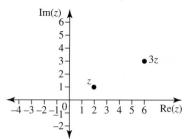

2. $-\dfrac{z}{2} = -2 + i$; $-\dfrac{z}{2}$ is in the opposite direction to z and half its length.

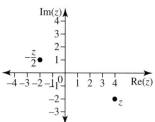

3. a. $u + v = 4$

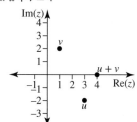

b. $u - v = 2 - 4i$

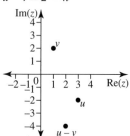

4. a $u + v = 4 + 2i$

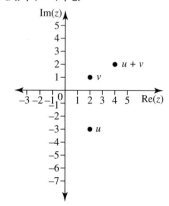

b $u - v = -4i$

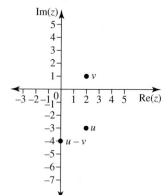

5. a. i. $u + v = 1 + i$

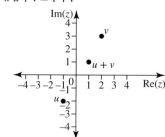

ii. $u - v = -3 - 5i$

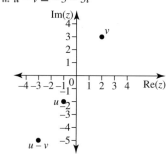

b. i. $u + v = 3 + i$

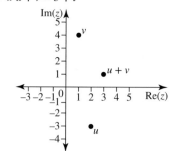

ii. $u - v = 1 - 7i$

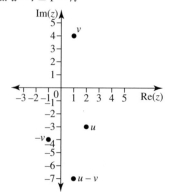

6. a.

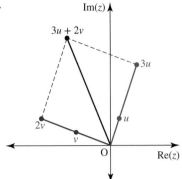

b.

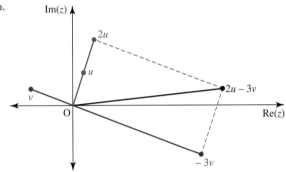

7. a.

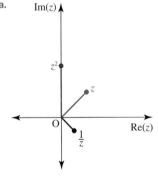

b. $\overrightarrow{AC} \neq k\overrightarrow{AB}$

$\overrightarrow{AB}$ and $\overrightarrow{AC}$ are not colinear thus A, B and C are not aligned.

8.

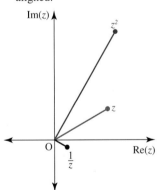

9. $\overrightarrow{AC} = \dfrac{3}{4}\overrightarrow{AB}$

Thus A, B and C are aligned.

10. a.

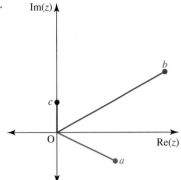

b.

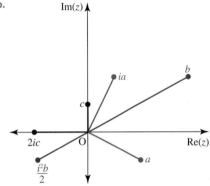

6.6 Conjugates and multiplication in the complex plane

6.6 Exercise

1. $\overline{u} = -3 + 2i$, $iu = 2 - 3i$, $\overline{u}$ is the reflection of the complex number in the real axis and iu is a rotation of 90° anticlockwise from u.

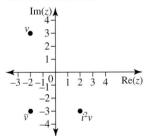

2. $\overline{v} = -2 - 3i$, $i^2v = 2 - 3i$, $\overline{v}$ is the reflection of the complex number in the real axis and i^2v is a rotation of 180° anti-clockwise from v.

3. a. $2z = 4 - 6i,\ \bar{z} = 2 + 3i,\ iz = 3 + 2i$

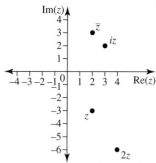

2z is dilated by a factor 2 in the same direction.
$\bar{z}$ is a reflection of z in the real axis.
iz is a rotation of 90° anticlockwise.

b. $-\dfrac{z}{2} = 2 + 2i,\ \bar{z} = -4 + 4i,\ iz = 4 - 4i$

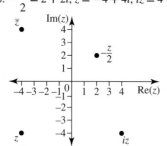

$-\dfrac{z}{2}$ is a rotation of 180°, followed by a dilation by a

factor $\dfrac{1}{2}$.

$\bar{z}$ is a reflection of z in the real axis.
iz is a rotation of 90° anticlockwise.

4. a. $z = 1 - i,\ i^2z = -1 + i,\ i^3z = -1 - i$

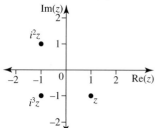

b. $z = -2 + 3i,\ i^2z = 2 - 3i,\ i^3z = 3 + 2i$

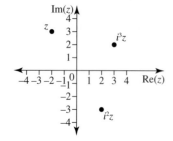

5. a. $u = 1 + i,\ \dfrac{1}{u} = \dfrac{1}{2} - \dfrac{1}{2}i$

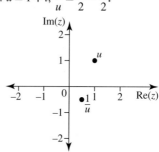

b. $u = -\sqrt{3} - i,\ \dfrac{1}{u} = -\dfrac{\sqrt{3}}{4} + \dfrac{1}{4}i$

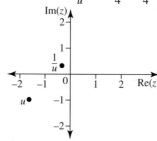

6. a. $uv = -4i,\ \dfrac{u}{v} = -2$

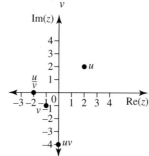

b. $uv = 2 + 2i,\ \dfrac{u}{v} = -1 + i$

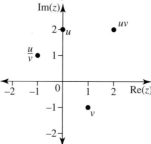

7. a. $z^2 = -2i,\ z^3 = -2 - 2i$

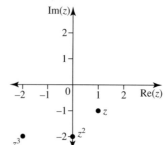

b. $z^2 = -2i, z^3 = 2 + 2i$

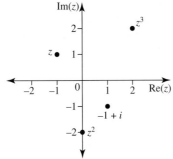

8. a. Sample responses can be found in the worked solutions in the online resources.

b. $(4, 3)$

9. Sample responses can be found in the worked solutions in the online resources.

10. $z' = 2 - \dfrac{3\sqrt{3}}{4} - \left(2\sqrt{3} + \dfrac{3}{4}\right)i$

6.7 Review

6.7 Exercise

1. $-1 - 2i$

2. $6 - 5i$

3. a. -48

    b. $a = -549, b = 296$

4. $\dfrac{12}{17} - \dfrac{14}{17}i$

5. C

6. D

7. B

8. B

9. $a = \sqrt{2}$

10.
$$z^8 = 16$$
$$z^{8n} = 16^n$$
$$\text{Im}\left(z^{8n}\right) = 0$$

11. Sample responses can be found in the worked solutions in the online resources.

12. a.

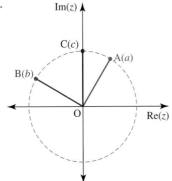

    b. $|\overrightarrow{OA}| = |\overrightarrow{OB}| = |\overrightarrow{OC}| = 2$ thus A, B and C are on a circle of centre 0 and radius 2.

13. a. $iz = -2 + 3i, i^2 z = -3 - 2i, i^3 z = 2 - 3i, i^4 z = 3 + 2i$

    b.

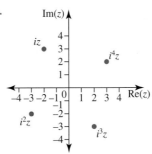

    c. Rotation of $\dfrac{\pi}{2}$ (90°) anticlockwise about the origin

14. a. Reflecting z in the real axis followed by a dilation of $\dfrac{1}{|z|}$

    b. Rotating z through $\dfrac{\pi}{2}$ (90°) in an anticlockwise direction

15. a.

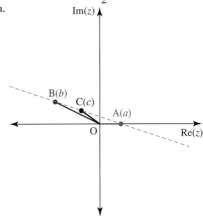

    b. $\overrightarrow{AC} = \dfrac{3}{5}\overrightarrow{AB}$

    $\overrightarrow{AB}$ and $\overrightarrow{AC}$ are colinear thus A, B and C are aligned.

16. • Rotation of 90° anticlockwise
    • Dilation of 5
    • Translation of $6i + 4$

17. a.

    b. $\theta = 45.0°$

18. a.

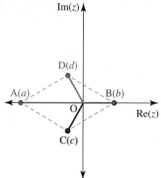

b. Sample responses can be found in the worked solutions in the online resources.

19. OAB is a right-angle triangle for $c = 18$.

20. Sample responses can be found in the worked solutions in the online resources.

7 Complex arithmetic and algebra

LESSON SEQUENCE

Fully worked solutions for this chapter are available online.

EXAM PREPARATION

Access exam-style questions in every lesson, available online.

on Resources

Solutions   Solutions — Chapter 7 (sol-0399)

Exam questions   Exam question booklet — Chapter 7 (eqb-0285)

Digital documents Learning matrix — Chapter 7 (doc-41928)
                       Chapter summary — Chapter 7 (doc-41929)

LESSON
7.1 Overview

7.1.1 Introduction

In Chapter 6, we discussed complex numbers in Cartesian form, complex-number arithmetic and the Argand plane, introducing transformations in the complex plane. In this topic, we will discuss the polar form of complex numbers and consider the geometric interpretation of the multiplication and division of complex numbers, making connections with transformations in the complex plane. We will also examine subsets of the Argand plane, and use complex numbers to solve quadratic equations.

Note that transformations in the plane will be covered in more depth in Chapter 10, matrices and transformations.

7.1.2 Syllabus links

Lesson	Lesson title	Syllabus links
7.2	**Complex numbers in polar form**	○ Use the modulus $\lvert z \rvert$ of a complex number z and the principal argument $\mathrm{Arg}(z)$ of a non-zero complex number z. • $\lvert z \rvert = \sqrt{a^2 + b^2}$ • $\mathrm{Arg}(z) = \theta$, $\tan(\theta) = \dfrac{b}{a}$, $-\pi < \theta \le \pi$, $a \ne 0$
		○ Understand the difference between the argument, $\arg(z)$, and the principal argument, $\mathrm{Arg}(z)$ of a non-zero complex number z. • $\arg(z) = \mathrm{Arg}(z) + 2n\pi$, $n \in \mathbb{Z}$
		○ Sketch and use complex numbers in polar form as polar coordinates.
		○ Convert between Cartesian form and polar form.
7.3	**Operations on complex numbers in polar form**	○ Express a complex number in Cartesian form and polar form. • $z = r(\cos(\theta) + i\sin(\theta))$ or $z = r\,\mathrm{cis}(\theta)$
		○ Convert between Cartesian form and polar form.
		○ Perform complex-number arithmetic: addition, subtraction, multiplication and division, with and without technology.
7.4	**Geometric interpretations of multiplication and division of complex numbers in polar form**	○ Understand and use multiplication, division of complex numbers in polar form and the geometric interpretation of these. • $z_1 z_2 = r_1 r_2 \mathrm{cis}(\theta_1 + \theta_2)$ • $\dfrac{z_1}{z_2} = \dfrac{r_1}{r_2}\mathrm{cis}(\theta_1 - \theta_2)$
7.5	**Subsets of the complex plane**	○ Identify and sketch subsets of the complex plane determined by straight lines and circles, e.g. $\lvert z - 3i \rvert < 4$, $\dfrac{\pi}{4} \le \mathrm{Arg}(z) \le \dfrac{3\pi}{4}$, $\mathrm{Re}(z) < \mathrm{Im}(z)$ and $\lvert z - 1 \rvert = 2\lvert z - i \rvert$.
7.6	**Roots of equations**	○ Determine complex conjugate solutions of real quadratic equations with real coefficients using factorisation, completing the square and the quadratic formula, with and without technology.
		○ Determine and use linear factors of quadratic polynomials with real coefficients that involve the complex conjugate root theorem, e.g. determine the coefficients of a real quadratic equation given one complex root.

Source: Specialist Mathematics Senior Syllabus 2024 © State of Queensland (QCAA) 2024; licensed under CC BY 4.0

LESSON
7.2 Complex numbers in polar form

7.2.1 The modulus of z

Like vectors, we can write a complex number in both Cartesian and polar form. The polar form of a complex number, like the polar form of a vector, requires the magnitude (modulus) of the complex number and the angle it makes, anticlockwise, with the positive direction of the real axis.

The **magnitude** (or **modulus** or **absolute value**) of the complex number $z = x + yi$ is the length of the line segment joining the origin to the point z. It is denoted by:

$$|z|,\ \ |x + yi|\ \text{or}\ \mathrm{mod}(z).$$

The modulus $|z|$ of z is calculated using Pythagoras' theorem.

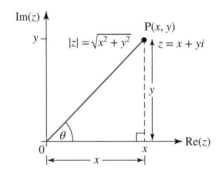

The magnitude (modulus) of a complex number

The magnitude of a complex number $z = x + yi$ is:

$$|z| = \sqrt{x^2 + y^2}$$

Recall that $z\bar{z} = x^2 + y^2$ and note that this is the square of the magnitude. Therefore, $|z|^2 = z\bar{z} = x^2 + y^2$.

WORKED EXAMPLE 1 Calculating the modulus of a complex number

Calculate the modulus of the complex number $z = 8 - 6i$.

THINK

Calculate the modulus by rule.
Because it forms the hypotenuse of a right-angled triangle, the modulus is always greater than or equal to Re(z) or Im(z).

WRITE

$$|z| = \sqrt{8^2 + (-6)^2}$$
$$= \sqrt{100}$$
$$= 10$$

TI | THINK	WRITE
1. On a Calculator page, press MENU, then select: 2: Number 9: Complex Number Tools 5: Magnitude. Complete the entry line as: $\|8 - 6i\|$ then press ENTER. *Note:* The symbol i can be found by pressing the π button.	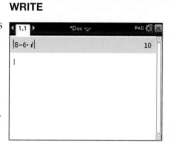
2. The answer appears on the screen.	$\|z\| = 10$

CASIO | THINK	WRITE
1. On a Run-Matrix screen, press OPTN, then select COMPLEX by pressing F3. Select Abs by pressing F2. Complete the entry line as: $\|8 - 6i\|$ then press EXE. *Note:* The symbol i can be found by pressing SHIFT 0.	
2. The answer appears on the screen.	$\|z\| = 10$

WORKED EXAMPLE 2 Calculating the modulus of the difference of complex numbers

If $z = 4 + 2i$ and $w = 7 + 6i$, represent the position of $w - z$ on an Argand diagram and calculate $|w - z|$.

THINK

1. Calculate $w - z$.

2. Represent it on an Argand diagram as a directed line segment OP.

3. Use Pythagoras theorem to determine the length of OP.

WRITE

$$w - z = 7 + 6i - (4 + 2i)$$
$$= 3 + 4i$$

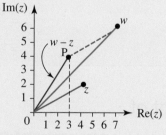

$$|OP|^2 = 3^2 + 4^2 = 25$$
$$|OP| = 5$$
So $|w - z| = 5$.

From Worked example 2, we can generalise that for affixes z(M) and w(N), the length of a segment $\overline{MN}$ on the Argand plane is given by $|w - z|$ (or $|z - w|$, which is the same).

This can be used to calculate lengths and areas on the Argand plane, as shown in Worked example 3.

WORKED EXAMPLE 3 Applications of modulus to areas represented by complex numbers

Represent the points $z_1 = 1 + 2i$, $z_2 = 4 - 2i$, $z_3 = -1 - 2i$ and $z_4 = -4 + 2i$ on the complex number plane and calculate the area of the rhombus formed when the four points are connected by straight line segments in the order z_1 to z_2 to z_3 3 to z_4.

THINK	WRITE
1. Sketch the shape on the complex number plane.	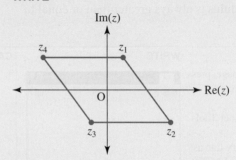
2. Recall the formula for the area of a rhombus with diagonals of lengths d and D is given by $A = \dfrac{dD}{2}$ and apply to this situation.	In this situation, $d = \|z_1 - z_3\|$ and $D = \|z_2 - z_4\|$. $$\therefore A = \frac{\|z_1 - z_3\| \times \|z_2 - z_4\|}{2}$$
3. Calculate $\|z_1 - z_3\|$ and $\|z_2 - z_4\|$.	$$\begin{aligned}\|z_1 - z_3\| &= \|1 + 2i - (-1 - 2i)\| \\ &= \|2 + 4i\| \\ &= \sqrt{2^2 + 4^2} \\ &= 2\sqrt{5}\end{aligned}$$ $$\begin{aligned}\|z_2 - z_4\| &= \|4 - 2i - (-4 + 2i)\| \\ &= \|8 - 4i\| \\ &= \sqrt{8^2 + 4^2} \\ &= 4\sqrt{5}\end{aligned}$$
4. Calculate the area of the rhombus.	$$\begin{aligned}A &= \frac{\|z_1 - z_3\| \times \|z_2 - z_4\|}{2} \\ &= \frac{2\sqrt{5} \times 4\sqrt{5}}{2} \\ &= 20 \text{ square units}\end{aligned}$$

7.2.2 The argument of z

The **argument** of z, $\arg(z)$, is the angle measurement anticlockwise from the positive real axis.

In the figure, $\arg(z) = \theta$, where $\tan(\theta) = \dfrac{y}{x}$.

Using this we can determine the argument for any complex number as

$$\theta = \tan^{-1}\left(\frac{y}{x}\right).$$

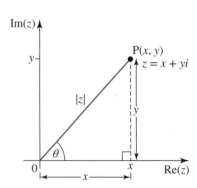

Note that there are an infinite number of arguments for any non-zero complex number z since adding or subtracting multiples of 2π radians (or $360°$) does not change the position of the complex number on the Argand plane.

The argument of a complex number

The argument of a complex number $z = x + yi$ is:

$$\arg(z) = \theta = \tan^{-1}\left(\frac{y}{x}\right), x \neq 0$$

Note that if $x = 0$, that is if z is is a pure imaginary number, then $\arg(z) = \pm\dfrac{\pi}{2}$

To ensure that there is only one value of θ corresponding to z, we refer to the *principal value* or principal argument of θ and denote it by $\text{Arg}(z)$. Note the capital A.

Assume all angles are now in radians, unless specified otherwise.

The principal argument of a complex number

The principal argument of a complex number $z = x + yi$ is:

$$\text{Arg}(z) = \theta, \tan(\theta) = \left(\frac{y}{x}\right), -\pi < \theta \leq \pi, x \neq 0$$

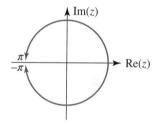

WORKED EXAMPLE 4 Converting arguments into principal arguments

Convert each of the following into principal arguments.

a. $\dfrac{7\pi}{4}$

b. $-\dfrac{5\pi}{2}$

THINK	WRITE
a. 1. Sketch the angle.	a.
2. Since the given angle is positive, subtract multiples of 2π until it lies in the range $(-\pi, \pi]$.	a. $\text{Arg}(z) = \dfrac{7\pi}{4} - 2\pi$ $= -\dfrac{\pi}{4}$

b. 1. Sketch the angle.

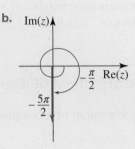

2. Since the given angle is negative, add multiples of 2π until it lies in the range $(-\pi, \pi]$.

$$\text{Arg}(z) = -\frac{5\pi}{2} + 2\pi$$
$$= -\frac{\pi}{2}$$

Exact values

It is useful to remember the exact values of the trigonometric ratios for the commonly used angles $\frac{\pi}{6}$, $\frac{\pi}{4}$ and $\frac{\pi}{3}$. To remember these we can sketch a couple of triangles that can be analysed to determine whichever exact value you require. The triangle on the left is an isosceles right-angled triangle with two sides of length 1. The triangle on the right is an equilateral triangle with sides of length 2. The vertical line in the middle creates right-angled triangles with angles $\frac{\pi}{3}$ and $\frac{\pi}{6}$. Using the definitions of the sine, cosine and tangent ratios, you will be able to determine the exact values for the angles $\frac{\pi}{6}$, $\frac{\pi}{4}$ and $\frac{\pi}{3}$.

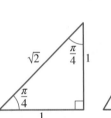 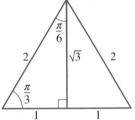

These exact values are summarised in the following table.

Angle	$\sin(\theta)$	$\cos(\theta)$	$\tan(\theta)$
$\dfrac{\pi}{6}$	$\dfrac{1}{2}$	$\dfrac{\sqrt{3}}{2}$	$\dfrac{1}{\sqrt{3}}$
$\dfrac{\pi}{4}$	$\dfrac{1}{\sqrt{2}}$	$\dfrac{1}{\sqrt{2}}$	1
$\dfrac{\pi}{3}$	$\dfrac{\sqrt{3}}{2}$	$\dfrac{1}{2}$	$\sqrt{3}$

WORKED EXAMPLE 5 Calculating the arguments of complex numbers

Calculate the argument of z for each of the following in the interval $(-\pi, \pi]$.

a. $z = 4 + 4i$ **b.** $z = 1 - \sqrt{3}i$

THINK

a. 1. Plot z.

WRITE

a.

Im(z)

θ, 4, 4, Re(z)

2. Sketch the triangle that has sides in this $1:1$ ratio.

From the diagram:

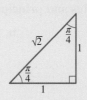

$$\theta = \frac{\pi}{4}$$

$$\therefore \text{Arg}(z) = \frac{\pi}{4}$$

3. This result can be verified using an inverse trigonometric ratio, $\theta = \tan^{-1}\left(\dfrac{y}{x}\right)$.

Check:

$$\theta = \tan^{-1}\left(\frac{4}{4}\right)$$

$$= \frac{\pi}{4}$$

b. 1. Plot z.

b.

2. Sketch the triangle that has sides in this ratio.

Remember: $-\pi < \theta \leq \pi$.

From the diagram:

$$\theta = -\frac{\pi}{3}$$

$$\therefore \text{Arg}(z) = -\frac{\pi}{3}$$

3. This result can be verified using an inverse trigonometric ratio, $\theta = \tan^{-1}\left(\dfrac{y}{x}\right)$.

Check:

$$\theta = \tan^{-1}\left(\frac{-\sqrt{3}}{1}\right)$$

$$= -\frac{\pi}{3}$$

**TI	THINK**	**WRITE**	**CASIO	THINK**	**WRITE**

a. 1. Put the calculator in Radian mode. On a Calculator page, press MENU, then select:
2: Number
9: Complex Number Tools
4: Polar Angle.
Complete the entry line as: angle $(4 + 4i)$ then press ENTER.
Note: The symbol i can be found by pressing the π button.

a. 1. Put the calculator in Radian mode. On a Run-Matrix screen, press OPTN, then select COMPLEX by pressing F3. Select Arg by pressing F3. Complete the entry line as: Arg$(4 + 4i)$ then press EXE.
Note: The symbol i can be found by pressing SHIFT 0.

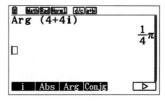

2. The answer appears on the screen.
Note: The calculator will give a decimal answer, not an exact answer.

$\text{Arg}(z) = 0.785$ radians (to three decimal places)

2. The answer appears on the screen.

$\text{Arg}(z) = \dfrac{\pi}{4}$

WORKED EXAMPLE 6 Determining the modulus and argument of complex numbers

Determine the modulus and principal argument for each of the following complex numbers.

a. $-\sqrt{3} + i$ b. $-\sqrt{2} - \sqrt{2}i$

THINK

WRITE

a. **1.** Plot z.

a.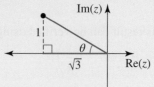

2. This triangle has sides in the same ratio as

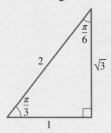

$|z| = \sqrt{\left(\sqrt{3}\right)^2 + (1)^2}$

$\quad = 2$

$\theta = \tan^{-1}\left(\dfrac{1}{\sqrt{3}}\right)$

$\quad = \dfrac{\pi}{6}$

$\operatorname{Arg}(z) = \pi - \dfrac{\pi}{6}$

$\qquad\quad = \dfrac{5\pi}{6}$

b. **1.** Plot z.

b.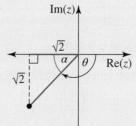

2. Determine the modulus.

$|z| = \sqrt{\left(-\sqrt{2}\right)^2 + \left(-\sqrt{2}\right)^2}$

$\quad = \sqrt{2 + 2}$

$\quad = \sqrt{4}$

$\quad = 2$

3. The triangle in the third quadrant will be used to determine α but the answer will be finally expressed as θ and $\operatorname{Arg}(z)$.

$\alpha = \tan^{-1}\left(\dfrac{y}{x}\right)$

$\quad = \tan^{-1}\left(\dfrac{-\sqrt{2}}{-\sqrt{2}}\right)$

$\quad = \dfrac{\pi}{4}$

$\operatorname{Arg}(z) = \theta = -\pi + \dfrac{\pi}{4}$

$\qquad\qquad\quad = -\dfrac{3\pi}{4}$

7.2.3 Expressing complex numbers in polar form

Now that we know how to calculate the modulus and arguments of complex numbers, we can express them in a new form. The **polar form** of a complex number can be used to express any complex number in terms of its modulus and argument.

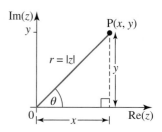

Suppose $z = x + yi$ is represented by the point P(x, y) on the complex plane using Cartesian coordinates. The modulus $|z|$ of z is also noted r.

Using the trigonometric properties of a right-angled triangle, z can also be expressed in polar coordinates as follows. We have:

$$\cos(\theta) = \frac{x}{r} \quad \text{or} \quad x = r\cos(\theta)$$

$$\sin(\theta) = \frac{y}{r} \quad \text{or} \quad y = r\sin(\theta)$$

where $|z| = r = \sqrt{x^2 + y^2}$ and $\theta = \text{Arg}(z)$.

Substituting these values, the point P(x, y) becomes the polar form P$\big(r\cos(\theta), \, r\sin(\theta)\big)$ as shown.

$z = x + yi$ in Cartesian form becomes:

$z = r\cos(\theta) + r\sin(\theta)i$

$\quad = r\big(\cos(\theta) + i\sin(\theta)\big)$

$\quad = r\,\text{cis}(\theta)$, where $\text{cis}(\theta)$ is the abbreviated form of $\cos(\theta) + i\sin(\theta)$.

(***Note:*** The acronym 'cis' is pronounced 'sis'.)

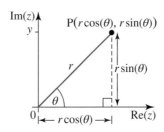

Polar form of complex numbers

The polar form of a complex number $z = x + yi$ is:

$$z = r\,\text{cis}(\theta) = r(\cos(\theta) + i\sin(\theta))$$

where $r = |z| = \sqrt{x^2 + y^2}$ and $\theta = \tan^{-1}\left(\dfrac{y}{x}\right), -\pi < \theta \leq \pi, x \neq 0$

To convert from Cartesian form to polar form, use the following formulas:

$$r = \sqrt{x^2 + y^2}$$
$$\theta = \text{Arg}(z)$$

You may find it useful to sketch the complex number on an Argand plane to help determine the principal argument.

WORKED EXAMPLE 7 Expressing complex numbers in polar form

Express each of the following in polar form, using the principal argument.

a. $z = 1 + i$ **b.** $z = 1 - \sqrt{3}i$

THINK

a. 1. Sketch z.

2. Calculate the value of r using $r = |z| = \sqrt{x^2 + y^2}$.

Determine θ from $\tan(\theta) = \left(\dfrac{y}{x}\right)$.

The angle θ is in the range $(-\pi, \pi]$, which is required.

3. Substitute the values of r and θ in $z = r \operatorname{cis}(\theta)$.

b. 1. Sketch z.

2. Calculate the value of r and θ.

3. Substitute the values of r and θ into $r \operatorname{cis}(\theta)$.

WRITE

a.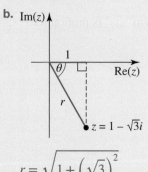

$r = \sqrt{1^2 + 1^2}$

$= \sqrt{2}$

$\tan(\theta) = \dfrac{1}{1}$

$\theta = \tan^{-1}(1)$

$\theta = \dfrac{\pi}{4}$

$z = \sqrt{2} \operatorname{cis}\left(\dfrac{\pi}{4}\right)$

b. Im(z) diagram with $z = 1 - \sqrt{3}i$

$r = \sqrt{1 + \left(\sqrt{3}\right)^2}$

$= 2$

$\tan(\theta) = -\dfrac{\sqrt{3}}{1}$

$\theta = \tan^{-1}\left(-\sqrt{3}\right)$

$\theta = -\dfrac{\pi}{3}$

$z = 2 \operatorname{cis}\left(-\dfrac{\pi}{3}\right)$

| TI | THINK | WRITE | CASIO | THINK | WRITE |

b. 1. Put the calculator in Radian mode. On a Calculator page, complete the entry line as: $1 - \sqrt{3}\,i$. Press MENU, then select:
2: Number
9: Complex Number Tools
6: Convert to Polar then press ENTER.
Note: The symbol i can be found by pressing the π button.

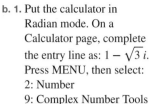

b. 1. Put the calculator in Radian mode. On a Run-Matrix screen, complete the entry line as: $1 - \sqrt{3}\,i$. Press OPTN, then select COMPLEX by pressing F3. Press F6 to scroll across to more menu options, then select $r\angle\theta$ by pressing F3. Press EXE.
Note: The symbol i can be found by pressing SHIFT 0.

2. The answer appears on the screen. Note: The calculator will give a decimal answer, not an exact answer.

The answer is given in the form $e^{i\theta} \cdot |z|$, where $\theta = -1.047$ and $|z| = 2$. Rewriting this in the form $|z|\operatorname{cis}\theta$ gives $z = 2\operatorname{cis}(-1.047)$.

2. The answer appears on the screen.

$$z = 2\operatorname{cis}\left(\frac{-\pi}{3}\right)$$

7.2.4 Polar coordinates

The Cartesian coordinates you are familiar with use the signed distance to the two coordinate axes to specify the location of a point P on the plane.

For instance, $(-3, 2)$ represents the point 3 units left of the y-axis and 2 units above the x-axis.

With polar coordinates, however, the position of a point P in the plane is specified by its distance r from the origin O, and the angle θ made between the horizontal x-axis, and the line segment OP.

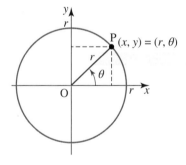

Thus, the polar form of complex numbers, which uses the modulus $|z| = r$ and principal argument θ of a complex number, is well suited to represent complex numbers on the Argand plane using the polar coordinates.

Note the similarities between the Cartesian form and the polar form of a vector and of a complex number.

Note that if the argument of a complex number is $0 \pm \pi$, then its imaginary part is equal to zero, it is a real number.

Similarly, if the argument of a complex number is $\pm\dfrac{\pi}{2}$, then its real part is equal to zero, it is a pure imaginary number.

In summary, for $z = r\cos(\theta) + r\sin(\theta)i$ and $k \in \mathbb{Z}$

If $\theta = 0 \pm k\pi$, $\operatorname{Im}(z) = r\sin(0) = 0$

If $\theta = \dfrac{\pi}{2} \pm k\pi$, $\operatorname{Re}(z) = \cos\left(\dfrac{\pi}{2}\right) = 0$

7.2.5 Converting from polar form to Cartesian form

To convert from polar form to Cartesian form, use the following formulas:

$$x = r\cos(\theta)$$
$$y = r\sin(\theta)$$

WORKED EXAMPLE 8 Converting from polar form to Cartesian form

Express $3\operatorname{cis}\left(\dfrac{\pi}{4}\right)$ in Cartesian (or standard $a + bi$) form.

THINK

1. Sketch z.

2. Express $3\operatorname{cis}\left(\dfrac{\pi}{4}\right)$ in Cartesian form.

3. Simplify using exact values from the following triangle:

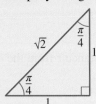

WRITE

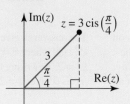

$$3\operatorname{cis}\left(\frac{\pi}{4}\right) = 3\cos\left(\frac{\pi}{4}\right) + 3\sin\left(\frac{\pi}{4}\right)i$$

$$= 3 \times \frac{1}{\sqrt{2}} + 3 \times \frac{1}{\sqrt{2}}i$$

$$= \frac{3}{\sqrt{2}} + \left(\frac{3}{\sqrt{2}}\right)i$$

TI | THINK

1. Put the calculator in Radian mode. On a Calculator page, complete the entry line as: $3\angle\dfrac{\pi}{4}$
Press MENU, then select:
2: Number
9: Complex Number Tools
7: Convert to Rectangular
then press ENTER.
Note: The $\angle$ symbol can be found by pressing CTRL, then the Catalogue button.

WRITE

2. The answer appears on the screen.
Note: The calculator will give a decimal answer, not an exact answer.

$3\operatorname{cis}\left(\dfrac{\pi}{4}\right) = 2.121 + 2.121i$

(to three decimal places)

CASIO | THINK

1. Put the calculator in Radian mode. On a Run-Matrix screen, complete the entry line as: $3\angle\dfrac{\pi}{4}$ Press OPTN, then select COMPLEX by pressing F3. Press F6 to scroll across to more menu options, then select $a + bi$ by pressing F4. Press EXE.
Note: The $\angle$ symbol can be found by pressing SHIFT, then the Variables button.

WRITE

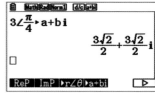

2. The answer appears on the screen.

$3\operatorname{cis}\left(\dfrac{\pi}{4}\right) = \dfrac{3\sqrt{2}}{2} + \dfrac{3\sqrt{2}}{2}i$

When not using technology to convert between the Cartesian and the polar forms, you might find the following **unit circle**, with its summary of important angles, useful.

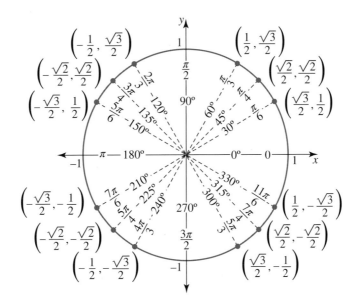

Exercise 7.2 Complex numbers in polar form

learn on

7.2 Exercise	7.2 Exam questions on

Simple familiar	Complex familiar	Complex unfamiliar
1, 2, 3, 4, 5, 6, 7, 8, 9, 10, 11, 12, 13, 14	15, 16, 17, 18, 19	20, 21

These questions are even better in jacPLUS!
- Receive immediate feedback
- Access sample responses
- Track results and progress

Find all this and MORE in jacPLUS ▶

Simple familiar

1. **WE1** Calculate the modulus of each of the following complex numbers.

   a. $z = 5 + 12i$

   b. $z = \sqrt{5} - 2i$

2. Calculate the modulus of each of the following complex numbers.

   a. $z = -4 + 7i$

   b. $z = -3 - 6i$

3. Calculate the modulus of each of the following complex numbers.

   a. $z = \sqrt{3} + \sqrt{2}i$

   b. $z = (2 + i)^2$

4. **WE2** If $z = 3 + i$, $w = 4 - 3i$ and $u = -2 + 5i$:

   i. represent each of the following on an Argand plane
   ii. calculate the magnitude in each case.

   a. $z - w$ b. $u + z$ c. $w - u$ d. $w + z$ e. $z + w - u$ f. z^2

5. **WE4** Convert each of the following into principal arguments.

   a. $\dfrac{3\pi}{2}$

   b. $-\dfrac{11\pi}{6}$

   c. $\dfrac{15\pi}{8}$

   d. $-\dfrac{5\pi}{4}$

6. Convert each of the following into principal arguments.

   a. $\dfrac{19\pi}{6}$

   b. $\dfrac{20\pi}{7}$

   c. $-\dfrac{18\pi}{5}$

   d. $-\dfrac{13\pi}{12}$

7. **WE5** Determine the principal argument of z for each of the following in the interval $(-\pi, \pi]$.

   a. $z = \sqrt{3} + i$ b. $z = 5 - 5i$ c. $z = -2 - 2\sqrt{3}i$

8. Determine the principal argument of z for each of the following in the interval $(-\pi, \pi]$.

   a. $z = 3i$ b. $z = -\sqrt{7}$ c. $z = -6i$ d. $z = 55$

9. **WE6** Determine the modulus and principal argument of each of the following complex numbers.

   a. $3 - 3i$ b. $-5 + 5i$ c. $-1 - \sqrt{3}i$ d. $4\sqrt{3} + 4i$

10. **WE7** Express each of the following in polar form, using the principal argument.

   a. $z = -1 + i$ b. $z = \sqrt{6} + \sqrt{2}i$ c. $z = -\sqrt{5} - \sqrt{5}i$

11. Express each of the following in polar form, using the principal argument.

   a. $z = \sqrt{5} - \sqrt{15}i$ b. $z = -\dfrac{1}{2} - \dfrac{\sqrt{3}}{2}i$ c. $z = -\dfrac{1}{4} + \dfrac{1}{4}i$

12. **WE8** Express each of the following complex numbers in Cartesian form.

   a. $2\operatorname{cis}\left(\dfrac{2\pi}{3}\right)$ b. $3\operatorname{cis}\left(\dfrac{\pi}{4}\right)$ c. $\sqrt{5}\operatorname{cis}\left(\dfrac{5\pi}{6}\right)$

13. Express each of the following complex numbers in Cartesian form.

   a. $4\operatorname{cis}\left(-\dfrac{\pi}{3}\right)$ b. $\sqrt{7}\operatorname{cis}\left(-\dfrac{7\pi}{4}\right)$ c. $8\operatorname{cis}\left(\dfrac{\pi}{2}\right)$

14. **WE3** a. Represent the points $z_1 = 3 + 3i$, $z_2 = 6 - 2i$, $z_3 = 1 - 5i$ and $z_4 = -2$ on the complex number plane.

   b. Calculate the area of the rhombus formed when the four points are connected by straight line segments in the order z_1 to z_2 to z_3 to z_4.

Complex familiar

15. a. Show the points $z = -1 + 3i$, $u = 3$ and $w = 3 + 12i$ on the complex number plane.
   b. Calculate the area of the triangle produced by joining the three points with straight line segments.

16. a. If the complex numbers $u = 3 - 4i$, $\bar{u}$, v and $\bar{v}$ form a square with an area of 64 square units, evaluate the complex number v.
   b. If the complex numbers $u = -2 + 5i$, $\bar{u}$, v and $\bar{v}$ form a rectangle with an area of 60 square units, evaluate the complex number v.
   c. If $a, b, c \in \mathbb{R}^+$, evaluate the area of the rectangle formed by the complex numbers $u = a + bi$, $\bar{u}$, $v = -c - bi$ and $\bar{v}$.

17. a. Evaluate the area of the triangle that is formed by the complex numbers $u = 4 + 3i$ and iu and the origin O.
   b. Evaluate the area of the triangle that is formed by the complex numbers $u = 12 + 5i$ and iu and the origin O.
   c. If $a, b \in \mathbb{R}^+$, evaluate the area of the triangle that is formed by the complex numbers $u = a + bi$, iu and the origin O.

18. **a.** Evaluate the area of the square that is formed by the complex numbers $u = 4 - 3i$, iu, $u + iu$ and the origin O.
    b. If the area of the square formed by the complex numbers $u = 6 + bi$, iu, $u + iu$ and the origin O is equal to 50 square units, evaluate the value of b.
    c. If $a, b \in \mathbb{R}^+$, evaluate the area of the square that is formed by the complex numbers $u = a + bi$, iu, $u + iu$ and the origin O.

19. **a.** Evaluate the equation and area of the circle that passes through the complex numbers $u = 3 - 4i$, iu and the origin O.
    b. Evaluate the equation and area of the circle that passes through the complex numbers $u = 5 + 12i$, iu and the origin O.

Complex unfamiliar

20. If $a, b \in \mathbb{R}^+$, evaluate the equation and area of the circle that passes through the complex numbers $u = a + bi$, iu and the origin O.

21. If $z_1 = 1 + i\sqrt{3}$, $z_2 = 1 + i$ and $Z = \dfrac{z_1}{z_2}$, use the Cartesian and polar forms of Z to determine the exact values of $\cos\left(\dfrac{\pi}{12}\right)$ and $\sin\left(\dfrac{\pi}{12}\right)$.

Fully worked solutions for this chapter are available online.

LESSON
7.3 Operations on complex numbers in polar form

SYLLABUS LINKS

- Express a complex number in Cartesian form and polar form.
  - $z = r(\cos(\theta) + i\sin(\theta))$ or $z = r\operatorname{cis}(\theta)$
- Convert between Cartesian form and polar form.
- Perform complex-number arithmetic: addition, subtraction, multiplication and division, with and without technology.

Source: Specialist Mathematics Senior Syllabus 2024 © State of Queensland (QCAA) 2024; licensed under CC BY 4.0

7.3.1 Addition and subtraction in polar form

In general there is no simple way to add or subtract complex numbers given in the polar form $r\operatorname{cis}(\theta)$. For addition or subtraction, the complex numbers need to be expressed in Cartesian form first. This is similar to vectors: the Cartesian form of vectors is more convenient for the addition and subtraction of vectors.

7.3.2 Multiplication, division and powers in polar form

The polar form of complex numbers is very useful for multiplying and dividing complex numbers. The product of two complex numbers can be calculated by multiplying their moduli and adding their arguments. It also applies to division, which is similar to multiplication but uses the inverse operations, division of moduli and subtraction of arguments.

> ## Multiplication and division of complex numbers in polar form.
>
> If $z_1 = r_1 \text{cis}(\theta_1)$ and $z_2 = r_2 \text{cis}(\theta_2)$, then
>
> $$z_1 \times z_2 = r_1 \times r_2 \text{cis}(\theta_1 + \theta_2)$$
>
> **and**
>
> $$\frac{z_1}{z_2} = \frac{r_1}{r_2} \text{cis}(\theta_1 - \theta_2).$$

Note that this applies to the multiplication or division by a scalar, k, which in polar form is $r \text{cis}(0 \pm \pi)$.

WORKED EXAMPLE 9 Calculating products of complex numbers in polar form (1)

Express $5 \text{cis}\left(\dfrac{\pi}{4}\right) \times 2 \text{cis}\left(\dfrac{5\pi}{6}\right)$ in the form $r \text{cis}(\theta)$ where $\theta \in (-\pi, \pi]$.

THINK

1. Recall the formula
   $z_1 \times z_2 = r_1 \times r_2 \text{cis}(\theta_1 + \theta_2)$.

2. Sketch this number.

3. Subtract 2π from θ to express the answer in the required form where $\theta \in (-\pi, \pi]$.

WRITE

$5 \text{cis}\left(\dfrac{\pi}{4}\right) \times 2 \text{cis}\left(\dfrac{5\pi}{6}\right) = (5 \times 2) \text{cis}\left(\dfrac{\pi}{4} + \dfrac{5\pi}{6}\right)$

$= 10 \text{cis}\left(\dfrac{13\pi}{12}\right)$

$10 \text{cis}\left(\dfrac{13\pi}{12}\right) = 10 \text{cis}\left(-\dfrac{11\pi}{12}\right)$

WORKED EXAMPLE 10 Calculating products of complex numbers in polar form (2)

Express $z_1 z_2$ in Cartesian form if $z_1 = \sqrt{2}\operatorname{cis}\left(\dfrac{5\pi}{6}\right)$ and $z_2 = \sqrt{6}\operatorname{cis}\left(-\dfrac{\pi}{3}\right)$.

THINK

1. Recall the formula
   $z_1 \times z_2 = r_1 \times r_2 \operatorname{cis}(\theta_1 + \theta_2)$.

2. Simplify and write the result in Cartesian form.

WRITE

$$z_1 \times z_2 = \sqrt{2}\operatorname{cis}\left(\dfrac{5\pi}{6}\right) \times \sqrt{6}\operatorname{cis}\left(-\dfrac{\pi}{3}\right)$$

$$= \left(\sqrt{2} \times \sqrt{6}\right)\operatorname{cis}\left(\dfrac{5\pi}{6} - \dfrac{\pi}{3}\right)$$

$$= 2\sqrt{3}\operatorname{cis}\left(\dfrac{\pi}{2}\right)$$

$$= 2\sqrt{3}\cos\left(\dfrac{\pi}{2}\right) + 2\sqrt{3}\sin\left(\dfrac{\pi}{2}\right)i$$

$$= 2\sqrt{3}\times 0 + 2\sqrt{3}\times 1i$$

$$= 2\sqrt{3}i$$

WORKED EXAMPLE 11 Converting to polar form and calculating products of complex numbers

If $z = 5\sqrt{3} + 5i$ and $w = 3 + 3\sqrt{3}i$, express the product zw in polar form.

THINK

1. Convert z to polar form.

WRITE

$$z = r_1 \operatorname{cis}(\theta_1)$$

$$r_1 = |z|$$

$$= \sqrt{\left(5\sqrt{3}\right)^2 + (5)^2}$$

$$= \sqrt{100}$$

$$= 10$$

$$\theta_1 = \tan^{-1}\left(\dfrac{5}{5\sqrt{3}}\right)$$

$$= \tan^{-1}\left(\dfrac{1}{\sqrt{3}}\right)$$

$$= \dfrac{\pi}{6}$$

$$z = 10\operatorname{cis}\left(\dfrac{\pi}{6}\right)$$

2. Convert w to polar form.

$$w = r_2 \operatorname{cis}(\theta_2)$$

$$r_2 = |w|$$

$$= \sqrt{(3)^2 + \left(3\sqrt{3}\right)^2}$$

$$= \sqrt{36}$$

$$= 6$$

$$\theta_2 = \tan^{-1}\left(\frac{3\sqrt{3}}{3}\right)$$

$$= \tan^{-1}\left(\sqrt{3}\right)$$

$$= \frac{\pi}{3}$$

$$w = 6\operatorname{cis}\left(\frac{\pi}{3}\right)$$

3. Recall the formula
$z_1 \times z_2 = r_1 \times r_2 \operatorname{cis}(\theta_1 + \theta_2)$.

$$zw = 10\operatorname{cis}\left(\frac{\pi}{6}\right) \times 6\operatorname{cis}\left(\frac{\pi}{3}\right)$$

$$= 60\operatorname{cis}\left(\frac{\pi}{6} + \frac{\pi}{3}\right)$$

$$= 60\operatorname{cis}\left(\frac{\pi}{2}\right)$$

$$= 60i$$

WORKED EXAMPLE 12 Calculating quotients of complex numbers in polar form

Express $10\operatorname{cis}\left(-\dfrac{\pi}{3}\right) \div 5\operatorname{cis}\left(\dfrac{5\pi}{6}\right)$ in the form $r\operatorname{cis}(\theta)$ where $\theta \in (-\pi,\ \pi]$.

THINK

1. Recall the formula

2. Sketch this number.

3. State θ, the principal argument.

4. State the result in polar form.

WRITE

$$10\operatorname{cis}\left(-\frac{\pi}{3}\right) \div 5\operatorname{cis}\left(\frac{5\pi}{6}\right) = 2\operatorname{cis}\left(-\frac{\pi}{3} - \frac{5\pi}{6}\right)$$

$$= 2\operatorname{cis}\left(-\frac{7\pi}{6}\right)$$

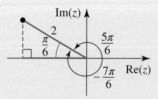

$$\operatorname{Arg}(z) = \frac{5\pi}{6}$$

$$z = 2\operatorname{cis}\left(\frac{5\pi}{6}\right)$$

TI	THINK	WRITE	CASIO	THINK	WRITE

1. In the Document Settings menu, change the Angle to Radian and the Calculation Mode to Polar, then select OK.

1. In the SETUP menu, change the Angle to Radian and the Complex Mode to $r\angle\theta$, then press EXIT.

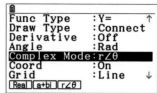

2. On a Calculator page, complete the entry line as: $\dfrac{10\angle\frac{-\pi}{3}}{5\angle\frac{5\pi}{6}}$ then press ENTER. *Note:* The $\angle$ symbol can be found by pressing CTRL, then pressing the Catalogue button.

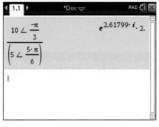

2. On a Run-Matrix screen, complete the entry line as: $\dfrac{10\angle\frac{-\pi}{3}}{5\angle\frac{5\pi}{6}}$ then press EXE. *Note:* The $\angle$ symbol can be found by pressing SHIFT, then the Variables button.

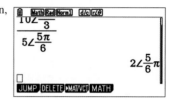

3. The answer appears on the screen. *Note:* The calculator will give a decimal answer, not an exact answer.

The answer is given in the form $e^{i\theta}\cdot|z|$, where $\theta = 2.618$ and $|z| = 2$. Rewriting this in the form $|z|\operatorname{cis}\theta$ gives $z = 2\operatorname{cis}(2.618)$.

3. The answer appears on the screen.

$z = 2\operatorname{cis}\left(\dfrac{5\pi}{6}\right)$

Index powers of z

A logical consequence of the multiplication and division of complex numbers in polar form, **de Moivre's theorem**, involves the calculation of a power of z.

$z^n = z \times z \times \ldots \times z$ (n times)

> ### de Moivre's theorem (powers of z)
>
> **If $z = r\operatorname{cis}(\theta)$, then $z^n = r^n\operatorname{cis}(n\theta)$.**
>
> **In other words,**
>
> $$\left(r\operatorname{cis}(\theta)\right)^n = r^n\operatorname{cis}(n\theta)$$

The proofs required to establish these rules are outside the scope of this course and are not included here. You will learn more about the applications of de Moivre's theorem in Unit 3.

WORKED EXAMPLE 13 Calculating powers of complex numbers in polar form

If $z = 2\operatorname{cis}\left(\dfrac{5\pi}{6}\right)$, calculate z^3, write your final answer in Cartesian form.

THINK

1. Recall de Moivre's theorem for index powers.

2. Substitute the values into de Moivre's theorem and simplify.

WRITE

$z^n = r^n\operatorname{cis}(n\theta)$

$z^3 = 2^3\operatorname{cis}\left(3 \times \dfrac{5\pi}{6}\right)$

$\quad = 8\operatorname{cis}\left(\dfrac{5\pi}{2}\right)$

3. Remember $-\pi < \text{Arg}(z) \le \pi$.	$\text{Arg}\left(\dfrac{5\pi}{2}\right) = \dfrac{\pi}{2}$
4. Write the answer.	$z^3 = 8\,\text{cis}\left(\dfrac{\pi}{2}\right)$ $= 8i$

Negative powers of z

Your earlier studies have shown that $z^{-1} = \dfrac{1}{z}$. Similarly, $z^{-3} = \dfrac{1}{z^3}$.

WORKED EXAMPLE 14 Calculating negative powers of complex numbers in polar form

Evaluate $(1-i)^{-4}$.

THINK	WRITE
1. Convert $(1-i)$ to polar form.	$\|1-i\| = \sqrt{1^2 + (-1)^2}$ $= \sqrt{2}$ $\text{Arg}(1-i) = \tan^{-1}(-1)$ $= -\dfrac{\pi}{4}$ $1-i = \sqrt{2}\,\text{cis}\left(-\dfrac{\pi}{4}\right)$
2. Write de Moivre's theorem and substitute in the values.	$z^n = r^n\,\text{cis}(n\theta)$ $(1-i)^{-4} = \left(\sqrt{2}\right)^{-4}\text{cis}\left(4 \times \dfrac{\pi}{4}\right)$
3. Simplify the expression, recalling the exact values of $\cos(\pi)$ and $\sin(\pi)$.	$= \dfrac{1}{4}\text{cis}(\pi)$ $= \dfrac{1}{4}(\cos(\pi) + i\sin(\pi))$
4. Write the final expression.	$(1-i)^{-4} = -\dfrac{1}{4}$

7.3.3 Powers of complex numbers in Cartesian form

Whole powers of z

As with real numbers, powers of complex numbers can be written as:

$$z^n = z \times z \times z \times z \times \dots \times z \text{ to } n \text{ factors.}$$

Since $z = a + bi$ is a binomial (containing two terms), we can express z^n using Pascal's triangle (see Chapter 1, Combinatorics) to generate the coefficients of each term.

$$
\begin{array}{ccccccccccc}
 & & & & & 1 & & & & & \\
 & & & & 1 & & 1 & & & & \\
 & & & 1 & & 2 & & 1 & & & \\
 & & 1 & & 3 & & 3 & & 1 & & \\
 & 1 & & 4 & & 6 & & 4 & & 1 & \\
\end{array}
$$

5th row $\rightarrow$ 1   5   10   10   5   1   and so on.

$(a+bi)^5$ can therefore be expanded using the elements of the fifth row of Pascal's triangle:

$$(a+bi)^5 = 1a^5 + 5a^4(bi)^1 + 10a^3(bi)^2 + 10a^2(bi)^3 + 5a(bi)^4 + (bi)^5$$
$$= 1a^5 + 5a^4bi + 10a^3b^2i^2 + 10a^2b^3i^3 + 5ab^4i^4 + b^5i^5$$
$$= 1a^5 + 5a^4bi - 10a^3b^2 - 10a^2b^3i + 5ab^4 + b^5i$$
$$= 1a^5 - 10a^3b^2 + 5ab^4 + 5a^4bi - 10a^2b^3i + b^5i$$
$$= 1a^5 - 10a^3b^2 + 5ab^4 + (5a^4b - 10a^2b^3 + b^5)\,i \text{ grouped into standard form.}$$

$$\text{Re}[(a+bi)^5] = 1a^5 - 10a^3b^2 + 5ab^4$$
$$\text{Im}[(a+bi)^5] = 5a^4b - 10a^2b^3 + b^5$$

WORKED EXAMPLE 15 Expanding powers of complex numbers in Cartesian form

Use Pascal's triangle to expand $(2-3i)^3$.

THINK	WRITE
1. Use the third row of Pascal's triangle to expand $(1 \quad 3 \quad 3 \quad 1)$. Use brackets to keep the negative sign of the second term.	$(2-3i)^3 = 1\left(2^3\right) + 3(2)^2\left(-3i\right) + 3(2)(-3i)^2 + (-3i)^3$
2. Simplify the expression.	$= 8 - 36i + 54i^2 - 27i^3$ $= 8 - 36i - 54 + 27i$ $= -46 - 9i$

7.3.4 Trigonometric proofs with complex numbers

In Chapter 9, you will explore trigonometric identities and the different ways to prove whether statements that are written as functions of one or more angles are true. It is possible to combine de Moivre's theorem and binomial expansion of complex numbers to prove some of these multi-angle identities. You will learn more about trigonometric proofs using de Moivre's theorem in Unit 3.

If $z = r\operatorname{cis}(\theta)$, then de Moivre's theorem tells us that:

$$z^n = r^n \operatorname{cis}(n\theta)$$
$$= r^n\left(\cos(n\theta) + i\sin(n\theta)\right) \quad [1]$$

We can also write:

$$z^n = r^n\left(\cos(\theta) + i\sin(\theta)\right)^n \quad [2]$$

Equating [1] and [2] gives us:

$$\cos(n\theta) + i\sin(n\theta) = \left(\cos(\theta) + i\sin(\theta)\right)^n$$

Equating the real and imaginary parts of equations [1] and [2] gives us the following identities.

Identities for $\sin(n\theta)$ and $\cos(n\theta)$

$$\sin(n\theta) = \text{Re}\left[\left(\cos(\theta) + i\sin(\theta)\right)^n\right]$$
$$\cos(n\theta) = \text{Im}\left[\left(\cos(\theta) + i\sin(\theta)\right)^n\right]$$

It may be necessary to use the Pythagorean identity $\sin^2(\theta) + \cos^2(\theta) = 1$ as part of your proof.

WORKED EXAMPLE 16 Using de Moivre's theorem to prove trigonometric identities

Use de Moivre's theorem to prove:

a. $\cos(3\theta) = 4\cos^3(\theta) - 3\cos(\theta)$

b. $\sin(3\theta) = 3\sin(\theta) - 4\sin^3(\theta)$

THINK	WRITE
1. Both proofs involve 3θ, so it will be necessary to determine $(\cos(\theta) + i\sin(\theta))^3$.	$(\cos(\theta) + i\sin(\theta))^3$
2. Determine the expansion using the third row (1 3 3 1) of Pascal's triangle.	$(\cos(\theta) + i\sin(\theta))^3 = 1(\cos(\theta))^3 + 3(\cos(\theta))^2(i\sin(\theta))$ $\qquad + 3(\cos(\theta))(i\sin(\theta))^2 + 1(i\sin(\theta))^3$ $= \cos^3(\theta) + 3i\cos^2(\theta)\sin(\theta)$ $\qquad + 3i^2\cos(\theta)\sin^2(\theta) + i^3\sin^3(\theta)$
3. Simplify the expression, remembering that $i^2 = -1$ and $i^3 = -i$.	$= \cos^3(\theta) + 3i\cos^2(\theta)\sin(\theta)$ $\qquad - 3\cos(\theta)\sin^2(\theta) - i\sin^3(\theta)$
4. Group the real and imaginary terms together.	$= \cos^3(\theta) - 3\cos(\theta)\sin^2(\theta)$ $\qquad + 3i\cos^2(\theta)\sin(\theta) - i\sin^3(\theta)$ $(\cos(\theta) + i\sin(\theta))^3 = \cos^3(\theta) - 3\cos(\theta)\sin^2(\theta)$ $\qquad + i(3\cos^2(\theta)\sin(\theta) - \sin^3(\theta))$
a. 1. State the rule for $\cos(n\theta)$.	$\cos(n\theta) = \text{Re}\left[(\cos(\theta) + i\sin(\theta))^n\right]$
2. Use the expansion of $(\cos(\theta) + i\sin(\theta))^3$ to determine $\cos(3\theta)$.	$\cos(3\theta) = \text{Re}\left[(\cos(\theta) + i\sin(\theta))^3\right]$ $= \cos^3(\theta) - 3\cos(\theta)\sin^2(\theta)$
3. We want to prove that $\cos(3\theta) = 4\cos^3(\theta) - 3\cos(\theta)$, so use the Pythagorean identity $\sin^2(\theta) = 1 - \cos^2(\theta)$ to replace $\sin^2\theta$.	$\cos(3\theta) = \cos^3(\theta) - 3\cos(\theta)(1 - \cos^2(\theta))$ $= \cos^3(\theta) - 3\cos(\theta) + 3\cos^3(\theta)$ $\therefore \cos(3\theta) = 4\cos^3(\theta) - 3\cos(\theta)$
b. 1. State the rule for $\sin(n\theta)$.	$\sin(n\theta) = \text{Im}\left[(\cos(\theta) + i\sin(\theta))^n\right]$
2. Use the expansion of $(\cos(\theta) + i\sin(\theta))^3$ to determine $\sin(3\theta)$.	$\sin(3\theta) = \text{Im}\left[(\cos(\theta) + i\sin(\theta))^3\right]$ $= 3\cos^2(\theta)\sin(\theta) - \sin^3(\theta)$
3. We want to prove that $\sin(3\theta) = 3\sin(\theta) - 4\sin^3(\theta)$, so use the Pythagorean identity $\cos^2(\theta) = 1 - \sin^2(\theta)$ to replace $\sin^2(\theta)$.	$\sin(3\theta) = 3(1 - \sin^2(\theta))\sin(\theta) - \sin^3(\theta)$ $= 3\sin(\theta) - 3\sin^3(\theta) - \sin^3(\theta)$ $\therefore \sin(3\theta) = 3\sin(\theta) - 4\sin^3(\theta)$

7.3 Exercise	7.3 Exam questions on	These questions are even better in jacPLUS!

Simple familiar	Complex familiar	Complex unfamiliar
1, 2, 3, 4, 5, 6, 7, 8, 9, 10	11, 12, 13, 14, 15, 16, 17, 18	19, 20

- Receive immediate feedback
- Access sample responses
- Track results and progress

Find all this and MORE in jacPLUS ▶

Simple familiar

1. **WE9** Express each of the following in the form $r\operatorname{cis}(\theta)$ where $\theta \in (-\pi, \pi]$.

   a. $2\operatorname{cis}\left(\dfrac{\pi}{4}\right) \times 3\operatorname{cis}\left(\dfrac{\pi}{2}\right)$

   b. $5\operatorname{cis}\left(\dfrac{2\pi}{3}\right) \times 4\operatorname{cis}\left(-\dfrac{\pi}{3}\right)$

   c. $6\operatorname{cis}\left(\dfrac{3\pi}{4}\right) \times \sqrt{5}\operatorname{cis}(\pi)$

   d. $\sqrt{3}\operatorname{cis}\left(-\dfrac{5\pi}{6}\right) \times \sqrt{2}\operatorname{cis}\left(-\dfrac{\pi}{2}\right)$

   e. $\sqrt{7}\operatorname{cis}\left(-\dfrac{7\pi}{12}\right) \times 2\operatorname{cis}\left(\dfrac{5\pi}{12}\right)$

2. **WE10** Express the resultant complex numbers in question 1 in Cartesian form.

3. **WE11** Express the following products in polar form.

   a. $(2+2i)\left(\sqrt{3}+i\right)$

   b. $\left(\sqrt{3}-3i\right)\left(2\sqrt{3}-2i\right)$

   c. $\left(-4+4\sqrt{3}i\right)(-1-i)$

4. **WE12** Express $12\operatorname{cis}\left(\dfrac{5\pi}{6}\right) \div 4\operatorname{cis}\left(\dfrac{\pi}{3}\right)$ in the form $r\operatorname{cis}(\theta)$ where $\theta \in (-\pi, \pi]$.

5. Express each of the following in the form $r\operatorname{cis}(\theta)$ where $\theta \in (-\pi, \pi]$.

   a. $36\operatorname{cis}\left(\dfrac{3\pi}{4}\right) \div 9\operatorname{cis}\left(-\dfrac{\pi}{6}\right)$

   b. $\sqrt{20}\operatorname{cis}\left(-\dfrac{\pi}{2}\right) \div \sqrt{5}\operatorname{cis}\left(-\dfrac{\pi}{5}\right)$

   c. $4\sqrt{3}\operatorname{cis}\left(\dfrac{4\pi}{7}\right) \div \sqrt{6}\operatorname{cis}\left(\dfrac{11\pi}{14}\right)$

   d. $3\sqrt{5}\operatorname{cis}\left(-\dfrac{7\pi}{12}\right) \div 2\sqrt{10}\operatorname{cis}\left(\dfrac{5\pi}{6}\right)$

6. **WE13** If $z = \sqrt{3}\operatorname{cis}\left(\dfrac{3\pi}{4}\right)$, calculate z^3, giving your answer in polar form.

7. If $w = 2\operatorname{cis}\left(-\dfrac{\pi}{4}\right)$, express each of the following in polar form.

   a. w^4

   b. w^5

8. **WE14** Evaluate $(1-i)^{-3}$.

9. If $z = 1-i$ and $w = -\sqrt{3}+i$, write the following in standard form.

   a. z^{-4} b. w^{-3} c. w^{-5} d. $\dfrac{z^3}{w^4}$ e. $z^2 w^3$

10. **WE15** Use Pascal's triangle to expand $(4+5i)^4$.

11. Determine $(2+2i)^2(1-\sqrt{3}i)^4$ in standard form.

12. Write $\dfrac{\left(\sqrt{3}-i\right)^6}{\left(2-2\sqrt{3}i\right)^3}$ in the form $x+yi$.

13. If $z=\sqrt{2}\operatorname{cis}\left(\dfrac{3\pi}{4}\right)$ and $w=\sqrt{3}\operatorname{cis}\left(\dfrac{\pi}{6}\right)$, determine the modulus and the argument of $\dfrac{z^6}{w^4}$.

14. If $z=4+i$ and $w=-3-2i$, determine $(z+w)^9$.

15. Evaluate z^6+w^4, if $z=\sqrt{2}-\sqrt{2}i$ and $w=2-2i$.

16. If $z_1=\sqrt{5}\operatorname{cis}\left(-\dfrac{2\pi}{5}\right)$, $z_2=2\operatorname{cis}\left(\dfrac{3\pi}{8}\right)$ and $z_3=\sqrt{10}\operatorname{cis}\left(\dfrac{\pi}{12}\right)$, determine the modulus and the argument of $\dfrac{z_1{}^2\times z_2{}^3}{z_3{}^4}$.

17. **WE16** Use de Moivre's theorem to prove:
   a. $\sin(2\theta)=2\sin(\theta)\cos(\theta)$
   b. $\cos(2\theta)=\cos^2(\theta)-\sin^2(\theta)$
   $\qquad\qquad = 2\cos^2(\theta)-1$

18. Use de Moivre's theorem to prove:
   a. $\sin(4\theta)=4\sin(\theta)\cos^3(\theta)-4\cos(\theta)\sin^3(\theta)$
   b. $\cos(4\theta)=\cos^4(\theta)-6\cos^2(\theta)\sin^2(\theta)+\sin^4(\theta)$
   $\qquad\qquad = 8\cos^4(\theta)-8\cos^2(\theta)+1$

19. Determine the Cartesian form of $z=\dfrac{(1+i)^{2000}}{\left(1-i\sqrt{3}\right)^{1000}}$.

20. Determine the values of the integer n for which $\left(1+i\sqrt{3}\right)^n$ is a real number greater than zero.

Fully worked solutions for this chapter are available online.

LESSON
7.4 Geometric interpretations of multiplication and division of complex numbers in polar form

SYLLABUS LINKS

- Understand and use multiplication, division of complex numbers in polar form and the geometric interpretation of these.
  - $z_1 z_2 = r_1 r_2 \text{cis}(\theta_1 + \theta_2)$
  - $\dfrac{z_1}{z_2} = \dfrac{r_1}{r_2} \text{cis}(\theta_1 - \theta_2)$

Source: Specialist Mathematics Senior Syllabus 2024 © State of Queensland (QCAA) 2024; licensed under CC BY 4.0

7.4.1 Geometrically multiplying or dividing complex numbers

As seen in section 7.3.2, to multiply two complex numbers in polar form, we multiply their moduli and add their arguments, and to divide two complex numbers in polar form, we divide their moduli and subtract their arguments.

This means that when multiplying or dividing two complex numbers together, the resulting complex number is dilated and rotated.

We can now generalise what we have seen in section 6.6.3 (see Worked example 17).

Geometric interpretation of complex multiplication and division

If z_1 and z_2 are two complex numbers, with $z_2 \neq 0$ then

$$|z_1 \times z_2| = |z_1| \times |z_2|$$

$$\left| \frac{z_1}{z_2} \right| = \frac{|z_1|}{|z_2|}$$

$$\arg(z_1 \times z_2) = \arg(z_1) + \arg(z_2)$$

$$\arg\left(\frac{z_1}{z_2} \right) = \arg(z_1) - \arg(z_2)$$

When multiplying z_1 by z_2, z_1 is dilated by a factor $|z_2|$ and rotated by θ_2 anticlockwise about the origin.

When dividing z_1 by z_2, z_1 is dilated by a factor $\dfrac{1}{|z_2|}$ and rotated by θ_2 clockwise about the origin.

The diagrams below illustrate this geometrically.

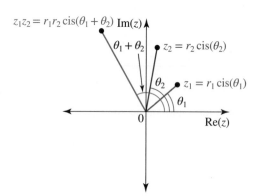

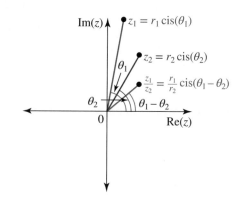

Let $z_1 = -3 + 3i$ and $z_2 = 1 - \sqrt{3}i$.

a. Express $z_1 z_2$ and $\dfrac{z_1}{z_2}$ in polar form.

b. Provide a geometric interpretation of $z_1 z_2$ and $\dfrac{z_1}{z_2}$ and represent z_1, z_2, $z_1 z_2$ and $\dfrac{z_1}{z_2}$ in the Argand plane.

THINK	WRITE
a. 1. Convert z_1 and z_2 from Cartesian form $z = x + yi$ to polar form $z = r\operatorname{cis}(\theta)$ using $r = \sqrt{x^2 + y^2}$ and $\theta = \tan^{-1}\left(\dfrac{y}{x}\right)$, $-\pi < \theta \leq \pi, x \neq 0$.	a. $\quad\|z_1\| = 3\sqrt{2}$ $\tan(\theta_1) = -1$ $\theta_1 = \dfrac{3\pi}{4}$ (2nd quadrant) $z_1 = 3\sqrt{2}\operatorname{cis}\left(\dfrac{3\pi}{4}\right)$ $\|z_2\| = 2$ $\tan(\theta_2) = -\sqrt{3}$ $\theta_1 = -\dfrac{\pi}{3}$ (4th quadrant) $z_1 = 2\operatorname{cis}\left(-\dfrac{\pi}{3}\right)$

2. Calculate $z_1 z_2$ and $\dfrac{z_1}{z_2}$

$$z_1 z_2 = 3\sqrt{2}\operatorname{cis}\left(\frac{3\pi}{4}\right) \times 2\operatorname{cis}\left(-\frac{\pi}{3}\right)$$

$$= 6\sqrt{2}\operatorname{cis}\left(\frac{3\pi}{4} + \left(-\frac{\pi}{3}\right)\right)$$

$$= 6\sqrt{2}\operatorname{cis}\left(\frac{5\pi}{12}\right)$$

$$\frac{z_1}{z_2} = \frac{3\sqrt{2}\operatorname{cis}\left(\dfrac{3\pi}{4}\right)}{2\operatorname{cis}\left(-\dfrac{\pi}{3}\right)}$$

$$= \frac{3\sqrt{2}}{2}\operatorname{cis}\left(\frac{3\pi}{4} - \left(-\frac{\pi}{3}\right)\right)$$

$$= \frac{3\sqrt{2}}{2}\operatorname{cis}\left(\frac{13\pi}{12}\right)$$

$$= \frac{3\sqrt{2}}{2}\operatorname{cis}\left(-\frac{11\pi}{12}\right)$$

b. 1. Recall that multiplying or dividing two complex numbers together correspond to a dilation and a rotation.

b. When multiplying z_1 by z_2, z_1 is dilated by a factor $|z_2| = 2$ and rotated by $\theta_2 = -\dfrac{\pi}{3}$ **anticlockwise** about the origin.

When dividing z_1 by z_2, z_1 is dilated by a factor $\dfrac{1}{|z_2|} = \dfrac{1}{2}$ and rotated by $\theta_2 = -\dfrac{\pi}{3}$ **clockwise** about the origin.

(Alternatively, when multiplying z_2 by z_1, z_2 is dilated by a factor $|z_1| = 3\sqrt{2}$ and rotated by $\theta_1 = \dfrac{3\pi}{4}$ anticlockwise about the origin.)

2. Represent z_1 and z_2 in the Argand plane, and use the geometric interpretation of $z_1 z_2$ and $\dfrac{z_1}{z_2}$ to represent them in the Argand plane.

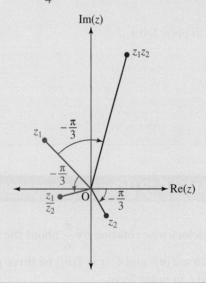

For linear transformations in the complex plane, complex numbers in polar form can conveniently be used for rotations, dilations, while complex numbers in Cartesian form are more convenient for translations.

WORKED EXAMPLE 18 Dilation and rotation about the origin

Let $z = 4\operatorname{cis}\left(\dfrac{\pi}{6}\right)$. z is rotated by $\dfrac{\pi}{4}$ clockwise about the origin and dilated by $\dfrac{1}{2}$, giving z'.
a. Represent z and z' on the Argand plane.
b. Express z' in polar form.

THINK

WRITE

a. 1. The modulus of z is 4 and its principal argument is $\dfrac{\pi}{6}$.
Represent z on a circle of radius 4 centered on the origin, and at an angle of $\dfrac{\pi}{6}$ anticlockwise from the real axis.

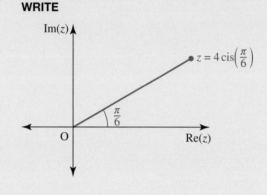

2. z is rotated by $\dfrac{\pi}{4}$ clockwise about the origin and dilated by $\dfrac{1}{2}$ thus z' is on a circled of radius $\dfrac{1}{2} \times 4$ centred on the origin, at an angle $\dfrac{\pi}{6} - \dfrac{\pi}{4} = -\dfrac{\pi}{12}$ to the real axis.

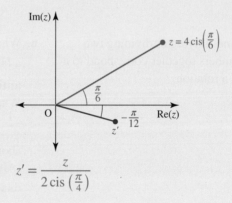

b. 1. Recall that a rotation about the origin by θ clockwise and dilation by $\dfrac{1}{r}$ is given by $\dfrac{z}{r\operatorname{cis}(\theta)}$.

$z' = \dfrac{z}{2\operatorname{cis}\left(\frac{\pi}{4}\right)}$

2. Express z' in polar form

$z' = \dfrac{4\operatorname{cis}\left(\frac{\pi}{6}\right)}{2\operatorname{cis}\left(\frac{\pi}{4}\right)}$

$= 2\operatorname{cis}\left(\dfrac{\pi}{6} - \dfrac{\pi}{4}\right)$

$= 2\operatorname{cis}\left(-\dfrac{\pi}{12}\right)$

WORKED EXAMPLE 19 Using transformations of the Complex plane.

Let T be the anticlockwise rotation by $\dfrac{\pi}{6}$ about the origin O and dilation of factor $\dfrac{\sqrt{3}}{2}$.
Let A($a = 6$), B ($b = T(a)$) and C ($c = T(b)$) be three points on the Complex plane.
a. Express b and c in polar form.
b. Represent the three points on the Argand plane.
c. Prove that OBC is a right-angle triangle.

THINK	WRITE

THINK

Express $T(z)$ the transformation of z in the Complex plane for an anticlockwise rotation by $\dfrac{\pi}{6}$ about the origin and dilation of factor $\dfrac{\sqrt{3}}{2}$.

a. Write b as $T(a)$ and c as $T(b)$.

b. Use the polar coordinates to represent A, B and C on the Complex plane.

c. 1. Determine the lengths of sides OB and OC using their polar form.

WRITE

$T(z) = \dfrac{\sqrt{3}}{2} \text{cis}\left(\dfrac{\pi}{6}\right) \times z.$

a.
$b = T(a)$

$= \dfrac{\sqrt{3}}{2} \text{cis}\left(\dfrac{\pi}{6}\right) \times a$

$= \dfrac{\sqrt{3}}{2} \text{cis}\left(\dfrac{\pi}{6}\right) \times 6$

$= 3\sqrt{3}\,\text{cis}\left(\dfrac{\pi}{6}\right)$

$c = T(b)$

$= \dfrac{\sqrt{3}}{2} \text{cis}\left(\dfrac{\pi}{6}\right) \times b$

$= \dfrac{\sqrt{3}}{2} \text{cis}\left(\dfrac{\pi}{6}\right) \times 3\sqrt{3}\,\text{cis}\left(\dfrac{\pi}{6}\right)$

$= \dfrac{9}{2}\,\text{cis}\left(\dfrac{\pi}{6} + \dfrac{\pi}{6}\right)$

$= \dfrac{9}{2}\,\text{cis}\left(\dfrac{\pi}{3}\right)$

b.

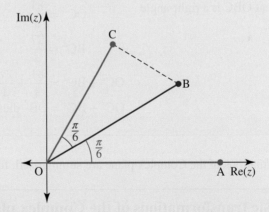

c.
$b = 3\sqrt{3}\,\text{cis}\left(\dfrac{\pi}{6}\right)$

$|b| = \text{OB} = 3\sqrt{3}$

$c = \dfrac{9}{2}\,\text{cis}\left(\dfrac{\pi}{3}\right)$

$|c| = \text{OC} = \dfrac{9}{2}$

2. Write b and c in Cartesian form to determine BC. (Remember that there is no easy way to add or subtract complex numbers in polar form).

$$c - b = \frac{9}{2}\operatorname{cis}\left(\frac{\pi}{3}\right) - 3\sqrt{3}\operatorname{cis}\left(\frac{\pi}{6}\right)$$

$$= \frac{9}{2}\left(\cos\left(\frac{\pi}{3}\right) + i\sin\left(\frac{\pi}{3}\right)\right)$$

$$-3\sqrt{3}\left(\cos\left(\frac{\pi}{6}\right) + i\sin\left(\frac{\pi}{6}\right)\right)$$

$$= \frac{9}{2}\left(\frac{1}{2} + i\frac{\sqrt{3}}{2}\right) - 3\sqrt{3}\left(\frac{\sqrt{3}}{2} + \frac{1}{2}i\right)$$

$$= \left(\frac{9}{4} + i\frac{9\sqrt{3}}{4}\right) - \left(\frac{9}{2} + \frac{3\sqrt{3}}{4}i\right)$$

$$= -\frac{9}{4} + \frac{3\sqrt{3}}{4}i$$

$$BC = |c - b|$$

$$= \left|-\frac{9}{4} + \frac{3\sqrt{3}}{4}i\right|$$

$$= \sqrt{\frac{81}{16} + \frac{27}{16}}$$

$$= \sqrt{\frac{108}{16}}$$

$$= \frac{3\sqrt{3}}{2}$$

3. Use Pythagoras' theorem to prove that OBC is a right-angle triangle.

$$OB^2 = 27$$

$$OC^2 = \frac{81}{4}$$

$$BC^2 = \frac{27}{4}$$

$$OC^2 + BC^2 = \frac{81}{4} + \frac{27}{4} = \frac{108}{4} = 27 = OB^2$$

$$OC^2 + BC^2 = OB^2 \text{ thus OBC is a right-angle triangle.}$$

The basic transformations of the Complex plane are summarised, for convenience, below.

Basic transformations of the Complex plane

Let $a = r\operatorname{cis}(\theta)$, b and z be complex numbers and k be a real number.

$T(z) = az + b$ is a linear transformation of z in the complex plane.

Dilation by a factor k: $T(z) = kz$

Translation by b: $T(z) = z + b$

Reflection in the real axis: $T(z) = \bar{z}$

Rotation by θ anticlockwise about the origin: $T(z) = z \times \operatorname{cis}(\theta)$

Rotation by θ clockwise about the origin: $T(z) = \dfrac{z}{\text{cis}(\theta)} = z \times \text{cis}(-\theta)$

Rotation about the origin by θ anticlockwise and dilation by r: $T(z) = az$

Rotation about the origin by θ clockwise and dilation by $\dfrac{1}{r}$: $T(z) = \dfrac{z}{a}$

7.4.2 Rotation about a point in the complex plane (Extension)

We have seen how to use complex division and multiplication to represent rotation about the origin. Can this be generalised to a rotation about any point in the complex plane?

Consider the three affixes z, z_1 and z_2 on the complex plane below where z_1 rotated by an angle θ anticlockwise about z gives z_2.

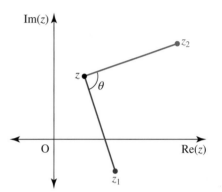

How can we relate z, z_1, z_2 and θ?

Consider the vectors $z_1 - z$ and $z_2 - z$ in the Complex plane.

$z_1 - z$ is rotated by θ anticlockwise to give $z_2 - z$.

This means, using the geometrical interpretation of complex multiplication, that $(z_1 - z) \times \text{cis}(\theta) = z_2 - z$.

$$\therefore \frac{z_1 - z}{z_2 - z} = \text{cis}(\theta)$$

This can be generalised to a rotation about a point and a dilation by a factor k.

Rotation about a point and dilation (Extension)

Let z, z_1 and z_2 be three complex numbers and k and θ be real numbers such that $\dfrac{|z_2 - z|}{|z_1 - z|} = k$ and the angle between vectors $z_1 - z$ and $z_2 - z$ in the Complex plane is θ.

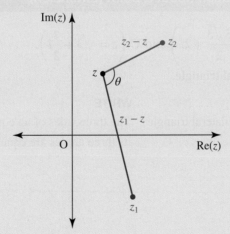

The relationship between z, z_1, z_2, k and θ is:

$$\frac{z_2 - z}{z_1 - z} = k\,\text{cis}(\theta) \text{ where } k = \frac{|z_2 - z|}{|z_1 - z|}$$

We now have an alternative method to prove that the triangle OBC from Worked Example 19 is a right-angle triangle, by proving that the angle $(\overrightarrow{CO}, \overrightarrow{CB})$ is $\dfrac{\pi}{2}$.

Let's prove that vector $\overrightarrow{CB}$ is a rotation by $\dfrac{\pi}{2}$ anticlockwise about $\mathbb{C}$ (and a dilation of factor k) of vector $\overrightarrow{CO}$ and thus that OBC is a right-angle triangle, that is, let's prove that $\dfrac{b-c}{0-c} = k \operatorname{cis}\left(\dfrac{\pi}{2}\right)$.

$$
\begin{aligned}
\frac{b-c}{0-c} &= \frac{\frac{3}{4}\left(3 - \sqrt{3}i\right)}{-\frac{9}{2}\left(1 + \sqrt{3}i\right)} \\
&= \frac{-3 + \sqrt{3}i}{6(1 + \sqrt{3}i)} \\
&= \frac{\left(-3 + \sqrt{3}i\right)\left(1 - \sqrt{3}i\right)}{12} \\
&= \frac{4\sqrt{3}i}{12} \\
&= \frac{\sqrt{3}i}{3} \\
&= \frac{\sqrt{3}}{3}\operatorname{cis}\left(\frac{\pi}{2}\right)
\end{aligned}
$$

$\overrightarrow{CB}$ is a rotation by $\dfrac{\pi}{2}$ anticlockwise about $\mathbb{C}$ (and a dilation of factor $\dfrac{\sqrt{3}}{3}$) of vector $\overrightarrow{CO}$.

$\therefore$ OBC is a right-angle triangle.

This can be used for instance to prove, using rotations, that 3 complex numbers form an equilateral triangle, as in Worked Example 20.

WORKED EXAMPLE 20 Rotation about a point (Extension)

Let $A\left(a = \sqrt{3} + \dfrac{1}{2}i\right)$, $B\left(b = \dfrac{5\sqrt{3}}{2} + 2i\right)$ and $C\left(c = \sqrt{3} + \dfrac{7}{2}i\right)$.

Prove that ABC is an equilateral triangle.

THINK	WRITE
1. Recall the properties of an equilateral triangle relating to its sides and angles.	The three sides of an equilateral triangle are equal and its three angles are equal to $\dfrac{\pi}{3}$.

2. Represent the three points on the Complex plane to help you visualise the rotations to consider.

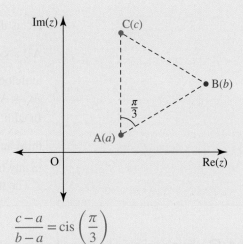

3. Write the condition for vector $\overrightarrow{AC}$ to be the rotation by $\dfrac{\pi}{3}$ anticlockwise of vector $\overrightarrow{AB}$ about point A.

$$\frac{c-a}{b-a} = \text{cis}\left(\frac{\pi}{3}\right)$$

4. Prove this equality.

$$\frac{c-a}{b-a} = \frac{\sqrt{3}+\frac{7}{2}i-\left(\sqrt{3}+\frac{1}{2}i\right)}{\frac{5\sqrt{3}}{2}+2i-\left(\sqrt{3}+\frac{1}{2}i\right)}$$

$$= \frac{3i}{\frac{3\sqrt{3}}{2}+\frac{3}{2}i}$$

$$= \frac{2i}{\sqrt{3}+i}$$

$$= \frac{2\,\text{cis}\left(\frac{\pi}{2}\right)}{2\,\text{cis}\left(\frac{\pi}{6}\right)}$$

$$= \text{cis}\left(\frac{\pi}{2}-\frac{\pi}{6}\right)$$

$$= \text{cis}\left(\frac{\pi}{3}\right)$$

5. Write the condition for vector $\overrightarrow{CB}$ to be the rotation by $\dfrac{\pi}{3}$ anticlockwise of vector $\overrightarrow{CA}$ about point $\mathbb{C}$.

$$\frac{b-c}{a-c} = \text{cis}\left(\frac{\pi}{3}\right)$$

6. Prove this equality.

$$\frac{b-c}{a-c} = \frac{\frac{5\sqrt{3}}{2}+2i-\left(\sqrt{3}+\frac{7}{2}i\right)}{\sqrt{3}+\frac{1}{2}i-\left(\sqrt{3}+\frac{7}{2}i\right)}$$

$$= \frac{\frac{3}{2}\sqrt{3}-\frac{3}{2}i}{-3i}$$

$$= \frac{\sqrt{3}-i}{-2i}$$

$$= \frac{2\,\text{cis}\left(-\frac{\pi}{6}\right)}{2\,\text{cis}\left(-\frac{\pi}{2}\right)}$$

$$= \text{cis}\left(-\frac{\pi}{6}+\frac{\pi}{2}\right)$$

$$= \text{cis}\left(\frac{\pi}{3}\right)$$

7. Conclude.

$\overrightarrow{AC}$ is the rotation by $\dfrac{\pi}{3}$ anticlockwise of vector $\overrightarrow{AB}$ about point A and $\overrightarrow{CB}$ is the rotation by $\dfrac{\pi}{3}$ anticlockwise of vector $\overrightarrow{CA}$ about point C. Thus $AC = AB$ and $CB = CA$. The three sides are equal. In addition, two of the angles are equal to $\dfrac{\pi}{3}$ thus the third angle is also equal to $\dfrac{\pi}{3}$ (The sum of the angles in any triangle is equal to π).

$\therefore$ The triangle ABC is equilateral.

Exercise 7.4 Geometric interpretations of multiplication and division of complex numbers in polar form

learn on

7.4 Exercise	7.4 Exam questions on

Simple familiar	Complex familiar	Complex unfamiliar
1, 2, 3, 4, 5, 6, 7, 8	9, 10	11, 12

Simple familiar

1. **WE17** Let $z_1 = -1 + \sqrt{3}i$ and $z_2 = 2 - 2i$.

   a. Express $z_1 z_2$ and $\dfrac{z_2}{z_1}$ in polar form.

   b. Provide a geometric interpretation of $z_1 z_2$ and $\dfrac{z_2}{z_1}$ and represent z_1, z_2, $z_1 z_2$ and $\dfrac{z_2}{z_1}$ in the Argand plane.

2. Let $z_1 = \text{cis}\left(\dfrac{6\pi}{7}\right)$, $z_2 = \text{cis}\left(\dfrac{\pi}{9}\right)$. A student represented the affixes z_1, z_2, $z_1 z_2$ and $\dfrac{z_2}{z_1}$ in the Argand plane, however they forgot to write down which affix, Z or W represents $z_1 z_2$.

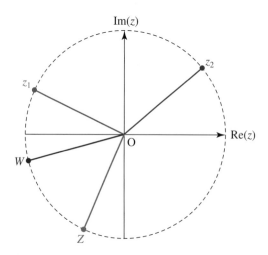

   The student thinks that $W = z_1 z_2$ and $Z = \dfrac{z_2}{z_1}$. Justify whether they are correct.

3. **WE18** Let $z = 2\text{cis}\left(\dfrac{5\pi}{6}\right)$. z is rotated by $\dfrac{3\pi}{4}$ anticlockwise about the origin and dilated by 3, giving z'.

   a. Represent z and z' on the Argand plane.
   b. Express z' in polar form.

4. **MC** Let $A\left(a = -\sqrt{3} + i\right)$ and $B(b = -4i)$ two complex numbers.

   Which of the following transformations of the Complex plane is such that $T(a) = b$?

   A. $T(z) = \dfrac{1}{2}\text{cis}\left(-\dfrac{5\pi}{6}\right)z$

   B. $T(z) = 2\text{cis}\left(\dfrac{5\pi}{6}\right)z$

   C. $T(z) = \dfrac{1}{2}\text{cis}\left(-\dfrac{2\pi}{3}\right)z$

   D. $T(z) = 2\text{cis}\left(\dfrac{2\pi}{3}\right)z$

5. Let $A(a = 1 - i)$ and $B\left(b = -2 - 2\sqrt{3}i\right)$ two complex numbers.

   a. Determine $Z = \dfrac{a}{b}$

      i. in Cartesian form
      ii. in polar form.

   b. **MC** Which of the following is the transformation of the Complex plane such that $T(a) = b$?

      A. Rotation by $\dfrac{\pi}{12}$ anticlockwise about the origin and dilated by $\dfrac{\sqrt{2}}{4}$.

      B. Rotation by $\dfrac{5\pi}{12}$ anticlockwise about the origin and dilated by $\dfrac{\sqrt{2}}{4}$.

      C. Rotation by $\dfrac{\pi}{12}$ clockwise about the origin and dilated by $\dfrac{1}{2}$.

      D. Rotation by $\dfrac{5\pi}{12}$ clockwise about the origin and dilated by $\dfrac{1}{2}$.

   c. Determine an exact value of $\sin\left(\dfrac{5\pi}{12}\right)$.

6. **WE19** Let $A\left(a = \dfrac{1}{2} + \dfrac{\sqrt{3}}{2}i\right)$, $B\left(b = a^2\right)$ and $C\left(c = \dfrac{1}{a}\right)$ be three points on the Complex plane.

   a. Express b and c in polar form.
   b. Represent the three points on the Argand plane.
   c. Prove that ABC is a right-angle triangle.

7. Consider the regular pentagon ABCDE inscribed in a circle centred on the origin and with radius 3.

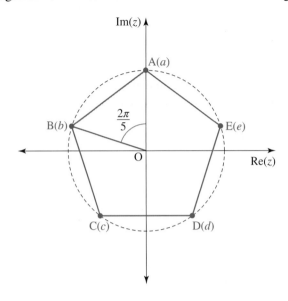

Express b in polar form.

8. **WE20** Let A($a = 2 - i$), B($b = 4 - 3i$) and C($c = -3i$) be three complex numbers.

a. Express $\dfrac{c - a}{b - a}$ in polar form.

b. Give a geometric interpretation of this result.

Complex familiar

9. Let T be the anticlockwise rotation by $\dfrac{2\pi}{3}$ about the origin O.

Let A$\left(a = 2\sqrt{3} - 2i\right)$, B($b = -8i$) and C($c = T(b)$) be three points on the Complex plane.
Show that A, B and C are aligned.

10. Let A($a = 1 + i$), B($b = -1 + i$), C($c = -1 - i$) and D($d = 1 - i$) be four points in the Complex plane.
Let T be the clockwise rotation by $\dfrac{\pi}{3}$ about point C.

a. Determine $e = T(b)$ and $f = T(d)$ in Cartesian form.
b. Represent the points A, B, C, D, E (e) and F (f) on the complex plane.
c. Show that the points A, E and F are aligned.

Complex unfamiliar

11. Let A($a = 4 - 2i$), B($b = -2 + 4i$) and C($k(1 + i)$) where $k \in \mathbb{R}$ be three points on the complex plane.
Determine k so that ABC is an equilateral triangle.

12. Let T be the clockwise rotation by $\dfrac{\pi}{3}$ about the origin O and a dilation of factor 2.

Let A$\left(a = 2\operatorname{cis}\left(\dfrac{2\pi}{3}\right)\right)$, B ($b = T(a)$), M($m$) be the middle of AB and C(k), where $k \in \mathbb{R}$, be four points on the complex plane.
Determine k so that OMC is a right-angle triangle where the right angle is at angle M. Give your answer to three decimal places.

Fully worked solutions for this chapter are available online.

LESSON
7.5 Subsets of the complex plane

So far in this chapter, complex numbers have been used to represent points on the Argand plane. If we consider z as a complex variable, we can sketch relations or regions of the Argand plane.

7.5.1 Lines

If $z = x + yi$, then $\text{Re}(z) = x$ and $\text{Im}(z) = y$. The equation $a\,\text{Re}(z) + b\,\text{Im}(z) = c$ where a, b and $c \in \mathbb{R}$ represents the line $ax + by = c$.

WORKED EXAMPLE 21 Sketching lines in the complex plane (1)

Determine the Cartesian equation and sketch the graph defined by $\{z: 2\,\text{Re}(z) - 3\,\text{Im}(z) = 6\}$.

THINK	WRITE
1. Consider the equation.	$2\,\text{Re}(z) - 3\,\text{Im}(z) = 6$ As $z = x + yi$, then $\text{Re}(z) = x$ and $\text{Im}(z) = y$. This is a straight line with the Cartesian equation $2x - 3y = 6$.
2. Evaluate the axial intercepts.	When $y = 0$, $2x = 6 \Rightarrow x = 3$. $(3, 0)$ is the intercept with the real axis. When $x = 0$, $-3y = 6 \Rightarrow y = -2$. $(0, -2)$ is the intercept with the imaginary axis.
3. Identify and sketch the equation.	The equation represents the line $2x - 3y = 6$. 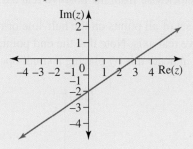

Lines in the complex plane can also be represented as a set of points that are equidistant from two other fixed points. The equations of a line in the complex plane can thus have multiple representations.

Determine the Cartesian equation and sketch the graph defined by $\{z: |z - 2i| = |z + 2|\}$.

THINK	WRITE
1. Consider the equation as a set of points.	$|z - 2i| = |z + 2|$ Substitute $z = x + yi$: $|x + yi - 2i| = |x + yi + 2|$
2. Group the real and imaginary parts together.	$|x + (y - 2)i| = |(x + 2) + yi|$
3. Use the definition of the modulus.	$\sqrt{x^2 + (y - 2)^2} = \sqrt{(x + 2)^2 + y^2}$
4. Square both sides, expand, and cancel like terms.	$x^2 + y^2 - 4y + 4 = x^2 + 4x + 4 + y^2$ $-4y = 4x$
5. Identify the required line.	$y = -x$
6. Identify the line geometrically.	The line is the set of points that is equidistant from the two points $(0, 2)$ and $(-2, 0)$.
7. Sketch the required line.	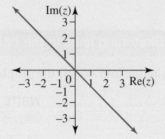

7.5.2 Rays

A ray is a half-line and is usually defined in terms of the argument of the complex number. When drawing rays, remember that $\text{Arg}(z) = \theta \in (-\pi, \pi]$. The argument of a complex number $z = x + yi$ is $\theta = \tan^{-1}\left(\dfrac{y}{x}\right)$, $x > 0$, where θ is measured anticlockwise from the positive real axis.

$\text{Arg}(z) = \theta$ represents the set of all points on the half-line or ray that has one end at the origin and makes an angle of θ with the positive real axis. Note that the end point, in this case the origin, is not included in the set since $x > 0$ and $y > 0$. We indicate this by placing a small open circle at this point.

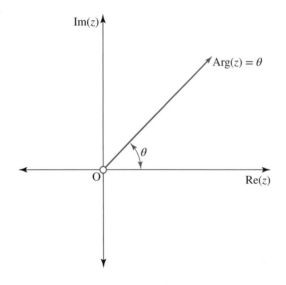

a. Describe and sketch the graph defined by $\left\{ z: \mathrm{Arg}(z) = \dfrac{-\pi}{4} \right\}$.

b. Determine the Cartesian equation and sketch the graph defined by $\left\{ z: \mathrm{Arg}(z - 1 + i) = \dfrac{-\pi}{4} \right\}$.

THINK	WRITE
a. 1. Consider the given equation.	**a.** $\mathrm{Arg}(z) = \dfrac{-\pi}{4}$
2. Recognise the equation as a ray.	The equation is of the form $\mathrm{Arg}(z) = \theta$.
3. Identify the point from which the ray starts.	The ray starts from the point $(0, 0)$, not including the point.
4. Determine the angle the ray makes.	The ray makes an angle of $-45°$ with the positive real axis.
5. Describe the ray.	The ray starts from $(0, 0)$, making an angle of $-45°$ with the positive real axis.
6. Sketch the required ray.	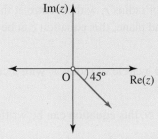
b. 1. Consider the given equation.	**b.** $\mathrm{Arg}(z - 1 + i) = \dfrac{-\pi}{4}$
2. Substitute $z = x + yi$.	$\mathrm{Arg}(x + yi - 1 + i) = \dfrac{-\pi}{4}$
3. Group the real and imaginary parts.	$\mathrm{Arg}((x - 1) + (y + 1)i) = \dfrac{-\pi}{4}$
4. Use the definition of the argument.	$\tan^{-1}\left(\dfrac{y+1}{x-1}\right) = \dfrac{-\pi}{4}$ for $x > 1$
5. Simplify.	$\dfrac{y+1}{x-1} = \tan\left(\dfrac{-\pi}{4}\right)$ for $x > 1$ $\dfrac{y+1}{x-1} = -1$ for $x > 1$
6. State the Cartesian equation of the ray.	$y + 1 = -(x - 1)$ for $x > 1$ $y = -x$ for $x > 1$
7. Identify the point from which the ray starts.	The ray starts from the point $(1, -1)$, not including the point
8. Determine the angle the ray makes with the positive real axis.	The ray makes an angle of $-45°$ with the positive real axis.

9. Sketch the ray.

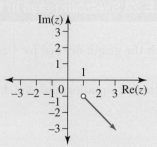

10. Alternatively, use translations from the answer of part **a**.

The ray from the origin making an angle of –45° with the positive real axis has been translated 1 unit to the right parallel to the real axis and 1 unit down parallel to the imaginary axis.

7.5.3 Circles

The equation of a circle of radius r, with a centre at the origin $(0, 0)$ is given by $x^2 + y^2 = r^2$. In the Argand plane, this equation can be written as $\sqrt{x^2 + y^2} = r$, or $|z| = r$, where $z = x + yi$.

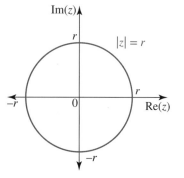

Similarly, the equation of a circle of radius r, with a centre at point (h, k) is given by $(x - h)^2 + (y - k)^2 = r^2$.

If $z = x + yi$ and $z_\Omega = h + ki$, this equation can be written as $|z - z_\Omega| = r$.

Geometrically, $|z| = r$ represents the set of points, or what is called the locus of points, in the Argand plane that are at r units from the origin.

Determine the Cartesian equation and sketch the graph of $\{z: |z + 2 - 3i| = 4\}$.

THINK	WRITE
1. Consider the equation.	$|z + 2 - 3i| = 4$ Substitute $z = x + yi$: $\quad |x + yi + 2 - 3i| = 4$
2. Group the real and imaginary parts.	$|(x + 2) + i(y - 3)| = 4$
3. Use the definition of the modulus.	$\sqrt{(x + 2)^2 + (y - 3)^2} = 4$
4. Square both sides.	$(x + 2)^2 + (y - 3)^2 = 16$

5. Sketch and identify the graph of the Argand plane.

The equation represents a circle with centre at $(-2, 3)$ and radius 4.

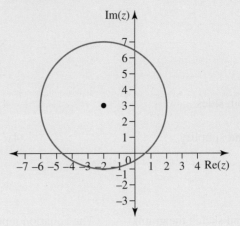

7.5.4 Ellipses (Extension)

The general form of an ellipse is $\dfrac{x^2}{a^2} + \dfrac{y^2}{b^2} = 1$.

The definition of an ellipse is that for all points on the curve the sum of the distances to two fixed points is a constant.

For example, the equation $|z - 1| + |z + 1| = 4$ would represent an ellipse where the sum of the distances from the variable point z to the points $(1, 0)$ and $(-1, 0)$ is equal to 4.

This is illustrated in the following Worked Example.

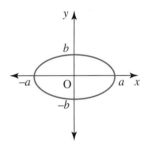

WORKED EXAMPLE 25 Sketching ellipses in the complex plane (Extension)

Determine the Cartesian equation and sketch the graph of $\{z: |z - 1| + |z + 1| = 4\}$.

THINK	WRITE
1. Consider the equation.	$|z - 1| + |z + 1| = 4$
2. Substitute $z = x + yi$.	$|x + yi - 1| + |x + yi + 1| = 4$
3. Group the real and imaginary parts.	$|(x - 1) + yi| + |(x + 1) + yi| = 4$
4. Use the definition of the modulus.	$\sqrt{(x - 1)^2 + y^2} + \sqrt{(x + 1)^2 + y^2} = 4$
5. Rearrange the equation.	$\sqrt{(x - 1)^2 + y^2} = 4 - \sqrt{(x + 1)^2 + y^2}$
6. Square both sides.	$(x - 1)^2 + y^2 = 16 - 8\sqrt{(x + 1)^2 + y^2}$ $+ (x + 1)^2 + y^2$

7. Expand and simplify.

$$x^2 - 2x + 1 + y^2 = x^2 + 2x + 1 + y^2 + 16$$
$$- 8\sqrt{(x+1)^2 + y^2}$$
$$8\sqrt{(x+1)^2 + y^2} = 16 + 4x$$
$$2\sqrt{(x+1)^2 + y^2} = 4 + x$$

8. Square both sides.

$$4\left((x+1)^2 + y^2\right) = 16 + 8x + x^2$$

9. Expand and simplify.

$$4x^2 + 8x + 4 + 4y^2 = 16 + 8x + x^2$$
$$3x^2 + 4y^2 = 12$$
$$\frac{x^2}{4} + \frac{y^2}{3} = 1$$

10. Identify and sketch the graph.

The equation represents an ellipse with centre at $(0,0)$, x-intercepts at $(2,0)$ and $(-2,0)$, and y-intercepts at $\left(0, \sqrt{3}\right)$ and $\left(0, -\sqrt{3}\right)$.

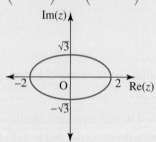

7.5.5 Regions in the complex plane

When graphing a linear inequality on a Cartesian plane, the solution is a region on one side of the line, called a half plane. Similarly, sets of points in the complex plane can be represented by regions. For example, $\{z: \text{Re}(z) \geq 4\}$ is the set of points on or to the right of the vertical line $x = 4$. The boundary line is a solid line if it is included or dotted if it is not included. Special care needs to be taken when the inequality involves the argument of the complex number. This is illustrated in the following worked example.

WORKED EXAMPLE 26 Sketching regions in the complex plane (1)

Sketch the region defined by $\left\{ z: \text{Arg}(z) \leq \frac{\pi}{4} \right\}$.

THINK	WRITE
1. Consider the given set.	$\text{Arg}(z) \leq \dfrac{\pi}{4}$
2. Determine the boundary line.	$\text{Arg}(z) = \dfrac{\pi}{4}$ is a ray from but not including the point $(0,0)$ making an angle of $45°$ with the positive real axis.
3. State the restrictions on θ for $\text{Arg}(z)$.	$\text{Arg}(z) = \theta, \theta \in (-\pi, \pi]$

4. Apply to the given inequality.

$$\text{Arg}(z) \le \frac{\pi}{4}$$
$$\therefore -\pi < \theta \le \frac{\pi}{4}$$

5. Sketch the region, dotting the negative real axis as it is not included. Show $(0, 0)$ as an open circle as it is also not included in the region.

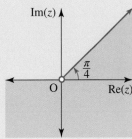

Regions may involve the intersection of two sets of points in the complex plane. This is illustrated in the following worked example.

WORKED EXAMPLE 27 Sketching regions in the complex plane (2)

Sketch the region defined by $\{z: 0 \le \text{Re}(z) \le 4\} \cap \{z: -1 \le \text{Im}(z) < 3\}$.

THINK	WRITE
1. Consider the first set, where $z = x + yi$.	$0 \le \text{Re}(z) \le 4$ $\therefore 0 \le x \le 4$
2. Determine the boundary lines and state the region.	The boundary lines are $x = 0$ and $x = 4$. The region is on and between the lines.
3. Sketch the region.	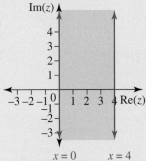
4. Consider the second set.	$-1 \le \text{Im}(z) < 3$ $\therefore -1 \le y < 3$
5. Determine the boundary lines and state the region.	The boundary lines are $y = -1$ and $y = 3$. The region includes $y = -1$ up to a dotted $y = 3$.
6. Sketch the region.	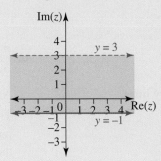

7. Shade the intersection, or overlap, of the two regions. Dot the boundary line $y = 3$ as it is not included. The points of intersection $(4, 3)$ and $(0, 3)$ are open circles.

The required region is:

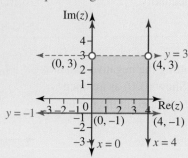

WORKED EXAMPLE 28 Sketching regions in the complex plane (3)

a. Sketch the region defined by $\{z: 2 < |z| \leq 4\}$.

b. Sketch the region defined by $\left\{ z: \dfrac{\pi}{3} \leq \text{Arg}(z) \leq \dfrac{2\pi}{3} \right\}$.

c. Hence, sketch the region defined by $\{z: 2 < |z| \leq 4\} \cap \left\{ z: \dfrac{\pi}{3} \leq \text{Arg}(z) \leq \dfrac{2\pi}{3} \right\}$.

THINK	WRITE
a. 1. Consider the given set, where $z = x + yi$.	**a.** $2 < |z| \leq 4$
2. Square each side.	$2 < \sqrt{x^2 + y^2} \leq 4$ $4 < x^2 + y^2 \leq 16$
3. Determine the boundary lines and state the region.	The boundary lines are two circles, centre $(0, 0)$. The circle with radius 2 is dotted. The circle with radius 4 is solid.
4. Identify and sketch the region.	The region lies between the two concentric circles of radius 2 and 4, not including the circumference of the circle with radius 2.

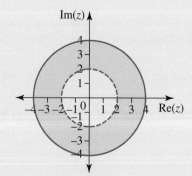

b. 1. Consider the given set.	**b.** $\dfrac{\pi}{3} \leq \text{Arg}(z) \leq \dfrac{2\pi}{3}$
2. Determine the boundary lines.	$\text{Arg}(z) = \dfrac{\pi}{3}$ and $\text{Arg}(z) = \dfrac{2\pi}{3}$ The boundary lines are rays from but not including $(0, 0)$ at angles of $60°$ and $120°$ from the positive real axis.

3. State the restrictions on θ for $\mathrm{Arg}(z)$. $\mathrm{Arg}(z) = \theta, \theta \in (-\pi, \pi]$

4. Apply to the given inequality.
$$\frac{\pi}{3} \le \mathrm{Arg}(z) \le \frac{2\pi}{3}$$
$$\therefore \frac{\pi}{3} \le \theta \le \frac{2\pi}{3}$$

5. Identify and sketch the region, shading between the two rays, with an open circle at $(0, 0)$.

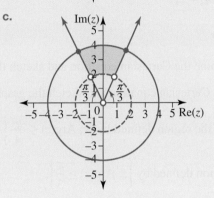

c. 1. On one Argand diagram, sketch both regions from parts **a** and **b**.
   2. Identify and sketch the required region, which is the common region. Open circles are needed where the rays intersect with the circle of radius 2 units.

c.

Exercise 7.5 Subsets of the complex plane

7.5 Exercise	7.5 Exam questions on

Simple familiar	Complex familiar	Complex unfamiliar
1, 2, 3, 4, 5, 6, 7, 8, 9, 10	11, 12, 13, 14, 15, 16	17, 18, 19, 20

Simple familiar

1. **WE21** Sketch and describe the region of the complex plane defined by $\{z: 4\,\mathrm{Re}(z) + 3\,\mathrm{Im}(z) = 12\}$.

2. Illustrate each of the following and describe the subset of the complex plane.
   a. $\{z: \mathrm{Im}(z) = 2\}$
   b. $\{z: \mathrm{Re}(z) + 2\,\mathrm{Im}(z) = 4\}$

3. **WE22** Sketch and describe the region of the complex plane defined by $\{z: |z + 3i| = |z - 3|\}$.

4. Sketch and describe the region of the complex plane defined by $\{z: |z - i| = |z + 3i|\}$.

5. Determine the Cartesian equation and sketch each of the following sets.
   a. $\{z: |z + 4| = |z - 2i|\}$
   b. $\{z: |z + 2 - 3i| = |z - 2 + 3i|\}$

6. **WE23** **a.** Describe and sketch the graph defined by $\left\{z: \text{Arg}(z) = \dfrac{\pi}{6}\right\}$.

   b. Determine the Cartesian equation and sketch the graph defined by $\left\{z: \text{Arg}(z-1) = \dfrac{\pi}{6}\right\}$.

7. **a.** Describe and sketch the graph defined by $\left\{z: \text{Arg}(z) = \dfrac{\pi}{4}\right\}$.

   b. Determine the Cartesian equation and sketch the graph defined by $\left\{z: \text{Arg}(z+2) = \dfrac{\pi}{4}\right\}$.

8. **WE24** Sketch and describe the region of the complex plane defined by $\{z: |z-3+2i| = 4\}$.

9. For each of the following, sketch and determine the Cartesian equation of the set, and describe the region.

   a. $\{z: |z| = 3\}$

   b. $\{z: |z| = 2\}$

10. For each of the following, sketch and determine the Cartesian equation of the set.

    a. $\{z: |z+2-3i| = 2\}$

    b. $\{z: |z-3+i| = 3\}$

Complex familiar

11. **WE25** Determine the Cartesian equation and sketch the graph of $\{z: |z-3| + |z+3| = 12\}$.

12. Determine the Cartesian equation and sketch the graph of $\{z: |z+2| + |z-2| = 6\}$.

13. **WE26** Sketch the region defined by $\left\{z: \text{Arg}(z) \le \dfrac{3\pi}{4}\right\}$.

14. Sketch the region defined by $\left\{z: \text{Arg}(z) \ge \dfrac{\pi}{3}\right\}$.

15. **WE27** Sketch the region defined by $\{z: -2 \le \text{Re}(z) \le 2\} \cap \{z: 3 \le \text{Im}(z) < 5\}$.

16. Sketch the region defined by $\{z: 0 < \text{Re}(z) < 6\} \cap \{z: 0 \le \text{Im}(z) \le 4\}$.

Complex unfamiliar

17. Describe and sketch the region defined by $\{z: 1 \le |z-2| \le 2\}$.

18. Describe and sketch the region defined by $\{z: |z| \le 3\} \cap \{z: |z-3| \le 3\}$.

19. **WE28** Sketch the region defined by $\{z: 3 < |z| \le 5\} \cap \left\{z: \dfrac{\pi}{4} \le \text{Arg}(z) \le \dfrac{3\pi}{4}\right\}$.

20. Sketch the region defined by $\{z: 3 \le |z| \le 6\} \cap \{z: \text{Re}(z) + \text{Im}(z) > 3\}$.

Fully worked solutions for this chapter are available online.

LESSON
7.6 Roots of equations

7.6.1 Linear factors of real quadratic polynomials

The solutions of an equation are also known as the **roots of the equation**. Geometrically, the roots of a function are the x-intercepts of the function.

Up to this point, linear factors of real quadratic polynomials have been limited to those over the set of real numbers. A Linear factor is a factor whose highest power of the variable is 1, such as $(2x - 3)$, $\left(5 - \dfrac{x}{3}\right)$ etc. Now, linear factors can be determined over the complex number field, such as $(3z - 2i)$ or $(1 + 2i - z)$.

WORKED EXAMPLE 29 Determining the linear factors of quadratic polynomials

Determine the linear factors over $\mathbb{C}$ for $z^2 - 4z + 29$.

THINK	WRITE
1. Complete the square. *Note:* The sum of two squares has no real factors.	$z^2 - 4z + 29$ $= z^2 - 4z + 4 + (29 - 4)$ $= (z - 2)^2 + 25$
2. Write the expression as the difference of two squares.	$= (z - 2)^2 - 25i^2 \text{ (using } i^2 = -1)$ $= (z - 2)^2 - (5i)^2$
3. Write the expression as a product of linear factors.	$= (z - 2 + 5i)(z - 2 - 5i)$
4. Write the answer.	The linear factors of $z^2 - 4z + 29$ are $(z - 2 + 5i)(z - 2 - 5i)$.

7.6.2 The general solution of real quadratic equations

Consider the quadratic equation $az^2 + bz + c = 0$, where the coefficients a, b and c are real. Recall that the roots depend upon the discriminant, $\Delta = b^2 - 4ac$.

- If $\Delta > 0$, the equation has two distinct real roots.
- If $\Delta = 0$, the equation has one real repeated root.
- If $\Delta < 0$, the equation has no real roots.

With the introduction of complex numbers, it can now be stated that if $\Delta 0 < 0$, then the equation has one pair of **complex conjugate roots** of the form $a \pm bi$.

Solve for z, given $z^2 - 6z + 25 = 0$.

THINK	WRITE
Method 1: Using completing the square	$z^2 - 6z + 25 = 0$
1. Complete the square.	$z^2 - 6z + 9 + (25 - 9) = 0$
	$(z - 3)^2 + 16 = 0$
	$(z - 3)^2 = -16$
2. Substitute -1 with i^2.	$(z - 3)^2 = 16i^2$
3. Solve for z.	$z - 3 = \pm 4i$
4. State the two solutions for z.	$z = 3 \pm 4i$
Method 2: Using the general quadratic formula	$z = \dfrac{-b \pm \sqrt{b^2 - 4ac}}{2a}$
1. Write the general quadratic formula.	
2. State the values for a, b and c.	$z^2 - 6z + 25 = 0$
	$a = 1, \ b = -6, \ c = 25$
3. Substitute the values into the formula and simplify.	$z = \dfrac{6 \pm \sqrt{(-6)^2 - 4(1)(25)}}{2}$
	$= \dfrac{6 \pm \sqrt{-64}}{2}$
	$= \dfrac{6 \pm \sqrt{64i^2}}{2}$
	$= \dfrac{6 \pm 8i}{2}$
	$= 3 \pm 4i$
4. State the two solutions.	$z = 3 \pm 4i$

7.6.3 The relationship between roots and coefficients

For a quadratic with real coefficients, we have seen that if the discriminant is negative, then the roots occur in *complex conjugate pairs*. Here we determine a relationship between the roots and the coefficients.

Given a quadratic $az^2 + bz + c = 0$ where $a \neq 0$, then:

$$z^2 + \frac{b}{a}z + \frac{c}{a} = 0$$

Let the roots of the quadratic equation be α and β, so the factors are $(z - \alpha)(z - \beta)$. Expanding the brackets, the expression becomes:

$$z^2 - (\alpha + \beta)z + \alpha\beta = 0 \text{ or}$$

$$z^2 - (\text{sum of roots})\, z + \text{product of roots} = 0$$

$$\text{so that } \alpha + \beta = -\frac{b}{a} \text{ and } \alpha\beta = \frac{c}{a}.$$

This gives us a relationship between the roots of a quadratic equation and its coefficients. Rather than formulating a problem as solving a quadratic equation, we now consider the reverse problem. That is forming a quadratic equation with real coefficients, given one of the roots.

WORKED EXAMPLE 31 Determining a quadratic given one of its roots

Determine the equation of the quadratic $P(z)$, with real coefficients given that $P(4 - 3i) = 0$.

THINK	WRITE
1. State the value of the second root, recalling that complex conjugate roots are of the form $a \pm bi$.	$P(4 - 3i) = 0$ $\Rightarrow 4 - 3i$ is a root of $P(z)$. $\Rightarrow 4 + 3i$ is also a root of $P(z)$. Let $\alpha = 4 - 3i$ and let $\beta = 4 + 3i$.
2. Determine the sum of the roots.	$\alpha + \beta = 8$
3. Determine the product of the roots.	$\alpha\beta = 16 - 9i^2 = 25$
4. State the quadratic equation.	$P(z) = (z - 4 + 3i)(z - 4 - 3i)$ $= z^2 - 8z + 25$

Exercise 7.6 Solving quadratic equations with complex roots   learn on

7.6 Exercise	7.6 Exam questions on

Simple familiar

1. **WE29** Determine the linear factors over $\mathbb{C}$ for $z^2 - 6z + 25$.

2. Determine the linear factors over $\mathbb{C}$ for $z^2 + 4z + 7$.

3. **WE30** Solve for z, given $z^2 - 4z + 29 = 0$.

4. Solve for z, given $z^2 + 2z + 26 = 0$.

5. Calculate the roots of each of the following.
   a. $z^2 + 2z + 17 = 0$
   b. $z^2 - 4z + 20 = 0$
   c. $z^2 - 6z + 13 = 0$
   d. $z^2 + 10z + 41 = 0$

6. **WE31** Determine the quadratic $P(z)$ with real coefficients given that $P(-5 - 2i) = 0$.

7. Determine the quadratic $P(z)$ with real coefficients given that $P(5i) = 0$.

8. Form a quadratic with integer coefficients for each of the following cases.
   a. -2 and $\dfrac{1}{3}$ are the roots.
   b. $2 - 6i$ is a root.
   c. $-2 + 3i$ is a root.
   d. $-4 - 5i$ is a root.

9. Form a quadratic with integer coefficients for each of the following cases.

   a. $\dfrac{1}{2}$ and $-\dfrac{1}{5}$ are the roots.

   b. $4 - \sqrt{3}i$ is a root.

   c. $-5 - \sqrt{7}i$ is a root.

   d. $-3 - \sqrt{8}i$ is a root.

Complex familiar

10. Solve for z in each of the following.

    a. $z(4 - z) = 8$
    c. $z(2 - z) = 26$

    b. $z(6 - z) = 10$
    d. $z(8 - z) = 41$

11. Solve for z in each of the following.

    a. $(z + 2)(z - 6) + 25 = 0$
    c. $(z + 2)(z - 4) + 12 = 0$

    b. $(z - 2)(z + 8) + 30 = 0$
    d. $(z + 6)(z + 4) + 7 = 0$

12. Determine the roots of each of the following.

    a. $4z^2 + 12z + 10 = 0$
    c. $4z^2 - 12z + 13 = 0$

    b. $4z^2 + 20z + 29 = 0$
    d. $9z^2 - 42z + 53 = 0$

Complex unfamiliar

13. Solve for z if $z^2 + (2 - i)z + (3 - i) = 0$.

14. Solve for z if $z^2 + (4 - 6i)z = 10$.

15. Solve for z if $z^2 + (4 - i)z + (5 - 2i) = 0$.

16. Solve for z if $z^2 + \sqrt{11}z + (1 + 6i) = 0$.

17. Let $w = 1 + 2\sqrt{2}i$.

    a. Determine z such that $z^2 = w$.

    b. Solve for z, given $z^2 + iz - \dfrac{1}{2} - i\dfrac{\sqrt{2}}{2} = 0$.

18. Solve for z given $iz^8 + iz^4 + 1 + i = 0$. Give your answers in polar form.

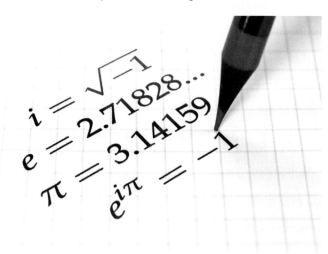

Fully worked solutions for this chapter are available online.

LESSON
7.7 Review

7.7.1 Summary

7.7 Exercise

learn on

7.7 Exercise	7.7 Exam questions on

Simple familiar	Complex familiar	Complex unfamiliar
1, 2, 3, 4, 5, 6, 7, 8, 9, 10, 11, 12	13, 14, 15, 16	17, 18, 19, 20

Simple familiar

1. **MC** The principal argument of $4\sqrt{3} - 4i$ is

   A. $\dfrac{\pi}{6}$
   B. $\dfrac{\pi}{3}$
   C. $\dfrac{5\pi}{6}$
   D. $-\dfrac{\pi}{6}$

2. **MC** In polar form, $5i$ is

   A. $\text{cis}(5\pi)$
   B. $\text{cis}\left(\dfrac{5\pi}{2}\right)$
   C. $5\,\text{cis}(5\pi)$
   D. $5\,\text{cis}\left(\dfrac{\pi}{2}\right)$

3. **MC** The Cartesian form of $\sqrt{3}\,\text{cis}\left(-\dfrac{7\pi}{6}\right)$ is

   A. $\dfrac{1}{2} + \dfrac{\sqrt{3}}{2}i$
   B. $-\dfrac{1}{2} + \dfrac{\sqrt{3}}{2}i$
   C. $-\dfrac{\sqrt{3}}{2} + \dfrac{1}{2}i$
   D. $-\dfrac{3}{2} + \dfrac{\sqrt{3}}{2}i$

4. **MC** $\sqrt{5}\,\text{cis}\left(-\dfrac{\pi}{3}\right) \times \sqrt{8}\,\text{cis}\left(-\dfrac{\pi}{6}\right)$ is equal to

   A. $6\sqrt{2}i$
   B. $-2\sqrt{10}i$
   C. $-6\sqrt{3}$
   D. $-6i$

5. Express $\left(\dfrac{-1+7i}{1+i} + \dfrac{7+i}{3-i}\right)$ in polar form.

6. Let $z = 2 - 3i$ and $w = 1 + 2i$.
   a. Express $\dfrac{w}{z}$ in polar form.
   b. Let T be the anticlockwise rotation by θ about the origin and dilation by k.
      Determine θ and k so that $w = T(z)$. Give your answers to three decimal places.

7. Evaluate $\text{Arg}(2 - 2i)$.

8. Determine the roots of the equation $4z^2 - 3z + 6 = 0$. Give your answers in Cartesian form, to three decimal places.

9. Solve for z, given $\dfrac{z+2}{z+1} = z + 3$.

10. Determine the Cartesian equation and sketch the graph of each of the following.

    a. $\{z: |z-4| = 2\,|z-1|\}$
    b. $\{z: |z-4| = |z-1|\}$

11. If $z_1 = 10\,\text{cis}\left(\dfrac{\pi}{4}\right)$ and $z_2 = 5\,\text{cis}\left(-\dfrac{\pi}{6}\right)$, evaluate $z_1 z_2$ in polar form.

12. Evaluate $12\sqrt{2}\,\text{cis}\left(\dfrac{3\pi}{4}\right) \div \left(3\,\text{cis}\left(-\dfrac{\pi}{2}\right)\right)$ in standard form.

Complex familiar

13. Sketch the following regions.

    a. $\{z: 1 < |z| \le 2\}$

    b. $\left\{z: \dfrac{\pi}{6} \le \text{Arg}\,(z) \le \dfrac{5\pi}{6}\right\}$

    c. Hence, sketch the region defined by $\{z: 1 < |z| \le 2\} \cap \left\{z: \dfrac{\pi}{6} \le \text{Arg}\,(z) \le \dfrac{5\pi}{6}\right\}$.

14. How many degrees apart are two consecutive roots of $z^8 = 1$ on the unit circle?

15. Let $P(z) = z^3 + iz^2 - iz + 1 + i$.

    a. Determine $P(-1-i)$
    b. Solve $P(z) = 0$

16. Let T be the anticlockwise rotation by $\dfrac{\pi}{3}$ about the origin and A(a), B(b) and C(c) the three points on the Complex plane such that $a = T(-4)$, $b = T(2)$ and $c = T(4)$.
    Show that A, B and C are aligned.

Complex unfamiliar

17. Explain why $z^5 + z + 1 = i$ does not have any real solutions.

18. Let $z = 2\,\text{cis}\left(\dfrac{\pi}{4}\right)$ and $w = T(z)$ where T is the anticlockwise rotation by $\dfrac{\pi}{6}$ about the origin.

    Show that $\tan\left(\dfrac{5\pi}{12}\right) = 2 + \sqrt{3}$.

19. Solve for z given $z^4 + z^2 + 1 = 0$. Give your answers in Cartesian form.

20. Let A(z_1) and B(z_2) the roots of $z^2 - 2\sqrt{3}z + 4 = 0$ and O the origin of the Complex plane.
    Show that OAC is an equilateral triangle.

Fully worked solutions for this chapter are available online.

Answers

Chapter 7 Complex arithmetic and algebra

7.2 Complex numbers in polar form

7.2 Exercise

1. a. 13 b. 3

2. a. $\sqrt{65}$ b. $3\sqrt{5}$

3. a. $\sqrt{5}$ b. 5

4. a. i.
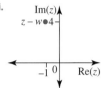
   ii. $\sqrt{17}$

b. i.

Im(z)
6 — •$u + z$

0 | 1   Re(z)

   ii. $\sqrt{37}$

c. i.

Im(z)
0   6   Re(z)
−8 — •$w - u$

   ii. 10

d. i.

Im(z)
0   7   Re(z)
−2 —
•$w + z$

   ii. $\sqrt{53}$

e. i.

Im(z)
0   9   Re(z)
−7 —
•$z + w - u$

   ii. $\sqrt{130}$

f. i.

Im(z)
6 — •z^2

0   8   Re(z)

   ii. 10

5. a. $-\dfrac{\pi}{2}$ b. $\dfrac{\pi}{6}$ c. $-\dfrac{\pi}{8}$ d. $\dfrac{3\pi}{4}$

6. a. $-\dfrac{5\pi}{6}$ b. $\dfrac{6\pi}{7}$ c. $\dfrac{2\pi}{5}$ d. $\dfrac{11\pi}{12}$

7. a. $\dfrac{\pi}{6}$ b. $-\dfrac{\pi}{4}$ c. $-\dfrac{2\pi}{3}$

8. a. $\dfrac{\pi}{2}$ b. π c. $-\dfrac{\pi}{2}$ d. 0

9. a. $3\sqrt{2}, -\dfrac{\pi}{4}$ b. $5\sqrt{2}, \dfrac{3\pi}{4}$

   c. $2, -\dfrac{2\pi}{3}$ d. $8, \dfrac{\pi}{6}$

10. a. $\sqrt{2}\operatorname{cis}\left(\dfrac{3\pi}{4}\right)$ b. $2\sqrt{2}\operatorname{cis}\left(\dfrac{\pi}{6}\right)$

   c. $\sqrt{10}\operatorname{cis}\left(-\dfrac{3\pi}{4}\right)$

11. a. $2\sqrt{5}\operatorname{cis}\left(-\dfrac{\pi}{3}\right)$ b. $\operatorname{cis}\left(-\dfrac{2\pi}{3}\right)$

   c. $\dfrac{\sqrt{2}}{4}\operatorname{cis}\left(\dfrac{3\pi}{4}\right)$

12. a. $-1 + \sqrt{3}\,i$ b. $\dfrac{3\sqrt{2}}{2} + \dfrac{3\sqrt{2}}{2}i$

   c. $-\dfrac{\sqrt{15}}{2} + \dfrac{\sqrt{5}}{2}i$

13. a. $2 - 2\sqrt{3}i$ b. $\dfrac{\sqrt{14}}{2} + \dfrac{\sqrt{14}}{2}i$

   c. $0 + 8i$

14. a.
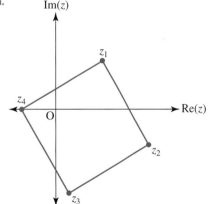

b. $A = 34$ square units

15. a.

Im(z)
12 — •w
10 —
8 —
6 —
4 —
z •
2 —
    •u
−4 −2 0   2 4 6 8   Re(z)

b. 24 square units

16. a. $v = -5 + 4i$ or $11 + 4i$

b. $v = 4 - 5i$ or $8 - 5i$

c. $2b(a + c)$

17. a. $\dfrac{25}{2}$ b. $\dfrac{169}{2}$ c. $\dfrac{1}{2}(a^2 + b^2)$

18. a. 25   b. $b = \pm\sqrt{14}$ c. $a^2 + b^2$

19. a. $x^2 - 7x + y^2 + y = 0$, $\dfrac{25\pi}{2}$

b. $x^2 + 7x + y^2 - 17y = 0$, $\dfrac{169\pi}{2}$

20. $x^2 - (a - b)x + y^2 - (a + b)y = 0$, $\dfrac{\pi}{2}(a^2 + b^2)$

21. $\cos\left(\dfrac{\pi}{12}\right) = \dfrac{\left(1 + \sqrt{3}\right)}{2\sqrt{2}}$

$\sin\left(\dfrac{\pi}{12}\right) = \dfrac{\left(\sqrt{3} - 1\right)}{2\sqrt{2}}$

7.3 Operations on complex numbers in polar form

7.3 Exercise

1. a. $6\operatorname{cis}\left(\dfrac{3\pi}{4}\right)$ b. $20\operatorname{cis}\left(\dfrac{\pi}{3}\right)$

c. $6\sqrt{5}\operatorname{cis}\left(-\dfrac{\pi}{4}\right)$ d. $\sqrt{6}\operatorname{cis}\left(\dfrac{2\pi}{3}\right)$

e. $2\sqrt{7}\operatorname{cis}\left(-\dfrac{\pi}{6}\right)$

2. a. $-3\sqrt{2} + 3\sqrt{2}i$ b. $10 + 10\sqrt{3}i$

c. $3\sqrt{10} - 3\sqrt{10}i$ d. $-\dfrac{\sqrt{6}}{2} + \dfrac{3\sqrt{2}}{2}i$

e. $\sqrt{21} - \sqrt{7}i$

3. a. $4\sqrt{2}\operatorname{cis}\left(\dfrac{5\pi}{12}\right)$

b. $8\sqrt{3}\operatorname{cis}\left(-\dfrac{\pi}{2}\right)$

c. $8\sqrt{2}\operatorname{cis}\left(-\dfrac{\pi}{12}\right)$

4. $3\operatorname{cis}\left(\dfrac{\pi}{2}\right)$

5. a. $4\operatorname{cis}\left(\dfrac{11\pi}{12}\right)$ b. $2\operatorname{cis}\left(-\dfrac{3\pi}{10}\right)$

c. $2\sqrt{2}\operatorname{cis}\left(-\dfrac{3\pi}{14}\right)$ d. $\dfrac{3\sqrt{2}}{4}\operatorname{cis}\left(\dfrac{7\pi}{12}\right)$

6. $3\sqrt{3}\operatorname{cis}\left(\dfrac{\pi}{4}\right)$

7. a. $16\operatorname{cis}(\pi)$ b. $32\operatorname{cis}\left(\dfrac{3\pi}{4}\right)$

8. $-\dfrac{1}{4} + \dfrac{1}{4}i$

9. a. $-\dfrac{1}{4} + 0i$

b. $0 - \dfrac{1}{8}i$

c. $\dfrac{\sqrt{3}}{64} - \dfrac{1}{64}i$

d. $\dfrac{\sqrt{2}}{8}\cos\left(\dfrac{\pi}{12}\right) - \dfrac{\sqrt{2}}{8}\sin\left(\dfrac{\pi}{12}\right)i$

e. $16 + 0i$

10. $-1519 - 720i$

11. $-64\sqrt{3} - 64i$

12. $1 + 0i$

13. $\dfrac{8}{9}, -\dfrac{\pi}{6}$

14. $16 - 16i$

15. $-64 + 64i$

16. $\dfrac{2}{5}, -\dfrac{\pi}{120}$

17. Sample responses can be found in the worked solutions in the online resources.

18. Sample responses can be found in the worked solutions in the online resources.

19. $z = \dfrac{1}{2} - \dfrac{\sqrt{3}}{2}i$

20. n must be a multiple of 6 for $\left(1 + i\sqrt{3}\right)^n$ to be a real number greater than zero.

7.4 Geometric interpretations of multiplication and division of complex numbers in polar form

7.4 Exercise

1. a. $z_1z_2 = 4\sqrt{2}\operatorname{cis}\left(\dfrac{5\pi}{12}\right)$ and $\dfrac{z_2}{z_1} = \sqrt{2}\operatorname{cis}\left(-\dfrac{11\pi}{12}\right)$

b. z_1 is dilated by a factor $2\sqrt{2}$ and rotated by $-\dfrac{\pi}{4}$ anticlockwise about the origin.

z_2 is dilated by a factor $\dfrac{1}{2}$ and rotated by $\dfrac{2\pi}{3}$ clockwise about the origin.

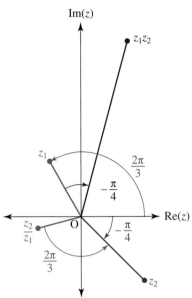

2. The student is correct.

3. a.

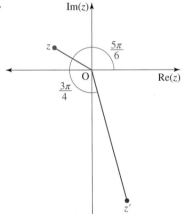

b. $z' = 6 \operatorname{cis}\left(-\dfrac{5\pi}{12}\right)$

4. D

5. a. i. $Z = \dfrac{-1 + \sqrt{3}}{8} + \dfrac{\left(1 + \sqrt{3}\right)}{8}i$

ii. $Z = \dfrac{\sqrt{2}}{4} \operatorname{cis}\left(\dfrac{5\pi}{12}\right)$

b. B

c. $\sin\left(\dfrac{5\pi}{12}\right) = \dfrac{\left(\sqrt{2} + \sqrt{6}\right)}{4}$

6. a. $b = \operatorname{cis}\left(\dfrac{2\pi}{3}\right)$

$c = \operatorname{cis}\left(-\dfrac{\pi}{3}\right)$

b.

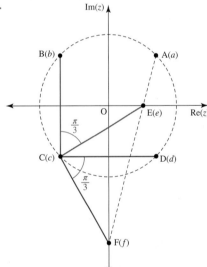

c. $\angle\left(\overrightarrow{AB}, \overrightarrow{AC}\right) = \left(\dfrac{\pi}{2}\right)$ (or $AC^2 + AB^2 = BC^2$)

7. $b = \operatorname{cis}\left(\dfrac{9\pi}{10}\right)$

8. a. $\dfrac{c - a}{b - a} = \operatorname{cis}\left(-\dfrac{\pi}{2}\right)$

b. The triangle ABC is a right-angle triangle where the right angle is at angle A.

9. $\left(\overrightarrow{AC}, \overrightarrow{AB}\right) = \operatorname{cis}(\pi)$ thus A, B and C are aligned.

10. a. $e = \sqrt{3} - 1$

$f = \left(-1 - \sqrt{3}\right)i$

b.

c. Show that $\left(\overrightarrow{EA}, \overrightarrow{EF}\right) = 0 \pm \pi$, that is $\dfrac{f - e}{a - e} \in \mathbb{R}$.

11. $k = 1 \pm 3\sqrt{3}$

12. $k = 14$

7.5 Exercise

1. The line $4x + 3y = 12$

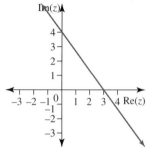

2. a. $y = 2$; line

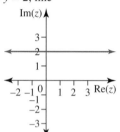

b. $x + 2y = 4$; line

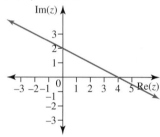

3. Line $y = -x$; the set of points equidistant from $(0, -3)$ and $(3, 0)$

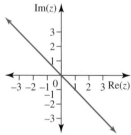

4. Line $y = -1$; the set of points equidistant from $(0, 1)$ and $(0, -3)$

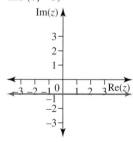

5. a. $y = -2x - 3$

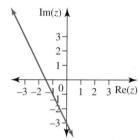

b. $y = \dfrac{2x}{3}$

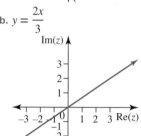

6. a. The ray makes an angle of 30° with the positive real axis, starting from but not including $(0, 0)$.

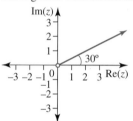

b. The ray starts from the point $(1, 0)$, not including the point, making an angle of 30° with the positive real axis.

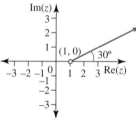

7. a. The ray makes an angle of 45° with the positive real axis, starting from but not including $(0, 0)$.

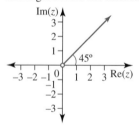

b. The ray starts from the point $(-2, 0)$, not including the point, making an angle of $45°$ with the positive real axis.

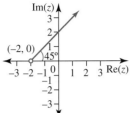

8. $(x-3)^2 + (y+2)^2 = 16$; circle with centre $(3, -2)$ and radius 4.

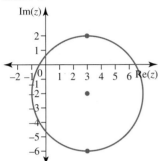

9. a. $x^2 + y^2 = 9$; circle with centre $(0, 0)$ and radius 3.

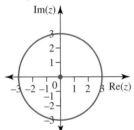

b. $x^2 + y^2 = 4$; circle with centre $(0, 0)$ and radius 2.

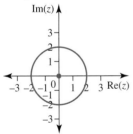

10. a. $(x+2)^2 + (y-3)^2 = 4$; circle with centre $(-2, 3)$ and radius 2.

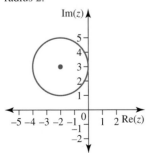

b. $(x-3)^2 + (y+1)^2 = 9$; circle with centre $(3, -1)$ and radius 3.

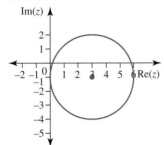

11. $\dfrac{x^2}{48} + \dfrac{y^2}{36} = 1$

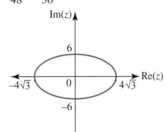

12. $\dfrac{x^2}{9} + \dfrac{y^2}{5} = 1$

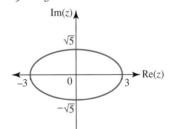

13.

14.

15.

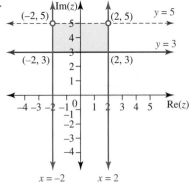

16.

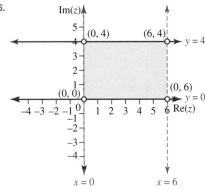

17. The region lies on and between the two circles centred at $(2, 0)$ with radii of 1 and 2 units.

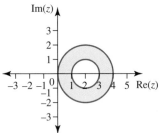

18. The required region is the overlap of these two circles, including points of intersection of the circumferences.

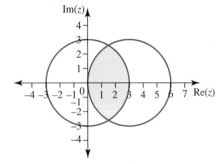

19.

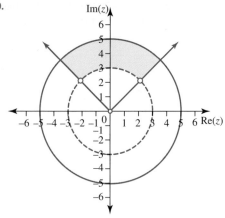

20.

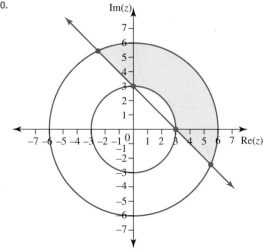

7.6 Roots of equations

7.6 Exercise

1. $(z - 3 + 4i)(z - 3 - 4i)$

2. $\left(z + 2 + \sqrt{3}i\right)\left(z + 2 - \sqrt{3}i\right)$

3. $2 \pm 5i$

4. $-1 \pm 5i$

5. a. $-1 \pm 4i$ b. $2 \pm 4i$
   c. $3 \pm 2i$ d. $-5 \pm 4i$

6. $z^2 + 10z + 29$

7. $z^2 + 25$

8. a. $3z^2 + 5z - 2$ b. $z^2 - 4z + 40$
   c. $z^2 + 4z + 13$ d. $z^2 + 8z + 41$

9. a. $10z^2 - 3z - 1$ b. $z^2 - 8z + 19$
   c. $z^2 + 10z + 32$ d. $z^2 + 6z + 17$

10. a. $2 \pm 2i$ **b.** $3 \pm i$

    c. $1 \pm 5i$ **d.** $4 \pm 5i$

11. a. $2 \pm 3i$ **b.** $-3 \pm \sqrt{5}i$

    c. $1 \pm \sqrt{3}i$ **d.** $-5 \pm \sqrt{6}i$

12. a. $-\dfrac{3}{2} \pm \dfrac{1}{2}i$ **b.** $-\dfrac{5}{2} \pm i$ **c.** $\dfrac{3}{2} \pm i$ **d.** $\dfrac{7}{3} \pm \dfrac{2}{3}i$

13. $-1 + 2i, \ -1 - i$

14. $-5 + 5i, \ 1 + i$

15. $-2 + \left(\dfrac{\sqrt{5}+1}{2}\right)i, \ -2 + \left(\dfrac{1-\sqrt{5}}{2}\right)i$

16. $-\left(\dfrac{\sqrt{11}+4}{2}\right) + \dfrac{3}{2}i, \ -\left(\dfrac{\sqrt{11}-4}{2}\right) - \dfrac{3}{2}i$

17. a. $z = \sqrt{2} + i$ or $z = -\sqrt{2} - i$

    b. $z = \dfrac{\sqrt{2}}{2}$ or $z = \dfrac{-\sqrt{2}}{2} - i$

18. $z = \text{cis}\left(\dfrac{\pi}{8} + \dfrac{k\pi}{2}\right), \ k = 0, 1, 2, 3$ and

    $z = 2^{\frac{1}{8}} \text{cis}\left(\dfrac{5\pi}{16} + \dfrac{k\pi}{2}\right), \ k = 0, 1, 2, 3$

7.7 Review

7.7 Exercise

1. D

2. D

3. D

4. B

5. $5\sqrt{2}\,\text{cis}\left(\dfrac{\pi}{4}\right)$

6. a. $\dfrac{w}{z} = \dfrac{-4 + 7i}{13}$

    b. $k = \dfrac{\sqrt{65}}{13} \approx 0.620$ and $\theta = \tan^{-1}\left(-\dfrac{7}{4}\right) = -60.255°$

7. $\text{Arg}(2 - 2i) = -\dfrac{\pi}{4}$

8. $z = \dfrac{3 \pm \sqrt{87}i}{8} = 0.375 \pm 1.166i$

9. $z = -2 \pm i$

10. a. Circle $x^2 + y^2 = 4$

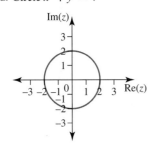

b. Line $x = \dfrac{5}{2}$

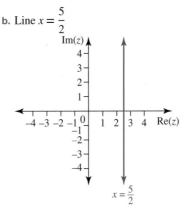

11. $50\,\text{cis}\left(\dfrac{\pi}{12}\right)$

12. $-4 - 4i$

13. a.

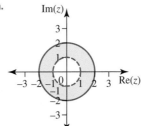

b.

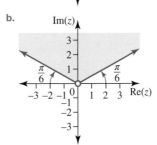

c.

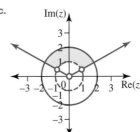

14. $45°$

15. $z = -1 - i, \ z = \dfrac{1 \mp \sqrt{3}i}{2}$

16. Show that $\dfrac{c - b}{a - b} = \text{cis}(0 \pm \pi)$. Sample responses can be found in the worked solutions in the online resources.

17. Sample responses can be found in the worked solutions in the online resources.

18. Sample responses can be found in the worked solutions in the online resources.

19. $z = \dfrac{1}{2} + \dfrac{\sqrt{3}}{2}i, \ -\dfrac{1}{2} - \dfrac{\sqrt{3}}{2}i, \ \dfrac{1}{2} - \dfrac{\sqrt{3}}{2}i, \ -\dfrac{1}{2} + \dfrac{\sqrt{3}}{2}i$

20. Sample responses can be found in the worked solutions in the online resources.

8 Circle and geometric proofs

Fully worked solutions for this chapter are available online.

EXAM PREPARATION

Access exam-style questions in every lesson, available online.

 Resources

Solutions	Solutions — Chapter 8 (sol-0400)
Exam questions	Exam question booklet — Chapter 8 (eqb-0286)
Digital documents	Learning matrix — Chapter 8 (doc-41930)
	Chapter summary — Chapter 8 (doc-41931)

LESSON
8.1 Overview

8.1.1 Introduction

Circles (from the Greek word *kirkos*, meaning 'ring') have been studied since ancient times. They are a useful shape because they enclose the largest possible area for a given perimeter (or have the smallest possible perimeter for a given area).

The study of circles has many applications, including the development and use of gyroscopes. A gyroscope is composed of a circular disc that can freely rotate to assume any orientation. This disc is mounted onto an axis in the centre of a larger and more stable wheel. As the disc spins, it wants to stay in the same orientation; any external force applied to change the orientation of the spin axis will be resisted, due to the conservation of angular momentum. Gyroscopes are used to maintain the stability of many machines, from bicycles, motorcycles and ships through to smart phones, virtual reality headsets, robots, aeroplanes and space telescopes.

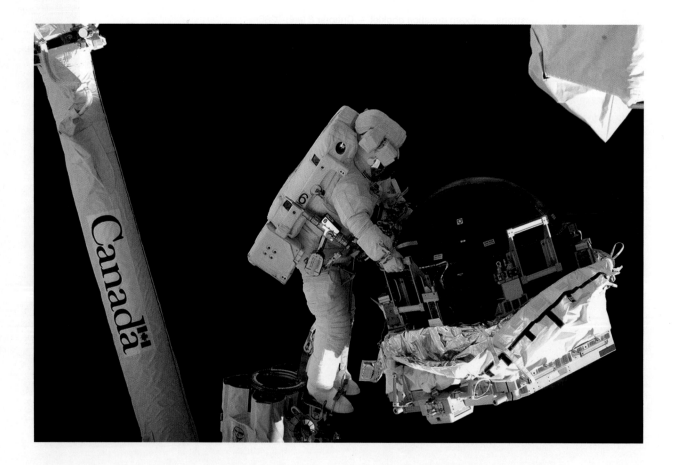

8.1.2 Syllabus links

Lesson	Lesson title	Syllabus links
8.2	**Congruent triangles and angle relationships**	Review of prerequisite concepts.
8.3	**Circle properties 1 – angles in a circle and chords**	○ Prove circle properties: • the angle at the centre subtended by an arc of a circle is twice the angle at the circumference subtended by the same arc • angles at the circumference of a circle subtended by the same arc are equal • an angle in a semicircle is a right angle. ○ Solve problems finding unknown angles and lengths and prove further results using circle properties.
8.4	**Circle properties 2 – tangents secants and segments**	○ Prove circle properties: • a tangent drawn to a circle is perpendicular to the radius at the point of contact and its converse • the alternate segment theorem. ○ Solve problems finding unknown angles and lengths and prove further results using circle properties.
8.5	**Circle properties 3 – cyclic quadrilaterals**	○ Prove circle properties: • the opposite angles of a cyclic quadrilateral are supplementary and its converse. ○ Solve problems finding unknown angles and lengths and prove further results using circle properties.
8.6	**Geometric proofs using vectors**	○ Prove the diagonals of a parallelogram meet at right angles if and only if it is a rhombus. ○ Prove midpoints of the sides of a quadrilateral join to form a parallelogram. ○ Prove the sum of the squares of the lengths of a parallelogram's diagonals is equal to the sum of the squares of the lengths of the sides. ○ Prove an angle in a semicircle is a right angle.

Source: Specialist Mathematics Senior Syllabus 2024 © State of Queensland (QCAA) 2024; licensed under CC BY 4.0

LESSON
8.2 Congruent triangles and angle relationships

8.2.1 Congruent triangle tests

You have already learned about tests that can be used to prove that triangles are congruent. These congruency tests are shown in the table below.

Description	Test	Abbreviation
Three sides of one triangle are congruent to the three sides of the other triangle.		SSS (side–side–side)
Two sides and the included angle of one triangle are congruent to two sides and the included angle of the other triangle.		SAS (side–angle–side)
Two angles and a side of one triangle are congruent to two angles and the corresponding side of the other triangle.		AAS (angle–angle–side)
The triangles are right-angled, and the hypotenuse and a side of one triangle are congruent to the hypotenuse and side of the other triangle.		RHS (right angle–hypotenuse–side)

The tests SSS, SAS and AAS are postulates, meaning that they are assumed to be true. The test RHS can be proven by using SSS and Pythagoras' theorem.

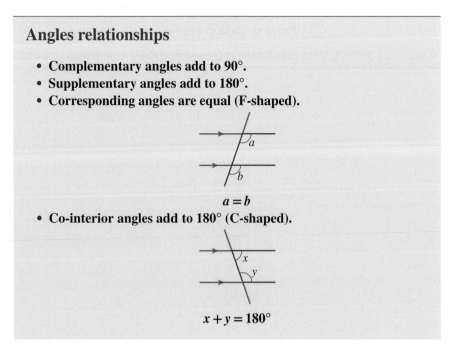

Angles relationships

- **Complementary angles add to 90°.**
- **Supplementary angles add to 180°.**
- **Corresponding angles are equal (F-shaped).**

$$a = b$$

- **Co-interior angles add to 180° (C-shaped).**

$$x + y = 180°$$

- **Alternate angles are equal (Z-shaped).**

$$a = b$$

- **Vertically opposite angles are equal (X-shaped).**

$$a = b$$

- **Two lines are perpendicular if they meet at right angles (90°).**

WORKED EXAMPLE 1 Proving that the diagonals of a rectangle bisect each other

Demonstrate that the diagonals of a rectangle bisect each other.

THINK	WRITE
1. Draw a diagram of a rectangle, marking the diagonals and labelling the points.	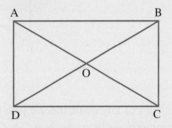
2. Opposite sides of a rectangle are equal and parallel.	Angle BAC = angle ACD (alternate) and angle BDC = angle ABD
3. Therefore, the triangles are congruent.	$\triangle$ABO $\cong$ $\triangle$CDO by ASA rule
4. The corresponding sides are equal.	AO = OC and BO = OD
5. The intersection is the midpoint of AC and BD.	The diagonals therefore bisect each other.

Exercise 8.2 Congruent triangles and angle relationships

learn on

8.2 Exercise	8.2 Exam questions on

Simple familiar	Complex familiar	Complex unfamiliar
1, 2, 3, 4, 5	6, 7, 8	9, 10

Simple familiar

1. **WE1** Demonstrate that a diagonal of a parallelogram divides it into two congruent triangles.

2. Demonstrate that the diagonals of a rectangle bisect each other.

3. Demonstrate that the opposite sides of a parallelogram are congruent.

4. For a rhombus, demonstrate that:
   a. the diagonals bisect the corner angles
   b. the diagonals bisect each other at right angles.

5. Prove that if a quadrilateral has one pair of opposite sides that are both congruent and parallel, it is a parallelogram.

Complex familiar

6. Prove that the diagonals of a kite intersect at right angles.

7. Triangle ABC is an equilateral triangle. The points X, Y and Z bisect $\overline{AB}$, $\overline{BC}$ and $\overline{CA}$ respectively. Prove that $\triangle XYZ$ is an equilateral triangle.

8. If $\overline{AE} \cong \overline{CD}$, prove that $\overline{AB}$ is parallel to $\overline{FD}$.

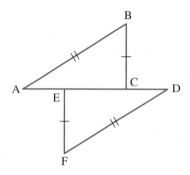

Complex unfamiliar

9. The midpoints of the sides of any quadrilateral join to form a parallelogram. Use geometry software such as GeoGebra to explore quadrilaterals, then prove that the statement is true. *Hint:* You will need to use similar triangles as part of your proof; the diagram shown may help you to see them.

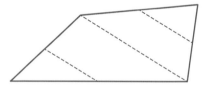

10. The sum of the squares of the lengths of the diagonals of a parallelogram is equal to the sum of the squares of the lengths of the four sides. Use geometry software such as GeoGebra to convince yourself that this is true. Use Pythagoras' theorem to prove that it is true. The diagram shown may help you.

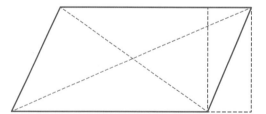

Fully worked solutions for this chapter are available online.

LESSON
8.3 Circle properties 1 — angles in a circle and chords

SYLLABUS LINKS

- Prove circle properties:
  - the angle at the centre subtended by an arc of a circle is twice the angle at the circumference subtended by the same arc
  - angles at the circumference of a circle subtended by the same arc are equal
  - an angle in a semicircle is a right angle.
- Solve problems finding unknown angles and lengths and prove further results using circle properties.

Source: Specialist Mathematics Senior Syllabus 2024 © State of Queensland (QCAA) 2024; licensed under CC BY 4.0

8.3.1 Useful definitions

Chord definitions

- **Circumference**: the distance around the circle (the perimeter), that is, the length of the circle if it was opened up and straightened into a line segment. (Circumference may also be used to refer to the edge of the circle.) A, B, D, E, F and G are all **points on the circle**.
- **Centre of the circle**: the middle point, equidistant from all points on the circumference; point O.
- **Chord**: an interval jointing two points on the circle. $\overline{AG}$, $\overline{AD}$ and $\overline{GD}$ are all chords.
- **Diameter**: a chord that passes through the centre. $\overline{GD}$ is also a diameter.
- **Radius**: any straight line joining a point on the circumference to the centre. $\overline{OG}$, $\overline{OD}$ and $\overline{OF}$ are all radii. Notice that every diameter consists of two radii.

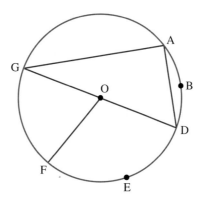

Angle definitions

- **Arc**: part of a circle defined by three points, for example arc $\overarc{ABD}$. A **minor arc** is smaller than a semicircle; a **major arc** is larger than a semicircle. The arc $\overarc{ABD}$ is said to **subtend** the angle $\angle AGD$ as G is on the circumference of the circle. $\angle AGD$ is called an **angle at the circumference** subtended by the arc $\overarc{ABD}$. The same angle is sometimes described as **standing** on chord $\overline{AD}$.
- **Angle at the centre**: the angle made by two radii intersecting the centre of the circle, for example $\angle FOD$, which is subtended by the arc $\overarc{FED}$.
- **Angle in a semicircle**: the angle standing on the diameter, for example, $\angle GAD$, as $\overline{GD}$ is a diameter.

8.3.2 Theorems 1, 2 and 3

There are a number of theorems involving chords and angles in circles. These can be proved by using the congruent triangle tests angle relationships, etc.

Theorem 1	Diagram	Symbol
Theorem 1: The angle at the centre of the circle is twice the angle at the circumference subtended on the same arc.	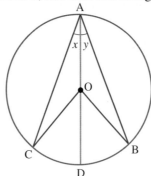 If the magnitude of the angle at the circumference is $x°$, then the magnitude of the angle at the centre subtended on the same arc is $2x°$.	

Proof of theorem 1

If a diameter is drawn passing through A and O, two isosceles triangles are formed.

As AO and CO are radii, $\triangle AOC$ is isosceles.
Determining the relationship between $\angle CAO$ and $\angle COD$:
Consider $\triangle AOC$.
Let $\angle CAO = x$.
$\angle ACO \cong \angle CAO$
$\angle ACO = x$

$\angle DOC = \angle ACO + \angle CAO$
$\qquad = x + x$
$\qquad = 2x$

Similarly, as AO and BO are radii, $\triangle AOB$ is isosceles.
Repeating the same steps as for $\triangle AOC$:
If $\angle BAO = y$, then $\angle DOB = 2y$
Thus
$\angle CAB = x + y$ and $\angle COB = 2x + 2y$
$\qquad\qquad\qquad\qquad\quad = 2(x + y)$
Therefore, the angle at the centre is twice the angle at the circumference subtended by the same arc.

Theorem 2	Diagram	Symbol
Theorem 2: The angle in a semicircle is a right angle.	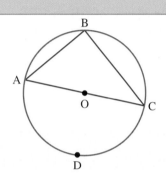 The angle in the semicircle is 90°.	

Proof of theorem 2

∠ABC and ∠AOC both stand on $\overset{\frown}{ADC}$

∠AOC = 2∠ABC

180° = 2∠ABC

∠ABC = 90°

Therefore, the angle subtended by a semicircle is a right angle.

Alternatively, using theorem 1:

The angle at the centre is 180° so it follows from Theorem 1 that the angle subtended by that arc on the circumference is half of 180°, which is 90°.

Theorem 3	Diagram	Symbol
Theorem 3: Angles at the circumference subtended by the same arc are congruent.	Same arc If the magnitude of the angle at the circumference is $x°$, then the magnitude of the angle at the circumference subtended by the same arc is also $x°$.	

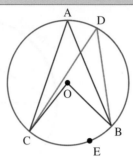

$\angle CAB$ and $\angle COB$ both stand on $\overset{\frown}{CEB}$

Therefore $\angle COB = 2\angle CAB$ [1]

Similarly, $\angle CDB$ and $\angle COB$ both stand on $\overset{\frown}{CEB}$

Therefore $\angle COB = 2\angle CDB$ [2]

Using equations [1] and [2], and matching equals, $\angle CAB = \angle CDB$

Therefore, angles at circumference subtended by the same arc are congruent.

Alternatively, using theorem 1:

The two angles are subtended by the same arc and therefore both have a magnitude half of the angle subtended at the centre: $\angle CAB = \angle CDB$.

These theorems can be used to calculate the values of unknown angles.

WORKED EXAMPLE 2 Using the angles at the circumference subtended by the same arc being congruent

Determine the values of a, b and c in the circle shown.

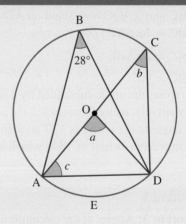

THINK

1. $\angle AOD$ and $\angle ABD$ are both standing on arc $\overset{\frown}{AED}$. By recognising the first theorem above, $\angle AOD$ is twice $\angle ABD$.

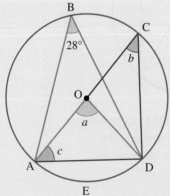

WRITE

$\angle ABD = 28°$

$\therefore a = 2 \times 28°$

$= 56°$

2. ∠ABD and ∠ACD are both standing on arc $\overset{\frown}{AED}$.
   By recognising the second theorem above,
   ∠ABD = ∠ACD.

∠ABD = 28°

∴ b = 28°

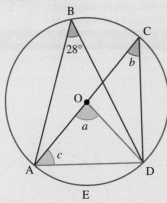

3. ∠ADC is in a semicircle. Recognising the third theorem above, ∠ADC = 90°, and recalling the sum of angles in a triangle is 180°, $b + c = 90°$.

∠ADC = 90°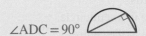

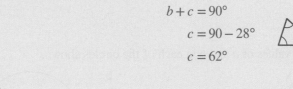

$b + c = 90°$

$c = 90 - 28°$

$c = 62°$

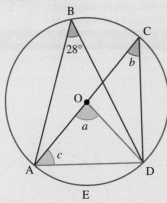

Worked example 2 uses theorems to calculate unknown values. As introduced in the previous chapter, postulates and previously proven theorems can also be used to prove other theorems.

 Resources

Interactivities	Circle theorem 1 (int-6218)
	Circle theorem 3 (int-6219)
	Circle theorem 2 (int-6220)

Exercise 8.3 Circle properties 1 — angles in a circle and chords

learnon

8.3 Exercise	8.3 Exam questions on

Simple familiar	Complex familiar	Complex unfamiliar
1, 2, 3, 4, 5	6, 7, 8, 9, 10, 11	12, 13

These questions are even better in jacPLUS!
• Receive immediate feedback
• Access sample responses
• Track results and progress

Find all this and MORE in jacPLUS ▶

Simple familiar

1. Refer to the diagram of the circle shown to answer the following questions.
   a. Identify the points on the circle.
   b. Identify the radii.
   c. Identify a major arc.
   d. Identify the chords.
   e. Identify an angle in a semicircle.
   f. Identify an angle at the centre.
   g. Identify two angles standing on arc $\overset{\frown}{AFE}$.

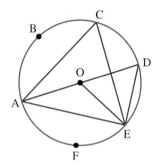

2. **WE2** Determine the values of a and b in each of the circles shown.

   a.
   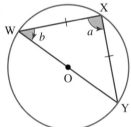

   b.
   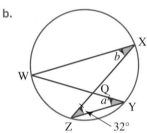

3. Prove that $\overline{AB}$ is parallel to $\overline{CD}$.

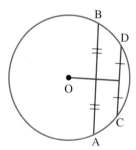

4. Prove that $\overline{AB}$ is congruent to $\overline{CD}$.

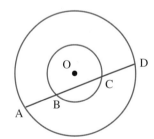

5. If AB = BC = 15, evaluate:

   a. ∠OBC   **b.** EC

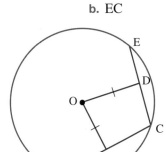

Complex familiar

6. a. If the radius of the circle shown is 5 units, how far apart are the chords $\overline{AB}$ and $\overline{CD}$?

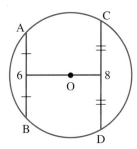

  b. If the radii of the circles shown are 13 and 15 units and the distance from the centre to $\overline{WZ}$ is 12 units, determine the length of $\overline{WX}$.

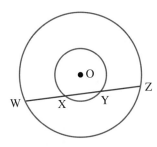

7. Prove the following: if a line from the centre bisects a chord, it is perpendicular to the chord.

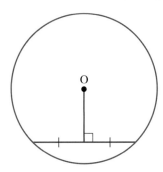

8. Prove the following: a line from the centre that is perpendicular to a chord will bisect the chord.

9. **a.** Prove the following: if chords are congruent, they are equidistant from the centre.

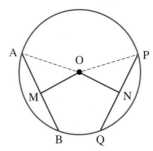

  b. Prove the converse: if chords are equidistant from the centre, they are congruent.

10. Prove each of the following. The diagram may assist you.

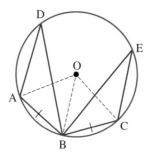

  a. Chords of equal length subtend equal angles at the centre and the converse.
  b. Angles at the circumference standing on chords of equal length are congruent.

11. Two congruent circles with centres O and C intersect at A and B. The centre of each circle is a point on the other circle. If $\overline{AD}$ and $\overline{AE}$ are diameters, prove that:

  a. $\triangle AED$ is an equilateral triangle
  b. AB is $\sqrt{3}$ times the length of the radius.

Complex unfamiliar

12. A regular pentagon has a side of 3 cm. If a circle is drawn so that the vertices of the pentagon lie on the circumference of the circle, what is the radius of the circle? Use suitable technology to verify your answer.

13. A second regular pentagon is formed by bisecting the sides of the pentagon from question **12**. A circle is drawn that passes through the vertices of the new pentagon. What is the ratio of the radius of the original circle to the radius of the new circle? Does this ratio remain the same if you continue to create pentagons by bisecting the sides of the previous pentagons? You may use suitable technology to explore your answer.

Fully worked solutions for this chapter are available online.

LESSON
8.4 Circle properties 2 — tangents, secants and segments

SYLLABUS LINKS

- Prove circle properties:
  - a tangent drawn to a circle is perpendicular to the radius at the point of contact and its converse
  - the alternate segment theorem.
- Solve problems finding unknown angles and lengths and prove further results using circle properties.

Source: Specialist Mathematics Senior Syllabus 2024 © State of Queensland (QCAA) 2024; licensed under CC BY 4.0

8.4.1 Useful definitions

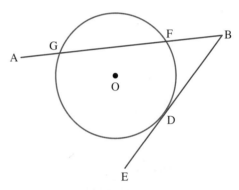

- **External** to the circle: outside the circle; for example points A, B and E
- **Secant**: a line that cuts a circle at two distinct points; for example $\overline{AB}$
- **Tangent**: a line that touches a circle at one point; for example $\overline{EB}$
- **Segment**: a region of a circle bounded by an arc and a chord passing through the endpoints of the arc; for example, chord $\overline{GF}$ divides the circle into two segments, a **minor segment**, and a **major segment**.

- Chord $\overline{AB}$ divides the circle into two segments. Notice that the minor segment, shaded in blue, is in the segment that is closest to the tangent $\overline{AT}$ here.
  The major segment, shaded in pink, is in the other segment, it is said to be its **alternate segment**.

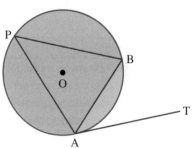

8.4.2 Theorems 4 and 5

Theorem 4	Diagram	Symbol
Theorem 4: A tangent and a radius intersect at right angles.	The tangent to the circle and the radius meet at 90°	

Proof of theorem 4

This is a proof by contradiction.

Let $\overline{OA}$ be the radius and AC be the tangent to the circle at A.

Let's assume that the radius and the tangent do NOT meet at 90°, then there must be another point, let's call it B, which does make a 90° angle with the tangent. Let OB pass through the circle at point D and meet the tangent at B.

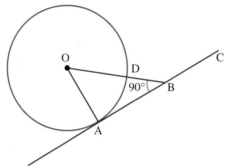

Now, triangle OBA is a right-angled triangle, with ∠OBA as the right-angle.

The hypothenuse of a right-angled triangle is always opposite to the right-angle so the hypothenuse is side $\overline{OA}$.

However, $\overline{OA}$ and $\overline{OD}$ are both radius lines, so they are of equal length.

$\overline{OA} = \overline{OD}$

But $\overline{OB} = \overline{OD} + \overline{DB}$, thus the side $\overline{OB}$ is longer than $\overline{OA}$.

$\overline{OB} > \overline{OA}$.

This is a contradiction as $\overline{OA}$ is the hypothenuse, the longest side.

This contradiction proves that no point B can exist away from point A, therefore the right-angle must be at point A, where the radius meets the tangent.

Therefore, the assumption is incorrect, and the tangent and radius meet at 90°.

Theorem 5	Diagram	Symbol
Theorem 5: The angle formed by a tangent and a chord is congruent to the angle in the alternate segment. This is known as the alternate segment theorem.	Angle in the alternate segment The angle formed by a tangent and a chord is congruent to the angle in the alternate segment.	

Proof of theorem 5

Angle in the alternate segment

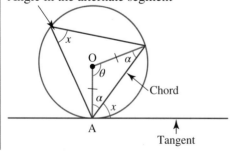

$\overline{OA}$ and $\overline{OC}$ are radius lines, and therefore of equal length.

$\triangle AOC$ is therefore an isosceles triangle, and angles $\angle OAC$ and $\angle OCA$ are therefore congruent. Let's call these angles' size α and the remaining angle, $\angle AOC$, θ.

The angles in a tringle add up to $180°$, thus, applying this to $\triangle AOC$:

$\theta + 2\alpha = 180°$

$\quad \therefore \theta = 180° - 2\alpha.$

Moreover, $\angle AOC$ is twice the size of the angle $\angle ABC$, of size y, subtended by the same arc at the circumference (Theorem 1).

$\theta = 2y.$

Thus, $180° - 2\alpha = 2y$

$\quad\quad \therefore y = 90° - \alpha.$

$\overline{OA}$ is a radius line, and the angle that a radius line makes with a tangent is $90°$, thus at A, $\alpha + x = 90°$.

Therefore, $\therefore y = 90° - \alpha = x$. Angle y is congruent to angle x.

WORKED EXAMPLE 3 Using radius and tangent intersecting at right angles

Determine the value of a in the diagram.

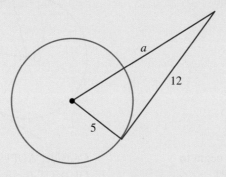

THINK

1. A tangent and radius meet at right angles.

WRITE

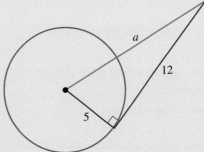

A tangent and radius meet at right angles The length of the hypotenuse is a.

2. Use Pythagoras' theorem to solve for a.

$a^2 = 5^2 + 12^2$
$\quad = 25 + 144$
$\quad = 169$
$a = 13$

WORKED EXAMPLE 4 Proving that two tangents drawn from an external point to a circle are congruent

Use the diagram to prove the following: two tangents drawn from an external point to a circle are congruent.

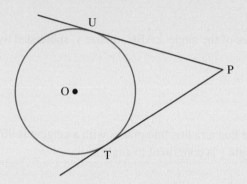

THINK

1. Draw radii $\overline{OU}$ and $\overline{OT}$ and the line $\overline{OP}$ so that congruent triangles can be found.

WRITE

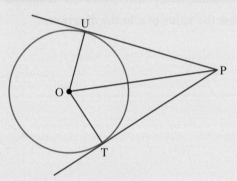

2. Identify the congruent sides to begin to demonstrate that $\triangle OPU \cong \triangle OPT$.

Consider $\triangle OPU$ and $\triangle OPT$.
$\overline{OU} \cong \overline{OT}$ (both radii) $\overline{OP}$ is common to both.

3. Use the theorem that the radius and tangent meet at right angles.

$\angle OUP = \angle OTP = 90°$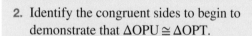

4. Conclude that the triangles are congruent using RHS.

Therefore, $\triangle OUP \cong \triangle OTP$ (RHS).

5. $\overline{UP}$ and $\overline{TP}$ are corresponding sides and are therefore congruent.

As $\overline{UP}$ and $\overline{TP}$ are corresponding sides, $\overline{UP} \cong \overline{TP}$.

6. Write the concluding statement.

Therefore, tangents drawn from an external point are congruent.

 Resources

Interactivities Circle theorem 4 (int-7260)
Circle theorem 5 (int-7263)

Exercise 8.4 Circle properties 2 — tangents, secants and segments

8.4 Exercise	8.4 Exam questions on

Simple familiar	Complex familiar	Complex unfamiliar
1, 2, 3, 4	5, 6, 7, 8, 9	10, 11

These questions are
even better in jacPLUS!
• Receive immediate feedback
• Access sample responses
• Track results and progress

Find all this and MORE in jacPLUS ▶

Simple familiar

1. Refer to the diagram to identify the following.
   a. Tangent
   b. Secant
   c. Major segment
   d. Angle in the alternate segment to ∠ABC

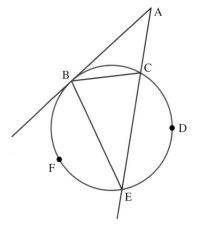

2. **WE3** Determine the value of a.

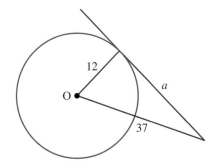

3. **WE4** Use the diagram to prove the following.

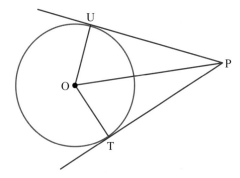

   a. $\overline{OP}$ bisects the angle made between the tangents drawn from P to the circle.
   b. $\overline{OP}$ bisects $\overline{TU}$.

4. Prove that $\overline{AB} \cong \overline{BC}$.

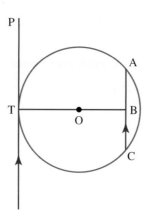

Complex familiar

5. If $\overline{AB}$ is a tangent and $\overline{BC}$ is a diameter, prove that $\angle ADB \cong \angle ABC$.

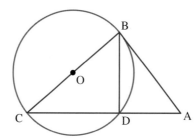

6. Use the alternate segment theorem and any other theorems you know to prove the following.

  a. $b = a + c$

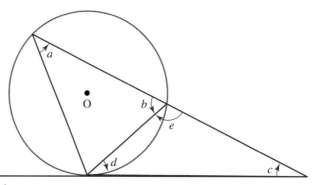

  b. $\triangle ABC$ is an isosceles triangle.

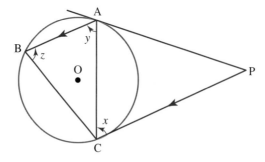

7. A proof by contradiction can be used to prove that the tangent drawn to a circle is perpendicular to the radius at the point of contact (theorem 6).

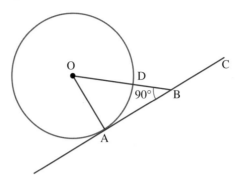

Let $\overline{OA}$ be the radius of the circle and $\overline{AC}$ be the tangent to the circle at A. Assume that the tangent and radius do not meet at 90°. This means that there must be a line from O that does meet the tangent at 90°. Let $\overline{OB}$ pass through the circle at D and intersect the tangent at point B at 90°.

a. What type of triangle is △AOB?
b. According to your answer in part a, what is the longest side of the triangle?
c. What do you know about $\overline{OA}$ and $\overline{OD}$? What does that tell you about OA and OB?
d. Use your answers to construct a proof by contradiction to prove that the tangent drawn to a circle is perpendicular to the radius at the point of contact.

8. Two circles intersect at A and B. If $\overline{AM}$ and $\overline{AN}$ are diameters of one of the circles and tangents to the other, prove that M, N and B are collinear.

9. Two circles intersect at points A and B. The tangents from point A intersect with the circles at C and D. $\overline{CD}$ also intersects with the circles at P and Q. Prove that $\overline{AP} \cong \overline{AQ}$.

Complex unfamiliar

10. Using suitable geometry software, construct a circle with a tangent from P touching the circle at T and a secant from P cutting the circle at A and B. Verify with your software that $PT^2 = PA \cdot PB$. Prove the relationship. *Note:* This is called the tangent–secant theorem (theorem 12).

11. Using suitable geometry software, construct a circle with two secants $\overline{AB}$ and $\overline{CD}$ that intersect at an external point P. Verify with your software that $PA \cdot PB = PC \cdot PD$. Without using the tangent–secant theorem, prove the relationship. *Note:* This is called the secant–secant theorem (theorem 13).

Fully worked solutions for this chapter are available online.

LESSON
8.5 Circle properties 3 — cyclic quadrilaterals

8.5.1 Useful definitions

- A **cyclic quadrilateral** is a quadrilateral whose vertices all lie on a circle.
- Quadrilateral ABCD is a cyclic quadrilateral as all vertices lie on the circumference of the circle.

  In this example, ∠CDE is an **exterior angle to a cyclic quadrilateral** and ∠ABC is the **interior opposite angle in a cyclic quadrilateral**.

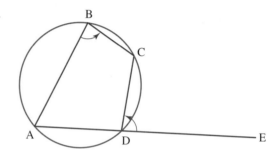

8.5.2 Opposite angles in a cyclic quadrilateral

Theorem 6	Diagram	Symbol
Theorem 6: The opposite angles of a cyclic quadrilateral are supplementary.	 ∠ABC + ∠CDA = 180° and ∠BCD + ∠DAB = 180°	
Proof of theorem 6		

Let's construct a radius to each of the four vertices.

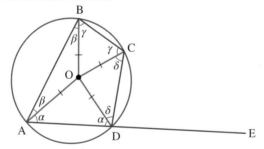

The radii of the circle are all congruent, thus △AOD, △AOB, △BOC and △COD are all isosceles triangles. The base angles of an isosceles triangle have the same measure, as illustrated by the labels in the figure above.

The sum of the angles in a triangle is equal to 180° and the sum of the angles around the centre of a circle is 360°, thus:

$(180° - 2\alpha) + (180° - 2\beta) + (180° - 2\gamma) + (180° - 2\delta) = 360°$

$$\therefore \alpha + \beta + \gamma + \delta = 180°$$

$\alpha + \beta$ is the measure of $\angle DAB$ and $\gamma + \delta$ is a measure of $\angle BCD$, its opposite angle.

$\alpha + \delta$ is the measure of $\angle ADC$ and $\beta + \gamma$ is a measure of $\angle ABC$, its opposite angle.

Thus, as $(\alpha + \beta) + (\gamma + \delta) = 180°$ and $(\alpha + \delta) + (\beta + \gamma) = 180°$, this proves that the opposite angles in a cyclic quadrilateral are supplementary.

Note that the converse is true.

WORKED EXAMPLE 5 Using the opposite angles in a cyclic quadrilateral being supplementary

Determine the values of a and b in the diagram.

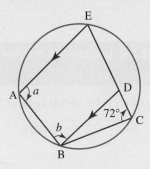

THINK

1. ABCE is a cyclic quadrilateral. Opposite angles of a cyclic quadrilateral are supplementary. $\angle EAB$ is opposite $\angle ECB$.

WRITE

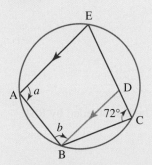

ABCE is a cyclic quadrilateral.
Therefore:

$\angle EAB + \angle ECB = 180°$

$a + 72° = 180°$

$a = 108°$

2. Co-interior angles are supplementary.

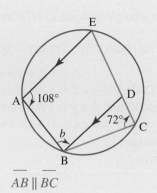

$$\overline{AB} \parallel \overline{BC}$$

$$\angle EAB + \angle ABD = 180°$$
$$108° + b = 180°$$
$$b = 72°$$

WORKED EXAMPLE 6 Proving that the opposites angles of a cyclic quadrilateral are supplementary

Prove the following the opposite angles of a cyclic quadrilateral are supplementary.

THINK

1. Draw a diagram of a cyclic quadrilateral and mark a pair of opposite angles.

2. ∠ROP and ∠RQP are both subtended by $\overset{\frown}{RSP}$.

WRITE

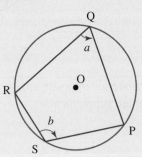

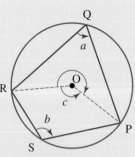

Let ∠RQP = a.
Let ∠ROP = c.
$\therefore c = 2a$

3. Reflex ∠ROP and ∠RSP are both subtended
   by R͡QP.

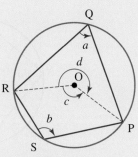

Let ∠RSP = b.
Let ∠ROP = d (reflex angle).
∴ $d = 2b$

4. The angles at the origin must add to 360°.

$c + d = 360°$

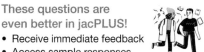

5. Use substitution for c and d.

$2a + 2b = 360°$
$a + b = 180°$

6. Write the concluding statement.

Therefore, the opposite angles of a cyclic
quadrilateral are supplementary.

Exercise 8.5 Circle properties 3 — cyclic quadrilaterals

learn on

8.5 Exercise	**8.5 Exam questions** on

Simple familiar	Complex familiar	Complex unfamiliar
1, 2, 3	4, 5, 6, 7, 8	9, 10, 11, 12

These questions are
even better in jacPLUS!
- Receive immediate feedback
- Access sample responses
- Track results and progress

Find all this and MORE in jacPLUS ▶

Simple familiar

1. **WE5** Determine the values of a and b in the diagram shown.

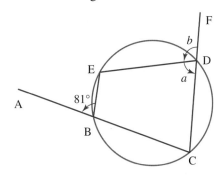

2. Calculate the values of a and b in the diagram shown.

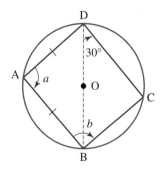

3. If one angle of a cyclic quadrilateral is a right angle, demonstrate that two of the points must be at either end of a diameter.

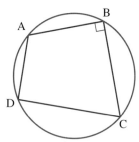

Complex familiar

4. **WE6** Prove the following: the exterior angle of a cyclic quadrilateral is congruent to the interior opposite angle. (You may use theorem 6, that opposite angles in a cyclic quadrilateral are supplementary, as part of your proof.)

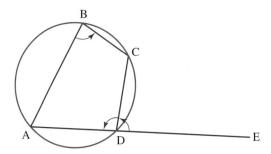

5. Prove that $a = b$.

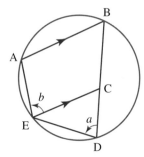

6. Prove that ∠EAB and ∠DCF are supplementary.

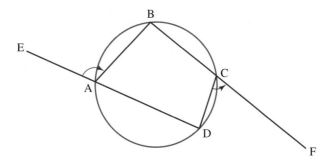

7. Prove that ∠FCD and ∠EDC are congruent.

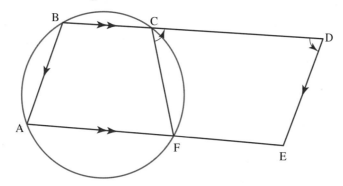

8. ABCDEF is a hexagon with vertices on the circumference of a circle.
   a. If $\overline{AB}\|\overline{ED}$ and $\overline{BC}\|\overline{FE}$, prove that ∠ABC ≅ ∠FED. (*Hint:* Join B and E.)
   b. By proving that ∠ADC ≅ ∠DAF, demonstrate that $\overline{CD}\|\overline{AF}$.

Complex unfamiliar

9. In a quadrilateral ABCD, prove that the bisectors of the angles intersect to form a cyclic quadrilateral.

10. If a parallelogram is inscribed inside a circle, prove that it is a rectangle.

11. Using suitable drawing software, draw a cyclic quadrilateral ABCD. Explore the relationship between ∠A, ∠B and ∠C if point D is moved to the centre of the circle. Prove this relationship.

12. Using suitable drawing software, determine the type of quadrilateral formed by the tangents to the points of a cyclic quadrilateral when the cyclic quadrilateral is a rectangle.

Fully worked solutions for this chapter are available online.

LESSON
8.6 Geometric proofs using vectors

Vectors can be used to prove theorems and other general statements in geometry.

8.6.1 Review of vectors

- A **vector** is an entity that has magnitude and direction.
  A **scalar** is an entity that only has magnitude (a number).
- If O is the origin, the vector connecting O to A can be written as $\overrightarrow{OA}$ or $\underset{\sim}{a}$. Similarly, $\overrightarrow{OB} = \underset{\sim}{b}$.
- The vector connecting A to B can be written as $\overrightarrow{AB} = \underset{\sim}{b} - \underset{\sim}{a}$.
- If two vectors $\overrightarrow{AB}$ and $\overrightarrow{CD}$ are parallel, then $\overrightarrow{AB} = n\overrightarrow{CD}$, where n is a scalar.
- If two vectors $\overrightarrow{AB}$ and $\overrightarrow{CD}$ are perpendicular, then $\overrightarrow{AB} \cdot \overrightarrow{CD} = 0$.
- The magnitude of a vector $\underset{\sim}{a}$ can be found using $|\underset{\sim}{a}|^2 = \underset{\sim}{a} \cdot \underset{\sim}{a}$, i.e. $|\underset{\sim}{a}|^2 = \underset{\sim}{a} \cdot \underset{\sim}{a}$.
- The dot product of two vectors $\underset{\sim}{a}$ and $\underset{\sim}{b}$ can be found using $\underset{\sim}{a} \cdot \underset{\sim}{b} = |\underset{\sim}{a}||\underset{\sim}{b}|\cos(\theta)$, where θ is the angle between $\underset{\sim}{a}$ and $\underset{\sim}{b}$.
- If two vectors $\overrightarrow{AB}$ and $\overrightarrow{CD}$ are equal, then $\overrightarrow{AB} \parallel \overrightarrow{CD}$ and $|\overrightarrow{AB}| = |\overrightarrow{CD}|$.
- If $\overrightarrow{AB} = n\overrightarrow{BC}$, then A, B and C are collinear.
- If M is the midpoint of $\overrightarrow{AB}$, $\overrightarrow{OM} = \overrightarrow{OA} + \overrightarrow{AM}$
$$= \overrightarrow{OA} + \frac{1}{2}\overrightarrow{AB}$$
$$\underset{\sim}{m} = \underset{\sim}{a} + \frac{1}{2}(\underset{\sim}{b} - \underset{\sim}{a})$$
$$= \frac{1}{2}\underset{\sim}{a} + \frac{1}{2}\underset{\sim}{b}$$

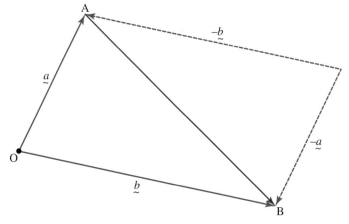

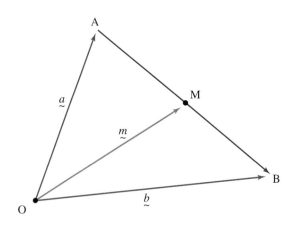

ABCD is a rectangle. Express the following in terms of $\underset{\sim}{a}$, $\underset{\sim}{b}$ and $\underset{\sim}{c}$.

a. $\overrightarrow{AB}$ b. $\overrightarrow{CB}$ c. $\overrightarrow{CD}$ d. $\overrightarrow{AD}$ e. $\underset{\sim}{e}$ f. $\underset{\sim}{f}$

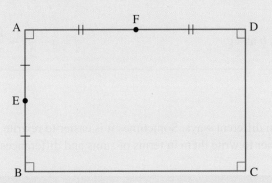

THINK	WRITE
a. $\overrightarrow{AB}$ means the vector connecting A to B. This can be expressed in terms of a and b.	$\overrightarrow{AB} = \underset{\sim}{b} - \underset{\sim}{a}$
b. $\overrightarrow{CB}$ means the vector connecting C to B. This can be expressed in terms of c and b.	$\overrightarrow{CB} = \underset{\sim}{b} - \underset{\sim}{c}$
c. 1. $\overrightarrow{CD}$ means the vector connecting C to D. As ABCD is a rectangle, $\overrightarrow{CD}$ is the reverse of the vector $\overrightarrow{AB}$.	$\overrightarrow{CD} = \overrightarrow{BA}$
2. As $\overrightarrow{CD} = \overrightarrow{BA}$, this is the reverse of $\overrightarrow{AB}$.	$\overrightarrow{CD} = \underset{\sim}{a} - \underset{\sim}{b}$
d. 1. $\overrightarrow{AD}$ means the vector connecting A to D. As ABCD is a rectangle, $\overrightarrow{AD}$ is equal to vector $\overrightarrow{BC}$.	$\overrightarrow{AD} = \overrightarrow{BC}$
2. Vector $\overrightarrow{AD}$ can be expressed in terms of b and c.	$\overrightarrow{AD} = \underset{\sim}{c} - \underset{\sim}{b}$
e. 1. E is the midpoint of $\overrightarrow{AB}$, so vector $\overrightarrow{AE}$ is half the magnitude of $\overrightarrow{AB}$.	$\overrightarrow{AE} = \dfrac{1}{2}\overrightarrow{AB}$
2. $\overrightarrow{AE}$ can be expressed in terms of a and e. $\overrightarrow{AB}$ can be expressed in terms of a and b.	$\overrightarrow{AE} = \dfrac{1}{2}\overrightarrow{AB}$ $\underset{\sim}{e} - \underset{\sim}{a} = \dfrac{1}{2}(\underset{\sim}{b} - \underset{\sim}{a})$ $\underset{\sim}{e} = \dfrac{1}{2}\underset{\sim}{b} - \dfrac{1}{2}\underset{\sim}{a} + \underset{\sim}{a}$
3. Express $\underset{\sim}{e}$ in terms of a and b.	$\underset{\sim}{e} = \dfrac{1}{2}\underset{\sim}{a} + \dfrac{1}{2}\underset{\sim}{b}$
f. 1. F is the midpoint of $\overrightarrow{AD}$, so vector $\overrightarrow{AF}$ is half the magnitude of $\overrightarrow{AD}$.	$\overrightarrow{AF} = \dfrac{1}{2}\overrightarrow{AD}$

2. $\overrightarrow{AF}$ can be expressed in terms of $\underset{\sim}{a}$ and $\underset{\sim}{f}$. $\overrightarrow{AD}$ can be expressed in terms of b and c (from part **d**).

$$\underset{\sim}{f} - \underset{\sim}{a} = \frac{1}{2}\overrightarrow{BC}$$

$$= \frac{1}{2}(\underset{\sim}{c} - \underset{\sim}{b})$$

3. Express $\underset{\sim}{f}$ in terms of $\underset{\sim}{a}$, $\underset{\sim}{b}$ and $\underset{\sim}{c}$.

$$\underset{\sim}{f} = \underset{\sim}{a} - \frac{1}{2}\underset{\sim}{b} + \frac{1}{2}\underset{\sim}{c}$$

Using vectors in proofs

In proofs, vectors can be used in different ways. Sometimes it is easier to rewrite the vectors using position vectors, and sometimes it is easier to write them in terms of sums and differences of other vectors.

WORKED EXAMPLE 8 Using vectors in proofs 1

AOB is a triangle. Point C is drawn on $\overline{AB}$ so that $AC = \frac{2}{3}AB$. The position vectors $\underset{\sim}{a}$, $\underset{\sim}{b}$ and $\underset{\sim}{c}$ are shown on the diagram. Demonstrate that $\underset{\sim}{c} = \frac{1}{3}(2\underset{\sim}{b} + \underset{\sim}{a})$.

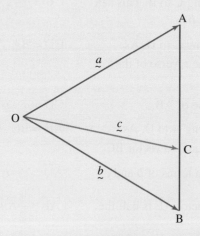

THINK

1. Express $\overrightarrow{AB}$ in terms of $\underset{\sim}{a}$ and $\underset{\sim}{b}$.

2. Express $\overrightarrow{AC}$ in terms of $\underset{\sim}{a}$ and $\underset{\sim}{c}$.

3. Express $\overrightarrow{AC}$ as a multiple of $\overrightarrow{AB}$.

4. Replace $\overrightarrow{AC}$ with $\underset{\sim}{c} - \underset{\sim}{a}$ and $\overrightarrow{AB}$ with $\underset{\sim}{b} - \underset{\sim}{a}$, and rearrange for $\underset{\sim}{c}$.

WRITE

$$\overrightarrow{AB} = \underset{\sim}{b} - \underset{\sim}{a}$$

$$\overrightarrow{AC} = \underset{\sim}{c} - \underset{\sim}{a}$$

$$\overrightarrow{AC} = \frac{2}{3}\overrightarrow{AB}$$

$$\underset{\sim}{c} - \underset{\sim}{a} = \frac{2}{3}(\underset{\sim}{b} - \underset{\sim}{a})$$

$$\underset{\sim}{c} = \frac{2}{3}\underset{\sim}{b} - \frac{2}{3}\underset{\sim}{a} + \underset{\sim}{a}$$

$$= \frac{2}{3}\underset{\sim}{b} + \frac{1}{3}\underset{\sim}{a}$$

$$= \frac{1}{3}(2\underset{\sim}{b} + \underset{\sim}{a})$$

In $\triangle ABC$, $AB = BC$. Using the properties of vectors only, show that the line $\overline{BD}$, drawn so that $\overline{BD}$ is perpendicular to $\overline{AC}$, divides $\overline{AC}$ in half. That is, $|\overrightarrow{AD}| = |\overrightarrow{DC}|$.

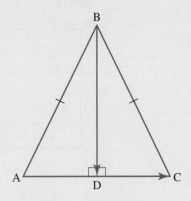

THINK	WRITE
1. Consider $\triangle ABD$	$\overrightarrow{AB} = \overrightarrow{AD} + \overrightarrow{DB}$
2. Use $|\overrightarrow{AB}|^2 = \overrightarrow{AB} \cdot \overrightarrow{AB}$ and determine the dot product of both sides.	$\overrightarrow{AB} \cdot \overrightarrow{AB} = \left(\overrightarrow{AD} + \overrightarrow{DB}\right) \cdot \left(\overrightarrow{AD} + \overrightarrow{DB}\right)$ $|\overrightarrow{AB}|^2 = \overrightarrow{AD} \cdot \overrightarrow{AD} + 2\overrightarrow{AD} \cdot \overrightarrow{DB} + \overrightarrow{DB} \cdot \overrightarrow{DB}$ $= |\overrightarrow{AD}|^2 + |\overrightarrow{DB}|^2 + 2\overrightarrow{AD} \cdot \overrightarrow{DB}$
3. As $\overrightarrow{AD}$ is perpendicular to $\overrightarrow{DB}$, $\overrightarrow{AD} \cdot \overrightarrow{DB} = 0$.	$\overrightarrow{AD} \perp \overrightarrow{DB}$, so $\overrightarrow{AD} \cdot \overrightarrow{DB} = 0$. Therefore, $|\overrightarrow{AB}|^2 = |\overrightarrow{AD}|^2 + |\overrightarrow{DB}|^2$.
4. Repeat using $\triangle CBD$.	$\overrightarrow{BC} = \overrightarrow{BD} + \overrightarrow{DC}$ $\overrightarrow{BC} \cdot \overrightarrow{BC} = \left(\overrightarrow{BD} + \overrightarrow{DC}\right) \cdot \left(\overrightarrow{BD} + \overrightarrow{DC}\right)$ $|\overrightarrow{BC}|^2 = \overrightarrow{BD} \cdot \overrightarrow{BD} + 2\overrightarrow{BD} \cdot \overrightarrow{DC} + \overrightarrow{DC} \cdot \overrightarrow{DC}$ $= |\overrightarrow{BD}|^2 + |\overrightarrow{DC}|^2 + 2\overrightarrow{BD} \cdot \overrightarrow{DC}$ $\overrightarrow{BD} \perp \overrightarrow{DC}$, so $\overrightarrow{BD} \cdot \overrightarrow{DC} = 0$. Therefore, $|\overrightarrow{BC}|^2 = |\overrightarrow{BD}|^2 + |\overrightarrow{DC}|^2$.
5. $\triangle ABC$ is isosceles and $|\overrightarrow{AB}|^2 = |\overrightarrow{BC}|^2$.	As $|\overrightarrow{AB}|^2 = |\overrightarrow{BC}|^2$ and $|\overrightarrow{AB}|^2 = |\overrightarrow{AD}|^2 + |\overrightarrow{BD}|^2$. $|\overrightarrow{AD}| = |\overrightarrow{DC}|$.
6. Write concluding statements.	$\overrightarrow{AC} = \overrightarrow{AD} + \overrightarrow{DC}$ and $|\overrightarrow{AD}| = |\overrightarrow{DC}|$. Therefore, $\overrightarrow{AD} = \dfrac{1}{2}\overrightarrow{AC}$, so D is the midpoint of $\overline{AC}$.

A useful method to demonstrate that lines bisect each other is to let the bisector of one line be one point and the bisector of the other line be a different point, then use position vectors to show that the points are in the same position.

Use a vector method to demonstrate that the diagonals of a rectangle bisect each other.

THINK	WRITE
1. Draw a rectangle ABCD.	

2. Observe that $\overrightarrow{AB} = \overrightarrow{DC}$ and write this in terms of the position vectors.

$$\overrightarrow{AB} = \overrightarrow{DC}$$
$$\underset{\sim}{b} - \underset{\sim}{a} = \underset{\sim}{c} - \underset{\sim}{d}$$
$$\underset{\sim}{b} + \underset{\sim}{d} = \underset{\sim}{a} + \underset{\sim}{c}$$

3. Let E be the midpoint of $\overline{AC}$.

Let E be the midpoint of $\overline{AC}$.

4. Write $\overrightarrow{AC}$ in terms of $\underset{\sim}{a}$ and $\underset{\sim}{c}$.

$$\overrightarrow{AC} = \underset{\sim}{c} - \underset{\sim}{a}$$

5. As E is the midpoint of $\overline{AC}$, $\overrightarrow{AE} = \frac{1}{2}\overrightarrow{AC}$.

$$\overrightarrow{AE} = \frac{1}{2}\overrightarrow{AC}$$

6. Replace with position vectors and rearrange to have $\underset{\sim}{e}$ as the subject.

$$\underset{\sim}{e} - \underset{\sim}{a} = \frac{1}{2}(\underset{\sim}{c} - \underset{\sim}{a})$$
$$\underset{\sim}{e} = \frac{1}{2}\underset{\sim}{c} - \frac{1}{2}\underset{\sim}{a} + \underset{\sim}{a}$$
$$\underset{\sim}{e} = \frac{1}{2}\underset{\sim}{c} + \frac{1}{2}\underset{\sim}{a}$$
$$= \frac{1}{2}(\underset{\sim}{a} + \underset{\sim}{c})$$

7. Let F be the midpoint of $\overline{BD}$. Repeat steps **3–6** to determine an expression for f.

Let F be the midpoint of $\overline{BD}$.
$$\overrightarrow{BD} = \underset{\sim}{d} - \underset{\sim}{b}$$
$$\overrightarrow{BF} = \frac{1}{2}\overrightarrow{BD}$$
$$\underset{\sim}{f} - \underset{\sim}{b} = \frac{1}{2}(\underset{\sim}{d} - \underset{\sim}{b})$$
$$\underset{\sim}{f} = \frac{1}{2}\underset{\sim}{d} + \frac{1}{2}\underset{\sim}{b}$$
$$= \frac{1}{2}(\underset{\sim}{b} + \underset{\sim}{d})$$

8. We observed in step **2** that $\underset{\sim}{b} + \underset{\sim}{d} = \underset{\sim}{a} + \underset{\sim}{c}$. Use this in the expression for f.

As $\underset{\sim}{b} + \underset{\sim}{d} = \underset{\sim}{a} + \underset{\sim}{c}$, $\underset{\sim}{f} = \frac{1}{2}(\underset{\sim}{a} + \underset{\sim}{c})$.

9. The expressions for $\underset{\sim}{e}$ and f are equal, so E and F are the same point. Write the concluding statement.

As $\underset{\sim}{e} = \frac{1}{2}(\underset{\sim}{a} + \underset{\sim}{c})$ and $\underset{\sim}{f} = \frac{1}{2}(\underset{\sim}{a} + \underset{\sim}{c})$, E and F are the same point, and they are the bisectors of the diagonals. Therefore, the diagonals of a rectangle bisect each other.

8.6.2 Proof that the diagonals of a parallelogram meet at right angles if and only if it is a rhombus

$$\underset{\sim}{a} \cdot \underset{\sim}{b} = |\underset{\sim}{a}||\underset{\sim}{b}| \cos(\theta)$$

$$\text{Diagonal } 1 = \underset{\sim}{a} + \underset{\sim}{b}$$

$$\text{Diagonal } 2 = \underset{\sim}{b} - \underset{\sim}{a}$$

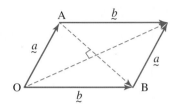

If the diagonals are perpendicular, the dot product will equal zero $\left(\cos \left(\pm \dfrac{\pi}{2} \right) = 0 \right)$.

That is

$$(\underset{\sim}{a} + \underset{\sim}{b}) \cdot (\underset{\sim}{b} - \underset{\sim}{a}) = 0$$

The dot product is distributive so

$$(\underset{\sim}{a} + \underset{\sim}{b}) \cdot (\underset{\sim}{b} - \underset{\sim}{a}) = \underset{\sim}{a}(\underset{\sim}{b} - \underset{\sim}{a}) + \underset{\sim}{b}(\underset{\sim}{b} - \underset{\sim}{a}) = \underset{\sim}{a} \cdot \underset{\sim}{b} - \underset{\sim}{a} \cdot \underset{\sim}{a} + \underset{\sim}{b} \cdot \underset{\sim}{b} - \underset{\sim}{b} \cdot \underset{\sim}{a}$$

The dot product is commutative so $\underset{\sim}{a} \cdot \underset{\sim}{b} - \underset{\sim}{b} \cdot \underset{\sim}{a} = 0$

Thus $(\underset{\sim}{a} + \underset{\sim}{b}) \cdot (\underset{\sim}{b} - \underset{\sim}{a}) = \underset{\sim}{b} \cdot \underset{\sim}{b} - \underset{\sim}{a} \cdot \underset{\sim}{a}$

$\underset{\sim}{b} \cdot \underset{\sim}{b} = |\underset{\sim}{b}|^2$ and $\underset{\sim}{a} \cdot \underset{\sim}{a} = |\underset{\sim}{a}|^2$

Thus $(\underset{\sim}{a} + \underset{\sim}{b}) \cdot (\underset{\sim}{b} - \underset{\sim}{a}) = |\underset{\sim}{b}|^2 - |\underset{\sim}{a}|^2$

Therefore, if the diagonals of a parallelogram meet at right angles, then

$$(\underset{\sim}{a} + \underset{\sim}{b}) \cdot (\underset{\sim}{b} - \underset{\sim}{a}) = |\underset{\sim}{b}|^2 - |\underset{\sim}{a}|^2 = 0$$
$$\therefore |\underset{\sim}{a}| = |\underset{\sim}{b}|$$

The adjacent sides of the parallelogram are equal in length, the parallelogram is a rhombus. If the adjacent sides of the parallelogram are equal in length, then

$$|\underset{\sim}{a}| = |\underset{\sim}{b}|$$
$$\therefore (\underset{\sim}{a} + \underset{\sim}{b}) \cdot (\underset{\sim}{b} - \underset{\sim}{a}) = 0$$

its diagonals intersect at right angles.

8.6.3 Proof that the midpoints of the sides of a quadrilateral join to form a parallelogram

Let vectors $\underset{\sim}{a}, \underset{\sim}{b}, \underset{\sim}{c}$ and $\underset{\sim}{d}$ be the four sides of the quadrilateral, oriented as shown in the diagram.

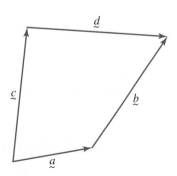

The vector sum of the four sides of the quadrilateral is $\underset{\sim}{0}$, that means $\underset{\sim}{a} + \underset{\sim}{b} - \underset{\sim}{d} - \underset{\sim}{c} = \underset{\sim}{0}$.

Thus

$\underset{\sim}{a} + \underset{\sim}{b} = \underset{\sim}{c} + \underset{\sim}{d}$

Now, since $\underset{\sim}{a} + \underset{\sim}{b} = \underset{\sim}{c} + \underset{\sim}{d}$, it follows that

$\underset{\sim}{c} - \underset{\sim}{a} = \underset{\sim}{b} - \underset{\sim}{d}$

$\therefore \dfrac{1}{2}(\underset{\sim}{c} - \underset{\sim}{a}) = \dfrac{1}{2}(\underset{\sim}{b} - \underset{\sim}{d})$

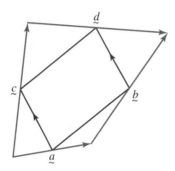

Now, the vector from the midpoint of side $\underset{\sim}{a}$ to the midpoint of side $\underset{\sim}{c}$ is $\dfrac{1}{2}(\underset{\sim}{c} - \underset{\sim}{a})$.

Similarly, the opposite vector, joining the midpoints of the other two sides is $\dfrac{1}{2}(\underset{\sim}{b} - \underset{\sim}{d})$.

Since $\dfrac{1}{2}(\underset{\sim}{c} - \underset{\sim}{a}) = \dfrac{1}{2}(\underset{\sim}{b} - \underset{\sim}{d})$, then $(\underset{\sim}{c} - \underset{\sim}{a}) = (\underset{\sim}{b} - \underset{\sim}{d})$:

the vector from the midpoint of side $\underset{\sim}{a}$ to the midpoint of side $\underset{\sim}{c}$ is equal to the vector from the midpoint of side $\underset{\sim}{b}$ to the midpoint of side $\underset{\sim}{d}$ (equal lengths and parallel).

Similarly, we can repeat those steps to show that the vector from the midpoint of side $\underset{\sim}{a}$ to the midpoint of side $\underset{\sim}{b}$ is equal to the vector from the midpoint of side $\underset{\sim}{c}$ to the midpoint of side $\underset{\sim}{d}$ (equal lengths and parallel).

Therefore, the quadrilateral joining the midpoints of the sides of the original quadrilateral has opposite sides that are parallel and equal: it is a parallelogram.

8.6.4 Proof that the sum of the squares of the lengths of a parallelogram's diagonals is equal to the sum of the squares of the lengths of the sides

The diagonals of this parallelogram are $(\underset{\sim}{a} + \underset{\sim}{b})$ and $(\underset{\sim}{a} - \underset{\sim}{b})$.

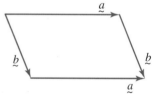

The sum of the squares of the diagonals $= (\underset{\sim}{a} + \underset{\sim}{b})^2 + (\underset{\sim}{a} - \underset{\sim}{b})^2$

$\qquad = |\underset{\sim}{a}|^2 + 2\underset{\sim}{a}\underset{\sim}{b} + |\underset{\sim}{b}|^2 + |\underset{\sim}{a}|^2 - 2\underset{\sim}{a}\underset{\sim}{b} + |\underset{\sim}{b}|^2$

$\qquad = 2\left(|\underset{\sim}{a}|^2 + |\underset{\sim}{b}|^2\right)$

$\qquad = \left(|\underset{\sim}{a}|^2 + |\underset{\sim}{b}|^2\right) + \left(|\underset{\sim}{a}|^2 + |\underset{\sim}{b}|^2\right)$

$\qquad = $ the sum of the squares of the sides.

$\qquad$ QED

8.6.5 Proof that an angle in a semicircle is a right-angle

Consider the following triangle ABC such that AB is a diameter of the circle with centre O.

Let's prove that if AB is a diameter, then $\angle ABC = 90°$.

$$\overrightarrow{BA} = \overrightarrow{BO} + \overrightarrow{OA}$$

and

$$\overrightarrow{BC} = \overrightarrow{BO} + \overrightarrow{OC} = \overrightarrow{BO} - \overrightarrow{OA}$$

$$\overrightarrow{BA} \cdot \overrightarrow{BC} = \left(\overrightarrow{BO} + \overrightarrow{OA}\right) \cdot \left(\overrightarrow{BO} - \overrightarrow{OA}\right) = |\overrightarrow{BO}|^2 - |\overrightarrow{OA}|^2$$

since $|\overrightarrow{BO}| = |\overrightarrow{OA}| = \text{radius}$, then

$$|\overrightarrow{BO}|^2 - |\overrightarrow{OA}|^2 = 0$$

Thus, $\overrightarrow{BA} \cdot \overrightarrow{BC} = 0$

$\therefore \angle ABC = 90°$

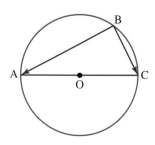

Exercise 8.6 Geometric proofs using vectors **learn**on

| **8.6 Exercise** | **8.6 Exam questions** on |

Simple familiar	Complex familiar	Complex unfamiliar
1, 2, 3	4, 5, 6, 7, 8, 9, 10, 11, 12, 13	14, 15, 16

These questions are even better in jacPLUS!
- Receive immediate feedback
- Access sample responses
- Track results and progress

Find all this and MORE in jacPLUS ▶

Simple familiar

1. **WE7** ABCD is a parallelogram. Express the following in terms of $\underset{\sim}{b}$ and $\underset{\sim}{c}$.

   a. $\overrightarrow{AB}$

   b. $\overrightarrow{CD}$

   c. $\overrightarrow{AD}$

   d. $\underset{\sim}{f}$

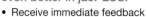
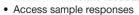

2. OPQRSTUV is a cube, where O is the origin. Express the following in terms of $\underset{\sim}{s}$, $\underset{\sim}{p}$ and $\underset{\sim}{z}$.

   a. $\overrightarrow{OU}$

   b. $\overrightarrow{OV}$

   c. $\overrightarrow{TP}$

   d. $\overrightarrow{RV}$

   e. $\overrightarrow{PS}$

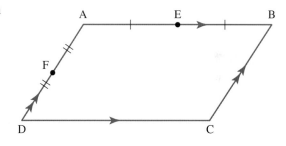
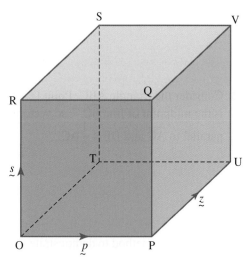

3. **WE8** AOB is a triangle. Point C is drawn on $\overline{AB}$ so that $AC = \frac{1}{4}AB$. The position vectors $\underset{\sim}{a}$, $\underset{\sim}{b}$ and $\underset{\sim}{c}$ are shown on the diagram. Demonstrate that $\underset{\sim}{c} = \frac{1}{4}(3\underset{\sim}{a} + \underset{\sim}{b})$.

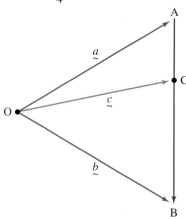

Complex familiar

4. In triangle ABC, O is the midpoint of $\overline{CB}$ and $\overline{OA}$ is perpendicular to $\overline{CB}$.

   a. If $\overrightarrow{OA} = \underset{\sim}{a}$ and $\overrightarrow{OB} = \underset{\sim}{b}$, determine in terms of $\underset{\sim}{a}$ and $\underset{\sim}{b}$ only:

      i. $\overrightarrow{OC}$ ii. $\overrightarrow{BA}$

      iii. $\overrightarrow{CA}$ iv. the dot product $\overrightarrow{BA} \cdot \overrightarrow{BA}$.

      v. the dot product $\overrightarrow{CA} \cdot \overrightarrow{CA}$.

   b. Hence, what can you say about AC and AB?

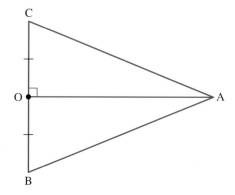

5. **WE9** ABC is an equilateral triangle. Using the properties of vectors only, show that the line $\overline{BD}$, drawn so that $\overline{BD}$ is perpendicular to $\overline{AC}$, divides $\overline{AC}$ in half. That is, $|\overrightarrow{AD}| = |\overrightarrow{DC}|$.

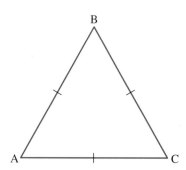

6. Consider the triangle ABC. Point D is the midpoint of line $\overline{AB}$ and E is the midpoint of line $\overline{BC}$. Use vectors to demonstrate that line $\overline{DE}$ is parallel to $\overline{AC}$ and $DE = \frac{1}{2}AC$.

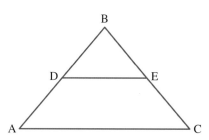

7. Use a vector method to demonstrate Pythagoras' theorem.

8. **WE10** Use a vector method to demonstrate that the diagonals of a parallelogram bisect each other.

9. Use vectors to demonstrate that the angle subtended by a diameter of a circle is a right angle.

10. Use vectors to demonstrate that joining the midpoints of the sides of a parallelogram results in a parallelogram.

11. Use vectors to demonstrate that joining the midpoints of the sides of a quadrilateral results in a parallelogram.

12. Prove that the diagonals of a rhombus intersect at 90°.

13. Prove that the diagonals of a parallelogram meet at 90° if and only if it is a rhombus.

Complex unfamiliar

14. Consider any two major diagonals of a cube. Use a vector method to prove that:
    a. the diagonals bisect each other
    b. the acute angle between the diagonals is 70.53°.

15. The median of a triangle is the line drawn from one vertex to the midpoint of the opposite side. Demonstrate that:

    a. the medians of a triangle are concurrent (pass through the same point)

    b. the distance from the vertex to the point of intersection is $\frac{2}{3}$ of the median length.

16. In any triangle ABC, the points D, E and F divide the line segments $\overline{BC}$, $\overline{CA}$ and $\overline{AB}$ respectively in the ratio 1 : 2.

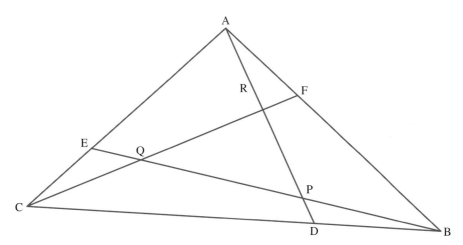

    a. Demonstrate that the line joining A and P is $\frac{6}{7}$ of the line joining A to D.

    b. Demonstrate that the line joining R and P is $\frac{3}{7}$ of the vector joining A to D.

    c. Determine the ratio of the area of ΔQRP to the area of ΔABC.
    d. If $\overline{BC}$, $\overline{CA}$ and $\overline{AB}$ are divided in the ratio 1 : n, determine the ratio of the area of ΔQRP to the area of ΔABC.

Fully worked solutions for this chapter are available online.

LESSON
8.7 Review

8.7.1 Summary

8.7 Exercise

learn on

8.7 Exercise	**8.7 Exam questions** on

Simple familiar	Complex familiar	Complex unfamiliar
1, 2, 3, 4, 5, 6, 7, 8, 9, 10, 11, 12	13, 14, 15, 16	17, 18, 19, 20

Simple familiar

1. If $\angle ABD = 32°$, determine the sizes of the following angles.

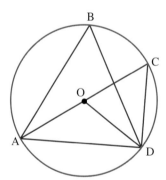

  a. $\angle AOD$ **b.** $\angle ACD$ **c.** $\angle COD$ **d.** $\angle CDA$

2. If $OB = 9$, determine the following.

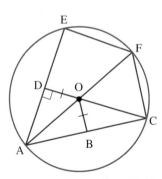

  a. $\angle AEF$ **b.** $\overline{EF}$

3. In the diagram, O and C are the centres of the circles.

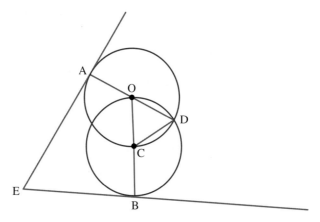

   a. Determine ∠COD.
   b. Determine ∠AEB.
   c. What type of quadrilateral is AOBE?
   d. Determine ∠OBD.

4. Prove that $\overline{OB}\|\overline{CD}$.

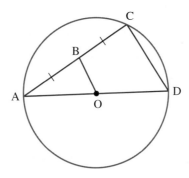

5. Is $\overline{AB}$ a diameter? Explain your reasoning.

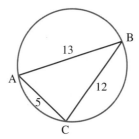

6. Establish the relationship between x and y.

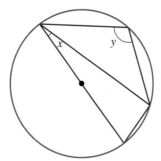

7. A smaller circle inside a larger circle touches the larger circle at A and passes through the centre of the larger circle. If line ABC passes through the smaller and larger circles at B and C respectively, prove that $\overline{AB} \cong \overline{BC}$.

8. Prove that a line from the centre bisects a chord if and only if it is perpendicular to the chord.

9. If $\triangle ABC$ is equilateral, prove that $\triangle PQR$ is also equilateral.

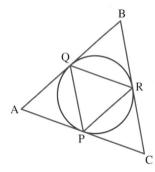

10. Prove that the angle at the centre of a circle is twice the angle at the circumference.

11. A circle is drawn through the vertices of a square ABCD. If the square has a side of 5 cm, calculate the area inside the circle but outside the square.

12. The diameter $\overline{AB}$ of a circle is extended through B to T. A tangent is drawn from T, touching the circle at C. $\overline{AC}$ is extended through C to D so that $\angle ATD = 90°$. Prove that $\overline{CT} \cong \overline{TD}$.

Complex familiar

13. Determine the magnitude of the angle at the circumference that is subtended by a chord of the same length as the radius of the circle.

14. Use a vector method to prove that a triangle can be constructed from the medians of a triangle without changing the lengths or slopes of the medians.

15. The point theorem says that if a line from point P cuts a circle at A and D and another line from P cuts the circle at C and B, then $PA \cdot PD = PC \cdot PB$.

    a Prove the point theorem if P is inside the circle.
    b Prove the point theorem if P is outside the circle.

16. If the four sides of a parallelogram touch a circle, prove that the parallelogram is a rhombus.

Complex unfamiliar

17. A circle of radius r is drawn so that its centre is on the hypotenuse of a right-angled triangle. If the circle touches the other sides of the triangle (lengths x and y), prove that $\dfrac{1}{r} = \dfrac{1}{x} + \dfrac{1}{y}$.

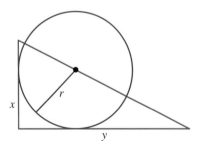

18. O is the centre of the circle shown, $\overline{OD} \cong \overline{CD}$ and $\overline{DB}$ is perpendicular to $\overline{AC}$. Evaluate $\angle CAB$.

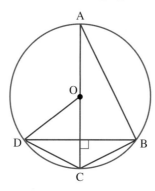

19. ABDC is a rhombus where A, B and D are points on a circle with centre C. If the area of the rhombus is $72\sqrt{3}$ units, what is the radius of the circle?

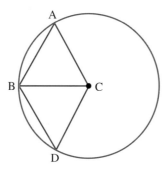

20. A tangent from a point P to a circle with centre O and diameter $\overline{AB}$ touches the circle at Q. If $\overline{QA} \parallel \overline{PO}$, demonstrate that $\overline{PB}$ is also a tangent.

Fully worked solutions for this chapter are available online.

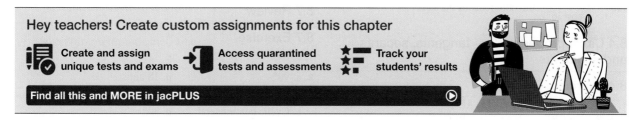

Answers

Chapter 8 Circle and geometric proofs

8.2 Congruent triangles and angle relationships

8.2 Exercises

1–10. Sample responses can be found in the worked solutions in the online resources.

8.3 Circle properties 1 — angles in a circle and chords

8.3 Exercise

1. a. A, B, C, D, E, F
   b. $\overline{OA}$, $\overline{OD}$, $\overline{OE}$
   c. Major arcs include $\overparen{ABE}$, $\overparen{ABF}$, $\overparen{CDA}$.
   d. $\overline{AC}$, $\overline{AD}$, $\overline{AE}$, $\overline{CE}$, $\overline{ED}$
   e. $\angle AED$
   f. $\angle AOE$, $\angle EOD$
   g. $\angle ACE$, $\angle ADE$
2. a. $a = 90°$, $b = 45°$ b. $a = 32°$, $b = 32°$
3–4. Sample responses can be found in the worked solutions in the online resources.
5. a. $\angle OBC = 90°$ b. $EC = 30$
6. a. 7 units   b. 4 units
7–11. Sample responses can be found in the worked solutions in the online resources.
12. 2.55 cm
13. $\dfrac{r_1}{r_2} = \dfrac{1}{\cos(36°)}$; the ratio remains the same.

8.4 Circle properties 2 — tangents, secants and segments

8.4 Exercise

1. a. $\overline{AB}$ b. $\overline{AE}$
   c. Region CBE   d. $\angle BEC$
2. 35
3–6. Sample responses can be found in the worked solutions in the online resources.
7. a. Right-angled triangle
   b. $\overline{OA}$
   c. They are radii. OB > OD, OB > OA
   d. A sample response can be found in the worked solutions in the online resources.
8–11. Sample responses can be found in the worked solutions in the online resources.

8.5 Circle properties 3 — cyclic quadrilaterals

8.5 Exercise

1. $a = 81°$, $b = 99°$
2. $a = 90°$, $b = 105°$

3–10. Sample responses can be found in the worked solutions in the online resources.
11. $\angle A + \angle C = \angle B$
    A sample response for the proof can be found in the worked solutions in the online resources.
12. Rhombus. A sample response can be found in the worked solutions in the online resources.

8.6 Geometric proofs using vectors

8.6 Exercise

1. a. $\underset{\sim}{b} - \underset{\sim}{a}$ b. $\underset{\sim}{a} - \underset{\sim}{b}$
   c. $\underset{\sim}{c} - \underset{\sim}{b}$ d. $\underset{\sim}{a} - \dfrac{1}{2}\underset{\sim}{b} + \dfrac{1}{2}\underset{\sim}{c}$
2. a. $\overrightarrow{OU} = \underset{\sim}{p} + \underset{\sim}{z}$
   b. $\overrightarrow{OV} = \underset{\sim}{p} + \underset{\sim}{z} + \underset{\sim}{s}$
   c. $\overrightarrow{TP} = \underset{\sim}{a} - \underset{\sim}{z}$
   d. $\overrightarrow{RV} = \underset{\sim}{p} + \underset{\sim}{z}$
   e. $\overrightarrow{PS} = \underset{\sim}{s} + \underset{\sim}{z} - \underset{\sim}{p}$
3, 4. Sample responses can be found in the worked solutions in the online resources.
5. a. i. $-\underset{\sim}{b}$ ii. $\underset{\sim}{a} - \underset{\sim}{b}$
      iii. $\underset{\sim}{a} + \underset{\sim}{b}$ iv. $\underset{\sim}{a} \cdot \underset{\sim}{a} + \underset{\sim}{b} \cdot \underset{\sim}{b}$
      v. $\underset{\sim}{a} \cdot \underset{\sim}{a} + \underset{\sim}{b} \cdot \underset{\sim}{b}$
   b. AC = AB
6–16. Sample responses can be found in the worked solutions in the online resources.
17. a, b. Sample responses can be found in the worked solutions in the online resources.
    c. 1 : 7
    d. $(n-1)^2 : n^2 + n + 1$

8.7 Review

8.7 Exercise

1. a. $64°$ b. $32°$ c. $116°$ d. $90°$
2. a. $90°$ b. 18 units
3. a. $60°$ b. $60°$
   c. Cyclic quadrilateral   d. $30°$
4. A sample response can be found in the worked solutions in the online resources.
5. Yes; a sample response can be found in the worked solutions in the online resources.
6. $y = 90° - x$
7–10. Sample responses can be found in the worked solutions in the online resources.
11. $\dfrac{50\pi}{4} - 25 = 14.3 \text{ cm}^2$ (to one decimal place)
12. A sample response can be found in the worked solutions in the online resources.
13. $30°$
14–17. Sample responses can be found in the worked solutions in the online resources.
18. $30°$
19. 12 units
20. A sample response can be found in the worked solutions in the online resources.

9 Trigonometry and functions

LESSON SEQUENCE

Fully worked solutions for this chapter are available online.

EXAM PREPARATION

Access exam-style questions in every lesson, available online.

on Resources

 Solutions   Solutions — Chapter 9 (sol-0401)

 Exam questions   Exam question booklet — Chapter 9 (eqb-0287)

 Digital documents Learning matrix — Chapter 9 (doc-41932)
                                              Chapter summary — Chapter 9 (doc-41933)

LESSON
9.1 Overview

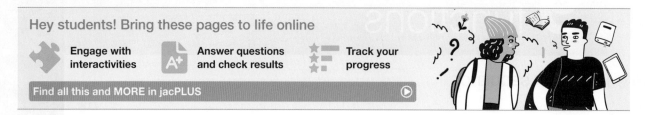
9.1.1 Introduction

Functions are used to model relationships between variables across a variety of disciplines, from forensic science and computer programming to climate science. For example, climate scientists use data about sea level, temperature, carbon dioxide and Arctic ice extent to define functions that allow them to create the complex models required to predict the effects of global climate change.

If we understand the relationship between two variables, it is possible to model their relationship. Knowing what a function looks like helps us understand why functions behave the way that they do. In this chapter you will explore the graphing of modulus functions, reciprocal functions and trigonometric functions.

Trigonometry is the branch of mathematics that deals with triangles and the relationship between the angles and sides of a triangle. Derived from *trigonon*, the ancient Greek word for triangle, trigonometry was originally devised around 1700 BCE as a tool for astronomers.

Hipparchus, the Greek astronomer and mathematician, is known as the father of trigonometry for compiling the first trigonometric tables around 150 BCE. From the 4th century onwards Indian mathematicians made significant contributions to trigonometry, with Bhaskara the First and Brahmagupta developing formulas for determining sine values. Islamic mathematicians built on the work of the Greeks and Indians, and by the 10th century were using all six trigonometric functions, with mathematician Abū al-Wafā' al-Būzjānī developing sine tables to eight decimal places. Islamic mathematicians were the first to use triangulation to determine distances between points.

The Russian mathematician Pafnuty Chebyshev (1821–1894) is better known for his work in the fields of probability, statistics, number theory and differential equations, but he also devised recurrence relations for trigonometric multiple angles that we explore later in this chapter.

9.1.2 Syllabus links

Lesson	Lesson title	Syllabus links
9.2	Sketching graphs	○ Use and apply the notation $\|x\|$ for the absolute value for the real number x and the graph of $y=\|x\|$.
		○ Understand and use the relationship between the graph of $y=f(x)$ and the graphs of $y=\dfrac{1}{f(x)}, y=\|f(x)\|$ and $y=f(\|x\|)$.
9.3	Reciprocal trigonometric functions	○ Define and use the reciprocal trigonometric functions to determine their simplified exact values and sketch their graphs.
9.4	Pythagorean identities	○ Prove and apply the Pythagorean identities. • $\sin^2(A)+\cos^2(A)=1$ • $\tan^2(A)+1=\sec^2(A)$ • $\cot^2(A)+1=\operatorname{cosec}^2(A)$
9.5	Compound angle identities	○ Prove and apply the angle sum, difference and double-angle identities for sines and cosines. • $\sin(A+B)=\sin(A)\cos(B)+\cos(A)\sin(B)$ • $\sin(A-B)=\sin(A)\cos(B)-\cos(A)\sin(B)$ • $\cos(A+B)=\cos(A)\cos(B)-\sin(A)\sin(B)$ • $\cos(A-B)=\cos(A)\cos(B)+\sin(A)\sin(B)$ • $\sin(2A)=2\sin(A)\cos(A)$ • $\cos(2A)=\cos^2(A)-\sin^2(A)$ $\qquad =1-2\sin^2(A)$ $\qquad =2\cos^2(A)-1$
9.6	Product identities	○ Prove and apply the identities for products of sines and cosines expressed as sums and differences. • $\sin(A)\sin(B)=\dfrac{1}{2}(\cos(A-B)-\cos(A+B))$ • $\cos(A)\cos(B)=\dfrac{1}{2}(\cos(A-B)+\cos(A+B))$ • $\sin(A)\cos(B)=\dfrac{1}{2}(\sin(A+B)+\sin(A-B))$ • $\sin(A)\cos(B)=\dfrac{1}{2}(\sin(A+B)+\sin(A-B))$
9.7	Applications of trigonometric identities	○ Convert sums $a\cos(x)+b\sin(x)$ to $R\cos(x\pm a)$ or $R\sin(x\pm a)$ and apply these to sketch graphs.
		○ Model and solve problems that involve equations of the form $a\cos(x)+b\sin(x)=c$.
		○ Prove and apply multi-angle trigonometric identities up to angles of $4x$ using the identities listed above, e.g. $\cos(4x)=8\cos^4(x)-8\cos^2(x)+1$ and $\operatorname{cosec}(2x)-\cot(2x)=\tan(x)$.

Source: Specialist Mathematics Senior Syllabus 2024 © State of Queensland (QCAA) 2024; licensed under CC BY 4.0

LESSON
9.2 Sketching graphs

9.2.1 Review of relations and function

As you have already learnt in Mathematical Methods, a mathematical **relation** is a set of ordered pairs. The pairs may be listed, for example $A = \{(-2, 4), (1, 5), (3, 4)\}$, or described by a rule such as $B = \{(x, y) : y = 2x\}$ or $C = \{(x, y) : y \leq 2x\}$. They may also be presented as a graph on coordinate axes.

For a set of ordered pairs (x, y), the **domain** is the set of all the x-values of the ordered pairs and the **range** is the set of all the y-values of the ordered pairs.

For $A = \{(-2, 4), (1, 5), (3, 4)\}$, the domain is $\{-2, 1, 3\}$ and the range is $\{4, 5\}$. For both $B = \{(x, y) : y = 2x\}$ and $C = \{(x, y) : y \leq 2x\}$, the domain is $\mathbb{R}$ and the range is $\mathbb{R}$, where $\mathbb{R}$ is the set of real numbers.

The graph of any polynomial relation normally has a domain of $\mathbb{R}$. However, restrictions can be placed on the values of the variables. If there are restrictions, the relation is defined on a **restricted domain**.

A **function** is a particular type of relation: in the set of ordered pairs, every x-value is paired to a unique y-value.

Functions can be recognised from their graphs by the **vertical line test**, as demonstrated in the graphs shown. If a vertical line can be drawn that cuts the graph at more than one place, the graph is not that of a function.

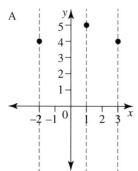

A vertical line cuts once.
$A = \{(-2, 4), (1, 5), (3, 4)\}$
is a function.

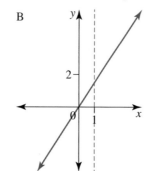

A vertical line cuts once.
$B = \{(x, y) : y = 2x\}$ is a function.

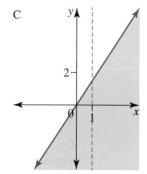

A vertical line
cuts many times.
$C = \{(x, y) : y \leq 2x\}$
is a relation.

9.2.2 The modulus function

Sometimes only the **magnitude** of a number (that is, the size of a number) is required. As you would remember from the vector chapter, this is referred to as the **absolute value** or **modulus** of a number.

Consider the numbers -3 and 3. They are both 3 units from the origin, as demonstrated in the line graph.

If x is a number, the absolute value of x is expressed as $|x|$. More formally, $|x| = \sqrt{x^2}$.

This means that when solving equations with magnitude involved, we need to consider the positive and negative answers carefully.

Consider 3 and -3 again. If $x = 3$, then $x^2 = 9$ and $\sqrt{9} = 3$. Additionally, if $x = -3$, then $x^2 = 9$ and $\sqrt{9} = 3$.

WORKED EXAMPLE 1 Solving simple equations using the modulus function (1)

Solve $|4x| = 16$.

THINK	WRITE
1. $|16| = 16$ and $|-16| = 16$, so there are 2 possible values for $4x$: 16 and -16.	$4x = 16$ or $4x = -16$
2. Solve the equations.	$x = 4$ or $x = -4$

WORKED EXAMPLE 2 Solving simple equations using the modulus function (2)

Solve $|x - 1| = 5$.

THINK	WRITE
1. $|5| = 5$ and $|-5| = 5$, so there are two possible values for $|x - 1|$: 5 and -5.	$x - 1 = 5$
	$x - 1 = -5$
2. Solve the equations	$x = 6$ or $x = -4$.

9.2.3 Sketching $y = |x|$

Consider the graph of $y = x$. When $x \geq 0$,

$y \geq 0$, and when $x < 0, y < 0$.

x	-2	-1	0	1	2
y	-2	-1	0	1	2

Now consider the graph $y = |x|$.

When $x \geq 0$, $y = x$ and $y = |x|$ are the same, but for $x < 0$, $y = |x|$ and $y = -x$ are the same.

x	-2	-1	0	1	2
$y = |x|$	2	1	0	1	2

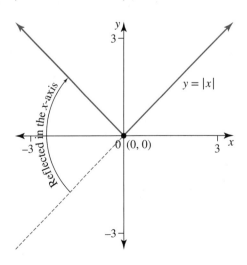

9.2.4 Graphing $y = |f(x)|$ from $y = f(x)$

Consider the graph $y = x + 1$.

The graph crosses the x-axis when $y = 0$, that is, when $x = -1$.
For $x \geq -1$, $y \geq 0$; for $x < -1$, $y < 0$.

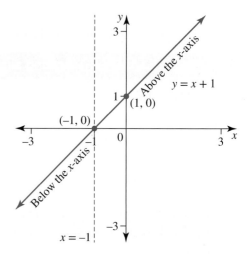

Now consider the graph $y = |x + 1|$.

For $x \geq -1$, $|x + 1| = x + 1$, but for $x < -1$, $|x + 1| = -(x + 1)$.

This results in a reflection in the x-axis at $x = -1$.

When graphing $y = |f(x)|$, it is useful to consider the original graph of $y = f(x)$ to determine significant features, then reflecting in the x-axis.

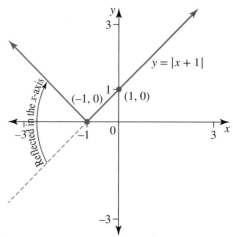

WORKED EXAMPLE 3 Sketching the modulus function for $y = |f(x)|$

Use the graph of $y = 2x + 1$ to do the following.
a. Sketch $y = |2x + 1|$ for the domain $-3 \leq x < 3$.
b. State the range of the function.

THINK

a. 1. Sketch $y = 2x + 1$ for $-3 \leq x < 3$.
   The gradient is 2 and the y-intercept is 1.

WRITE

a.

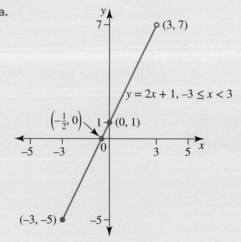

---

2. Identify where $2x + 1 \geq 0$.

$2x + 1 \geq 0$

$x \geq -\dfrac{1}{2}$

3. Sketch $y = |2x + 1|$ for $-3 \leq x < 3$.
Given the domain is restricted to $-3 \leq x < 3$,
the graph has end points — a solid point at
$x = -3$, as the restricted domain states $x \geq -3$,
and a hollow point at $x = 3$, as the restricted
domain states $x < 3$. The graph is reflected in
the x-axis at the x-intercept.

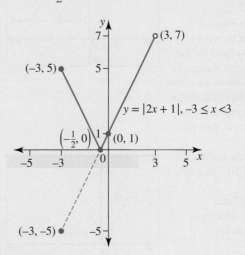

b. To determine the range, consider the end
points of the restricted domain. The minimum
value is $y = 0$. The larger value between 5
and 7 is 7; however, as it is a hollow point,
it cannot be included. Hence, the range is
$0 \leq y < 7$ or $[0, 7)$.

b. When $x = -3$,
$y = |2x + 1|$
$y = |2 \times -3 + 1|$
$y = 5$
When $x = 3$,
$y = |2x + 1|$
$y = |2 \times 3 + 1|$
$y = 7$
Hence, the range is $[0, 7)$.

TI | THINK

WRITE

a. 1. On a Graphs page,
complete the entry line
for function 1 as:
$f1(x) = \text{abs}(2x + 1)|$
$-3 \leq x \leq 3$
then press ENTER.
Note: The restricted
domain $-3 \leq x < 3$
doesn't include the point
at $x = 3$. However, on the
calculator, the point at
$x = 3$ needs to be
included so that the
calculator can find the
coordinates of this point.

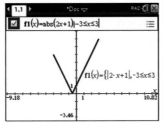

CASIO | THINK

WRITE

a. 1. On a Graph screen,
press SHIFT 4, then
use the down arrow to
scroll down to Abs, then
press EXE. Complete
the entry line for Y1 as:
Y1 = $|2X + 1|$, $[-3, 3]$
then press EXE.
Select DRAW by
pressing F6.

Note: The restricted
domain $-3 \leq x < 3$
doesn't include the
point at $x = 3$. However,
on the calculator, the
point at $x = 3$ needs to
be included so that the
calculator can find the
coordinates of this point.

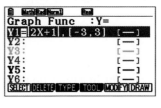

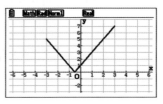

2. To find the coordinates of the vertex, press MENU, then select:
   6: Analyze Graph
   1: Zero.
   Move the cursor to the left of the vertex when prompted for the lower bound, then press ENTER. Move the cursor to the right of the vertex when prompted for the upper bound, then press ENTER.

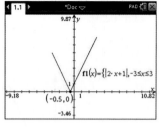

3. To label the end points, press MENU, then select:
   5: Trace
   1: Graph Trace
   Type '3', then press ENTER twice.

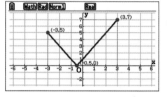

4. Sketch the graph.

Copy the graph from the screen, remembering that the end point at $x = -3$ is included, so it should be drawn with a solid point, and the end point at $x = 3$ is not included, so it should be drawn with a hollow point (as in **3a** above).

2. To find the coordinates of the vertex, select G-Solv by pressing F5. Select ROOT by pressing F1, then press EXE.

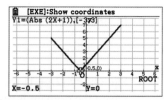

3. To label the end points, select Trace by pressing F1, then type '−3' and press EXE twice. Type '3', then press EXE twice.

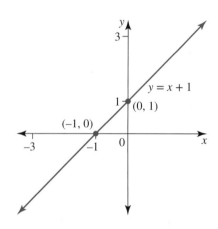

4. Sketch the graph.

Copy the graph from the screen, remembering that the end point at $x = -3$ is included, so it should be drawn with a solid point, and the end point at $x = 3$ is not included, so it should be drawn with a hollow point (as in **3a** above).

The patterns observed above are true for any graph involving the modulus function.

When graphing $y = |f(x)|$,

$$y = \begin{cases} f(x)\,, & f(x) \geq 0 \\ -f(x)\,, & f(x) < 0 \end{cases}$$

This results in a reflection in the x-axis at the x-intercept for $y < 0$.

9.2.5 Graphing $y = f(|x|)$ from $y = f(x)$

Consider again the graph of $y = x + 1$.

x	-2	-1	0	1	2
$y = x + 1$	-1	0	1	2	3

How would this change if we sketched $y = |x| + 1$?

For $x \geq 0$, $|x| = x$. Therefore, the graph would be unchanged for this interval, just the same as in the graph of $y = |x + 1|$.

x	-2	-1	0	1	2
$y = |x + 1|$	-1	0	1	2	3

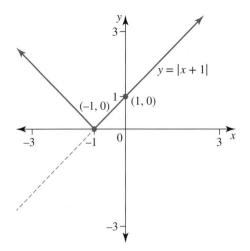

For $x \leq 0$, the graph $y = |x| + 1$ is reflected in the y-axis, as can be seen in the following table of values.

Note: This graph is a vertical shift of 1 unit from the graph of $y = |x|$.

x	-2	-1	0	1	2
$|x|$	2	1	0	1	2
$y = |x| + 1$	3	2	1	2	3

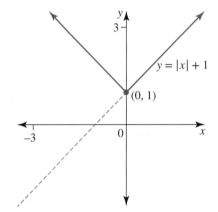

WORKED EXAMPLE 4 Sketching the modulus function for $y = f(|x|)$

Use the graph of $y = 2x + 1$ to do the following.
a. **Sketch $y = 2|x| + 1$ for the domain $-4 < x \leq 5$.**
b. **State the range of the function.**

THINK

a. 1. Sketch $y = 2x + 1$ for $-4 < x \leq 5$.
   The gradient is 2 and the y-intercept is 1.

WRITE

a.

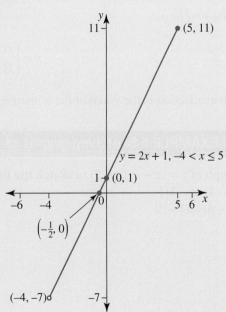

2. $|x| = x$ for $x \geq 0$; therefore, the graph is unchanged for $x \geq 0$.

The graph is symmetrical about the y-axis. Given the domain is restricted to $-4 < x \leq 5$, the graph has end points — a hollow point at $x = -4$, as the restricted domain states $x > 4$, and a solid point at $x = 5$, as the restricted domain states $x \leq 5$.

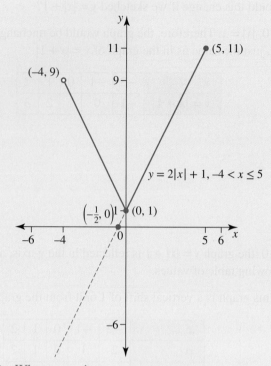

b. To determine the range, consider the end points of the restricted domain and the y-intercept.

b. When $x = -4$,
$y = 2|x| + 1$
$y = 2|-4| + 1$
$y = 9$
When $x = 5$,
$y = 2|x| + 1$
$y = 2|5| + 1$
$y = 11$
When $x = 0$, $y = 1$.
Hence, the range is $[1, 11]$.

When graphing $y = f(|x|)$,

$$y = \begin{cases} f(x), & x \geq 0 \\ f(-x), & x < 0 \end{cases}$$

This results in a reflection in the y-axis at the y-intercept for $x < 0$.

WORKED EXAMPLE 5 Sketching graphs of the form $y = f|x|)$ and $y = |f(x)|$

Use the graph of $y = (x - 4)(x - 2)$ to sketch the following.
a. $y = |(x - 4)(x - 2)|$
b. $y = (|x| - 4)(|x| - 2)$

THINK	WRITE

a. 1. Sketch $y = (x - 4)(x - 2)$. The x-intercepts are $x = 4$ and $x = 2$. The x-value for the turning point is the midpoint of the intercepts. This gives the turning point as $(3, -1)$. The y-intercept is $y = 8$.

a.

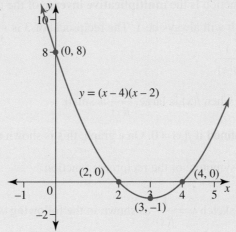

2. Identify where the graph is above and below the x-axis.

The graph is at or above the x-axis for $x \leq 2$ and $x \geq 4$.
The graph is below the x-axis for $2 < x < 4$.

3. Sketch $y = |(x - 4)(x - 2)|$ by reflecting the graph for the interval $(2, 4)$ in the x-axis.

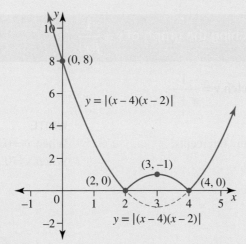

b. The graph for $y = (|x| - 4)(|x| - 2)$ is the same as the graph of $y = (x - 4)(x - 2)$ for $x \geq 0$. For $x < 0$, reflect the graph in the y-axis.

b.

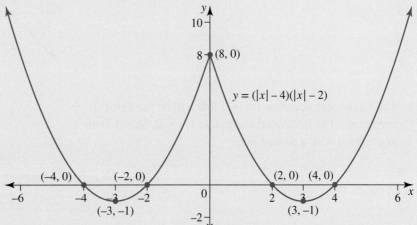

9.2.6 Graphing reciprocal functions

The **reciprocal** of any number or function is the **multiplicative inverse** of the original. When the original is multiplied by its reciprocal, the result will always be 1. The reciprocal of 3 is $\frac{1}{3}$, that is, $3 \times \frac{1}{3} = 1$. Similarly, for a function $f(x)$, the reciprocal is $\frac{1}{f(x)}$.

When $f(x)$ is small, $\frac{1}{f(x)}$ is large and when $f(x)$ is large, $\frac{1}{f(x)}$ is small.

The reciprocal of a function is **undefined** if $f(x) = 0$. On a graph, this is shown as a vertical **asymptote**.

As $\frac{1}{f(x)} \neq 0$, $y = 0$ is the horizontal asymptote of the reciprocal function.

The graph of $y = f(x)$ can be used to sketch $y = \frac{1}{f(x)}$ as shown in the following worked example.

The y-intercept of $y = f(x)$ is $f(0)$. Therefore, the y-intercept of $\frac{1}{f(x)}$ is $\frac{1}{f(0)}$.

WORKED EXAMPLE 6 Sketching the graph of $y = \dfrac{1}{f(x)}$

Use the graph of $y = x + 2$ to sketch $y = \dfrac{1}{x+2}$.

THINK

1. Sketch $y = x + 2$ and identify any intercepts.

2. The x-intercept occurs at $x = -2$; this will be the vertical asymptote. The horizontal asymptote is $y = 0$. Sketch both asymptotes with a dashed line.

3. The y-intercept of $y = x + 2$ is $y = 2$. Therefore, the y-intercept of $y = \dfrac{1}{x+2}$ is $y = \dfrac{1}{2}$.

WRITE

When $x = 0$, $y = 2$.
When $y = 0$, $x = -2$.

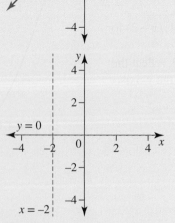

The y-intercept of $y = \dfrac{1}{x+2}$ is $y = \dfrac{1}{2}$.

---

4. Sketch the reciprocal functions, considering the following.

$y = x + 2$ is below the x-axis for $x < -2$. Therefore,
$y = \dfrac{1}{x+2}$ is also below the x-axis for $x < -2$.

$y = x + 2$ is above the x-axis for $x > -2$. Therefore,
$y = \dfrac{1}{x+2}$ is also above the x-axis for $x > -2$.

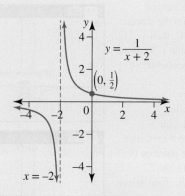

TI	THINK	WRITE

1. On a Graphs page, complete the entry line for function 1 as:
$f1(x) = \dfrac{1}{x+2}$ then press ENTER.

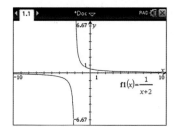

2. The position of the horizontal asymptote can be identified from the graph.

The equation of the horizontal asymptote is $y = 0$.

3. To find the position of the vertical asymptote, press MENU, then select:
7: Table
1: Split-screen Table.
Use the up/down arrows to scroll until a cell containing #undef is found.

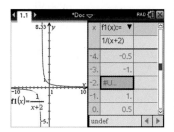

4. The position of the vertical asymptote can be read from the table.

The equation of the vertical asymptote is $x = -2$.

CASIO	THINK	WRITE

1. On a Graph screen, complete the entry line for Y1 as: Y1 = $1 \div (X + 2)$ then press EXE.
Select DRAW by pressing F6.

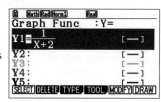

2. The position of the horizontal asymptote can be identified from the graph.

The equation of the horizontal asymptote is $y = 0$.

3. To find the position of the vertical asymptote, select Trace by pressing F1, then use the left/right arrows to trace along the graph until the screen reads Y = ERROR.

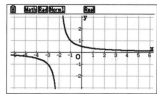

4. The position of the vertical asymptote can be read from the screen.

The equation of the vertical asymptote is $x = -2$.

5. To draw the vertical asymptote on the screen, press MENU, then select 2. Table 1. Remove table then press MENU, then select:
8: Geometry
4: Construction
1: Perpendicular.
Click on the x-axis, then click on the point on the x-axis where $x = 2$.
Press MENU, then select:
1: Actions
4: Attributes.
Click on the vertical line, then press the down arrow, then the right arrow to change the line style to dotted. Press ENTER.

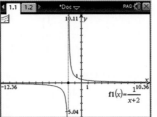

5. To draw the vertical asymptote on the screen, select DRAW by pressing F4. Press F6 to scroll across to more options, then select Vertical by pressing F4. Use the left arrow to move the vertical line to $x = -2$. Press SHIFT 5 to format the line, then select Broken for the Line Style. Press EXIT.

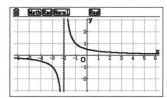

Compare the graphs of $y = x + 2$ and $y = \dfrac{1}{x + 2}$.

For the graph of $y = x + 2$, the x-intercept occurs at $(-2, 0)$, that is, when $y = 0$.

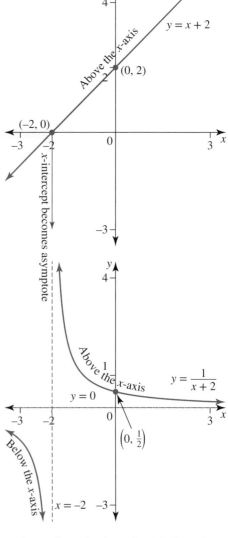

The reciprocal of $y = x + 2$, $y = \dfrac{1}{x + 2}$, is undefined at $y = \dfrac{1}{0}$, that is at $x = -2$. Hence, the x-intercept of the original $y = x + 2$ becomes the asymptote of $y = \dfrac{1}{x + 2}$.

The same process as shown above can be used to graph the reciprocal functions of quadratic and cubic functions.

Use the graph of $y = (x + 2)(x - 4)$ to sketch $y = \dfrac{1}{(x + 2)(x - 4)}$.

THINK

1. Identify the key points of $y = (x + 2)(x - 4)$.

2. Sketch $y = (x + 2)(x - 4)$.

WRITE

y-intercept: $x = 0$, $y = -8$
x-intercepts: $y = 0$, $x = -2$ or $x = 4$
Turning point: $x = 1$, $y = -9$

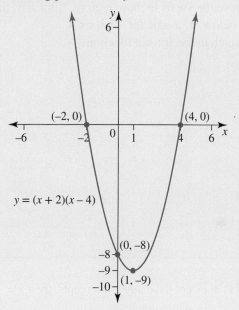

3. The x-intercepts occur at $x = -2$ and $x = 4$, so these become the vertical asymptotes. The horizontal asymptote is $y = 0$. Include all asymptotes by sketching with a dashed line.

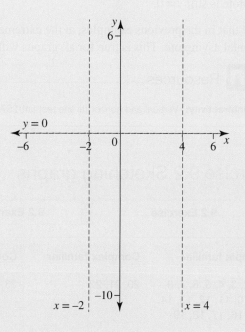

4. The y-intercept of $y = (x + 2)(x - 4)$ is -8. Therefore, the y-intercept of $y = \dfrac{1}{(x + 2)(x - 4)}$ is $\dfrac{-1}{8}$.

The y-intercept is $\left(0, \dfrac{-1}{8} \right)$.

5. The turning point of $y = (x+2)(x-4)$ is $(1, -9)$.
   Therefore, the turning point of
   $$y = \frac{1}{(x+2)(x-4)} \text{ is } \left(1, \frac{-1}{9}\right).$$

   The turning point is $\left(1, \frac{-1}{9}\right)$.

6. The graph of $y = (x+2)(x-4)$ is above the x-axis
   for $x < -2$ and $x > 4$. Therefore, $y = \frac{1}{(x+2)(x-4)}$ is

   above the x-axis in those regions. Similarly, the graph
   is below the x-axis for $-2 < x < 4$.
   Sketch the reciprocal function.

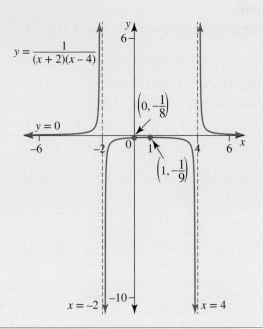

If $y = f(x)$ does not have any x-intercepts, then $y = \frac{1}{f(x)}$ does not have any vertical asymptotes. The horizontal
asymptote is still $y = 0$.

Notice that in the previous examples, at the extreme values of x ($x \to \infty$ and $x \to -\infty$) the graph approaches the
horizontal asymptote. This is true for all graphs with a horizontal asymptote.

 Resources

 Interactivity: Vertical and horizontal line test (int-2570)

Exercise 9.2 Sketching graphs

learn

9.2 Exercise	9.2 Exam questions on

These questions are
even better in jacPLUS!
- Receive immediate feedback
- Access sample responses
- Track results and progress

Find all this and MORE in jacPLUS ▶

Simple familiar	Complex familiar	Complex unfamiliar
1, 2, 3, 4, 5, 6, 7, 8, 9, 10, 11, 12, 13, 14, 15, 16, 17, 18, 19	20, 21, 22	23, 24, 25

Simple familiar

1. Identify whether the following statements are true or false.

   a. $|-5| = 5$ b. $|6| = -6$ c. $\sqrt{9} = \pm 3$ d. $|7| = 7$

2. **WE1** Solve $|3x| = 12$.

3. Solve $|-2a| = 6$.

4. **WE2** Solve $|3 - x| = 2$.

5. Solve $|x + 1| = 3$.

6. Use the graph of $y = 2x - 3$ to sketch the following, and state the range of each.
    a. **WE3** $y = |2x - 3|$ for the domain $-4 \leq x \leq 2$
    b. **WE4** $y = 2|x| - 3$ for the domain $-2 < x \leq 4$

7. Use the graph of $y = 1 - 2x$ to sketch the following.
    a. $y = |1 - 2x|$
    b. $y = 1 - 2|x|$

8. Use the graph of $y = 3x + 2$ to sketch the following.
    a. $y = |3x + 2|$
    b. $y = 3|x| + 2$

9. Use the graph of $y = \dfrac{x}{4} + 1$ to sketch the following.
    a. $y = \left|\dfrac{x}{4} + 1\right|$
    b. $y = \dfrac{|x|}{4} + 1$

10. **WE5** Use the graph of $y = (x - 1)(x + 3)$ to sketch the following.
    a. $y = |(x - 1)(x + 3)|$
    b. $y = (|x| - 1)(|x| + 3)$

11. Use the graph of $y = (1 - x)(x + 5)$ to sketch the following.
    a. $y = |(1 - x)(x + 5)|$
    b. $y = (1 - |x|)(|x| + 5)$

12. **WE6** a. Sketch $y = x - 2$.

    b. Use your sketch of $y = x - 2$ to identify the asymptotes for $y = \dfrac{1}{x - 2}$.

    c. Identify the y-intercept of $y = x - 2$ and use this to identify the y-intercept of $y = \dfrac{1}{x - 2}$.

    d. Use your sketch of $y = x - 2$ to sketch $y = \dfrac{1}{x - 2}$.

13. a. Sketch $y = x + 1$.

    b. Use your sketch of $y = x + 1$ to identify the asymptotes for $y = \dfrac{1}{x + 1}$.

    c. Identify the y-intercept of $y = x + 1$ and use this to identify the y-intercept of $y = \dfrac{1}{x + 1}$.

    d. Use your sketch of $y = x + 1$ to sketch $y = \dfrac{1}{x + 1}$.

14. Use the graph of $y = 3 - x$ to sketch $y = \dfrac{1}{3 - x}$.

15. Use the graph of $y = 2x - 1$ to sketch $y = \dfrac{1}{2x - 1}$.

16. **a.** Sketch $y = (x+1)(x-3)$.

   b. Use your sketch of $y = (x+1)(x-3)$ to identify the asymptotes for $y = \dfrac{1}{(x+1)(x-3)}$.

   c. Identify the y-intercept and turning point of $y = (x+1)(x-3)$ and use these to identify the y-intercept and turning point of $y = \dfrac{1}{(x+1)(x-3)}$.

   d. Use your sketch of $y = (x+1)(x-3)$ to sketch $y = \dfrac{1}{(x+1)(x-3)}$.

17. Use the graph of $y = (x+1)(x-3)$ to sketch the following.

   a. $y = |(x+1)(x-3)|$ **b.** $y = (|x|+1)(|x|-3)$

18. Use the graph of $y = (3-x)(5-x)$ to sketch the following.

   a. $y = |(3-x)(5-x)|$ **b.** $y = (3-|x|)(5-|x|)$

19. Use the graph of $y = \dfrac{1}{x}$ to sketch:

   a. $y = \left|\dfrac{1}{x}\right|$ **b.** $y = \dfrac{1}{|x|}$

   What do you notice about your answers for parts **a** and **b**?

Complex familiar

20. **WE7** Use the graph of $y = (x-4)(x-3)$ to sketch $y = \dfrac{1}{(x-4)(x-3)}$.

21. Use the graph of $y = (x-1)(x+3)$ to sketch $y = \dfrac{1}{(x-1)(x+3)}$.

22. Use the graph of $y = (x-2)^2$ to sketch the following.

   a. $y = |(x-2)^2|$ **b.** $y = (|x|-2)^2$

   What do you notice about your answer for part **a**?

Complex unfamiliar

23. Use sketches of $y = |2x+1|$ and $y = |1-2x|$ to determine what the graph of $y = |2x+1| + |1-2x|$ looks like. You must justify your prediction.

24. The graph of $y = \dfrac{x}{x+1}$ is shown. Use your understanding of reciprocal functions to sketch $y = \dfrac{x+1}{x}$. You must justify the decisions you have made. You may use technology to confirm your decisions.

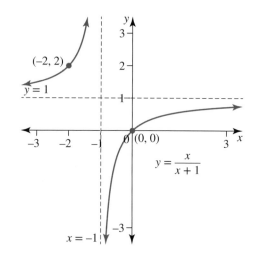

25. The graph of $y = \sqrt{x}$ is shown. Use your understanding of reciprocal functions to sketch $y = \dfrac{1}{\sqrt{x}}$. You must justify the decisions you have made. You may use technology to confirm your decisions.

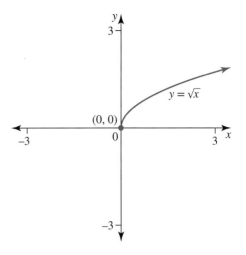

Fully worked solutions for this chapter are available online.

LESSON
9.3 Reciprocal trigonometric functions

SYLLABUS LINKS

- Define and use the reciprocal trigonometric functions to determine their simplified exact values and sketch their graphs.

Source: Specialist Mathematics Senior Syllabus 2024 © State of Queensland (QCAA) 2024; licensed under CC BY 4.0

9.3.1 Naming the reciprocal trigonometric functions

The reciprocal of the sine function is called the **cosecant function**; it may be abbreviated to cosec or csc. It is defined as $\operatorname{cosec}(x) = \dfrac{1}{\sin(x)}$, $\sin(x) \neq 0$.

The reciprocal of the cosine function is called the **secant function**, often abbreviated to sec. It is defined as $\sec(x) = \dfrac{1}{\cos(x)}$, $\cos(x) \neq 0$.

The reciprocal of the tangent function is called the **cotangent function**, often abbreviated to cot. It is defined as $\cot(x) = \dfrac{1}{\tan(x)} = \dfrac{\cos(x)}{\sin(x)}$, $\sin(x) \neq 0$.

Note that the reciprocal functions are not the inverse trigonometric functions. (eg. $\operatorname{cosec}(\theta) \neq \sin^{-1}(\theta)$)

These functions can also be defined in terms of right-angled triangles.

$$\csc(x) = \frac{1}{\sin(x)} = \frac{1}{\frac{\text{opposite}}{\text{hypotenuse}}} = \frac{\text{hypotenuse}}{\text{opposite}} = \frac{c}{b}$$

$$\sec(x) = \frac{1}{\cos(x)} = \frac{1}{\frac{\text{adjacent}}{\text{hypotenuse}}} = \frac{\text{hypotenuse}}{\text{adjacent}} = \frac{c}{a}$$

$$\cot(x) = \frac{1}{\tan(x)} = \frac{1}{\frac{\text{opposite}}{\text{adjacent}}} = \frac{\text{adjacent}}{\text{opposite}} = \frac{a}{b}$$

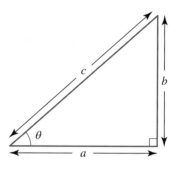

9.3.2 Determining exact values

The exact values for the reciprocal trigonometric functions for angles that are multiples of $\frac{\pi}{6}$ and $\frac{\pi}{4}$ (or 30° and 45°) can be found from the corresponding trigonometric values by determining the reciprocal. Note that we may need to simplify the resulting expression or rationalise the denominator.

WORKED EXAMPLE 8 Determining exact values of reciprocal trigonometric functions (1)

Determine the exact value of $\csc\left(\dfrac{5\pi}{4}\right)$.

THINK

1. Rewrite $\csc\left(\dfrac{5\pi}{4}\right)$ by recalling $\csc(\theta) = \dfrac{1}{\sin(\theta)}$.

2. $\dfrac{5\pi}{4}$ is in quadrant 3. Sine is negative in that quadrant.

WRITE

$$\csc\left(\frac{5\pi}{4}\right) = \frac{1}{\sin\left(\frac{5\pi}{4}\right)}$$

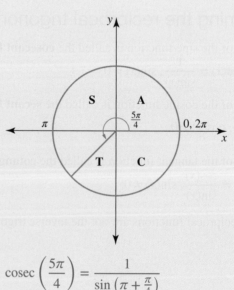

$$\csc\left(\frac{5\pi}{4}\right) = \frac{1}{\sin\left(\pi + \frac{\pi}{4}\right)}$$

$$= \frac{1}{-\sin\left(\frac{\pi}{4}\right)}$$

3. Use $\sin\left(\dfrac{\pi}{4}\right) = \dfrac{1}{\sqrt{2}}$ and simplify the fraction.

$$\csc\left(\frac{5\pi}{4}\right) = \frac{1}{-\dfrac{1}{\sqrt{2}}}$$

$$= 1 \times \frac{-\sqrt{2}}{1}$$

$$= -\sqrt{2}$$

In a circle of radius 1, for an acute angle θ, it is possible to locate lengths that are $\sin(\theta)$, $\cos(\theta)$, $\tan(\theta)$, $\sec(\theta)$, $\csc(\theta)$ and $\cot(\theta)$.

Through the use of right-angled triangles and the unit circle, if one trigonometric ratio for an angle is known, it is possible to calculate all of the ratios.

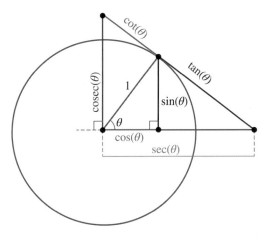

WORKED EXAMPLE 9 Determining exact values of reciprocal trigonometric functions (2)

If $\csc(\theta) = \dfrac{7}{4}$ and $\dfrac{\pi}{2} \le \theta \le \pi$, determine the exact value of $\cot(\theta)$.

THINK

1. Recall the definition $\csc(\theta) = \dfrac{1}{\sin(\theta)}$ and solve for $\sin(\theta)$.

2. Recall that $\sin(\theta) = \dfrac{\text{opposite}}{\text{hypotenuse}}$.
   Draw a right-angled triangle and mark an angle with an opposite side length of 4 and a hypotenuse of length 7. Label the unknown side length as x.

WRITE

$$\csc(\theta) = \frac{1}{\sin(\theta)}$$

$$\frac{7}{4} = \frac{1}{\sin(\theta)}$$

$$\sin(\theta) = \frac{4}{7}$$

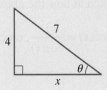

3. Use Pythagoras' theorem to determine the unknown side length.

$$x = \sqrt{7^2 - 4^2}$$
$$= \sqrt{49 - 16}$$
$$= \sqrt{33}$$

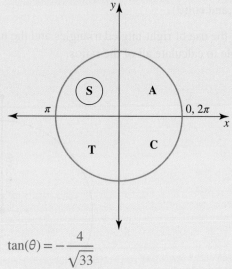

4. Given that $\dfrac{\pi}{2} < \theta < \pi$, θ is in the second quadrant, and although $\sin(\theta)$ is positive, $\tan(\theta)$ is negative. Determine $\tan(\theta)$ using the value of x from step 3.

$$\tan(\theta) = -\dfrac{4}{\sqrt{33}}$$

5. Use $\tan(\theta)$ to calculate $\cot(\theta)$.

$$\cot(\theta) = \dfrac{1}{\tan(\theta)}$$

$$= \dfrac{1}{-\dfrac{4}{\sqrt{33}}}$$

$$= -\dfrac{\sqrt{33}}{4}$$

$$\therefore \cot(\theta) = -\dfrac{\sqrt{33}}{4}$$

9.3.3 Sketching the graphs of reciprocal trigonometric functions

In the previous section, we looked at how the graph of $f(x)$ can be used to determine the graph of $\dfrac{1}{f(x)}$. This method will be used to graph $\sec(x) = \dfrac{1}{\cos(x)}$, $\operatorname{cosec}(x) = \dfrac{1}{\sin(x)}$ and $\cot(x) = \dfrac{1}{\tan(x)}$.

The graph of $y = \sec(x)$

Consider the graph of $y = \cos(x)$.

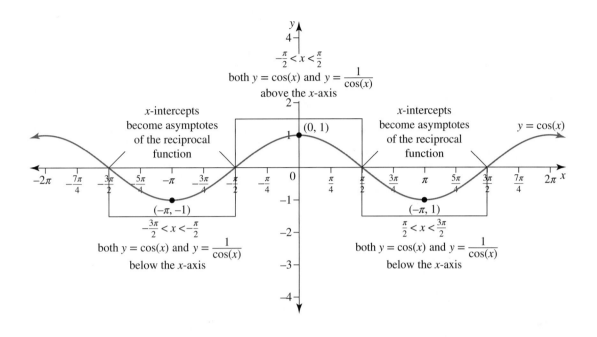

The graph of $y = \csc(x)$

In a similar fashion, the graph of $y = \sin(x)$ can be used to determine the graph of $y = \dfrac{1}{\sin(x)}$ (also known as $y = \csc(x)$).

Consider the graph of $y = \sin(x)$.

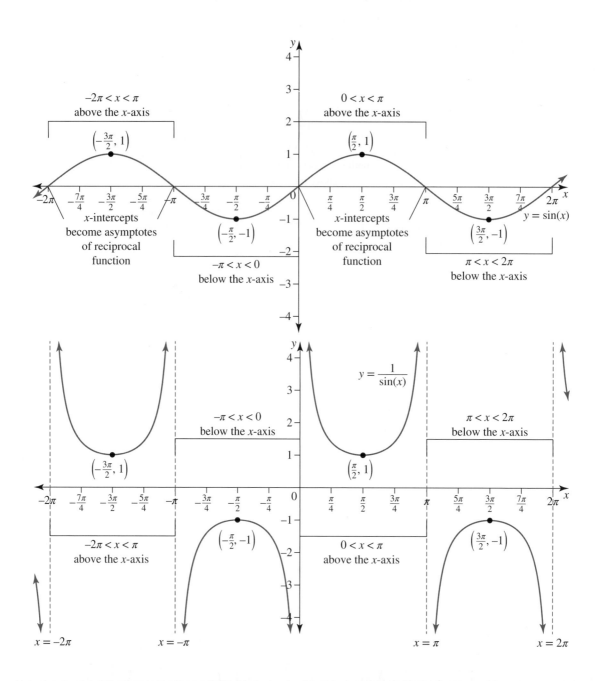

WORKED EXAMPLE 10 Sketching the graphs of reciprocal trigonometric functions (1)

Use the graph of $y = 2\cos(x)$ to sketch $y = \dfrac{1}{2}\sec(x)$ over the domain $-2\pi \leq x \leq 2\pi$.

THINK

1. Rewrite $y = \dfrac{1}{2}\sec(x)$ in terms of $\cos(x)$.

WRITE

$$y = \frac{1}{2}\sec(x)$$

$$= \frac{1}{2} \times \frac{1}{\cos(x)}$$

$$= \frac{1}{2\cos(x)}$$

2. Sketch $y = 2\cos(x)$ by identifying the amplitude, period, horizontal shift and vertical shift. Identify the end points at $x = -2\pi$ and $x = 2\pi$.

$y = 2\cos(x)$
Amplitude: 2   Period: 2π
Horizontal shift: 0   Vertical shift: 0
$2\cos(-2\pi) = 2\cos(2\pi) = 2$

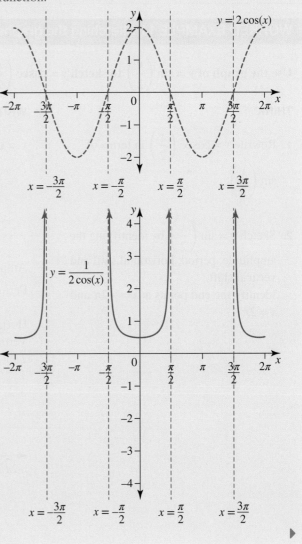

3. Determine the x-intercepts and hence the vertical asymptotes for the reciprocal graph.

The x-intercepts occur at
$x = -\dfrac{3\pi}{2}, x = -\dfrac{\pi}{2}, x = \dfrac{\pi}{2}$ and $x = \dfrac{3\pi}{2}$.
These will be the vertical asymptotes for the reciprocal function.

4. The graph of $y = 2\cos(x)$ is above the x-axis in the regions $-2\pi \le x < -\dfrac{-3\pi}{2}$, $-\dfrac{\pi}{2} < x < \dfrac{\pi}{2}$ and $\dfrac{3\pi}{2} < x \le 2\pi$. The graph of $y = \dfrac{1}{2\cos(x)}$ will also be above the x-axis in these regions. A maximum value of $y = 2$ is reached in the original graph, meaning that a minimum of $y = \dfrac{1}{2}$ will be reached in the reciprocal function.

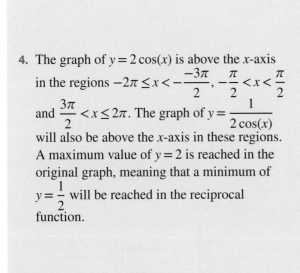

5. The graph of $y = 2\cos(x)$ is below the x-axis in the regions $-\dfrac{3\pi}{3} < x < -\dfrac{\pi}{2}$ and $\dfrac{\pi}{2} < x < \dfrac{3\pi}{2}$.

Therefore, $y = \dfrac{1}{2\cos(x)}$ is also below the x-axis in these regions. The minimum of $y = -2$ will become a maximum of $y = -\dfrac{1}{2}$.

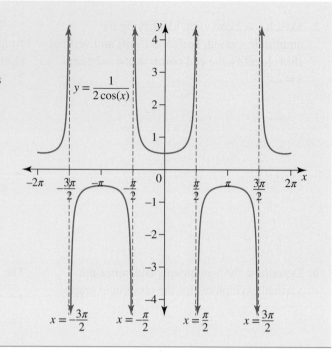

WORKED EXAMPLE 11 Sketching the graphs of reciprocal trigonometric functions (2)

Use the graph of $y = \sin\left(\dfrac{x}{2}\right)$ to sketch $y = \operatorname{cosec}\left(\dfrac{x}{2}\right)$ over the domain $-2\pi \le x \le 2\pi$.

THINK	WRITE
1. Rewrite $y = \operatorname{cosec}\left(\dfrac{x}{2}\right)$ in terms of $\sin\left(\dfrac{x}{2}\right)$.	$y = \operatorname{cosec}\left(\dfrac{x}{2}\right)$ $\quad = \dfrac{1}{\sin\left(\frac{x}{2}\right)}$
2. Sketch $y = \sin\left(\dfrac{x}{2}\right)$ by identifying the amplitude, period, horizontal shift and vertical shift. Identify the end points at $x = -2\pi$ and $x = 2\pi$.	$y = \sin\left(\dfrac{x}{2}\right)$ Amplitude: 1 Period: $\dfrac{2\pi}{\frac{1}{2}} = 4\pi$ Horizontal shift: 0 Vertical shift: 0 $2\sin(-2\pi) = 2\sin(2\pi) = 0$

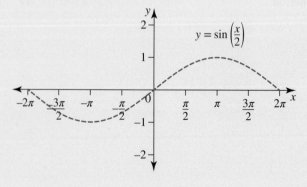

442   Jacaranda Maths Quest 11 Specialist Mathematics Units 1 & 2 for Queensland Second Edition

3. Determine the *x*-intercepts and hence the vertical asymptotes for the reciprocal graph.

The *x*-intercepts occur at $x = -2\pi$, $x = 0$ and $x = 2\pi$. These will be the vertical asymptotes for the reciprocal function.

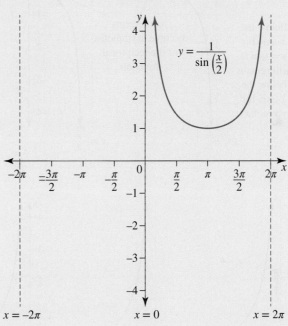

4. The graph of $y = \sin\left(\dfrac{x}{2}\right)$ is above the *x*-axis in the region $0 < x \le 2\pi$. The graph of $y = \dfrac{1}{\sin\left(\frac{x}{2}\right)}$ will also be above the *x*-axis in these regions.

A maximum value of $y = 1$ is reached in the original graph, meaning that a minimum of $y = 1$ will be reached in the reciprocal function.

5. The graph of $y = \sin\left(\dfrac{x}{2}\right)$ is below the *x*-axis in the region $-2\pi \le x < 0$.

Therefore, $y = \dfrac{1}{\sin\left(\frac{x}{2}\right)}$ is also below the *x*-axis in these regions. The minimum of $y = -1$ will become a maximum of $y = -1$.

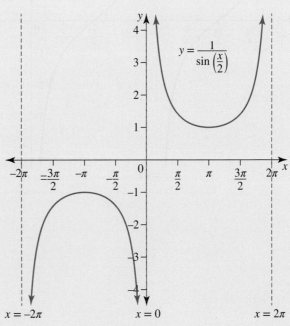

The graph of $y = \cot(x)$

The graph of $y = \tan(x)$ can also be used to determine the graph of $y = \dfrac{1}{\tan(x)}$ (also known as $y = \cot(x)$).

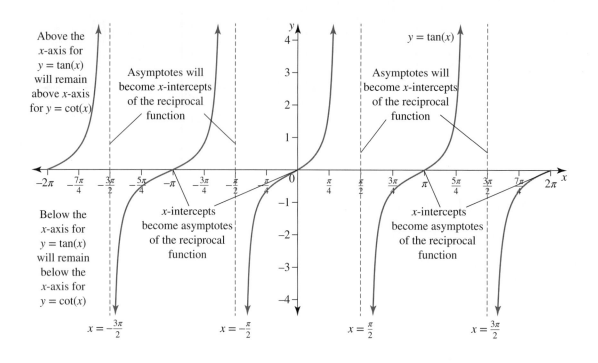

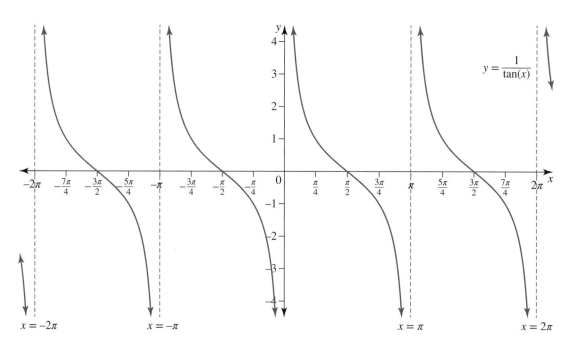

Use the graph of $y = \tan\left(\dfrac{x}{2}\right)$ to sketch $y = \cot\left(\dfrac{x}{2}\right)$ over the domain $-2\pi \leq x \leq 2\pi$.

THINK

1. Rewrite $y = \cot\left(\dfrac{x}{2}\right)$ in terms of $\tan\left(\dfrac{x}{2}\right)$.

2. Sketch $y = \tan\left(\dfrac{x}{2}\right)$ by identifying the period, horizontal shift and vertical shift. Identify the end points at $x = -2\pi$ and $x = 2\pi$.

3. The asymptotes of $y = \tan\left(\dfrac{x}{2}\right)$ will become the x-intercepts of $y = \cot\left(\dfrac{x}{2}\right)$.

4. Determine the x-intercepts and hence the vertical asymptotes for the reciprocal graph.

WRITE

$y = \cot\left(\dfrac{x}{2}\right)$

$\quad = \dfrac{1}{\tan\left(\frac{x}{2}\right)}$

$y = \tan\left(\dfrac{x}{2}\right)$ Period: $\dfrac{\pi}{\frac{1}{2}} = 2\pi$

Horizontal shift: 0 Vertical shift: 0

$\tan(-2\pi) = \tan 2\pi = 0$

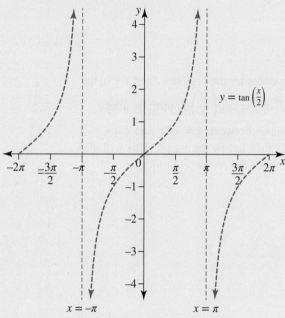

The x-intercepts of $y = \cot\left(\dfrac{x}{2}\right)$ will be $x = \pm\pi$.

The x-intercepts of $y = \tan\left(\dfrac{x}{2}\right)$ are $x = -2\pi$, $x = 0$ and $x = 2\pi$. These will become the vertical asymptotes of $y = \cot\left(\dfrac{x}{2}\right)$.

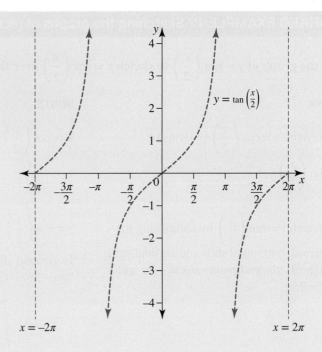

$y = \tan\left(\dfrac{x}{2}\right)$

$x = -2\pi$ $x = 2\pi$

5. If we consider the region $-2\pi \leq x < 0$, the graph of $y = \tan\left(\dfrac{x}{2}\right)$ is initially above the x-axis between $x = -2\pi$ and $x = -\pi$ and is then below the x-axis. This will also be true for the reciprocal function.

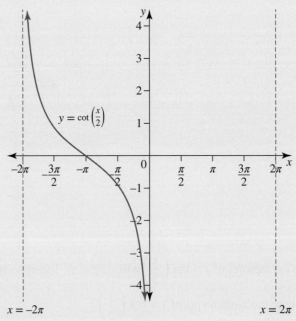

$y = \cot\left(\dfrac{x}{2}\right)$

$x = -2\pi$ $x = 2\pi$

6. In a similar fashion, the graph for $x = 0$ to $x = 2\pi$ can be obtained.

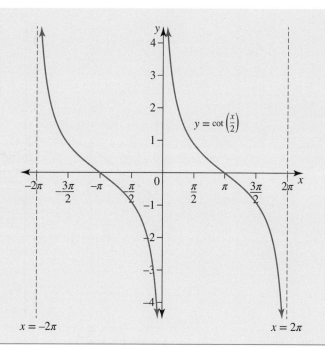

9.3.4 Transformations of the graphs of the reciprocal functions

Transformations can also be applied to the reciprocal functions.

> **WORKED EXAMPLE 13** Applying transformations to the graphs of reciprocal trigonometric functions
>
> Sketch the graph of $y = \sec\left(x + \dfrac{\pi}{4}\right) + 1$ over the domain $[-\pi, 2\pi]$.

THINK	WRITE
1. Sketch the graph of $y = \cos\left(x + \dfrac{\pi}{4}\right)$ to determine the graph of $y = \sec\left(x + \dfrac{\pi}{4}\right)$.	$y = \cos\left(x + \dfrac{\pi}{4}\right)$ Amplitude: 1 Period: 2π Horizontal shift: $\dfrac{\pi}{4}$ Left Vertical shift: 0 End points: $\cos\left(-\pi + \dfrac{\pi}{4}\right) = \cos\left(-\dfrac{3\pi}{4}\right)$ $\qquad\qquad\qquad = -\dfrac{\sqrt{2}}{2}$ $\cos\left(2\pi + \dfrac{\pi}{4}\right) = \cos\left(\dfrac{\pi}{4}\right)$ $\qquad\qquad\qquad = \dfrac{\sqrt{2}}{2}$

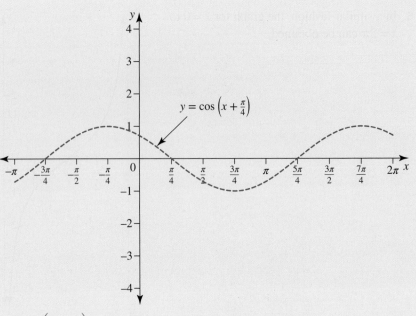

$$y = \cos\left(x + \frac{\pi}{4}\right)$$

2. Now consider $y = \sec\left(x + \frac{\pi}{4}\right)$. The x-intercepts of $y = \cos\left(x + \frac{\pi}{4}\right)$ will become the asymptotes. Sections of the graph above the x-axis will remain above the x-axis, and sections below the x-axis will remain below the x-axis.

$$y = \sec\left(x + \frac{\pi}{4}\right)$$

The asymptotes will be $x = -\frac{\pi}{4}$, $x = \frac{3\pi}{4}$ and $x = \frac{7\pi}{4}$.

The graph of $y = \cos\left(x + \frac{\pi}{4}\right)$ and $y = \sec\left(x + \frac{\pi}{4}\right)$:

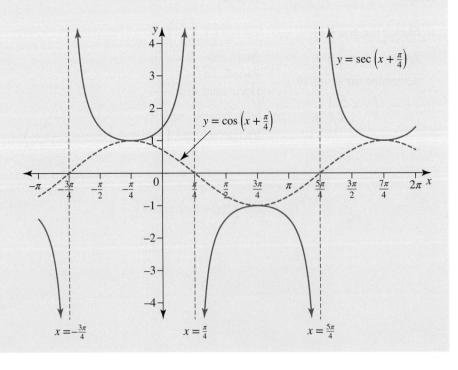

$$x = -\frac{3\pi}{4} \qquad x = \frac{\pi}{4} \qquad x = \frac{5\pi}{4}$$

3. To graph
$$y = \sec\left(x + \frac{\pi}{4}\right) + 1,$$
move $y = \sec\left(x + \frac{\pi}{4}\right)$
up 1.

The graph of $y = \sec\left(x + \frac{\pi}{4}\right) + 1$:

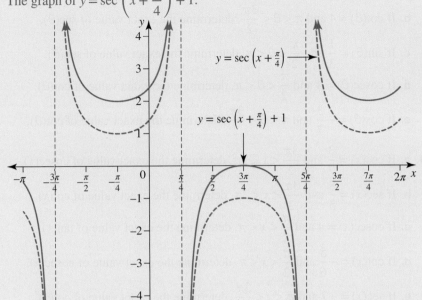

Exercise 9.3 Reciprocal trigonometric functions

9.3 Exercise	9.3 Exam questions on

Simple familiar	Complex familiar	Complex unfamiliar
1, 2, 3, 4, 5, 6, 7, 8, 9, 10	11, 12, 13	14, 15

These questions are even better in jacPLUS!
- Receive immediate feedback
- Access sample responses
- Track results and progress

Find all this and MORE in jacPLUS ▶

Simple familiar

1. **WE8** Determine the exact value of each of the following.

   a. $\operatorname{cosec}\left(\dfrac{2\pi}{3}\right)$ b. $\operatorname{cosec}\left(\dfrac{\pi}{3}\right)$ c. $\operatorname{cosec}\left(\dfrac{5\pi}{6}\right)$ d. $\operatorname{cosec}\left(\dfrac{7\pi}{4}\right)$

2. Determine the exact value of each of the following.

   a. $\sec\left(\dfrac{\pi}{6}\right)$ b. $\sec\left(-\dfrac{7\pi}{6}\right)$ c. $\sec\left(\dfrac{4\pi}{3}\right)$ d. $\sec\left(-\dfrac{7\pi}{4}\right)$

3. a. WE9 If $\operatorname{cosec}(\theta) = \dfrac{5}{2}$ and $\dfrac{\pi}{2} < \theta < \pi$, determine the exact value of $\cot(\theta)$.

b. If $\cot(\theta) = 4$ and $\pi < \theta < \dfrac{3\pi}{2}$, determine the exact value of $\sec(\theta)$.

c. If $\sin(\theta) = \dfrac{1}{3}$ and $\dfrac{\pi}{2} < \theta < \pi$, determine the exact value of $\sec(\theta)$.

d. If $\operatorname{cosec}(\theta) = 4$ and $\dfrac{\pi}{2} < \theta < \pi$, determine the exact value of $\cot(\theta)$.

e. If $\cos(\theta) = -\dfrac{3}{7}$ and $\pi < \theta < \dfrac{3\pi}{2}$, determine the exact value of $\cot(\theta)$.

4. a. If $\cos(x) = \dfrac{3}{7}$ and $\dfrac{3\pi}{2} < x < 2\pi$, determine the exact value of $\operatorname{cosec}(x)$.

b. If $\sec(x) = \dfrac{8}{5}$ and $\dfrac{3\pi}{2} < x < 2\pi$, determine the exact value of $\cot(x)$.

c. If $\operatorname{cosec}(x) = 4$ and $\dfrac{\pi}{2} < x < \pi$, determine the exact value of $\tan(x)$.

d. If $\cot(x) = -\dfrac{5}{6}$ and $\dfrac{\pi}{2} < x < \pi$, determine the exact value of $\operatorname{cosec}(x)$.

e. If $\sec(x) = -7$ and $\pi < x < \dfrac{3\pi}{2}$, determine the exact value of $\cot(x)$.

5. a. WE10 Use the graph of $y = 4\cos(x)$ to sketch $y = \dfrac{1}{4}\sec(x)$ over the domain $-2\pi \le x \le 2\pi$.

b. Use the graph of $y = 2\sin(x)$ to sketch $y = \dfrac{1}{2}\operatorname{cosec}(x)$ over the domain $-2\pi \le x \le 2\pi$.

6. Sketch $y = \dfrac{1}{4}\operatorname{cosec}(x)$ over the domain $[-\pi, \pi]$.

7. WE11 Use the graph of $y = \sin(2x)$ to sketch $y = \operatorname{cosec}(2x)$ over the domain $-2\pi \le x \le 2\pi$.

8. Sketch $y = \sec\left(\dfrac{x}{2}\right)$ over the domain $[-\pi, \pi]$.

9. WE12 Use the graph of $y = \tan(2x)$ to sketch $y = \cot(2x)$ over the domain $-2\pi \le x \le 2\pi$.

10. Sketch $y = \cot\left(\dfrac{x}{3}\right)$ over the domain $[-3\pi, 3\pi]$.

Complex familiar

11. a. WE13 Sketch the graph of $y = \dfrac{1}{2}\sec\left(x + \dfrac{\pi}{4}\right) - 1$ over the domain $[-\pi, 2\pi]$.

b. Sketch the graph of $y = \cot\left(x + \dfrac{\pi}{4}\right) + 1$ over the domain $[-\pi, 2\pi]$.

12. Sketch the following over the domain of $[-\pi, \pi]$.

a. $y = 2\sec(x) - 1$ **b.** $y = \dfrac{2}{\sin\left(x + \frac{\pi}{4}\right)}$ **c.** $y = 0.25\operatorname{cosec}\left(x - \dfrac{\pi}{4}\right)$

13. Use the graph of $y = \sin(x) + 2$ to sketch $y = \dfrac{1}{\sin(x) + 2}$ over the domain $\left[-\dfrac{5\pi}{2}, \dfrac{5\pi}{2}\right]$.

14. Sketch $y = 3 \sec\left(2x + \dfrac{\pi}{2}\right) - 2$ over the domain $[-\pi, \pi]$.

15. If $\operatorname{cosec}(x) = \dfrac{p}{q}$ where $p, q \in \mathbb{R}^+$ and $\dfrac{\pi}{2} < x < \pi$, determine $\sec(x) - \cot(x)$.

Fully worked solutions for this chapter are available online.

LESSON
9.4 Pythagorean identities

SYLLABUS LINKS

- Prove and apply the Pythagorean identities.
  - $\sin^2(A) + \cos^2(A) = 1$
  - $\tan^2(A) + 1 = \sec^2(A)$
  - $\cot^2(A) + 1 = \operatorname{cosec}^2(A)$

Source: Specialist Mathematics Senior Syllabus 2024 © State of Queensland (QCAA) 2024; licensed under CC BY 4.0

9.4.1 Prove the Pythagorean Identity

An **identity** is a relationship that hold true for all possible values of the variable or variables. This section explores trigonometric identities — identities that use functions of one or more angles. You have already used the following trigonometric identity:

$$\tan(A) = \frac{\sin(A)}{\cos(A)}$$

Identities are useful for simplifying expressions involving trigonometric functions, which becomes important in the study of calculus.

Consider a right-angled triangle drawn in the first quadrant of the unit circle, with a hypotenuse of one unit. The x-coordinate of a point on the unit circle can be described as the cosine of the angle and the y-coordinate of the point on the unit circle can be described as the sine of the angle.

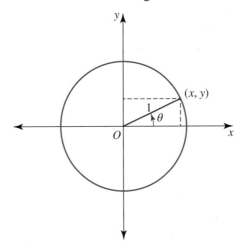

$$\cos(\theta) = \frac{\text{adjacent}}{\text{hypotenuse}} = \frac{x}{1}$$

$$x = \cos(\theta)$$

$$\sin(\theta) = \frac{\text{opposite}}{\text{hypotenuse}} = \frac{y}{1}$$

$$y = \sin(\theta)$$

The relationship between x and y, as a point on the unit circle is the equation of a circle.

$$x^2 + y^2 = 1$$

It follows that

$$\cos^2(\theta) + \sin^2(\theta) = 1$$

This is known as the **Pythagorean identity**.

9.4.2 Solve simple trigonometric unknowns

The Pythagorean identity is usually written as $\sin^2(A) + \cos^2(A) = 1$ and is used to solve simple trigonometric unknowns.

WORKED EXAMPLE 14 Using the Pythagorean identity to solve simple trigonometric unknowns (1)

If $\sin(A) = 0.4$ and $0° < A < 90°$, determine $\cos(A)$ correct to three decimal places.

THINK	WRITE
a. 1. Use the identity $\sin^2(A) + \cos^2(A) = 1$.	a. $\sin^2(A) + \cos^2(A) = 1$
2. Substitute 0.4 for $\sin(A)$.	$(0.4)^2 + \cos^2(A) = 1$
3. Solve the equation for $\cos(A)$ correct to three decimal places.	$\cos^2(A) = 1 - 0.16$ $= 0.84$ $\cos(A) = \pm\sqrt{0.84}$ $= 0.917 \text{ or } -0.917$
4. Retain the positive answer only as cosine is positive in the first quadrant.	For $0° < A < 90°$, cos is positive so $\cos(A) = 0.917$.

TI	THINK	WRITE	CASIO	THINK	WRITE

TI | THINK

1. Put the calculator in DEGREE mode.
On a Calculator page, select MENU, then select:
3: Algebra
1: Numerical Solve
Complete the entry line as:
nSolve(sin(a) = 0.4, a)
then press ENTER.

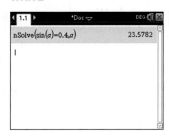

2. To store this answer as the value for a, press ctrl then var, then type "a". Press ENTER.

3. Complete the next entry line as:
cos(a)
then press ENTER.

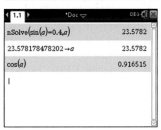

4. The answer appears on the screen.
$\cos(A) = 0.917$ (to three decimal places).

CASIO | THINK

1. Put the calculator in DEGREE mode.
On an Equation screen, select Solver by pressing F3.
Complete the entry line for the equation as: sin(A) = 0.4 then press EXE.
Set the lower bound to 0 and the upper bound to 90.

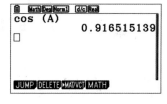

2. Select SOLVE by pressing F6.

3. On a Run-Matrix screen, complete the entry line as: cos(A) then press EXE.

4. The answer appears on the screen.
$\cos(A) = 0.917$ (to three decimal places).

WORKED EXAMPLE 15 Using the Pythagorean identity to solve simple trigonometric unknowns (2)

Determine all possible values of sin(A) over the domain $0 \le A \le 2\pi$ if cos(A) = 0.75.

THINK	WRITE
1. Use the identity $\sin^2(A) + \cos^2(A) = 1$.	$\sin^2(A) + \cos^2(A) = 1$
2. Substitute 0.75 for $\cos(A)$.	$\sin^2(A) + (0.75)^2 = 1$
3. Solve the equation for $\sin(A)$ correct to three decimal places.	$\sin^2(A) = 1 - 0.5625$ $= 0.4375$ $\sin(A) = \pm\sqrt{0.4375}$
4. Retain both the positive and negative solutions, since the angle could be in either the first or fourth quadrants.	$= 0.661$ or -0.661

TI	THINK	WRITE	CASIO	THINK	WRITE

TI | THINK

1. Put the calculator in RADIAN mode.
   On a Calculator page, select MENU, then select:
   3: Algebra
   1: Numerical Solve
   Complete the entry line as:
   $nSolve(\cos(a) = 0.75, a)$
   then press ENTER.
   This gives the solution found in the 1st quadrant.

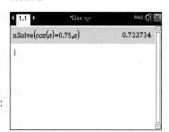

WRITE

CASIO | THINK

1. Put the calculator in RADIAN mode.
   On a Run-Matrix screen, press OPTN, then select CALC by pressing F4.
   Select SolveN by pressing F5.
   Complete the entry line as:
   $SolveN(\cos(A) = 0.75, A, 0, 2\pi)$
   Press b, then press OPTN.
   Select LIST by pressing F1, then select List by pressing F1 again.
   Type '1', then press EXE.
   This will store all solutions in List1.

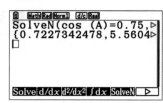

WRITE

2. To find the solution found in the 4th quadrant, complete the next entry line as:
   $2\pi - ans$
   then press ENTER.
   Note: Use the up arrow to highlight a previous answer and press ENTER to paste it onto the current entry line.

2. Complete the next entry line as:
   $\sin(List\ 1)$
   then press EXE.
   This calculates sine of all values in List 1.

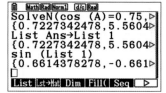

3. Complete the next entry line as:
   $\sin(0.722734)$
   then press ENTER.
   Complete the next entry line as:
   $\sin(5.56045)$
   then press ENTER.

3. The answers appear on the screen.

$\sin(A) = 0.661, -0.661$

4. The answers appear on the screen. $\sin(A) = 0.661, -0.661$

9.4.3 Other Pythagorean identities

The Pythagorean identity can be manipulated to demonstrate other relationships, which can then be used to simplify more complex trigonometric expressions.

Remember that $\tan(A) = \dfrac{\sin(A)}{\cos(A)}$, $\cot(A) = \dfrac{\cos(A)}{\sin(A)}$, $\sec(A) = \dfrac{1}{\cos(A)}$ and $\operatorname{cosec}(A) = \dfrac{1}{\sin(A)}$.

Note that cancelling down fractions is a key skill that will be especially useful in this section.

Divide each term in the original Pythagorean identity, $\sin^2(A) + \cos^2(A) = 1$, by $\cos^2(A)$:

$$\frac{\sin^2(A)}{\cos^2(A)} + \frac{\cos^2(A)}{\cos^2(A)} = \frac{1}{\cos^2(A)}$$
$$\tan^2(A) + 1 = \sec^2(A)$$

Divide each term in the original Pythagorean identity, $\sin^2(A) + \cos^2(A) = 1$, by $\sin^2(A)$:

$$\frac{\sin^2(A)}{\sin^2(A)} + \frac{\cos^2(A)}{\sin^2(A)} = \frac{1}{\sin^2(A)}$$
$$1 + \cot^2(A) = \mathrm{cosec}^2(A)$$

Pythagorean identities

$$\sin^2(A) + \cos^2(A) = 1$$
$$\tan^2(A) + 1 = \sec^2(A)$$
$$\cot^2(A) + 1 = \mathrm{cosec}^2(A)$$

WORKED EXAMPLE 16 Simplifying expressions involving trigonometric functions using Pythagorean identities (1)

Use a Pythagorean identity to simplify $\cos^2(A)\tan^2(A)$.

THINK	WRITE
1. Use the identity $\tan^2(A) + 1 = \sec^2(A)$.	Using $\tan^2(A) = \sec^2(A) - 1$, $\cos^2(A)\tan^2(A) = \cos^2(A)\left(\sec^2(A) - 1\right)$
	$= \cos^2(A)\sec^2(A) - \cos^2(A)$
2. As $\sec(A) = \dfrac{1}{\cos(A)}$, $\cos^2(A)\sec^2(A) = 1$.	$= 1 - \cos^2(A)$
3. $\sin^2(A) + \cos^2(A) = 1$ so $\sin^2(A) = 1 - \cos^2(A)$.	$= \sin^2(A)$

Sometimes it is easier to simplify expressions involving $\tan(A)$, $\cot(A)$, $\sec(A)$ and $\mathrm{cosec}(A)$ by rewriting them in terms of $\sin(A)$ and $\cos(A)$.

WORKED EXAMPLE 17 Simplifying expressions involving trigonometric functions using Pythagorean identities (2)

Simplify $\dfrac{\tan^2(A) - 1}{\tan^2(A) + 1}$.

THINK

1. Substitute $\tan(A) = \dfrac{\sin(A)}{\cos(A)}$.

2. To assist with simplifying the fraction, multiply by $\dfrac{\cos^2(A)}{\cos^2(A)}$.

3. Use $\sin^2(A) + \cos^2(A) = 1$ to simplify the denominator.

WRITE

$$\frac{\tan^2(A) - 1}{\tan^2(A) + 1} = \frac{\frac{\sin^2(A)}{\cos^2(A)} - 1}{\frac{\sin^2(A)}{\cos^2(A)} + 1}$$

$$= \frac{\frac{\sin^2(A)}{\cos^2(A)} - 1}{\frac{\sin^2(A)}{\cos^2(A)} + 1} \times \frac{\cos^2(A)}{\cos^2(A)}$$

$$= \frac{\frac{\sin^2(A)\,\cancel{\cos^2(A)}}{\cancel{\cos^2(A)}} - \cos^2(A)}{\frac{\sin^2(A)\,\cancel{\cos^2(A)}}{\cancel{\cos^2(A)}} + \cos^2(A)}$$

$$= \frac{\sin^2(A) - \cos^2(A)}{\sin^2(A) + \cos^2(A)}$$

$$= \frac{\sin^2(A) - \cos^2(A)}{1}$$

$$= \sin^2(A) - \cos^2(A)$$

9.4.4 Solve quadratic trigonometric equations

In your studies so far, you have factorised quadratic expressions and solved quadratic equations. These skills can also be used to factorise and solve equations when the quadratic is written in terms of a **trigonometric function**. For instance, in Worked example 18, we create the variable $a = \sin(A)$.

WORKED EXAMPLE 18 Solving quadratic trigonometric functions (1)

Solve the equation $2\sin^2(A) = \sin(A)$ for A over the domain $0 \leq A \leq 2\pi$.

THINK

1. Write the equation.

2. Move $\sin(A)$ to the left of the equation to make the equation equal to zero.

3. This is a quadratic in terms of $\sin(A)$. It can be easier to identify the factors if you rewrite the quadratic in terms of a variable.

4. Rewrite the equation in terms of $\sin(A)$.

5. Using the Null Factor Law, solve each factor equal to 0.

WRITE

$2\sin^2(A) = \sin(A)$

$2\sin^2(A) - \sin(A) = 0$

Let $a = \sin(A)$.
$2a^2 - a = 0$
$a(2a - 1) = 0$

$\sin(A)(2\sin(A) - 1) = 0$

$\sin(A) = 0$ or $2\sin(A) - 1 = 0$

6. Solve $\sin(A) = 0$.

$$\sin(A) = 0 \qquad \sin(A) = \frac{1}{2}$$
$$A = 0, \pi, 2\pi$$

7. Solve $\sin(A) = \frac{1}{2}$, remembering that $\sin(A)$ will be positive in quadrants 1 and 2.

$\sin\left(\dfrac{\pi}{6}\right) = \dfrac{1}{2}$ is a known trigonometric ratio.

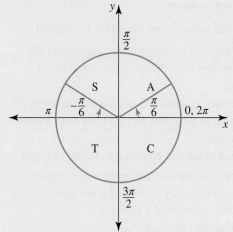

$$\sin(A) = \frac{1}{2}$$

The quadrant 1 solution becomes:
$$A = \frac{\pi}{6}$$

The quadrant 2 solution becomes:
$$A = \frac{5\pi}{6}$$

8. Combine all 5 solutions to the equation.

$$A = \left\{0, \frac{\pi}{6}, \frac{5\pi}{6}, \pi, 2\pi\right\}$$

Sometimes it is necessary to use the Pythagorean identities to rewrite the expression as a quadratic in terms of one trigonometric function.

WORKED EXAMPLE 19 Solving quadratic trigonometric functions (2)

Use a Pythagorean identity to factorise $\operatorname{cosec}^2(A) - \cot(A) - 3$.

THINK

1. Use the identity $1 + \cot^2(A) = \operatorname{cosec}^2(A)$ and then collect like terms.

2. This is a quadratic in terms of $\cot(A)$. It can be easier to identify the factors if you rewrite the quadratic in terms of a variable.

3. Rewrite the factors in terms of $\cot(A)$.

WRITE

$$\operatorname{cosec}^2(A) - \cot(A) - 3 = 1 + \cot^2(A) - \cot(A) - 3$$
$$= \cot^2(A) - \cot(A) - 2$$

Let $a = \cot(A)$.
$$\cot^2(A) - \cot(A) - 2 = a^2 - a - 2$$
$$= (a - 2)(a + 1)$$

$$\operatorname{cosec}^2(A) - \cot(A) - 3 = (\cot(A) - 2)(\cot(A) + 1)$$

Solve the equation $2\sin^2(A) = \cos(A) + 1$ for A over the domain $0 \leq \theta \leq 2\pi$.

THINK	WRITE
1. Write the equation.	$2\sin^2(A) = \cos(A) + 1$
2. Make the substitution $\sin^2(A) = 1 - \cos^2(A)$.	$2\left(1 - \cos^2(A)\right) = \cos(A) + 1$
3. Form a quadratic equation by expanding the brackets and then bringing all of the terms to one side. Before we factorise a quadratic, we normally rearrange to write the square term with a positive coefficient. This is a quadratic equation in terms of $\cos(A)$.	$2 - 2\cos^2(A) = \cos(A) + 1$ $0 = \cos(A) + 1 + 2\cos^2(A) - 2$ $0 = 2\cos^2(A) + \cos(A) - 1$
4. Factors of quadratic trigonometric equations can be more easily identified if the quadratic is rewritten in terms of a variable.	Let $a = \cos(A)$. $2a^2 + a - 1 = 0$
5. Factorise the equation. 6. Rewrite the factors in terms of $\cos(A)$.	$(2a - 1)(a + 1) = 0$ $(2\cos(A) - 1)(\cos(A) + 1) = 0$
7. Using the Null Factor Law, solve each factor equal to 0.	$2\cos(A) - 1 = 0$ or $\cos(A) + 1 = 0$ $\cos(A) = \dfrac{1}{2}$ $\qquad$ $\cos(A) = -1$
8. Solve $\cos(A) = \dfrac{1}{2}$, remembering that $\cos(A)$ will be positive in quadrants 1 and 4. $\cos\left(\dfrac{\pi}{3}\right) = \dfrac{1}{2}$ is a known trigonometric ratio.	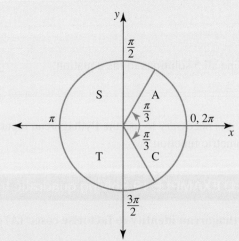 $\cos\left(\dfrac{\pi}{3}\right) = \dfrac{1}{2}$ The quadrant 1 solution becomes: $A = \dfrac{\pi}{3}$ The quadrant 4 solution becomes: $A = 2\pi - \dfrac{\pi}{3}$ $\quad = \dfrac{5\pi}{3}$

9. Solve $\cos(A) = -1$.
    $\cos(\pi) = -1$ is a known trigonometric ratio.

    $\cos(A) = -1$
    $A = \pi$

10. Combine all solutions.

    $A = \left\{ \dfrac{\pi}{3}, \pi, \dfrac{5\pi}{3} \right\}$

WORKED EXAMPLE 21 Solving quadratic trigonometric functions (4)

Determine the general solution to $2\sin^2(2A) - \sin(2A) - 1 = 0$.

THINK

1. This is a quadratic in terms of $\sin(2A)$. Factors of quadratic trigonometric equations can be more easily identified if the quadratic is rewritten in terms of a variable.

2. Factorise the equation.

3. Rewrite the factors in terms of $\sin(2A)$.

4. Using the Null Factor Law, solve each factor equal to 0.

5. Solve $\sin(2A) = -\dfrac{1}{2}$, remembering that $\sin(2A)$ will be negative in quadrants 3 and 4, and $\sin\left(\dfrac{\pi}{6}\right) = \dfrac{1}{2}$ is a known trigonometric ratio.

WRITE

Let $a = \sin(2A)$.
$2a^2 - a - 1 = 0$

$(2a + 1)(a - 1) = 0$

$(2\sin(2A) + 1)(\sin(2A) - 1) = 0$

$2\sin(2A) + 1 = 0 \quad$ or $\quad \sin(2A) - 1 = 0$
$\quad\quad \sin(2A) = -\dfrac{1}{2} \quad\quad\quad \sin(2A) = 1$

$\sin(2A) = -\dfrac{1}{2}$

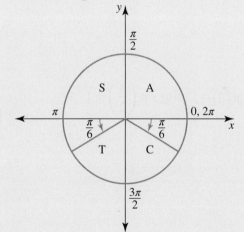

As a general solution is needed, each solution for $2A$ will require $2n\pi$, $n \in \mathbb{Z}$.

The quadrant 3 solution becomes:
$$2A = -\pi + \frac{\pi}{6} + 2n\pi$$
$$= -\frac{5\pi}{6} + 2n\pi$$
$$A = -\frac{5\pi}{12} + n\pi$$
$$= \pi\left(-\frac{5}{12} + n\right)$$
$$= \pi\left(\frac{-5 + 12n}{12}\right)$$
$$= \frac{\pi}{12}(12n - 5)$$

The quadrant 4 solution becomes:
$$2A = -\frac{\pi}{6} + 2n\pi$$
$$A = -\frac{\pi}{12} + n\pi$$
$$= \pi\left(-\frac{1}{12} + n\right)$$
$$= \pi\left(\frac{12n - 1}{12}\right)$$
$$= \frac{\pi}{12}(12n - 1)$$

6. Solve $\sin(2A) = 1$ using $\sin\left(\dfrac{\pi}{2}\right) = 1$.

$\sin(2A) = 1$
The general solution becomes:
$$2A = \frac{\pi}{2} + 2n\pi$$
$$A = \frac{\pi}{4} + n\pi$$
$$= \pi\left(\frac{1}{4} + n\right)$$
$$= \pi\left(\frac{1 + 4n}{4}\right)$$
$$= \frac{\pi}{4}(1 + 4n)$$

7. Combine all solutions.

$$A = \left\{ \frac{\pi}{12}(12n - 5), \frac{\pi}{12}(12n - 1), \frac{\pi}{4}(4n + 1), n \in \mathbb{Z} \right\}$$

9.4 Exercise	9.4 Exam questions `on`

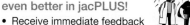

Simple familiar	Complex familiar	Complex unfamiliar
1, 2, 3, 4, 5, 6, 7, 8, 9, 10, 11	12, 13	14, 15, 16, 17, 18

Simple familiar

1. Copy and complete the table, correct to three decimal places:

A	30°	81°	129°	193°	260°	350°	−47°
$\sin^2(A)$							
$\cos^2(A)$							
$\sin^2(A) + \cos^2(A)$							

2. **WE14** If $\sin(A) = 0.8$ and $0° < A < 90°$, calculate, correct to three decimal places:
   a. $\cos(A)$ b. $\tan(A)$

3. If $\cos(A) = 0.3$ and $0° < A < 90°$, calculate, correct to three decimal places:
   a. $\sin(A)$ b. $\tan(A)$

4. **WE15** Determine all possible values of the following, correct to three decimal places.
   a. $\cos(x)$ if $\sin(x) = 0.4$ b. $\cos(x)$ if $\sin(x) = -0.7$
   c. $\sin(x)$ if $\cos(x) = 0.24$ d. $\sin(x)$ if $\cos(x) = -0.9$

5. Use the diagram to calculate the exact value of:
   a. $\sin(x)$ b. $\cos(x)$

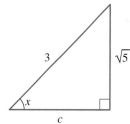

6. Use the diagram to determine the exact value of:
   a. $\cos(x)$ b. $\tan(x)$

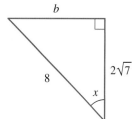

7. Given that $\sin(A) = \dfrac{\sqrt{5}}{4}$ and $\dfrac{3\pi}{2} < A < 2\pi$, calculate:

   a. the exact value of $\cos(A)$
   b. the exact value of $\tan(A)$.

8. **WE16** Use the Pythagorean identity to simplify $\tan^2(A) - \sin^2(A)\tan^2(A)$.

9. **WE17** Simplify $\dfrac{\operatorname{cosec}(A)\cos^2(A)}{1 + \operatorname{cosec}(A)}$.

10. **WE18** Solve each of the following equations over the domain $0 \le x \le 2\pi$.

   a. $\sin^2(x) - \sin(x) = 0$ b. $\cos^2(x) + \cos(x) = 0$

   c. $2\sin^2(x) + \sqrt{3}\sin(x) = 0$ d. $2\cos^2(x) + \cos(x) - 1 = 0$

11. **WE19** Use a Pythagorean identity to factorise each of the following.

   a. $1 + \sin(A) - 2\cos^2(A)$ b. $\sec^2(x) - \tan(x) - 3$

   c. $2\cot^2(\alpha) - \operatorname{cosec}(\alpha) + 1$ d. $2\cos^2(3x) - 2\sin^2(3x) + 1$

Complex familiar

12. **WE20** Solve each of the following equations over the domain $0 \le A \le 2\pi$.

   a. $2\cos^2(A) = 1 + \sin(A)$ b. $2\sin^2(A) + \sin(A) - 1 = 0$

   c. $2\sin^2(A) - 1 = 0$ d. $1 + \cos(A) = 2\sin^2(A)$

13. **WE21** Determine the general solution to each of the following equations.

   a. $2\sin^2(2A) - 3\sin(2A) + 1 = 0$ b. $2\cos^2(2A) + \cos(2A) - 1 = 0$

   c. $2\sin^2(4x) + \sin(4x) = 0$ d. $\cos^2(4x) - \cos(4x) = 0$

Complex unfamiliar

14. Determine the range of $\sin^4(x) + \cos^2(x)$.

15. Determine the general solution to each of the following equations.

   a. $\tan^2(x) + \left(\sqrt{3} + 1\right)\tan(x) + \sqrt{3} = 0$

   b. $\tan^2(x) + \left(\sqrt{3} - 1\right)\tan(x) - \sqrt{3} = 0$

16. Determine the general solution to each of the following equations.

   a. $2\sin^3(x) + \sin^2(x) - 2\sin(x) - 1 = 0$

   b. $2\cos^3(x) - \cos^2(x) - 2\cos(x) + 1 = 0$

17. Solve the following equation for A.

$$\frac{1}{\sin^2(A)} - \frac{1}{\cos^2(A)} - \frac{1}{\tan^2(A)} - \frac{1}{\cot^2(A)} - \frac{1}{\sec^2(A)} - \frac{1}{\operatorname{cosec}^2(A)} = -3,\ A \in [0,\ 2\pi]$$

18. If $\sin^{16}(A) = \dfrac{1}{9}$, solve the following equation for A. $\dfrac{1}{\cos^2(A)} + \dfrac{1}{1 + \sin^2(A)} + \dfrac{2}{1 + \sin^4(A)} + \dfrac{4}{1 + \sin^8(A)}$.

Fully worked solutions for this chapter are available online.

LESSON
9.5 Compound angle identities

SYLLABUS LINKS

- Prove and apply the angle sum, difference and double-angle identities for sines and cosines.
  - $\sin(A + B) = \sin(A)\cos(B) + \cos(A)\sin(B)$
  - $\sin(A - B) = \sin(A)\cos(B) - \cos(A)\sin(B)$
  - $\cos(A + B) = \cos(A)\cos(B) - \sin(A)\sin(B)$
  - $\cos(A - B) = \cos(A)\cos(B) + \sin(A)\sin(B)$
  - $\sin(2A) = 2\sin(A)\cos(A)$
  - $\cos(2A) = \cos^2(A) - \sin^2(A)$
    $$= 1 - 2\sin^2(A)$$
    $$= 2\cos^2(A) - 1$$

Source: Specialist Mathematics Senior Syllabus 2024 © State of Queensland (QCAA) 2024; licensed under CC BY 4.0

9.5.1 Prove angle sum and angle difference identities

Consider a rectangle PQSU with right-angled triangles PQR and PRT as shown. The length of PT is 1.

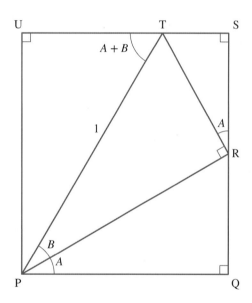

Consider $\triangle$PRT.

$$\cos(B) = \frac{PR}{PT} \qquad\qquad \sin(B) = \frac{RT}{PT}$$
$$= \frac{PR}{1} \qquad\qquad\qquad = \frac{RT}{1}$$
$$PR = \cos(B) \qquad\qquad RT = \sin(B)$$

Consider $\triangle PQR$.

$$\cos(A) = \frac{PQ}{PR}$$
$$= \frac{PQ}{\cos(B)}$$
$$PQ = \cos(A)\cos(B)$$

$$\sin(A) = \frac{RQ}{PR}$$
$$= \frac{RQ}{\cos(B)}$$
$$RQ = \sin(A)\cos(B)$$

Consider $\triangle RST$.

$$\sin(A) = \frac{ST}{RT}$$
$$= \frac{ST}{\sin(B)}$$
$$ST = \sin(A)\sin(B)$$

$$\cos(A) = \frac{RS}{RT}$$
$$= \frac{RS}{\sin(B)}$$
$$RS = \cos(A)\sin(B)$$

Consider $\triangle PTU$.

As $UT \parallel PQ$, $\angle UTP = \angle TPQ$ (Z)

$$= A + B$$

$$\cos(A + B) = \frac{UT}{PT}$$
$$= \frac{UT}{1}$$
$$UT = \cos(A + B)$$

$$\sin(A + B) = \frac{PU}{PT}$$
$$= \frac{PU}{1}$$
$$PU = \sin(A + B)$$

Adding these lengths gives us the following diagram.

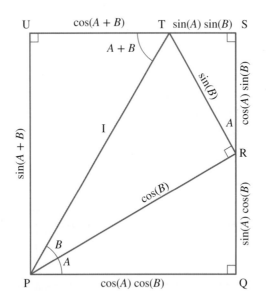

In the rectangle, $PU = QS$. Therefore, $\sin(A + B) = \sin(A)\cos(B) + \cos(A)\sin(B)$.

As $PQ = US$, $\cos(A)\cos(B) = \cos(A + B) + \sin(A)\sin(B)$
$$\cos(A + B) = \cos(A)\cos(B) - \sin(A)\sin(B)$$

If we consider $A + (-B)$ and remember that $\cos(-B) = \cos(B)$ and $\sin(-B) = -\sin(B)$, then the identities become:

$$\sin(A + (-B)) = \sin(A)\,\cos(-B) + \cos(A)\,\sin(-B)$$
$$\sin(A - B) = \sin(A)\,\cos(B) - \cos(A)\,\sin(B)$$

$$\cos(A + (-B)) = \cos(A)\,\cos(-B) - \sin(A)\,\sin(-B)$$
$$\cos(A - B) = \cos(A)\,\cos(B) + \sin(A)\,\sin(B)$$

These are known as the **angle sum** and **angle difference identities**, or compound angle identities.

The angle sum identities

$$\sin(A + B) = \sin(A)\,\cos(B) + \cos(A)\,\sin(B)$$
$$\cos(A + B) = \cos(A)\,\cos(B) - \sin(A)\,\sin(B)$$

The angle difference identities

$$\sin(A - B) = \sin(A)\,\cos(B) - \cos(A)\,\sin(B)$$
$$\cos(A - B) = \cos(A)\,\cos(B) + \sin(A)\,\sin(B)$$

Each of these rules can be used in both ways, depending on the question.

WORKED EXAMPLE 22 Evaluating expressions using the compound angle identities

Evaluate $\sin(22°)\cos(38°) + \cos(22°)\sin(38°)$, giving your answer as an exact value.

THINK	WRITE
1. The identity $\sin(A + B) = \sin(A)\,\cos(B) + \cos(A)\,\sin(B)$ can be used.	Let $A = 22°$ and $B = 38°$. $\sin(A + B) = \sin(A)\,\cos(B) + \cos(A)\,\sin(B)$ $\sin(22°)\cos(38°) + \cos(22°)\sin(38°) = \sin(22° + 38°)$
2. Simplify the expression	$= \sin(60°)$
3. If the exact value is known, then simplify further.	$= \dfrac{\sqrt{3}}{2}$

The compound angle identities can be used to simply expressions, particularly when the sum or difference involves an angle with a known trigonometric ratio.

WORKED EXAMPLE 23 Expanding expressions using the compound angle identities

Expand $2\cos\left(\theta + \dfrac{\pi}{3}\right)$.

THINK	WRITE
1. The identity $\cos(A + B) = \cos(A)\,\cos(B) - \sin(A)\,\sin(B)$ can be used.	Let $A = \theta$ and $B = \dfrac{\pi}{3}$. $\cos(A + B) = \cos(A)\,\cos(B) - \sin(A)\,\sin(B)$ $2\cos\left(\theta + \dfrac{\pi}{3}\right) = 2\left(\cos(\theta)\cos\left(\dfrac{\pi}{3}\right) - \sin(\theta)\sin\left(\dfrac{\pi}{3}\right)\right)$

▶

2. Exact values $\cos\left(\dfrac{\pi}{3}\right) = \dfrac{1}{2}$ and $\qquad\qquad\qquad$ $= 2\left(\dfrac{1}{2}\cos(\theta) - \dfrac{\sqrt{3}}{2}\sin(\theta)\right)$

$\sin\left(\dfrac{\pi}{3}\right) = \dfrac{\sqrt{3}}{2}$ are known, so use these
to simplify.

3. Simplify. $\qquad\qquad\qquad\qquad\qquad\qquad$ $= \cos(\theta) - \sqrt{3}\sin(\theta)$

4. State the solution. $\qquad\qquad$ $2\cos\left(\theta + \dfrac{\pi}{3}\right) = \cos(\theta) - \sqrt{3}\sin(\theta)$

WORKED EXAMPLE 24 Simplifying expressions using the compound angle identities

Use one of the compound angle identities to simplify $\cos\left(\dfrac{3\pi}{2} - \theta\right)$.

THINK

1. The identity
$\cos(A - B) = \cos(A)\cos(B) + \sin(A)\sin(B)$
can be used.

2. Exact values $\cos\left(\dfrac{3\pi}{2}\right) = 0$ and

$\sin\left(\dfrac{3\pi}{2}\right) = -1$ are known, so use these

to simplify.

3. Simplify.

4. State the solution.

WRITE

Let $A = \dfrac{3\pi}{2}$ and $B = \theta$.

$\cos(A - B) = \cos(A)\cos(B) + \sin(A)\sin(B)$

$\cos\left(\dfrac{3\pi}{2} - \theta\right) = \cos\left(\dfrac{3\pi}{2}\right)\cos(\theta) + \sin\left(\dfrac{3\pi}{2}\right)\sin(\theta)$

$= 0 \times \cos(\theta) - 1 \times \sin(\theta)$

$= -\sin(\theta)$

$\cos\left(\dfrac{3\pi}{2} - \theta\right) = -\sin(\theta)$

9.5.2 Determining exact values

In previous work, we have used the exact values for $30°$ and $45°$

$\left(\text{or } \dfrac{\pi}{6} \text{ and } \dfrac{\pi}{4}\right)$ and multiples of these values.

As $45° - 30° = 15°$ $\left(\text{or } \dfrac{\pi}{4} - \dfrac{\pi}{6} = \dfrac{\pi}{12}\right)$, it is possible to use the
compound angle identities to evaluate exact values for multiples of

$15°$ $\left(\dfrac{\pi}{12}\right)$.

Calculate the exact value of $\sin\left(\dfrac{13\pi}{12}\right)$.

THINK

1. Express $\dfrac{13\pi}{12}$ in terms of multiples of $\dfrac{\pi}{4}$ and $\dfrac{\pi}{6}$.

2. The compound angle identities
   $\sin(A+B) = \sin(A)\cos(B) + \cos(A)\sin(B)$
   can be used.

3. As $\dfrac{5\pi}{6}$ is in quadrant 2, $\sin\left(\dfrac{5\pi}{6}\right)$ is positive and $\cos\left(\dfrac{5\pi}{6}\right)$ is negative.

4. $\dfrac{\pi}{4}$ is in quadrant 1, so both $\sin\left(\dfrac{\pi}{4}\right)$ and $\cos\left(\dfrac{\pi}{4}\right)$ will be positive.

5. Substitute values for $\sin\left(\dfrac{5\pi}{6}\right)$, $\cos\left(\dfrac{5\pi}{6}\right)$, $\sin\left(\dfrac{\pi}{4}\right)$, $\cos\left(\dfrac{\pi}{4}\right)$ and simplify.

WRITE

$\dfrac{13\pi}{12} = \dfrac{10\pi}{12} + \dfrac{3\pi}{12}$

$= \dfrac{5\pi}{6} + \dfrac{\pi}{4}$

Let $A = \dfrac{5\pi}{6}$ and $B = \dfrac{\pi}{4}$.

$\sin(A+B) = \sin(A)\cos(B) + \cos(A)\sin(B)$

$\sin\dfrac{13\pi}{12} = \sin\left(\dfrac{5\pi}{6} + \dfrac{\pi}{4}\right)$

$= \sin\left(\dfrac{5\pi}{6}\right)\cos\left(\dfrac{\pi}{4}\right) + \cos\left(\dfrac{5\pi}{6}\right)\sin\left(\dfrac{\pi}{4}\right)$

$\sin\left(\dfrac{5\pi}{6}\right) = \sin\left(\dfrac{\pi}{6}\right) \qquad \cos\left(\dfrac{5\pi}{6}\right) = -\cos\left(\dfrac{\pi}{6}\right)$

$\qquad\qquad = \dfrac{1}{2} \qquad\qquad\qquad\qquad = -\dfrac{\sqrt{3}}{2}$

$\sin\left(\dfrac{\pi}{4}\right) = \cos\left(\dfrac{\pi}{4}\right) = \dfrac{\sqrt{2}}{2}$

$\sin\left(\dfrac{13\pi}{12}\right) = \dfrac{1}{2} \times \dfrac{\sqrt{2}}{2} - \dfrac{\sqrt{3}}{2} \times \dfrac{\sqrt{2}}{2}$

$= \dfrac{\sqrt{2}}{4} - \dfrac{\sqrt{6}}{4}$

$= \dfrac{\sqrt{2} - \sqrt{6}}{4}$

9.5.3 Applying compound angle identities

Sometimes it is necessary to use the Pythagorean identities to find the unknown ratios before it is possible to use the compound angle identities.

WORKED EXAMPLE 26 Calculating exact values using the compound angle identities (2)

If $\cos(A) = \dfrac{12}{13}$ and $\sin(B) = \dfrac{7}{25}$, where $0 < A < \dfrac{\pi}{2}$ and $\dfrac{\pi}{2} < B < \pi$, calculate the exact value of $\sin(A - B)$.

THINK	WRITE
1. To calculate $\sin(A - B)$ we will need to calculate $\sin(A)$ and $\cos(B)$. The Pythagorean identity can be used.	$\sin^2(A) + \cos^2(A) = 1$ $\sin^2(A) + \left(\dfrac{12}{13}\right)^2 = 1$ $\sin^2(A) = 1 - \dfrac{144}{169}$ $\qquad = \dfrac{25}{169}$ $\sin(A) = \pm\dfrac{5}{13}$ As $0 < A < \dfrac{\pi}{2}$, $\sin(A) > 0$, $\sin(A) = \dfrac{5}{13}$ $\sin^2(B) + \cos^2(B) = 1$ $\left(\dfrac{7}{25}\right)^2 + \cos^2(B) = 1$ $\cos^2(B) = 1 - \dfrac{49}{625}$ $\qquad = \dfrac{576}{625}$ $\cos(B) = \pm\dfrac{24}{25}$ As $\dfrac{\pi}{2} < B < \pi$, $\cos(B) < 0$, $\cos(B) = -\dfrac{24}{25}.$

2. Use the compound angle identity
$\sin(A - B) = \sin(A) \cos(B) - \cos(A) \sin(B)$ to simplify and solve.

$$\sin(A - B) = \sin(A) \cos(B) - \cos(A) \sin(B)$$

$$= \frac{5}{13} \times -\frac{24}{25} - \frac{12}{13} \times \frac{7}{25}$$

$$= -\frac{120}{325} - \frac{84}{325}$$

$$= -\frac{204}{325}$$

TI	THINK	WRITE

1. Put the calculator in RADIAN mode.
On a Calculator page, select MENU, then select:
3: Algebra
1: Numerical Solve.
Complete the entry line as:
$nSolve(\cos(a) = \frac{12}{13}, a)|0 < a < \frac{\pi}{2}$
Press [ctrl], [var] and A to store the result as a. Press ENTER.

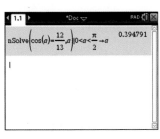

2. Select MENU, then select:
3: Algebra
1: Numerical Solve.
Complete the entry line as:
$nSolve(\sin(b) = \frac{7}{25}, b)|\frac{\pi}{2} < b < \pi$
Press [ctrl], [var] and B to store the result as b. Press ENTER.

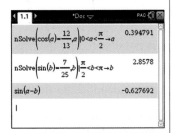

3. Complete the next entry line as:
$\sin(a - b)$
Then press ENTER.

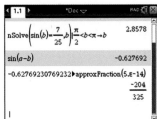

4. Press MENU, then select:
2: Number
2: Approximate to Fraction.
Then press ENTER.

5. The answer appears on the screen.
$\sin(A - B) = -\dfrac{204}{325}$

CASIO	THINK	WRITE

1. Form an expression for A.
For $0 < A < \dfrac{\pi}{2}$, if $\cos(A) = \dfrac{12}{13}$,
then $A = \cos^{-1}\left(\dfrac{12}{13}\right)$.[1]

2. Form an expression for B.
$\sin(B) = \dfrac{7}{25}$ For $\dfrac{\pi}{2} < B < \pi$,
$\sin(\pi - B) = \dfrac{7}{25}$ and
$B = \pi - \sin^{-1}\left(\dfrac{7}{25}\right)$ [2]

3. Form an expression for $\sin(A - B)$.
Substituting [1] and [2] into $\sin(A - B)$ gives:
$$\sin(A - B) = \sin\left(\cos^{-1}\left(\dfrac{12}{13}\right) - \left(\pi - \sin^{-1}\left(\dfrac{7}{25}\right)\right)\right)$$

4. Put the calculator in RADIAN mode.
On a Run-Matrix screen, complete the entry line as:
$$\sin\left(\cos^{-1}\left(\dfrac{12}{13}\right) - \left(\pi - \sin^{-1}\left(\dfrac{7}{25}\right)\right)\right)$$
Then press EXE.

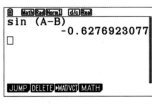

5. The answer appears on the screen.
$\sin(A - B) = -0.627... = -\dfrac{204}{325}$

The compound angle identities can be used to prove other identities.

Prove that $\sin(A+B)\sin(A-B) = \sin^2(A) - \sin^2(B)$.

THINK	WRITE
1. When trying to decide where to begin, it is often best to begin with the more complicated side (the side with more terms or with the compound angles). In this case, it is the left-hand side.	$\text{LHS} = \sin(A+B)\sin(A-B)$
2. Substitute the compound angle identities.	$= (\sin(A)\cos(B) + \cos(A)\sin(B))$ $\times (\sin(A)\cos(B) - \cos(A)\sin(B))$
3. Expand the brackets. Notice in this equation the difference of two squares pattern can be used.	$= \sin^2(A)\cos^2(B) - \cos^2(A)\sin^2(B)$
4. The right-hand side of the identity is written in terms of sine. Use the Pythagorean identity to replace the cosines.	$= \sin^2(A)\left(1 - \sin^2(B)\right) - \left(1 - \sin^2(A)\right)\sin^2(B)$
5. Simplify. As it has been demonstrated that the left-hand side is equal to the right-hand side, the identity is proven.	$= \sin^2(A) - \sin^2(A)\sin^2(B) - \sin^2(B) + \sin^2(A)\sin^2(B)$ $= \sin^2(A) - \sin^2(B)$ $= \text{RHS}$ $\therefore \sin(A+B)\sin(A-B) = \sin^2(A) - \sin^2(B)$

9.5.4 Double angle identities

If the trigonometric ratio of a particular angle is known, it is possible to find the trigonometric ratio of multiples of that angle.

Using the compound angle identities, it is possible to determine identities for double angles.

Using $\sin(A+B) = \sin(A)\cos(B) + \cos(A)\sin(B)$, we can determine an expression for $\sin(2A)$.

$$\begin{aligned} \sin(2A) &= \sin(A+A) \\ &= \sin(A)\cos(A) + \cos(A)\sin(A) \\ &= 2\sin(A)\cos(A) \end{aligned}$$

Similarly, using $\cos(A+B) = \cos(A)\cos(B) - \sin(A)\sin(B)$:

$$\begin{aligned} \cos(2A) &= \cos(A+A) \\ &= \cos(A)\cos(A) - \sin(A)\sin(A) \\ &= \cos^2(A) - \sin^2(A) \end{aligned}$$

Using $\cos^2(A) = 1 - \sin^2(A)$: $\cos(2A) = \left(1 - \sin^2(A)\right) - \sin^2(A)$
$$= 1 - 2\sin^2(A)$$

Using $\sin^2(A) = 1 - \cos^2(A)$: $\cos(2A) = \cos^2(A) - \left(1 - \cos^2(A)\right)$

$$= 2\cos^2(A) - 1$$

These are known as the **double angle identities**.

Double angle identities

$$\sin(2A) = 2\sin(A)\cos(A)$$

$$\cos(2A) = \cos^2(A) - \sin^2(A)$$

$$= 2\cos^2(A) - 1$$

$$= 1 - 2\sin^2(A)$$

The double angle identities express $\sin(2A)$ and $\cos(2A)$ in terms of $\sin(A)$ and $\cos(A)$. Using these identities, it is also possible to express $\sin(A)$ and $\cos(A)$ in terms of $\sin\left(\dfrac{A}{2}\right)$ and $\cos\left(\dfrac{A}{2}\right)$, which are the **half-angle identities**.

Half-angle identities

$$\sin(A) = 2\sin\left(\frac{A}{2}\right)\cos\left(\frac{A}{2}\right)$$

$$\cos(A) = \cos^2\left(\frac{A}{2}\right) - \sin^2\left(\frac{A}{2}\right)$$

$$= 2\cos^2\left(\frac{A}{2}\right) - 1$$

$$= 1 - 2\sin^2\left(\frac{A}{2}\right)$$

9.5.5 Determining exact values

The double angle identities can be used to determine exact values of expressions. The identities may be used in either direction.

WORKED EXAMPLE 28 Evaluating using double angle identities (1)

Determine the exact value of $\sin\left(\dfrac{7\pi}{12}\right)\cos\left(\dfrac{7\pi}{12}\right)$.

THINK	WRITE
1. As the expression is a multiple of sine and cosine of the same angle, the identity $\sin(2A) = 2\sin(A)\cos(A)$ can be used.	$2\sin(A)\cos(A) = \sin(2A)$ $\sin(A)\cos(A) = \dfrac{1}{2}\sin(2A)$

2. Substitute $A = \dfrac{7\pi}{12}$ and simplify.

$$\sin\left(\dfrac{7\pi}{12}\right)\cos\left(\dfrac{7\pi}{12}\right) = \dfrac{1}{2}\sin\left(2 \times \dfrac{7\pi}{12}\right)$$
$$= \dfrac{1}{2}\sin\left(\dfrac{7\pi}{6}\right)$$

3. $\dfrac{7\pi}{6}$ is in quadrant 3; $\sin\left(\dfrac{\pi}{6}\right) = \dfrac{1}{2}$ is a known trigonometric ratio and is negative in quadrant 3.

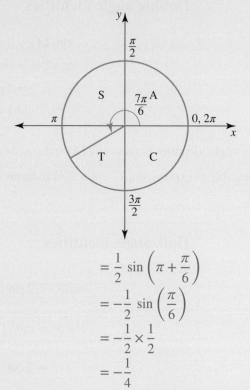

$$= \dfrac{1}{2}\sin\left(\pi + \dfrac{\pi}{6}\right)$$
$$= -\dfrac{1}{2}\sin\left(\dfrac{\pi}{6}\right)$$
$$= -\dfrac{1}{2} \times \dfrac{1}{2}$$
$$= -\dfrac{1}{4}$$

TI | THINK	WRITE	CASIO | THINK	WRITE
1. Put the calculator in RADIAN mode. On a Calculator page, complete the entry line as: $\sin\left(\dfrac{7\pi}{12}\right) \times \cos\left(\dfrac{7\pi}{12}\right)$ Then press ENTER.		**1.** Put the calculator in RADIAN mode. On a Run-Matrix screen, complete the entry line as: $\sin\left(\dfrac{7\pi}{12}\right) \times \cos\left(\dfrac{7\pi}{12}\right)$ Then press EXE.	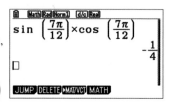
2. Type '–0.25', then press MENU and select: 2: Number 2: Approximate to Fraction. Then press ENTER.	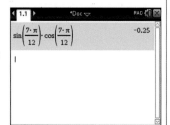	**2.** The answer appears on the screen.	$\sin\left(\dfrac{7\pi}{12}\right) \times \cos\left(\dfrac{7\pi}{12}\right) = -\dfrac{1}{4}$
3. The answer appears on the screen.	$\sin\left(\dfrac{7\pi}{12}\right) \times \cos\left(\dfrac{7\pi}{12}\right) = -\dfrac{1}{4}$		

WORKED EXAMPLE 29 Evaluating using double angle identities (2)

If $0 \leq A \leq \dfrac{\pi}{2}$ and $\cos(A) = \dfrac{1}{4}$, determine the exact values of:

a. $\cos(2A)$ **b.** $\sin(2A)$ **c.** $\tan(2A)$

THINK	WRITE
a. 1. As we know the value for $\cos(A)$, use $\cos(2A) = 2\cos^2(A) - 1$.	$\cos(2A) = 2\cos^2(A) - 1$
2. Substitute for $\cos(A)$.	$= 2 \times \left(\dfrac{1}{4}\right)^2 - 1$ $= \dfrac{2}{16} - 1$ $= -\dfrac{14}{16}$ $= -\dfrac{7}{8}$
b. 1. As $\sin(2A) = 2\sin(A)\cos(A)$, it is necessary to calculate $\sin(A)$. The Pythagorean identity can be used.	$\sin^2(A) + \cos^2(A) = 1$ $\sin^2(A) + \left(\dfrac{1}{4}\right)^2 = 1$ $\sin^2(A) + \dfrac{1}{16} = 1$ $\sin^2(A) = \dfrac{15}{16}$ $\sin(A) = \pm\dfrac{\sqrt{15}}{4}$
2. $0 \leq A \leq \dfrac{\pi}{2}$ so $\sin(A) > 0$.	As $0 \leq A \leq \dfrac{\pi}{2}$, $\sin(A) = \dfrac{\sqrt{15}}{4}$.
3. Use the identity to determine $\sin(2A)$.	$\sin(2A) = 2\sin(A)\cos(A)$ $= 2 \times \dfrac{\sqrt{15}}{4} \times \dfrac{1}{4}$ $= \dfrac{\sqrt{15}}{8}$
c. Use the identity $\tan(2A) = \dfrac{\sin(2A)}{\cos(2A)}$.	$\tan(2A) = \dfrac{\sin(2A)}{\cos(2A)}$ $= \dfrac{\frac{\sqrt{15}}{8}}{-\frac{7}{8}}$ $= \dfrac{\sqrt{15}}{8} \times -\dfrac{8}{7}$ $= -\dfrac{\sqrt{15}}{7}$

9.5.6 Applying the double angle identities

Previously, the Pythagorean identity was used to rewrite expressions using a single trigonometric ratio. In a similar fashion, the double angle identities can be used to rewrite the expressions in terms of the same angle.

WORKED EXAMPLE 30 Solving trigonometric equations using double angle identities

If $\sin(2x) + \sqrt{3}\cos(x) = 0$, solve for $x \in [0, 2\pi]$.

THINK

1. The equation involves trigonometric ratios of both x and $2x$. Use the double angle identity to rewrite $\sin(2x)$ in terms of $\sin(x)$ and $\cos(x)$.

2. Factorise by identifying the common factor of $\cos(x)$.

3. Solve each factor equal to 0.

4. Solve $\cos(x) = 0$.

5. Solve $\sin(x) = -\dfrac{\sqrt{3}}{2}$, remembering that $\sin(x)$ is negative in quadrants 3 and 4. $\sin\left(\dfrac{\pi}{3}\right) = \dfrac{\sqrt{3}}{2}$ is a known trigonometric ratio.

6. Combine the solutions.

WRITE

$$\sin(2x) + \sqrt{3}\cos(x) = 0$$
$$2\sin(x)\cos(x) + \sqrt{3}\cos(x) = 0$$

$$\cos(x)\left(2\sin(x) + \sqrt{3}\right) = 0$$

$$\cos(x) = 0 \text{ or } 2\sin(x) + \sqrt{3} = 0$$
$$\sin(x) = -\frac{\sqrt{3}}{2}$$

$$\cos(x) = 0$$
$$x = \frac{\pi}{2}, \frac{3\pi}{2}$$

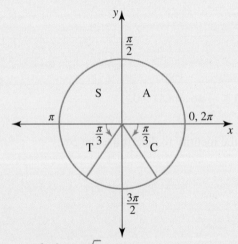

$$\sin\left(\frac{\pi}{3}\right) = \frac{\sqrt{3}}{2}$$

$$x = \pi + \frac{\pi}{3} \quad \text{or} \quad x = 2\pi - \frac{\pi}{3}$$
$$= \frac{4\pi}{3} \qquad\qquad = \frac{5\pi}{3}$$

$$x = \left\{\frac{\pi}{2}, \frac{4\pi}{3}, \frac{3\pi}{2}, \frac{5\pi}{3}\right\}$$

The double angle identities can be combined with the other identities covered in this chapter and the Pythagorean identities to prove trigonometric identities. When deciding where to start, it is often best to begin with the more complex side and work to simplify it.

WORKED EXAMPLE 31 Proving trigonometric identities using double angle identities

Prove the identity $\dfrac{\cos(2\alpha)\cos(\alpha) + \sin(2\alpha)\sin(\alpha)}{\sin(3\alpha)\cos(\alpha) - \cos(3\alpha)\sin(\alpha)} = \dfrac{1}{2}\text{cosec}(\alpha).$

THINK

1. The left-hand side is the more complex side, so begin with that side.

2. Use the angle difference identities. The numerator can be simplified using $\cos(A - B) = \cos(A)\cos(B) + \sin(A)\sin(B)$, and the denominator can be simplified using $\sin(A - B) = \sin(A)\cos(B) - \cos(A)\sin(B)$.

3. Simplify.

4. Substitute $\sin(2\alpha) = 2\sin(\alpha)\cos(\alpha)$ for the denominator.

5. Simplify the fraction and use $\text{cosec}(A) = \dfrac{1}{\sin(A)}$.

6. As it has been demonstrated that the left-hand side is equal to the right-hand side, the identity is proven.

WRITE

$\text{LHS} = \dfrac{\cos(2\alpha)\cos(\alpha) + \sin(2\alpha)\sin(\alpha)}{\sin(3\alpha)\cos(\alpha) - \cos(3\alpha)\sin(\alpha)}$

$= \dfrac{\cos(2\alpha - \alpha)}{\sin(3\alpha - \alpha)}$

$= \dfrac{\cos(\alpha)}{\sin(2\alpha)}$

$= \dfrac{\cos(\alpha)}{2\sin(\alpha)\cos(\alpha)}$

$= \dfrac{1}{2\sin(\alpha)}$

$= \dfrac{1}{2}\text{cosec}(\alpha)$

$= \text{RHS}$

$\therefore \dfrac{\cos(2\alpha)\cos(\alpha) + \sin(2\alpha)\sin(\alpha)}{\sin(3\alpha)\cos(\alpha) - \cos(3\alpha)\sin(\alpha)} = \dfrac{1}{2}\text{cosec}(\alpha)$

WORKED EXAMPLE 32 Proving trigonometric identities using half-angle identities

Prove the identity $\text{cosec}(A) - \cot(A) = \tan\left(\dfrac{A}{2}\right).$

THINK

1. The left-hand side is the more complex side, so begin with that side.

2. Remember that $\text{cosec}(A) = \dfrac{1}{\sin(A)}$ and $\cot(A) = \dfrac{\cos(A)}{\sin(A)}$ and make the substitutions.

WRITE

$\text{LHS} = \text{cosec}(A) - \cot(A)$

$= \dfrac{1}{\sin(A)} - \dfrac{\cos(A)}{\sin(A)}$

3. As the fractions have a common denominator, simplify the expression by adding the fractions.

$$= \frac{1 - \cos(A)}{\sin(A)}$$

4. On the right-hand side, the angle is $\frac{A}{2}$, so use the half angle identities for $\cos(A)$ and $\sin(A)$. There are three options to use as a substitution for $\cos(A)$. In this instance, as $\tan\left(\frac{A}{2}\right) = \frac{\sin\left(\frac{A}{2}\right)}{\cos\left(\frac{A}{2}\right)}$, writing $\cos(A)$ in terms of $\sin\left(\frac{A}{2}\right)$ would be the first option to try.

$$= \frac{1 - \left(1 - 2\sin^2\left(\frac{A}{2}\right)\right)}{2\sin\left(\frac{A}{2}\right)\cos\left(\frac{A}{2}\right)}$$

5. Simplify the fraction.

$$= \frac{1 - 1 + 2\sin^2\left(\frac{A}{2}\right)}{2\sin\left(\frac{A}{2}\right)\cos\left(\frac{A}{2}\right)}$$

$$= \frac{2\sin^2\left(\frac{A}{2}\right)}{2\sin\left(\frac{A}{2}\right)\cos\left(\frac{A}{2}\right)}$$

$$= \frac{\sin\left(\frac{A}{2}\right)}{\cos\left(\frac{A}{2}\right)}$$

$$= \tan\left(\frac{A}{2}\right)$$

$$= \text{RHS}$$

6. As it has been demonstrated that the left-hand side is equal to the right-hand side, the identity is proven.

$$\therefore \operatorname{cosec}(A) - \cot(A) = \tan\left(\frac{A}{2}\right)$$

Exercise 9.5 Compound angle identities

9.5 Exercise	9.5 Exam questions on

Simple familiar	Complex familiar	Complex unfamiliar
1, 2, 3, 4, 5, 6, 7, 8, 9, 10	11, 12, 13, 14, 15, 16, 17, 18	19, 20, 21, 22, 23, 24, 25

Simple familiar

1. **WE22** Evaluate the following, giving your answers as exact values.
   a. $\sin(27°)\cos(33°) + \cos(27°)\sin(33°)$
   b. $\cos(47°)\cos(43°) - \sin(47°)\sin(43°)$
   c. $\cos(76°)\cos(16°) + \sin(76°)\sin(16°)$

2. **WE23** Expand the following.

   a. $\sqrt{2}\sin\left(\theta - \dfrac{\pi}{4}\right)$

   b. $2\sin\left(\theta + \dfrac{\pi}{3}\right)$

   c. $2\cos\left(\theta - \dfrac{\pi}{6}\right)$

   d. $\sqrt{2}\cos\left(\theta + \dfrac{\pi}{4}\right)$

3. **WE24** Use one of the compound angle identities to simplify each of the following.

   a. $\sin\left(\dfrac{\pi}{2} - \theta\right)$
   b. $\cos\left(\dfrac{\pi}{2} - \theta\right)$
   c. $\sin(\pi + \theta)$
   d. $\cos(\pi - \theta)$

4. Use one of the compound angle identities to simplify each of the following.

   a. $\sin\left(\dfrac{3\pi}{2} - \theta\right)$
   b. $\cos\left(\dfrac{3\pi}{2} + \theta\right)$
   c. $\tan(\pi - \theta)$
   d. $\tan(\pi + \theta)$

5. Simplify each of the following.

   a. $\sin\left(x + \dfrac{\pi}{3}\right) - \sin\left(x - \dfrac{\pi}{3}\right)$

   b. $\cos\left(\dfrac{\pi}{3} + x\right) - \cos\left(\dfrac{\pi}{3} - x\right)$

6. **WE25** Calculate the exact values of the following.

   a. $\cos\left(\dfrac{7\pi}{12}\right)$
   b. $\sin\left(\dfrac{11\pi}{12}\right)$
   c. $\tan\left(\dfrac{\pi}{12}\right)$
   d. $\tan\left(\dfrac{5\pi}{12}\right)$

7. **WE28** Calculate the exact values of the following.

   a. $\sin\left(\dfrac{\pi}{8}\right)\cos\left(\dfrac{\pi}{8}\right)$

   b. $\cos^2(112°30') - \sin^2(112°30')$

8. **WE29** Given that $\sec(A) = \dfrac{8}{3}$ and $0 \le A \le \dfrac{\pi}{2}$, calculate the exact values of the following.

   a. $\cos(2A)$
   b. $\sin(2A)$
   c. $\tan(2A)$
   d. $\cot(2A)$

9. **WE30** Solve each of the following equations for $x \in [0, 2\pi]$.

   a. $\sin(2x) - \cos(x) = 0$
   b. $\sin(x) - \sin(2x) = 0$
   c. $\cos(2x) - \cos(x) = 0$
   d. $\sin(x) - \cos(2x) = 0$

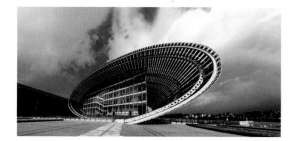

10. Solve each of the following equations for $x \in [0, 2\pi]$.

   a. $\tan(x) = \sin(2x)$
   b. $\sin(2x) = \sqrt{3}\cos(x)$

Complex familiar

11. **WE26** Given that $\cos(A) = \dfrac{4}{5}$ and $\sin(B) = \dfrac{12}{13}$ and that A and B are both acute angles, determine the exact values of:

   a. $\cos(A - B)$
   b. $\tan(A + B)$

12. Given that $\sin(A) = \dfrac{5}{13}$ and $\tan(B) = \dfrac{24}{7}$ and A is obtuse and B is acute, determine the exact values of:

   a. $\sin(A + B)$
   b. $\cos(A + B)$

13. Given that $\sec(A) = \dfrac{7}{2}$ and $\csc(B) = \dfrac{3}{2}$ and A is acute but B is obtuse, calculate the value exact values of:

   a. $\cos(A + B)$

   b. $\sin(A - B)$

14. Given that $\csc(A) = \dfrac{1}{a}$ and $\sec(B) = \dfrac{1}{b}$ and A and B are both acute, evaluate $\tan(A + B)$.

15. **WE27** Prove that $\cos(A + B)\cos(A - B) = \cos^2(A) - \sin^2(B)$.

16. **WE31** Prove the following identities.

   a. $\dfrac{\sin(2A)\cos(A) - \cos(2A)\sin(A)}{\cos(2A)\cos(A) + \sin(2A)\sin(A)} = \tan(A)$

   b. $\dfrac{\cos(2A)\cos(A) + \sin(2A)\sin(A)}{\sin(2A)\cos(A) - \cos(2A)\sin(A)} = \cot(A)$

   c. $\dfrac{\sin(3A)}{\sin(A)} - \dfrac{\cos(3A)}{\cos(A)} = 2$

   d. $\dfrac{\cos(3A)}{\sin(A)} + \dfrac{\sin(3A)}{\cos(A)} = 2\cot(2A)$

17. Prove the following identities.

   a. $\dfrac{\tan(3A) + \tan(A)}{\tan(3A) - \tan(A)} = 2\cos(2A)$

   b. $\dfrac{\tan(A) - \tan(B)}{\tan(A) + \tan(B)} = \dfrac{\sin(A - B)}{\sin(A + B)}$

   c. $\dfrac{\sin(A) - \cos(A)}{\sin(A) + \cos(A)} - \dfrac{\sin(A) + \cos(A)}{\sin(A) - \cos(A)} = 2\tan(2A)$

18. **WE32** Prove the following identities.

   a. $\dfrac{\sin(A)}{1 - \cos(A)} = \cot\left(\dfrac{A}{2}\right)$

   b. $\dfrac{\sin(A)}{1 + \cos(A)} = \tan\left(\dfrac{A}{2}\right)$

Complex unfamiliar

19. If ABCD is a cyclic quadrilateral, show that $\cos(A) + \cos(B) + \cos(C) + \cos(D) = 0$.

20. If $\tan(x) - \tan(y) = m$ and $\cot(y) - \cot(x) = n$, prove that $\dfrac{1}{m} + \dfrac{1}{n} = \cot(x - y)$.

21. If $\tan(B) = \dfrac{\sin(A)\cos(A)}{2 + \cos^2(A)}$, prove that $3\tan(A - B) = 2\tan(A)$.

22. Prove the following identities.

   a. $\dfrac{1 - \cos(A) + \sin(A)}{1 + \cos(A) + \sin(A)} = \tan\left(\dfrac{A}{2}\right)$

   b. $\dfrac{\sin\left(\frac{A}{2}\right) + \sin(A)}{1 + \cos(A) + \cos\left(\frac{A}{2}\right)} = \tan\left(\dfrac{A}{2}\right)$

23. Prove the following identities.

   a. $\sin(A + B)\sin(A - B) = \sin^2(A) - \sin^2(B)$

   b. $\tan(A + B)\tan(A - B) = \dfrac{\tan^2(A) - \tan^2(B)}{1 - \tan^2(A)\tan^2(B)}$

   c. $\cot(A + B) = \dfrac{\cot(A)\cot(B) - 1}{\cot(A) + \cot(B)}$

24. Prove the following identities.

   a. $\sin(2A) = \dfrac{2\tan(A)}{1 + \tan^2(A)}$

   b. $\cos(2A) = \dfrac{1 - \tan^2(A)}{1 + \tan^2(A)}$

   c. $\dfrac{\cos^3(A) - \sin^3(A)}{\cos(A) - \sin(A)} = 1 + \dfrac{1}{2}\sin(2A)$

25. In a triangle ABC with side lengths a, b and c, where C is a right angle, c is the hypotenuse, and a is opposite A, show that:

a. $\sin(2A) = \dfrac{2ab}{c^2}$

b. $\cos(2A) = \dfrac{b^2 - a^2}{c^2}$

c. $\tan(2A) = \dfrac{2ab}{b^2 - a^2}$

d. $\sin\left(\dfrac{A}{2}\right) = \sqrt{\dfrac{c-b}{2c}}$

e. $\cos\left(\dfrac{A}{2}\right) = \sqrt{\dfrac{c+b}{2c}}$

f. $\tan\left(\dfrac{A}{2}\right) = \sqrt{\dfrac{c-b}{c+b}}$

Fully worked solutions for this chapter are available online.

LESSON
9.6 Product identities

SYLLABUS LINKS

- Prove and apply the identities for products of sines and cosines expressed as sums and differences.
  - $\sin(A)\sin(B) = \dfrac{1}{2}(\cos(A-B) - \cos(A+B))$

  - $\cos(A)\cos(B) = \dfrac{1}{2}(\cos(A-B) + \cos(A+B))$

  - $\sin(A)\cos(B) = \dfrac{1}{2}(\sin(A+B) + \sin(A-B))$

  - $\cos(A)\sin(B) = \dfrac{1}{2}(\sin(A+B) - \sin(A-B))$

Source: Specialist Mathematics Senior Syllabus 2024 © State of Queensland (QCAA) 2024; licensed under CC BY 4.0

9.6.1 Prove product-sum and difference identities

The angle sum and angle difference identities can be written as products of the functions, allowing a sum to be expressed as a product and vice versa.

Starting with the angle sum and angle difference identities:

$$\sin(A+B) = \sin(A)\cos(B) + \cos(A)\sin(B) \quad [1]$$

$$\sin(A-B) = \sin(A)\cos(B) - \cos(A)\sin(B) \quad [2]$$

$[1] + [2]$: $\sin(A+B) + \sin(A-B) = 2\sin(A)\cos(B)$

$$\sin(A)\cos(B) = \frac{1}{2}[\sin(A+B) + \sin(A-B)]$$

$[1] - [2]$: $\sin(A+B) - \sin(A-B) = 2\cos(A)\sin(B)$

$$\cos(A)\sin(B) = \frac{1}{2}[\sin(A+B) - \sin(A-B)]$$

In a similar fashion,

$$\cos(A+B) = \cos(A)\cos(B) - \sin(A)\sin(B) \quad [3]$$

$$\cos(A-B) = \cos(A)\cos(B) + \sin(A)\sin(B) \quad [4]$$

$[3]+[4]:$ $\quad \cos(A+B) + \cos(A-B) = 2\cos(A)\cos(B)$

$$\cos(A)\cos(B) = \frac{1}{2}[\cos(A+B) + \cos(A-B)]$$

$[3]-[4]:$ $\quad \cos(A+B) - \cos(A-B) = -2\sin(A)\sin(B)$

$$\sin(A)\sin(B) = -\frac{1}{2}[\cos(A+B) - \cos(A-B)] \text{ or}$$

$$\sin(A)\sin(B) = \frac{1}{2}[\cos(A-B) - \cos(A+B)]$$

For the trigonometric functions sine and cosine, to write products of the functions as sums of the functions, the following identities can be used.

Product–sum identities

$$\sin(A)\sin(B) = \frac{1}{2}[\cos(A-B) - \cos(A+B)]$$

$$\cos(A)\cos(B) = \frac{1}{2}[\cos(A+B) + \cos(A-B)]$$

$$\sin(A)\cos(B) = \frac{1}{2}[\sin(A+B) + \sin(A-B)]$$

$$\cos(A)\sin(B) = \frac{1}{2}[\sin(A+B) - \sin(A-B)]$$

For the purposes of the proof of the sum identities, let $\alpha = A+B$ and $\beta = A-B$. Then:

$$\alpha + \beta = 2A \qquad \text{and} \qquad \alpha - \beta = 2B.$$

$$A = \frac{\alpha+\beta}{2} \qquad\qquad\qquad B = \frac{\alpha-\beta}{2}$$

These substitutions can be used to express the sums of trigonometric functions as products of trigonometric functions.

$$\sin\left(\frac{\alpha+\beta}{2}\right)\cos\left(\frac{\alpha-\beta}{2}\right) = \frac{1}{2}[\sin(\alpha) + \sin(\beta)]$$

$$2\sin\left(\frac{\alpha+\beta}{2}\right)\cos\left(\frac{\alpha-\beta}{2}\right) = \sin(\alpha) + \sin(\beta)$$

$$\cos\left(\frac{\alpha+\beta}{2}\right)\sin\left(\frac{\alpha-\beta}{2}\right) = \frac{1}{2}[\sin(\alpha) - \sin(\beta)]$$

$$2\cos\left(\frac{\alpha+\beta}{2}\right)\sin\left(\frac{\alpha-\beta}{2}\right) = \sin(\alpha) - \sin(\beta)$$

$$\cos\left(\frac{\alpha+\beta}{2}\right)\cos\left(\frac{\alpha-\beta}{2}\right) = \frac{1}{2}[\cos(\alpha) + \cos(\beta)]$$

$$2\cos\left(\frac{\alpha+\beta}{2}\right)\cos\left(\frac{\alpha-\beta}{2}\right) = \cos(\alpha) + \cos(\beta)$$

$$\sin\left(\frac{\alpha+\beta}{2}\right)\sin\left(\frac{\alpha-\beta}{2}\right) = -\frac{1}{2}[\cos(\alpha)-\cos(\beta)]$$

$$-2\sin\left(\frac{\alpha+\beta}{2}\right)\sin\left(\frac{\alpha-\beta}{2}\right) = \cos(\alpha)-\cos(\beta)$$

$$2\sin\left(\frac{\alpha+\beta}{2}\right)\sin\left(\frac{\alpha-\beta}{2}\right) = \cos(\beta)-\cos(\alpha)$$

For consistency with the other identities, these identities are expressed in terms of A and B as follows.

Sum–product identities

$$\cos(B) - \cos(A) = 2\sin\left(\frac{A+B}{2}\right)\sin\left(\frac{A-B}{2}\right)$$

$$\cos(A) + \cos(B) = 2\cos\left(\frac{A+B}{2}\right)\cos\left(\frac{A-B}{2}\right)$$

$$\sin(A) + \sin(B) = 2\sin\left(\frac{A+B}{2}\right)\cos\left(\frac{A-B}{2}\right)$$

$$\sin(A) - \sin(B) = 2\cos\left(\frac{A+B}{2}\right)\sin\left(\frac{A-B}{2}\right)$$

Note that these identities are not included in the syllabus, and thus will not be in the QCAA formula book for Specialist students.

9.6.2 Determining exact values

By rewriting products as sums or sums as products, it may be possible to rewrite expressions in terms of angles that the trigonometric ratios are known for.

WORKED EXAMPLE 33 Using product identities to simplify expressions

Evaluate $\dfrac{\sin(85°) + \sin(35°)}{\cos(85°) + \cos(35°)}$, giving your answer as an exact value.

THINK

1. Simplify the numerator using
$\sin(A) + \sin(B) =$
$2\sin\left(\dfrac{A+B}{2}\right)\cos\left(\dfrac{A-B}{2}\right)$ and
the denominator using
$\cos(A) + \cos(B) =$
$2\cos\left(\dfrac{A+B}{2}\right)\cos\left(\dfrac{A-B}{2}\right)$.

2. Simplify the terms and cancel any common factors.

WRITE

Let $A = 85°$ and $B = 35°$.

$$\frac{\sin(85°) + \sin(35°)}{\cos(85°) + \cos(35°)} = \frac{2\sin\left(\frac{85°+35°}{2}\right)\cos\left(\frac{85°-35°}{2}\right)}{2\cos\left(\frac{85°+35°}{2}\right)\cos\left(\frac{85°-35°}{2}\right)}$$

$$= \frac{2\sin(60°)\cos(25°)}{2\cos(60°)\cos(25°)}$$

$$= \frac{\sin(60°)}{\cos(60°)}$$

$$= \tan(60°)$$

$$= \sqrt{3}$$

WORKED EXAMPLE 34 Using product identities

Demonstrate that $\cos(70°) - \sqrt{2}\cos(25°) + \cos(20°) = 0$.

THINK

1. Using the $\cos(A) + \cos(B) = 2\cos\left(\dfrac{A+B}{2}\right)\cos\left(\dfrac{A-B}{2}\right)$ sum-product identity, it is possible to rewrite $\cos(70°) + \cos(20°)$.

2. Substitute $\cos(45°) = \dfrac{\sqrt{2}}{2}$.

3. As it has been demonstrated that the left-hand side is equal to the right-hand side, the identity is demonstrated.

WRITE

$\text{LHS} = \cos(70°) - \sqrt{2}\cos(25°) + \cos(20°)$

$= (\cos(70°) + \cos(20°)) - \sqrt{2}\cos(25°)$

$= \left(2\cos\left(\dfrac{70° + 20°}{2}\right)\cos\left(\dfrac{70° - 20°}{2}\right)\right)$
$\quad - \sqrt{2}\cos(25°)$

$= 2\cos(45°)\cos(25°) - \sqrt{2}\cos(25°)$

$= 2 \times \dfrac{\sqrt{2}}{2}\cos(25°) - \sqrt{2}\cos(25°)$

$= \sqrt{2}\cos(25°) - \sqrt{2}\cos(25°)$

$= 0$

$= \text{RHS}$

$\therefore \cos(70°) - \sqrt{2}\cos(25°) + \cos(20°) = 0$

9.6.3 Applying product sum identities

Like the identities introduced earlier in the chapter, the product as sum identities and the sum as product identities can be used to prove other identities.

WORKED EXAMPLE 35 Using product identities in trigomometric proofs

Prove the identity $\dfrac{\sin(5\alpha) - \sin(3\alpha)}{\cos(5\alpha) + \cos(3\alpha)} = \tan(\alpha)$.

THINK

1. The left-hand side is the more complex side, so begin with that side.

2. The numerator can be rewritten using $\sin(A) - \sin(B) = 2\cos\left(\dfrac{A+B}{2}\right)\sin\left(\dfrac{A-B}{2}\right)$ and the denominator can be rewritten using $\cos(A) + \cos(B) = 2\cos\left(\dfrac{A+B}{2}\right)\cos\left(\dfrac{A-B}{2}\right)$.

WRITE

$\text{LHS} = \dfrac{\sin(5\alpha) - \sin(3\alpha)}{\cos(5\alpha) + \cos(3\alpha)}$

Let $A = 5\alpha$ and $B = 3\alpha$.

$\text{LHS} = \dfrac{2\cos\left(\frac{5\alpha + 3\alpha}{2}\right)\sin\left(\frac{5\alpha - 3\alpha}{2}\right)}{2\cos\left(\frac{5\alpha + 3\alpha}{2}\right)\cos\left(\frac{5\alpha - 3\alpha}{2}\right)}$

3. Simplify the terms, noting that $\dfrac{5A + 3A}{2} = \dfrac{8A}{2}$,

$$= 4A$$

$$\dfrac{5A - 3A}{2} = \dfrac{2A}{2} \text{ and } \tan(A) = \dfrac{\sin(A)}{\cos(A)}.$$

$$= A$$

$$= \dfrac{2\cos(4\alpha)\sin(\alpha)}{2\cos(4\alpha)\cos(\alpha)}$$

$$= \dfrac{\sin(\alpha)}{\cos(\alpha)}$$

$$= \tan(\alpha)$$

$$= \text{RHS}$$

4. As it has been demonstrated that the left-hand side is equal to the right-hand side, the identity is demonstrated.

$$\therefore \dfrac{\sin(5\alpha) - \sin(3\alpha)}{\cos(5\alpha) + \cos(3\alpha)} = \tan(\alpha)$$

By rewriting sums of trigonometric ratios as products, it is possible to use the Null Factor Law to solve equations.

WORKED EXAMPLE 36 Solving trigonometric equations using product identities

Determine the general solution for x of $\cos(4x) + \cos(2x) = 0$.

THINK

1. Rewriting the left-hand side as a product allows the Null Factor Law to be used.

2. Set each factor to zero and solve.

3. Solve $\cos(x) = 0$. Note that $\cos\left(\dfrac{\pi}{2}\right) = 0$ and $\cos\left(\dfrac{3\pi}{2}\right) = 0$. These solutions are π apart, so the general solution is $\dfrac{\pi}{2} + n\pi, n \in \mathbb{Z}$.

WRITE

Use

$$\cos(A) + \cos(B) = 2\cos\left(\dfrac{A+B}{2}\right)\cos\left(\dfrac{A-B}{2}\right)$$

where $A = 4x$ and $B = 2x$.

$$\cos(4x) + \cos(2x) = 0$$

$$2\cos\left(\dfrac{4x + 2x}{2}\right)\cos\left(\dfrac{4x - 2x}{2}\right) = 0$$

$$2\cos(3x)\cos(x) = 0$$

$$\cos(3x) = 0 \text{ or } \cos(x) = 0$$

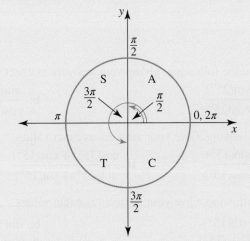

$$\cos(x) = 0$$
$$x = \frac{\pi}{2} + n\pi, \; n \in \mathbb{Z}$$
$$= \pi\left(\frac{1}{2} + n\right)$$
$$= \pi\left(\frac{1 + 2n}{2}\right)$$
$$= \frac{\pi(2n + 1)}{2}$$

4. Solve $\cos(3x) = 0$.

$$\cos(3x) = 0$$
Using the previous solution:
$$3x = \frac{\pi(2n + 1)}{2}, \; n \in \mathbb{Z}$$
$$x = \frac{\pi(2n + 1)}{6}$$

5. Combine the solutions.

$$x = \frac{\pi(2n + 1)}{2}, \frac{\pi(2n + 1)}{6}, \; n \in \mathbb{Z}$$

Exercise 9.6 Product identities

learn **on**

9.6 Exercise	9.6 Exam questions **on**

Simple familiar	Complex familiar	Complex unfamiliar
1, 2, 3, 4, 5, 6, 7, 8	9, 10, 11, 12, 13	14, 15, 16

These questions are even better in jacPLUS!
- Receive immediate feedback
- Access sample responses
- Track results and progress

Find all this and MORE in jacPLUS

Simple familiar

1. **WE33** Evaluate the following, giving your answers as exact values.

   a. $\dfrac{\sin(95°) + \sin(25°)}{\cos(95°) + \cos(25°)}$

   b. $\dfrac{\sin(95°) - \sin(25°)}{\cos(95°) - \cos(25°)}$

2. Evaluate the following. Give your answers as exact values.

   a. $\dfrac{\sin(75°) - \sin(15°)}{\cos(75°) + \cos(15°)}$

   b. $\dfrac{\cos(75°) + \cos(15°)}{\sin(75°) + \sin(15°)}$

   c. $\dfrac{\cos(75°) - \cos(15°)}{\sin(75°) + \sin(15°)}$

3. Evaluate the following. Give your answers as exact values.

   a. $\sin(105°) + \sin(15°)$

   b. $\sin(105°) - \sin(15°)$

   c. $\cos(105°) + \cos(15°)$

   d. $\cos(105°) - \cos(15°)$

4. **WE34** Demonstrate the following.

   a. $\cos(80°) - \sqrt{3}\cos(50°) + \cos(20°) = 0$

   b. $\sin(80°) - \sqrt{3}\sin(50°) + \sin(20°) = 0$

5. Demonstrate the following.
   a. $\cos(10°) + \cos(110°) + \cos(130°) = 0$
   b. $\cos(20°) + \cos(100°) + \cos(220°) = 0$
   c. $\cos(40°) - \sqrt{2}\cos(5°) + \cos(50°) = 0$

6. **WE35** Prove the identity $\dfrac{\sin(5A) - \sin(3A)}{\cos(5A) - \cos(3A)} = -\cot(4A)$.

7. Prove the identity $\dfrac{\cos(4A) + \cos(2A)}{\sin(4A) - \sin(2A)} = \cot A$.

8. Prove the identity $\dfrac{\sin(2A) + \sin(4A)}{\cos(2A) + \cos(4A)} = \tan(3A)$

Complex familiar

9. Prove the following identities.
   a. $\dfrac{\sin(2A) + \sin(2B)}{\sin(2A) - \sin(2B)} = \dfrac{\tan(A + B)}{\tan(A - B)}$
   b. $\dfrac{\cos(2A) + \cos(2B)}{\cos(2A) - \cos(2B)} = -\cot(A + B)\cot(A - B)$

10. Demonstrate that the following are true.
    a. $\sin(2A) + \sin(6A) + \sin(8A) = 4\cos(A)\cos(3A)\sin(4A)$
    b. $\sin(2A) + \sin(4A) + \sin(6A) = 4\cos(A)\cos(2A)\sin(3A)$

11. Demonstrate that the following is true $\dfrac{\sin(A) + 2\sin(3A) + \sin(5A)}{\sin(3A) + 2\sin(5A) + \sin(7A)} = \dfrac{\sin(3A)}{\sin(5A)}$

12. **WE36** Determine the general solutions of the following.
    a. $\cos(2x) + \cos(3x) = 0$
    b. $\sin(2x) + \sin(4x) = 0$

13. Determine the general solutions to the following.
    a. $\sin(x) + \sin(3x) = 0$
    b. $\cos(x) + \cos(3x) = 0$

Complex unfamiliar

14. Determine the general solutions to the following.
    a. $\sin(x) + \sin(2x) + \sin(3x) = 0$
    b. $\cos(x) + \cos(2x) + \cos(3x) = 0$
    c. $\sin(x) + \cos(x) + \sin(3x) = 0$

15. Prove the following.
    a. $\dfrac{\sin(A) + \sin(B) - \sin(A + B)}{\sin(A) + \sin(B) + \sin(A + B)} = \tan\left(\dfrac{A}{2}\right)\tan\left(\dfrac{B}{2}\right)$
    b. $\sin(A) + \sin(B) + \sin(C) - \sin(A + B + C) = 4\sin\left(\dfrac{A + B}{2}\right)\sin\left(\dfrac{B + C}{2}\right)\sin\left(\dfrac{A + C}{2}\right)$

16. a. Prove that $\dfrac{\sin(3A) + \sin(5A)}{\cos(3A) + \cos(5A)} = \tan(4A)$.
    b. Prove that $\dfrac{\sin(3A) + \sin(5A) + \sin(7A)}{\cos(3A) + \cos(5A) + \cos(7A)} = \tan(5A)$.
    c. Prove that $\dfrac{\sin(3A) + \sin(5A) + \sin(7A) + \sin(9A)}{\cos(3A) + \cos(5A) + \cos(7A) + \cos(9A)} = \tan(6A)$.
    d. Simplify the following.
    $\dfrac{\sin(3A) + \sin(5A) + \sin(7A) + \sin(9A) + \sin(11A) + \sin(13A)}{\cos(3A) + \cos(5A) + \cos(7A) + \cos(9A) + \cos(11A) + \cos(13A)}$.

Fully worked solutions for this chapter are available online.

LESSON
9.7 Applications of trigonometric identities

The trigonometric identities presented so far can be rearranged, extended and used in real world applications. These identities, and calculus more broadly, are applied across all fields of engineering.

9.7.1 Convert $a\cos(x) + b\sin(x)$ to $R\cos(x \pm \alpha)$ or $R\sin(x \pm \alpha)$

In section 9.3.3, you reviewed how to graph trigonometric functions. Using the compound angle identities, it is possible to rewrite expressions of the form $a\cos(x) + b\sin(x)$ as either $R\cos(x \pm \alpha)$ or $R\sin(x \pm \alpha)$. This makes it easier to visualise what the graph might look like and to identify the maximum and minimum values.

WORKED EXAMPLE 37 Converting $a\cos(x) + b\sin(x)$ to $R\cos(x - \alpha)$

Express $\sqrt{3}\cos(x) + \sin(x)$ in the form $R\cos(x - \alpha)$.

THINK	WRITE
1. Let $\sqrt{3}\cos(x) + \sin(x)$ equal $R\cos(x - \alpha)$ and use the compound angle identities to expand $R\cos(x - \alpha)$.	Let $\sqrt{3}\cos(x) + \sin(x) = R\cos(x - \alpha)$. $\begin{aligned}\sqrt{3}\cos(x) + \sin(x) &= R\cos(x - \alpha)\\ &= R(\cos(x)\cos(\alpha) + \sin(x)\sin(\alpha))\\ &= R\cos(x)\cos(\alpha) + R\sin(x)\sin(\alpha)\end{aligned}$
2. Equate the coefficients of $\cos(x)$ and $\sin(x)$.	Equating the coefficients of $\cos(x)$ and $\sin(x)$: $\sqrt{3} = R\cos(\alpha) \quad [1]$ $1 = R\sin(\alpha) \quad\quad [2]$
3. As R is a factor of both equations, it can be eliminated by calculating $\dfrac{[2]}{[1]}$. This will also allow us to use $\tan(\alpha) = \dfrac{\sin(\alpha)}{\cos(\alpha)}$.	$\begin{aligned}\dfrac{[2]}{[1]}: \dfrac{1}{\sqrt{3}} &= \dfrac{R\sin(\alpha)}{R\cos(\alpha)}\\ &= \dfrac{\sin(\alpha)}{\cos(\alpha)}\\ &= \tan(\alpha)\end{aligned}$
4. From equations [1] and [2], both $\sin(\alpha)$ and $\cos(\alpha)$ are positive, so α is in quadrant 1. Remember $\tan\left(\dfrac{\pi}{6}\right) = \dfrac{1}{\sqrt{3}}$. Solve for α.	$\alpha = \dfrac{\pi}{6}$

5. Use either equation [1] or [2] to solve for R.

$$[1]: \sqrt{3} = R\cos\left(\frac{\pi}{6}\right)$$

$$= R \times \frac{\sqrt{3}}{2}$$

$$R = \sqrt{3} \times \frac{2}{\sqrt{3}}$$

$$= 2$$

6. State the final result.

$$\sqrt{3}\cos(x) + \sin(x) = 2\cos\left(x - \frac{\pi}{6}\right)$$

9.7.2 Sketching the graphs

To make it easier to visualise the graph and to interpret the function, R is normally a positive number and α is generally in the first quadrant. Note that $-R\cos(x) = R\cos(x + \pi)$ and $-R\sin(x) = R\sin(x + \pi)$. Additionally, $R\cos(x) = R\sin\left(x + \frac{\pi}{2}\right)$ and $R\sin(x) = R\cos\left(x - \frac{\pi}{2}\right)$. This means that it is possible to write $a\cos(x) + b\sin(x)$ as either a sine or a cosine function, and the magnitude of α will vary depending on the function chosen.

For R to be positive and α to be in quadrant 1, the following transformations are used:

> **Sum conversion transformations**
>
> $a\cos(x) \pm b\sin(x)$ in the form $R\cos(x \mp \alpha)$
>
> $a\sin(x) \pm b\cos(x)$ in the form $R\sin(x \pm \alpha)$

WORKED EXAMPLE 38 Using $R\sin(x \pm \alpha)$ and $R\cos(x \pm \alpha)$ to sketch functions

Given the function $f(x) = 2\sin(x) - 2\sqrt{3}\cos(x)$, $x \in [0, 2\pi]$:
a. express $f(x)$ in the form $R\sin(x - \alpha)$
b. determine the coordinates of the maximum and minimum of $f(x)$
c. sketch $f(x)$
d. solve $f(x) = 2$.

THINK

a. 1. Let $2\sin(x) - 2\sqrt{3}\cos(x)$ equal $R\sin(x - \alpha)$ and use the compound angle identities to expand $R\sin(x - \alpha)$.

2. Equate the coefficients of $\cos(x)$ and $\sin(x)$.

WRITE

a. Let $2\sin(x) - 2\sqrt{3}\cos(x) = R\sin(x - \alpha)$.

$$2\sin(x) - 2\sqrt{3}\cos(x) = R\sin(x - \alpha)$$
$$= R(\sin(x)\cos(\alpha) - \cos(x)\sin(\alpha))$$
$$= R\sin(x)\cos(\alpha) - R\cos(x)\sin(\alpha)$$

Equating the coefficients of $\cos(x)$ and $\sin(x)$:

$$2 = R\cos(\alpha) \qquad [1]$$
$$2\sqrt{3} = R\sin(\alpha) \qquad [2]$$

3. As R is a factor of both equations, it can be eliminated by calculating $\dfrac{[2]}{[1]}$. This will also allow us to use $\tan(\alpha) = \dfrac{\sin(\alpha)}{\cos(\alpha)}$.

$\dfrac{[2]}{[1]}: \dfrac{2\sqrt{3}}{2} = \dfrac{R\sin(\alpha)}{R\cos(\alpha)}$

$\sqrt{3} = \dfrac{\sin(\alpha)}{\cos(\alpha)}$

$= \tan(\alpha)$

4. From equations [1] and [2], both $\sin(\alpha)$ and $\cos(\alpha)$ are positive, so α is in quadrant 1. Remember $\tan\left(\dfrac{\pi}{3}\right) = \sqrt{3}$. Solve for α.

$\alpha = \dfrac{\pi}{3}$

5. Use either equation [1] or [2] to solve for R.

$[1]: 2 = R\cos\left(\dfrac{\pi}{3}\right)$

$= R \times \dfrac{1}{2}$

$R = 2 \times 2$

$= 4$

6. State the final result.

$2\sin(x) - 2\sqrt{3}\cos x = 4\sin\left(x - \dfrac{\pi}{3}\right)$

b. 1. R will be the amplitude of the graph. There is no vertical shift.

b. $R = 4$; therefore, the amplitude of the graph is 4. This means that the maximum is 4 and the minimum is -4.

2. Identify where the maximum occurs. (The largest possible value of a sine function is 1 and $\sin\left(\dfrac{\pi}{2}\right) = 1$.) Note we are looking for $x \in [0, 2\pi]$.

The maximum will occur when

$\sin\left(x - \dfrac{\pi}{3}\right) = 1$

$x - \dfrac{\pi}{3} = \dfrac{\pi}{2}$

$x = \dfrac{\pi}{2} + \dfrac{\pi}{3}$

$= \dfrac{3\pi}{6} + \dfrac{2\pi}{6}$

$= \dfrac{5\pi}{6}$

The maximum point is $\left(\dfrac{5\pi}{6}, 4\right)$.

3. Identify where the minimum occurs. (The smallest possible value of a sine function is -1 and $\sin\left(\dfrac{3\pi}{2}\right) = 1$.) Note we are looking for $x \in [0, 2\pi]$.

The minimum will occur when

$$\sin\left(x - \dfrac{\pi}{3}\right) = -1$$

$$x - \dfrac{\pi}{3} = \dfrac{3\pi}{2}$$

$$x = \dfrac{3\pi}{2} + \dfrac{\pi}{3}$$

$$= \dfrac{9\pi}{6} + \dfrac{2\pi}{6}$$

$$= \dfrac{11\pi}{6}$$

The Minimum point is $\left(\dfrac{11\pi}{6}, -4\right)$.

c. Use the maximum and minimum points to sketch the function over the required domain.

c. $f(x) = 2\sin(x) - 2\sqrt{3}\cos(x),\ x \in [0, 2\pi]$

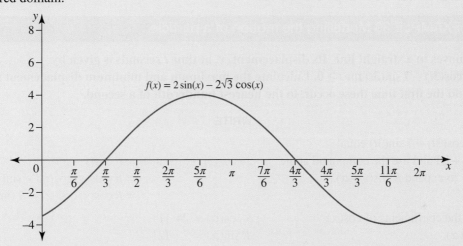

$f(x) = 2\sin(x) - 2\sqrt{3}\cos(x)$

d. 1. To solve $f(x)$, use the form $R\sin(x - \alpha)$.

d. $2\sin(x) - 2\sqrt{3}\cos(x) = 2$

$$4\sin\left(x - \dfrac{\pi}{3}\right) = 2$$

2. Solve for x. Remember that $\sin\left(\dfrac{\pi}{6}\right) = \dfrac{1}{2}$ and sine is positive in quadrants 1 and 2.

$$\sin\left(x - \dfrac{\pi}{3}\right) = \dfrac{1}{2}$$

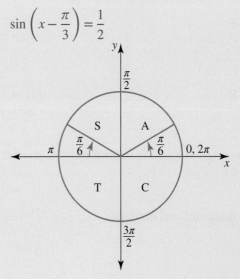

$$\sin\left(\frac{\pi}{6}\right) = \frac{1}{2}$$

Therefore: $x - \dfrac{\pi}{3} = \dfrac{\pi}{6}$ or $\quad x - \dfrac{\pi}{3} = \pi - \dfrac{\pi}{6}$

$$x = \frac{\pi}{6} + \frac{2\pi}{6} \qquad\qquad x = \pi - \frac{\pi}{6} + \frac{\pi}{3}$$

$$= \frac{\pi}{6} + \frac{2\pi}{6} \qquad\qquad = \frac{6\pi}{6} - \frac{\pi}{6} + \frac{2\pi}{6}$$

$$= \frac{3\pi}{6} \qquad\qquad\qquad = \frac{7\pi}{6}$$

$$= \frac{\pi}{2}$$

3. Combine the solutions.

$$x = \frac{\pi}{2} \text{ or } x = \frac{7\pi}{6}$$

9.7.3 Model and solve problems

WORKED EXAMPLE 39 Modelling the motion of a particle

A particle moves in a straight line. Its displacement, x, at time t seconds is given by
$x = 30 + 24\cos(3t) - 7\sin(3t)$ for $t \geq 0$. Calculate the maximum and minimum displacement from the
origin, O, and the first time these occur, to the nearest hundredth of a second.

THINK	WRITE
1. Let $24\cos(3t) - 7\sin(3t)$ equal $R\cos(y + \alpha)$ and use the compound angle identity to expand $R\cos(y + \alpha)$.	$24\cos(3t) - 7\sin(3t) = R\cos(y + \alpha)$ $= R(\cos(y)\cos(\alpha) - \sin(y)\sin(\alpha))$ $= R\cos(y)\cos(\alpha) - R\sin(y)\sin(\alpha)$
2. Equate the coefficients of $\cos(\alpha)$ and $\sin(\alpha)$.	$R\cos(\alpha) = 24$ [1] $R\sin(\alpha) = 7$ [2]
3. Eliminate R.	$\dfrac{[2]}{[1]} : \dfrac{R\sin(\alpha)}{R\cos(\alpha)} = \dfrac{7}{24}$ $\tan\alpha = 0.291\dot{6}$
4. Solve for α.	$\alpha = \tan^{-1}(0.291\dot{6})$ $= 0.284\,\text{rad}$ $(= 16.3°)$
5. Solve for R.	$[1]^2 + [2]^2 : R^2\cos^2(\alpha) + R^2\sin^2(\alpha) = 24^2 + 7^2$ $R^2 = 625$ $R = 25$
6. Present the original equation in the form $R\cos(y + \alpha)$.	$f(y) = 25\cos(y + 0.284)$ $x = f(3t) + 30$ $= 25\cos(3t + 0.284) + 30$
7. Calculate maximum and minimum.	The amplitude of this function $f(y)$ is 25. This means that the maximum is $25 + 30 = 55$ and the minimum is $30 - 25 = 5$.

8. Identify when the maximum occurs for the first time.	The maximum of 55 occurs for the first time when $\cos(3t + 0.284) = 1$ for the first time, that is when $(3t + 0.284) = 0 \pm 2k\pi, k \in \mathbb{Z}, t \geq 0$ for the first time. $\therefore t = \dfrac{2\pi - 0.284}{3} \simeq 2.00\,\text{s}.$
9. Identify when the minimum occurs for the first time.	The minimum of 5 occurs for the first time when $\cos(3t + 0.284) = -1$ for the first time, that is when $(3t + 0.284) = \pi \pm 2k\pi, k \in \mathbb{Z}, t \geq 0$ for the first time. $\therefore t = \dfrac{\pi - 0.284}{3} \simeq 0.95\,\text{s}$
10. Write the answer.	The minimum displacement from the origin, O, is 5 and occurs for the first time at $t = 0.95$ s, and the maximum displacement is 55 and occurs for the first time at $t = 2.00$ s.

9.7.4 Prove and apply multi-angle trigonometric identities

When the multiple of the angle is greater than 2, the identities are referred to as **multiple angle identities**.

These identities were proved using complex numbers and de Moivre's theorem in Chapter 7, in Worked example 13 Calculating powers of complex numbers in polar form.

Proof of these multiple angle identities may also be demonstrated using the compound angle identities and the Pythagorean identities. In the worked example and the exercise that follow, these multiple angle proofs will be demonstrated using both the compound angle identities and de Moivre's theorem (as seen in section 7.3.2). You will learn more about the applications of de Moivre's theorem in Unit 3.

In Worked example 40, the use of de Moivre's formula is provided as an example of the applications you will see in Unit 3.

> ### Multiple angle formulas
>
> $$\sin(3A) = 3\sin(A) - 4\sin^3(A)$$
> $$\cos(3A) = 4\cos^3(A) - 3\cos(A)$$
> $$\sin(4A) = \cos(A)\left(4\sin(A) - 8\sin^3(A)\right)$$
> $$\cos(4A) = 8\cos^4(A) - 8\cos^2(A) + 1$$

WORKED EXAMPLE 40 Proving multiple angle identities

Prove the multiple angle identity $\cos(3A) = 4\cos^3(A) - 3\cos(A)$:
a. **using the compound angle identities**
b. **using de Moivre's theorem.**

THINK	WRITE
a. 1. Although the right-hand side has more terms, by using the compound angle identities and then the double angle identities, it is possible to rewrite $\cos(3A)$ in terms of $\sin(A)$ and $\cos(A)$. So begin with the left-hand side.	LHS $= \cos(3A)$

2. Use the multiple angle identity $\cos(A + B)$ $= \cos(A)\cos(B) - \sin(A)\sin(B)$ by using $3A = 2A + A$.

$= \cos(2A + A)$
$= \cos(2A)\cos(A) - \sin(2A)\sin(A)$

3. Use the double angle identities to rewrite $\cos(2A)$ and $\sin(2A)$. As the right-hand side is written in terms of $\cos(A)$, use that option for $\cos(2A)$.

$= \left(2\cos^2(A) - 1\right)\cos(A) - (2\sin(A)\cos(A))\sin(A)$
$= 2\cos^3(A) - \cos(A) - 2\sin^2(A)\cos(A)$

4. Substitute $\sin^2(A) = 1 - \cos^2(A)$.

$= 2\cos^3(A) - \cos(A) - 2\left(1 - \cos^2(A)\right)\cos(A)$
$= 2\cos^3(A) - \cos(A) - 2\cos(A) + 2\cos^3(A)$
$= 4\cos^3(A) - 3\cos(A)$

5. As it has been demonstrated that the left-hand side is equal to the right-hand side, the identity is proven.

$= \text{RHS}$
$\therefore \cos(3A) = 4\cos^3(A) - 3\cos(A)$

b. 1. The proof involves $(3A)$ so it will be necessary to determine $(\cos(A) + i\sin(A))^3$.

$(\cos(A) + i\sin(A))^3$

2. Determine the expansion using the third row $(1\,3\,3\,1)$ of Pascal's triangle.

$(\cos(A) + i\sin(A))^3 = 1(\cos(A))^3 + 3(\cos(A))^2(i\sin(A))$
$\qquad + 3(\cos(A))(i\sin(A))^2 + 1(i\sin(A))^3$
$= \cos^3(A) + 3i\cos^2(A)\sin(A)$
$\qquad + 3i^2\cos(A)\sin^2(A) + i^3\sin^3(A)$

3. Simplify the expression, remembering that $i^2 = -1$ and $i^3 = -i$.

$= \cos^3(A) + 3i\cos^2(A)\sin(A)$
$\quad - 3\cos(A)\sin^2(A) - i\sin^3(A)$

4. Group the real and imaginary terms together.

$= \cos^3(A) - 3\cos(A)\sin^2(A)$
$\quad + 3i\cos^2(A)\sin(A) - i\sin^3(A)$
$(\cos(A) + i\sin(A))^3 = \cos^3(A) - 3\cos(A)\sin^2(A)$
$\qquad + i\left(3\cos^2(A)\sin(A) - \sin^3(A)\right)$

5. State the rule for $\cos(n\theta)$.

$\cos(nA) = \text{Re}(\cos(A) + i\sin(A))^n$

6. Use the expansion of $(\cos(\theta) + i\sin(\theta))^3$ to simplify the left-hand side term, $\cos(3\theta)$.

$\text{LHS} = \cos(3A)$
$\quad = \text{Re}(\cos(A) + i\sin(A))^3$
$\quad = \cos^3(A) - 3\cos(A)\sin^2(A)$

7. We want to prove that $\cos(3\theta) = 4\cos^3(\theta) - 3\cos(\theta)$ so use the Pythagorean identity $\sin^2(\theta) = 1 - \cos^2(\theta)$ to replace $\sin^2(\theta)$.

$\cos(3A) = \cos^3(A) - 3\cos(A)\left(1 - \cos^2(A)\right)$
$\quad = \cos^3(A) - 3\cos(A) + 3\cos^3(A)$
$\quad = 4\cos^3(A) - 3\cos(A)$
$\quad = \text{RHS}$
$\therefore \cos(3A) = 4\cos^3(A) - 3\cos(A)$

Exercise 9.7 Applications of trigonometric identities

9.7 Exercise	9.7 Exam questions on

Simple familiar	Complex familiar	Complex unfamiliar
1, 2, 3, 4, 5, 6	7, 8, 9, 10	11, 12, 13

These questions are even better in jacPLUS!
- Receive immediate feedback
- Access sample responses
- Track results and progress

Find all this and MORE in jacPLUS ▶

Simple familiar

1. **WE37** Express $\cos(x) + \sin(x)$ in the form $R\cos(x - \alpha)$.

2. Express $\sqrt{3}\cos(x) - \sin(x)$ in the form $R\cos(x + \alpha)$.

3. Express the following in the form $R\sin(x + \alpha)$.

   a. $\sin(x) + \cos(x)$

   b. $\sqrt{3}\sin(x) + \cos(x)$

4. Express the following in the form $R\sin(x - \alpha)$.

   a. $\sin(x) - \cos(x)$

   b. $\sqrt{3}\sin(x) - \cos(x)$

5. Express the following in the form $R\cos(x - \alpha)$.

   a. $4\sqrt{2}\cos(x) + 4\sqrt{2}\sin(x)$

   b. $7\cos(x) + 24\sin(x)$

6. **WE40** Prove the multiple angle identities $\sin(3A) = 3\sin(A) - 4\sin^3(A)$:

   a. using the compound angle identities
   b. using de Moivre's theorem.

Complex familiar

7. **WE38** Given the function $f(x) = \sqrt{2}\sin(x) - \sqrt{2}\cos(x), x \in [0, 2\pi]$:

   a. express $f(x)$ in the form $R\sin(x - \alpha)$
   b. determine the coordinates of the maximum and minimum of $f(x)$
   c. sketch $f(x)$
   d. solve $f(x) = 1$.

8. Consider the function $f(x) = 5\sin(x) - 12\cos(x), x \in [0, 2\pi]$.
   In your answer use three decimal places where necessary.

   a. Express $f(x)$ in the form $R\cos(x + \alpha)$.
   b. Determine the coordinates of the maximum and minimum of $f(x)$.
   c. Sketch $f(x)$.
   d. Solve $f(x) = 6.5$.

9. Consider the function $f(x) = 3\sin(x) - 4\cos(x), x \in [0°, 360°]$.
   In your answer, use two decimal places where necessary.

   a. Express $f(x)$ in the form $R\sin(x - \alpha)$.
   b. Determine the coordinates of the maximum and minimum of $f(x)$.
   c. Sketch $f(x)$.
   d. Solve $f(x) = 2.5$.

10. Prove each of the following multiple angle identities.

   i. using the compound angle identities

   ii. using de Moivre's theorem.

   a. $\sin(4A) = \cos(A)\left(4\sin(A) - 8\sin^3(A)\right)$

   b. $\cos(4A) = 8\cos^4(A) - 8\cos^2(A) + 1$

Complex unfamiliar

11. Chebyshev, the Russian mathematician mentioned in the overview, developed a number of recurrence relationships for trigonometric multiple angles.

   a. Prove the relationship $\cos(nx) = 2\cos(x)\cos[(n-1)x] - \cos[(n-2)x]$.

   b. Use this relationship and the identity for $\cos(2x)$ to demonstrate that $\cos(5x) = 16\cos^5(x) - 20\cos^3(x) + 5\cos(x)$.

12. Calculate the maximum and minimum values of each of the following.

   a. $\dfrac{4}{5 + 2.4\cos(x) - 3.2\sin(x)}$

   b. $\dfrac{\sqrt{3}}{2 - 4\cos(x) - 4\sqrt{3}\sin(x)}$

13. a. **WE39** At time t, the current I in a circuit is given by $I = 40\cos(300t) + 9\sin(300t)$. Express the current in the form $R\cos(nt - \alpha)$ where α is in decimal degrees, correct to two decimal places.

   b. At a time t, the current I in a circuit is given by $I = 11\cos(800t) - 60\sin(800t)$. Express the current in the form $R\cos(nt + \alpha)$ where α is in decimal degrees, correct to two decimal places.

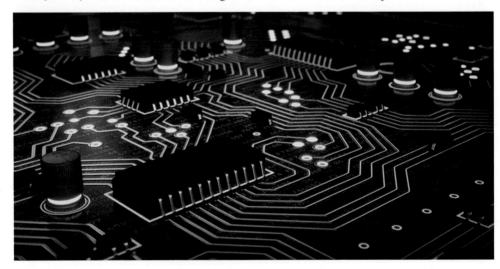

Fully worked solutions for this chapter are available online.

LESSON
9.8 Review

9.8.1 Summary

doc-41933

Hey students! Now that it's time to revise this chapter, go online to:

Access the chapter summary

Review your results

Practise exam questions

Find all this and MORE in jacPLUS

9.8 Exercise

learn on

9.8 Exercise	**9.8 Exam questions** on

Simple familiar	Complex familiar	Complex unfamiliar
1, 2, 3, 4, 5, 6, 7, 8, 9, 10, 11, 12	13, 14, 15, 16	17, 18, 19, 20

These questions are even better in jacPLUS!
- Receive immediate feedback
- Access sample responses
- Track results and progress

Find all this and MORE in jacPLUS

Simple familiar

1. Solve the following.
   a. $|2x| = 10$
   b. $|x + 1| = 7$

2. Use the diagram to determine the exact values of
   a. $\cos(x)$
   b. $\sin(x)$
   c. $\tan(x)$
   d. $\sec(x)$
   e. $\operatorname{cosec}(x)$
   f. $\cot(x)$

3. Determine the exact value of the following.
   a. $\sin\left(\dfrac{7\pi}{12}\right)$
   b. $\cos\left(\dfrac{\pi}{12}\right)$

4. Determine the exact value of each of the following.
   a. $\cot\left(\dfrac{\pi}{6}\right)$
   b. $\cot\left(\dfrac{2\pi}{3}\right)$
   c. $\cot\left(\dfrac{7\pi}{4}\right)$
   d. $\cot\left(\dfrac{11\pi}{6}\right)$

5. Answer the following.
   a. If $\sec(\theta) = -\dfrac{5}{2}$ and $\pi < \theta < \dfrac{3\pi}{2}$, determine the exact value of $\operatorname{cosec}(\theta)$.

   b. If $\cot(x) = 4$ and $\pi < x < \dfrac{3\pi}{2}$, determine the exact value of $\operatorname{cosec}(x)$.

6. a. Use the graph of $y = \cos 2(x)$ to sketch $y = \sec(2x)$ over the domain $-2\pi \leq x \leq 2\pi$.
   b. Use the graph of $y = \tan(3x)$ to sketch $y = \cot(3x)$ over the domain $-2\pi \leq x \leq 2\pi$.

7. Solve each of the following equations over the domain $0 \le x \le 2\pi$.

   a. $\sin^2(x) + 3\sin(x) - 4 = 0$

   b. $\sin^2(x) - \sin(x) - 1 = 0$

8. Use a Pythagorean identity to factorise each of the following.

   a. $\mathrm{cosec}^2(3A) - 2\mathrm{cosec}(3A) + 2\cot^2(3A) + 1$

   b. $\tan^2(2\beta) - \sec(2\beta) - 1$

9. Solve each of the following equations over the domain $0 \le A \le 2\pi$.

   a. $\sin^2(A) = 1 + \cos(A)$

   b. $2\cos^2(A) = 5 + 5\sin(A)$

10. Determine the general solution to each of the following equations.

    a. $2\cos^2(3x) + \sqrt{3}\cos(3x) = 0$

    b. $2\sin^2(3x) - \sqrt{3}\sin(3x) = 0$

11. a. Sketch $y = 25 - x^2$ and use your sketch to answer parts b–d.

    b. Sketch $y = |25 - x^2|$.

    c. Sketch $y = 25 - |x|^2$.

    d. Sketch $y = \dfrac{1}{25 - x^2}$.

12. Sketch $y = \dfrac{x^2 - 1}{x - 1}$.

Complex familiar

13. a. Sketch $y = |(1 - x)(x + 5)|$.

    b. Sketch $y = (1 - |x|)(|x| + 5)$.

    c. Sketch $y = \dfrac{1}{(1 - x)(x + 5)}$.

14. If $\sec(x) = \dfrac{a}{b}$ where $a, b \in \mathbb{R}^+$ and $\dfrac{3\pi}{2} < x < 2\pi$, determine $\cot(x) - \mathrm{cosec}(x)$.

15. Given the function $f(x) = \sqrt{2}\sin(x) - \sqrt{2}\cos(x)$, $x \in [0, 2\pi]$:

    a. express $f(x)$ in the form $R\sin(x - \alpha)$

    b. determine the coordinates of the maximum and minimum of $f(x)$

    c. sketch $f(x)$

    d. solve $f(x) = 1$.

16. For the function $f(x) = 12\cos(x) + 5\sin(x)$, $x \in [0, 2\pi]$. In your answer, use two decimal places where necessary.

    a. Express $f(x)$ in the form $R\cos(x - \alpha)$.

    b. Determine the coordinates of the maximum and minimum of $f(x)$.

    c. Sketch $f(x)$.

    d. Solve $f(x) = 6.5$ and hence determine the general solution of $12\cos(x) + 5\sin(x) = 6.5$.

Complex unfamiliar

17. Chebyshev's recurrence formula for multiple angles of the sine function is given by
    $$\sin(nx) = 2\cos(x)\sin[(n - 1)x] - \sin[(n - 2)x].$$

    a. Prove Chebyshev's recurrence formula for multiple angles of the sine function.

    b. Demonstrate that $\sin(6x) = \cos(x)\left(32\sin^5(x) - 32\sin^3(x) + 6\sin(x)\right)$.

18. Sketch the following.

   a. Sketch $y = \left| \dfrac{x+1}{x+2} \right|$.

   b. Sketch $y = \dfrac{|x|+1}{|x|+2}$.

   c. Sketch $y = \dfrac{x+2}{x+1}$.

19. Determine the general solution to each of the following equations.

   a. $\tan^3(x) - \tan^2(x) - \tan(x) + 1 = 0$

   b. $\tan^4(x) - 4\tan^2(x) + 3 = 0$

20. The depth of water near a pier changes with the tides according to the rule

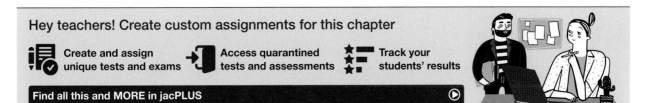

$h(t) = 5 + 2\sin\left(\dfrac{\pi t}{12}\right) + 2\sqrt{3}\cos\left(\dfrac{\pi t}{12}\right)$ where t hours is the time, for $t \geq 0$, and h is depth in metres.

Sketch the graph of the depth of the water, showing one cycle, labelling extremums.

Fully worked solutions for this chapter are available online.

Answers

Chapter 9 Trigonometry and functions

9.2 Sketching graphs

9.2 Exercise

1. **a.** True   **b.** False
   c. False   **d.** True

2. $x = \pm 4$

3. $a = \pm 3$

4. $x = 1, x = 5$

5. $x = 2, x = -4$

6. **a.**

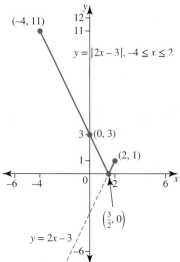

The range is $[0, 11]$

b.

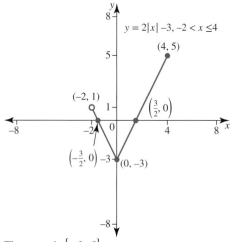

The range is $[-3, 5]$

7. **a.**

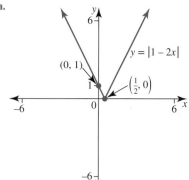

b.

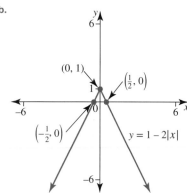

8. **a.**

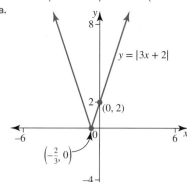

b.

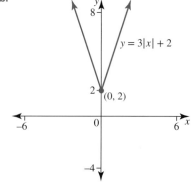

9. a.

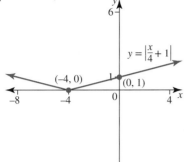

b.

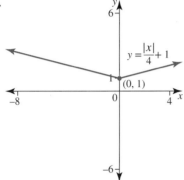

10. a.

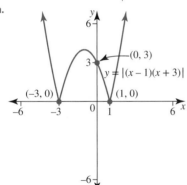

b.

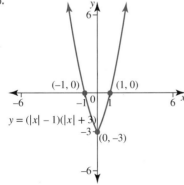

11. a.

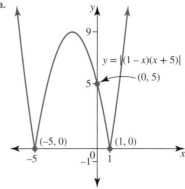

b.

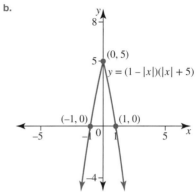

12. a.

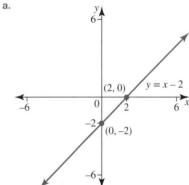

b. $x = 2, y = 0$

c. $\left(0, -\dfrac{1}{2}\right)$

d.

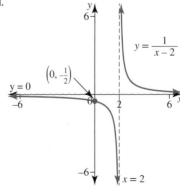

13. a.

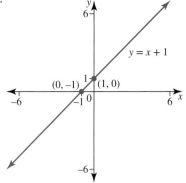

$y = x + 1$

$(0, -1)$　$(1, 0)$

b. $x = -1, y = 0$

c. $(0, 1)$

d.

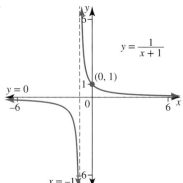

$y = \dfrac{1}{x+1}$

$(0, 1)$

$y = 0$

$x = -1$

14.

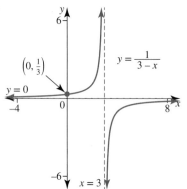

$\left(0, \dfrac{1}{3}\right)$

$y = \dfrac{1}{3-x}$

$y = 0$

$x = 3$

15.

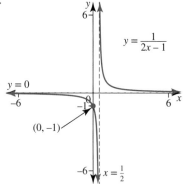

$y = \dfrac{1}{2x-1}$

$y = 0$

$(0, -1)$

$x = \dfrac{1}{2}$

16. a.

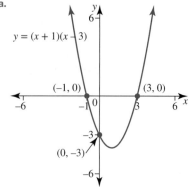

$y = (x+1)(x-3)$

$(-1, 0)$　$(3, 0)$

$(0, -3)$

b. $x = -1, x = 3, y = 0$

c. y-intercept $\left(0, -\dfrac{1}{3}\right)$, turning point $\left(1, -\dfrac{1}{4}\right)$

d.

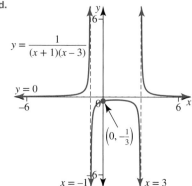

$y = \dfrac{1}{(x+1)(x-3)}$

$y = 0$

$\left(0, -\dfrac{1}{3}\right)$

$x = -1$　$x = 3$

17. a.

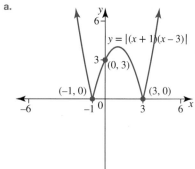

$y = |(x+1)(x-3)|$

$(0, 3)$

$(-1, 0)$　$(3, 0)$

b.

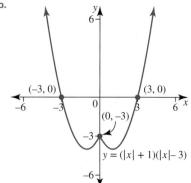

$(-3, 0)$　$(3, 0)$

$(0, -3)$

$y = (|x| + 1)(|x| - 3)$

18. a.

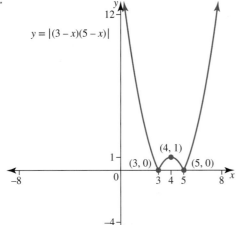

$y = |(3 - x)(5 - x)|$

(4, 1)
(3, 0)   (5, 0)

b.

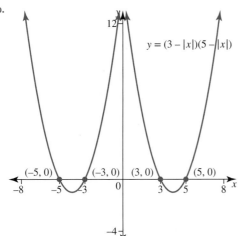

$y = (3 - |x|)(5 - |x|)$

(−5, 0)   (−3, 0)   (3, 0)   (5, 0)

19. a, b The graphs for **a** and **b** are identical.

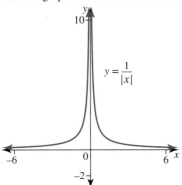

$y = \dfrac{1}{|x|}$

20.

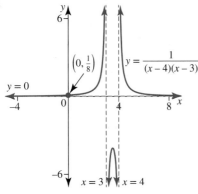

$\left(0, \frac{1}{8}\right)$ $y = \dfrac{1}{(x - 4)(x - 3)}$

$y = 0$

$x = 3$ $x = 4$

21.

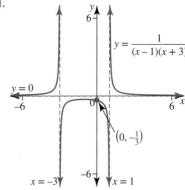

$y = \dfrac{1}{(x - 1)(x + 3)}$

$y = 0$

$\left(0, -\frac{1}{3}\right)$

$x = -3$ $x = 1$

22. a.

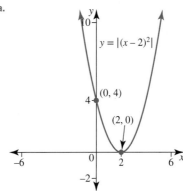

$y = |(x - 2)^2|$

(0, 4)
(2, 0)

b.

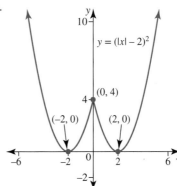

$y = (|x| - 2)^2$

(0, 4)
(−2, 0)   (2, 0)

The graph for **a** is identical to the original graph.

23.

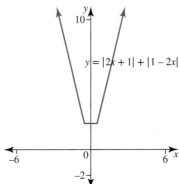

$y = |2x + 1| + |1 - 2x|$

24.

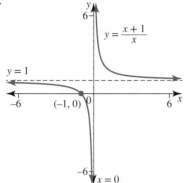

$y = \dfrac{x+1}{x}$

$y = 1$

$(-1, 0)$

$x = 0$

25.

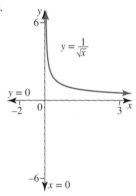

$y = \dfrac{1}{\sqrt{x}}$

$y = 0$

$x = 0$

9.3 Exercise

1. a. $\dfrac{2\sqrt{3}}{3}$ **b.** $\dfrac{2\sqrt{3}}{3}$ **c.** 2 **d.** $-\sqrt{2}$

2. a. $\dfrac{2\sqrt{3}}{3}$ **b.** $-\dfrac{2\sqrt{3}}{3}$ **c.** -2 **d.** $\sqrt{2}$

3. a. $-\dfrac{\sqrt{21}}{2}$ **b.** $-\dfrac{\sqrt{17}}{4}$ **c.** $-\dfrac{3\sqrt{8}}{8}$

  d. $-\sqrt{15}$ **e.** $\dfrac{3\sqrt{40}}{40}$

4. a. $-\dfrac{7\sqrt{40}}{40}$ **b.** $-\dfrac{5\sqrt{39}}{39}$ **c.** $-\dfrac{\sqrt{15}}{15}$

  d. $\dfrac{\sqrt{61}}{6}$ **e.** $\dfrac{\sqrt{3}}{12}$

5. a. See figure at the bottom of the page*

  b. See figure at the bottom of the page**

---

*5. a.

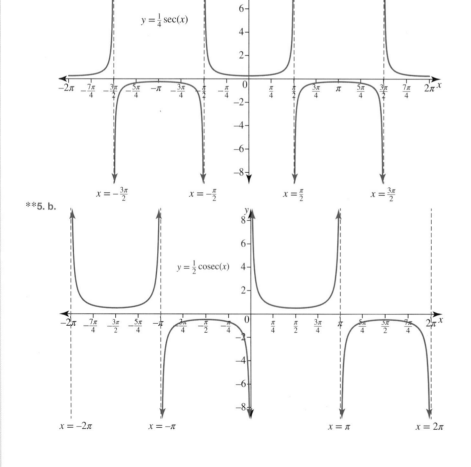

$y = \frac{1}{4}\sec(x)$

$x = -\frac{3\pi}{2}$ $\qquad$ $x = -\frac{\pi}{2}$ $\qquad$ $x = \frac{\pi}{2}$ $\qquad$ $x = \frac{3\pi}{2}$

**5. b.

$y = \frac{1}{2}\csc(x)$

$x = -2\pi$ $\qquad$ $x = -\pi$ $\qquad$ $x = \pi$ $\qquad$ $x = 2\pi$

6.

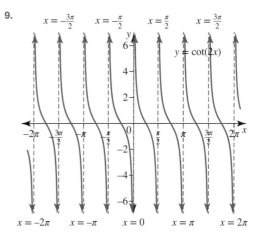

7. See figure at the bottom of the page*

8.

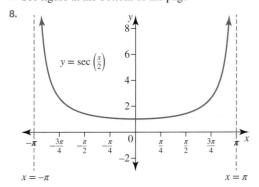

9.

10.

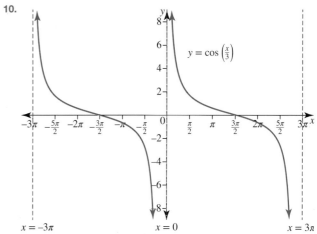

*7.

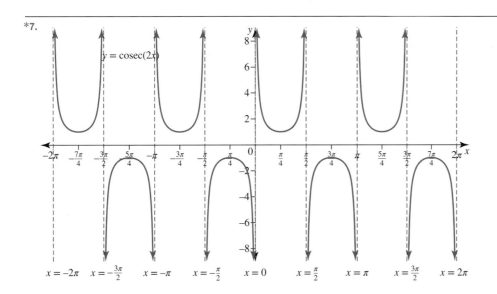

11. a. See figure at the bottom of the page*

    b. See figure at the bottom of the page**

b.

12. a.

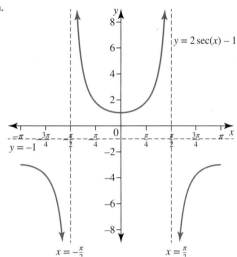

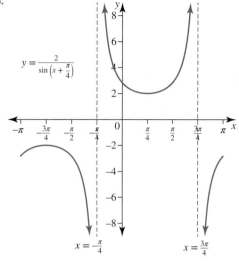

***11. a.**

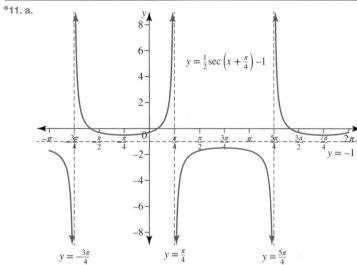

****11. b.**

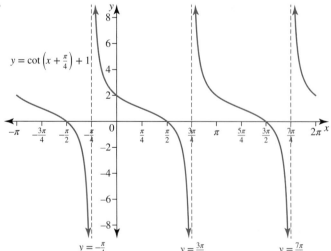

c.

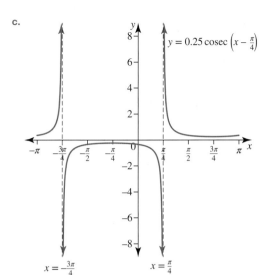

$$y = 0.25 \operatorname{cosec}\left(x - \frac{\pi}{4}\right)$$

$$x = -\frac{3\pi}{4} \qquad x = \frac{\pi}{4}$$

13.

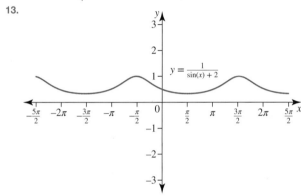

$$y = \frac{1}{\sin(x) + 2}$$

14.

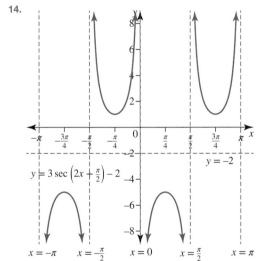

$$y = 3\sec\left(2x + \frac{\pi}{2}\right) - 2$$

$$y = -2$$

$$x = -\pi \qquad x = -\frac{\pi}{2} \qquad x = 0 \qquad x = \frac{\pi}{2} \qquad x = \pi$$

15. $\dfrac{p^2 - q^2 - pq}{q\sqrt{p^2 - q^2}}$

9.4 Pythagorean identities

9.4 Exercise

1. See table at the bottom of the page*
2. **a.** 0.600   **b.** 1.333
3. **a.** 0.954   **b.** 3.180
4. **a.** ± 0.917 **b.** ± 0.714
   c. ± 0.971 **d.** ± 0.436
5. **a.** $\dfrac{\sqrt{5}}{3}$ **b.** $\dfrac{2}{3}$
6. **a.** $\dfrac{\sqrt{7}}{4}$ **b.** $\dfrac{3}{\sqrt{7}}$
7. **a.** $\dfrac{\sqrt{11}}{4}$ **b.** $-\sqrt{\dfrac{5}{11}}$
8. $\sin^2(A)$
9. $1 - \sin(A)$
10. **a.** $0, \dfrac{\pi}{2}, \pi, 2\pi$ **b.** $\dfrac{\pi}{2}, \pi, \dfrac{3\pi}{2}$
    c. $0, \pi, \dfrac{4\pi}{3}, \dfrac{5\pi}{3}, 2\pi$ **d.** $\dfrac{\pi}{3}, \pi, \dfrac{5\pi}{3}$
11. **a.** $(2\sin(A) - 1)(\sin(A) + 1)$
    b. $(\tan(x) - 2)(\tan(x) + 1)$
    c. $(2\operatorname{cosec}(\alpha) + 1)(\operatorname{cosec}(\alpha) - 1)$
    d. $(1 - 2\sin(3x))(1 + 2\sin(3x))$ or
    $(2\cos(3x) + 1)(2\cos(3x) - 1)$
12. **a.** $\dfrac{\pi}{6}, \dfrac{5\pi}{6}, \dfrac{3\pi}{2}$ **b.** $\dfrac{\pi}{6}, \dfrac{5\pi}{6}, \dfrac{3\pi}{2}$
    c. $\dfrac{\pi}{4}, \dfrac{3\pi}{4}, \dfrac{5\pi}{4}, \dfrac{7\pi}{4}$ **d.** $\dfrac{\pi}{3}, \pi, \dfrac{5\pi}{3}$
13. **a.** $\dfrac{\pi}{12}(12n + 1), \dfrac{\pi}{12}(12n + 5), \dfrac{\pi}{4}(4n + 1), n \in \mathbb{Z}$
    b. $\dfrac{\pi}{6}(6n \pm 1), \dfrac{\pi}{2}(2n + 1), n \in \mathbb{Z}$
    c. $\dfrac{\pi}{24}(12n - 5), \dfrac{\pi}{24}(12n - 1), \dfrac{n\pi}{4}, n \in \mathbb{Z}$
    d. $\dfrac{n\pi}{2}, \dfrac{\pi}{8}(2n + 1), n \in \mathbb{Z}$
14. $\left[\dfrac{3}{4}, 1\right]$
15. **a.** $\dfrac{\pi}{3}(3n - 1), \dfrac{\pi}{4}(4n - 1), n \in \mathbb{Z}$
    b. $\dfrac{\pi}{3}(3n - 1), \dfrac{\pi}{4}(4n + 1), n \in \mathbb{Z}$

*1.

A°	30°	81°	129°	193°	260°	350°	−47°
$\sin^2(A)$	0.25	0.976	0.604	0.051	0.970	0.030	0.535
$\cos^2(A)$	0.75	0.024	0.396	0.949	0.030	0.970	0.465
$\sin^2(A) + \cos^2(A)$	1.00	1.000	1.000	1.000	1.000	1.000	1.000

16. a. $\dfrac{\pi}{6}(12n+1), \dfrac{\pi}{6}(12n+5), \dfrac{\pi}{2}(2n+1), n \in \mathbb{Z}$

  b. $\dfrac{\pi}{3}(6n \pm 1), n\pi, n \in \mathbb{Z}$

17. $\dfrac{\pi}{4}, \dfrac{3\pi}{4}, \dfrac{5\pi}{4}, \dfrac{7\pi}{4}$

18. 9

9.5 Compound angle identities

9.5 Exercise

1. a. $\dfrac{\sqrt{3}}{2}$ **b.** 0 **c.** $\dfrac{1}{2}$

2. a. $\sin(\theta) - \cos(\theta)$ **b.** $\sin(\theta) + \sqrt{3}\cos(\theta)$
  c. $\sqrt{3}\cos(\theta) + \sin(\theta)$ **d.** $\cos(\theta) - \sin(\theta)$

3. a. $\cos(\theta)$ **b.** $\sin(\theta)$
  c. $-\sin(\theta)$ **d.** $-\cos(\theta)$

4. a. $-\cos(\theta)$ **b.** $\sin(\theta)$
  c. $-\tan(\theta)$ **d.** $\tan(\theta)$

5. a. $\sqrt{3}\cos(x)$ **b.** $-\sqrt{3}\sin(x)$

6. a. $\dfrac{\sqrt{2} - \sqrt{6}}{4}$ **b.** $\dfrac{\sqrt{6} - \sqrt{2}}{4}$
  c. $2 - \sqrt{3}$ **d.** $2 + \sqrt{3}$

7. a. $\dfrac{\sqrt{2}}{4}$ **b.** $-\dfrac{\sqrt{2}}{2}$
  c. $-\dfrac{\sqrt{2}}{2}$ **d.** $\dfrac{\sqrt{3} + 2}{4}$

8. a. $-\dfrac{23}{32}$ **b.** $\dfrac{3\sqrt{55}}{32}$
  c. $-\dfrac{3\sqrt{55}}{23}$ **d.** $-\dfrac{23\sqrt{55}}{165}$

9. a. $\dfrac{\pi}{6}, \dfrac{\pi}{2}, \dfrac{5\pi}{6}, \dfrac{3\pi}{2}$
  b. $0, \dfrac{\pi}{3}, \pi, \dfrac{5\pi}{3}, 2\pi$

10. a. $0, \dfrac{\pi}{4}, \dfrac{3\pi}{4}, \pi, \dfrac{5\pi}{4}, \dfrac{7\pi}{4}, 2\pi$
  b. $\dfrac{\pi}{3}, \dfrac{\pi}{2}, \dfrac{2\pi}{3}, \dfrac{3\pi}{2}$

11. a. $\dfrac{56}{65}$ **b.** $-\dfrac{63}{16}$

12. a. $-\dfrac{253}{325}$ **b.** $-\dfrac{204}{325}$

13. a. $-\dfrac{8\sqrt{5}}{21}$ **b.** $-\dfrac{19}{21}$

14. $\dfrac{ab + \sqrt{1 - a^2}\sqrt{1 - b^2}}{b\sqrt{1 - a^2} - a\sqrt{1 - b^2}}$

15-25. Sample responses can be found in the worked solutions in the online resources.

9.6 Product identities

9.6 Exercise

1. a. $\sqrt{3}$ **b.** $-\dfrac{\sqrt{3}}{3}$

2. a. $\dfrac{\sqrt{3}}{3}$ **b.** 1 **c.** $-\dfrac{\sqrt{3}}{3}$

3. a. $\dfrac{\sqrt{6}}{2}$ **b.** $\dfrac{\sqrt{2}}{2}$ **c.** $\dfrac{\sqrt{2}}{2}$ **d.** $-\dfrac{\sqrt{6}}{2}$

4-11. Sample responses can be found in the worked solutions in the online resources.

12. a. $\dfrac{\pi}{5}(2n + 1), \pi(2n + 1), n \in \mathbb{Z}$
  b. $\dfrac{n\pi}{3}, \dfrac{\pi}{2}(2n + 1), n \in \mathbb{Z}$

13. a. $\dfrac{n\pi}{2}, \dfrac{\pi}{2}(2n + 1), n \in \mathbb{Z}$
  b. $\dfrac{\pi}{4}(2n + 1), \dfrac{\pi}{2}(2n + 1), n \in \mathbb{Z}$

14. a. $\dfrac{n\pi}{2}, \dfrac{\pi}{3}(6n \pm 2), n \in \mathbb{Z}$
  b. $\dfrac{\pi}{4}(2n + 1), \dfrac{\pi}{3}(6n \pm 2), n \in \mathbb{Z}$
  c. $\dfrac{\pi}{2}(2n + 1), \dfrac{\pi}{12}(12n - 1), \dfrac{\pi}{12}(12n - 5), n \in \mathbb{Z}$

15-16. Sample responses can be found in the worked solutions in the online resources.

9.7 Applications of trigonometric identities

9.7 Exercise

1. $\sqrt{2}\cos\left(x - \dfrac{\pi}{4}\right)$

2. $2\cos\left(x + \dfrac{\pi}{6}\right)$

3. a. $\sqrt{2}\sin\left(x + \dfrac{\pi}{4}\right)$ **b.** $2\sin\left(x + \dfrac{\pi}{6}\right)$

4. a. $\sqrt{2}\sin\left(x - \dfrac{\pi}{4}\right)$ **b.** $2\sin\left(x - \dfrac{\pi}{6}\right)$

5. a. $8\cos\left(x - \dfrac{\pi}{4}\right)$ **b.** $25\cos(x - 1.287)$

6. Sample responses can be found in the worked solutions in the online resources.

7. a. $f(x) = 2\sin\left(x - \dfrac{\pi}{4}\right)$
  b. Maximum $\left(\dfrac{3\pi}{4}, 2\right)$, minimum $\left(\dfrac{7\pi}{4}, -2\right)$

c.

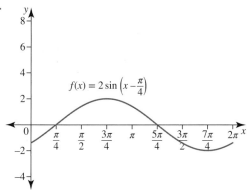

$f(x) = 2\sin\left(x - \dfrac{\pi}{4}\right)$

d. $\dfrac{5\pi}{12}, \dfrac{13\pi}{12}$

8. a. $f(x) = 13\cos(x - 2.747)$

b. Maximum $(2.747, 13)$, minimum $(5.888, -13)$

c.

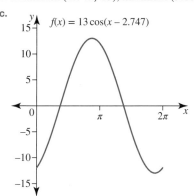

$f(x) = 13\cos(x - 2.747)$

d. $1.7, \ 3.794$

9. a. $f(x) = 5\sin(x° - 53.13°)$

b. Maximum $(143.13°, 5)$, minimum $(323.13°, -5)$

c. See figure at the bottom of the page*

d. $83.13°, 203.13°$

10-11. Sample responses can be found in the worked solutions in the online resources.

12. a. Maximum 4, minimum $\dfrac{4}{9}$

b. Maximum $\dfrac{\sqrt{3}}{10}$, minimum $-\dfrac{\sqrt{3}}{6}$

13. a. $I = 41\cos(300t - 12.68°)$

b. $I = 61\cos(800t - 79.61°)$

9.8 Review

9.8 Exercise

1. a. $x = 5 \text{ or } x = -5$ **b.** $x = 6 \text{ or } x = -8$

2. a. $\cos(x) = \dfrac{3}{\sqrt{58}}$ **b.** $\sin(x) = \dfrac{7}{\sqrt{58}}$

c. $\tan(x) = \dfrac{7}{3}$ **d.** $\sec(x) = \dfrac{\sqrt{58}}{3}$

e. $\operatorname{cosec}(x) = \dfrac{\sqrt{58}}{7}$ **f.** $\cot(x) = \dfrac{3}{7}$

3. a. $\sin\left(\dfrac{7\pi}{12}\right) = \dfrac{\sqrt{6} + \sqrt{2}}{4}$

b. $\cos\left(\dfrac{\pi}{12}\right) = \dfrac{\sqrt{6} + \sqrt{2}}{4}$

4. a. $\sqrt{3}$ **b.** $-\dfrac{\sqrt{3}}{3}$ **c.** -1 **d.** $-\sqrt{3}$

5. a. $-\dfrac{5\sqrt{21}}{21}$ **b.** $-\sqrt{17}$

---

*9. c.

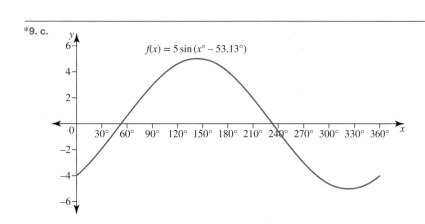

$f(x) = 5\sin(x° - 53.13°)$

6. a. See figure at the bottom of the page*

b.
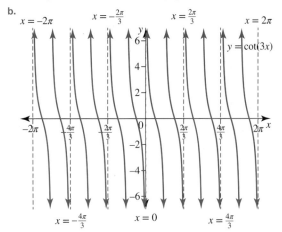

7. a. $\dfrac{\pi}{2}$

  b. $\dfrac{\pi}{2}, \dfrac{7\pi}{6}, \dfrac{11\pi}{6}$

8. a. $(\operatorname{cosec}(3A) - 1)(3\operatorname{cosec}(3A) + 1)$

  b. $(\sec(2\beta) + 1)(\sec(2\beta) - 2)$

9. a. $\dfrac{\pi}{2}, \pi, \dfrac{3\pi}{2}$ **b.** $\dfrac{3\pi}{2}$

10. a. $\dfrac{\pi}{18}(12n + 5), \dfrac{\pi}{18}(12n - 5), \dfrac{\pi}{6}(2n + 1), n \in \mathbb{Z}$

  b. $\dfrac{\pi}{9}(6n + 1), \dfrac{2\pi}{9}(3n + 1), \dfrac{n\pi}{3}, n \in \mathbb{Z}$

11. a.
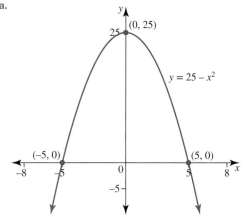

b.
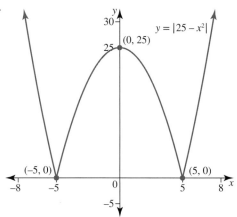

*6. a.
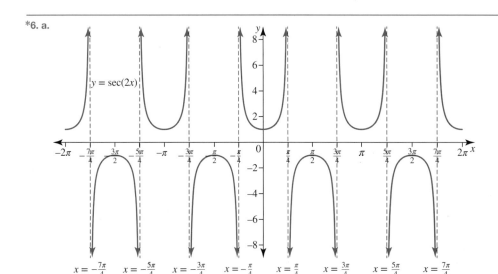

c.

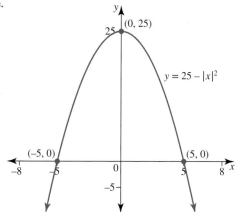

d.

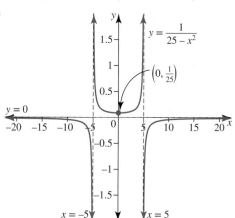

12. a.

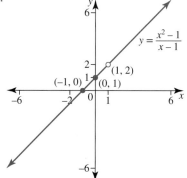

b.

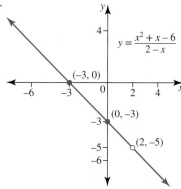

13. a.

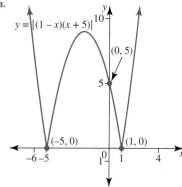

b.

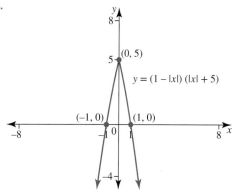

c.

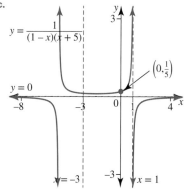

14. $\dfrac{\sqrt{a^2 - b^2}}{a + b}$

15. a. $f(x) = 2 \sin\left(x - \dfrac{\pi}{4}\right)$

    b. Maximum $\left(\dfrac{3\pi}{4}, 2\right)$, minimum $\left(\dfrac{7\pi}{4}, -2\right)$

c.

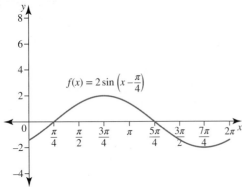

$f(x) = 2\sin\left(x - \dfrac{\pi}{4}\right)$

d. $\dfrac{5\pi}{12}, \dfrac{13\pi}{12}$

16. a. $f(x) = 13\cos(x - 0.395)$

b. Maximum $(0.395, 13)$, minimum $(3.537, -13)$

c.

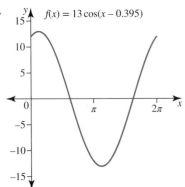

$f(x) = 13\cos(x - 0.395)$

d. $x = 1.442, 5.631$; general solution:
$x = 1.442 + 2n\pi, 5.631 + 2n\pi, n \in \mathbb{Z}$

17. Sample responses can be found in the worked solutions in the online resources.

18. a.

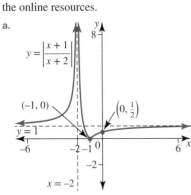

$y = \left|\dfrac{x+1}{x+2}\right|$

b.

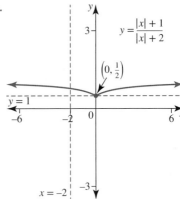

$y = \dfrac{|x|+1}{|x|+2}$

c.

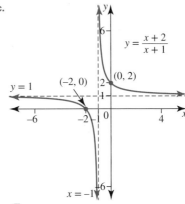

$y = \dfrac{x+2}{x+1}$

19. a. $\dfrac{\pi}{4}(2n+1), n \in \mathbb{Z}$

b. $\dfrac{\pi}{4}(2n+1), \dfrac{\pi}{3}(3n \pm 1), n \in \mathbb{Z}$

20. See figure at the bottom of the page*

***20.**

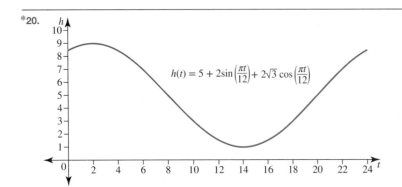

$h(t) = 5 + 2\sin\left(\dfrac{\pi t}{12}\right) + 2\sqrt{3}\cos\left(\dfrac{\pi t}{12}\right)$

10 Matrices and transformations

LESSON SEQUENCE

Fully worked solutions for this chapter are available online.

EXAM PREPARATION

Access exam-style questions in every lesson, available online.

 Resources

Solutions	Solutions — Chapter 10 (sol-0402)
Exam questions	Exam question booklet — Chapter 10 (eqb-0288)
Digital documents	Learning matrix — Chapter 10 (doc-41934)
	Chapter summary — Chapter 10 (doc-41935)

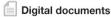

LESSON
10.1 Overview

10.1.1 Introduction

Following on from the introduction to matrices in Chapter 5, we will now investigate geometric transformations with matrices. A transformation is a movement or change in a geometric shape through translation, reflection, rotation and dilation. A translation can be thought of as a 'slide', a reflection as a 'mirror image', a rotation as a 'spin' about a point and a dilation as an enlargement or reduction of a shape or figure.

One of the most visually arresting examples of the manipulation of images through transformations is in the fascinating art of Maurits Cornelis Escher. Although this Dutch artist was not formally mathematically trained, he drew great inspiration from the mathematical ideas he read about.

Early animations, such as cartoons, were originally drawn by hand — a very time-consuming process. Today the transformation of points, and hence images, through matrix transformations is the basis of computer based animation. In this chapter, you will see how matrix transformations of reflection, dilation and rotation can be combined into a single new matrix. This combination of the movements of several individual matrices into one operation allows movement of every vertex in three-dimensional space. When matrix transformations are combined with the imagination of animators, designers and storytellers, images are created that have the power to shock, awe and inspire.

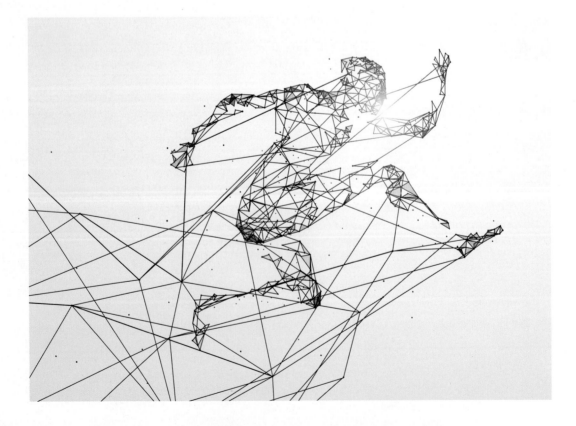

10.1.2 Syllabus links

Lesson	Lesson title	Syllabus links
10.2	**Translations**	○ Understand translations and their representation as column vectors.
		○ Apply these transformations to points in the plane and polygons.
10.3	**Reflections and rotations**	○ Use basic linear transformations: rotations about the origin and reflection in a line that passes through the origin, and the representations of these transformations by 2×2 matrices. • rotation of angle θ anticlockwise about the origin: $\begin{bmatrix} \cos(\theta) & -\sin(\theta) \\ \sin(\theta) & \cos(\theta) \end{bmatrix}$ • reflection in the line $y = \tan(\theta)$: $\begin{bmatrix} \cos(2\theta) & \sin(2\theta) \\ \sin(2\theta) & -\cos(2\theta) \end{bmatrix}$
		○ Apply these transformations to points in the plane and polygons.
10.4	**Dilations**	○ Use basic linear transformations: dilations of the form $(x, y) \rightarrow (ax, by)$ and the representations of these transformations by 2×2 matrices. • dilation of factor a parallel to the x-axis and factor b parallel to the y-axis: $\begin{bmatrix} a & 0 \\ 0 & b \end{bmatrix}$
		○ Apply these transformations to points in the plane and polygons.
10.5	**Combinations of transformations**	○ Use basic linear transformations: dilations of the form $(x, y) \rightarrow (ax, by)$, rotations about the origin and reflection in a line that passes through the origin, and the representations of these transformations by 2×2 matrices. • dilation of factor a parallel to the x-axis and factor b parallel to the y-axis: $\begin{bmatrix} a & 0 \\ 0 & b \end{bmatrix}$ • rotation of angle θ anticlockwise about the origin: $\begin{bmatrix} \cos(\theta) & -\sin(\theta) \\ \sin(\theta) & \cos(\theta) \end{bmatrix}$
		○ reflection in the line $y = \tan(\theta)$: $\begin{bmatrix} \cos(2\theta) & \sin(2\theta) \\ \sin(2\theta) & -\cos(2\theta) \end{bmatrix}$
		○ Apply these transformations to points in the plane and polygons.
		○ Understand and use composition of linear transformations and the corresponding matrix products.
		○ Understand and use inverses of linear transformations and the relationship with the matrix inverse.
		○ Understand and use the relationship between the determinant and the effect of a linear transformation on area.
		○ Determine geometric results by matrix multiplications, e.g. showing that the combined effect of two reflections in lines through the origin is a rotation.

Source: Specialist Mathematics Senior Syllabus 2024 © State of Queensland (QCAA) 2024; licensed under CC BY 4.0

LESSON
10.2 Translations

10.2.1 Matrix transformations

A **transformation** is a function which maps a set of points, called the original or the **pre-image** points to a new set of points called the **image** points. It can be a change of position of points, lines, curves or shapes in a plane, or a change in shape due to an enlargement or reduction by a scale factor.

Each point of the plane is transformed or mapped to another point.

Matrix transformations

The transformation, T, is written as:

$$T : \begin{bmatrix} x \\ y \end{bmatrix} \rightarrow \begin{bmatrix} x' \\ y' \end{bmatrix}$$

The matrix $\begin{bmatrix} x \\ y \end{bmatrix}$ is the column matrix representing the coordinates of the point $P(x, y)$, the pre-image or original point. The matrix $\begin{bmatrix} x' \\ y' \end{bmatrix}$ represents the coordinates of the point $P'(x', y')$, the image of $P(x, y)$ after a transformation.

Any transformation that can be represented by a 2×2 matrix, $\begin{bmatrix} a & b \\ c & d \end{bmatrix}$, is called a **linear transformation**. The origin never moves under a linear transformation.

An **invariant point** or **fixed point** is a point of the domain of the function which is mapped onto itself after a transformation, that is, the pre-image point is the same as the image point and is unchanged by the transformation.

$$\begin{bmatrix} x \\ y \end{bmatrix} = \begin{bmatrix} x' \\ y' \end{bmatrix} \Rightarrow x' = x \text{ and } y' = y$$

For example, a reflection in the line $y = x$ leaves every point on the line $y = x$ unchanged.

The transformations which will be studied in this chapter are:
- translations
- reflections
- rotations
- dilations.

10.2.2 Translations

A **translation** is a transformation where each point in the plane is moved a given distance in a horizontal and/or vertical direction.

Consider a marching band marching in perfect formation. As the leader of the marching band moves from a position $P(x, y)$ to a new position $P'(x', y')$, which is a steps across and b steps up, all members of the band will also move to a new position $P'(x + a, y + b)$. Their new position could be defined as $P'(x', y') = P'(x + a, y + b)$ where a represents the horizontal translation and b represents the vertical translation and P' is the image of P.

Translation matrix equation

The matrix equation for a translation can be given as:

$$\begin{array}{ccc} \mathbf{P'} & \mathbf{P} & \mathbf{T} \\ \begin{bmatrix} x' \\ y' \end{bmatrix} = & \begin{bmatrix} x \\ y \end{bmatrix} + & \begin{bmatrix} a \\ b \end{bmatrix} \end{array}$$

where a represents the horizontal translation and b the vertical translation.

The matrix $\begin{bmatrix} a \\ b \end{bmatrix}$ is called the **translation matrix** or column vector and is denoted by T. It represents the horizontal and vertical displacement, as shown in the diagram:

$$x' = x + a$$
$$y' = y + b$$

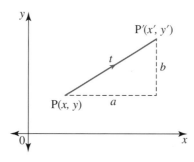

Each x-coordinate is moved a units parallel to the x-axis and each y-coordinate in moved b units parallel to the y-axis.

Note t (lower case) denotes the translation itself and T (upper case) denotes the matrix of the translation.

A cyclist in a race needs to move from the front position at (8, 1) across 2 positions, to the left, so that the other cyclists can pass. Write the translation matrix and determine the cyclist's new position.

THINK	WRITE
1. Write down the translation matrix, T, using the information given.	The cyclist moves across to the left by 2 units. Translating 2 units to the left means each x-coordinate decreases by 2. $$T = \begin{bmatrix} -2 \\ 0 \end{bmatrix}$$
2. Recall and apply the matrix transformation for a translation equation.	$\quad$ P′ $\quad$ P $\quad$ T $$\begin{bmatrix} x' \\ y' \end{bmatrix} = \begin{bmatrix} x \\ y \end{bmatrix} + \begin{bmatrix} -2 \\ 0 \end{bmatrix} = \begin{bmatrix} x - 2 \\ y + 0 \end{bmatrix}$$
3. Substitute the pre-image point into the matrix equation.	The pre-image point is (8, 1). P′ $$\begin{bmatrix} x' \\ y' \end{bmatrix} = \begin{bmatrix} 8 - 2 \\ 1 + 0 \end{bmatrix}$$
4. State the cyclist's new position by calculating the coordinates of the image point from the matrix equation. *Note:* The y-coordinate has not changed; that is, the cyclist is still in front, but to the left.	$\quad$ P′ $\quad$ P $\quad$ T $$\begin{bmatrix} x' \\ y' \end{bmatrix} = \begin{bmatrix} 8 \\ 1 \end{bmatrix} + \begin{bmatrix} -2 \\ 0 \end{bmatrix} = \begin{bmatrix} 6 \\ 1 \end{bmatrix}$$ The new position is (6, 1).

10.2.3 Translations of an object

Matrix addition can be used to determine the Cartesian coordinates of a translated object when an object is moved or translated from one location to another on the coordinate plane without changing its size or orientation.

Consider the triangle ABC with coordinates $A(-1, 3), B(0, 2)$ and $C(-2, 1)$. It is to be moved 3 units to the right and 1 unit down. To calculate the coordinates of the vertices of the translated $\Delta A'B'C'$, we can use matrix addition.

Note that you could translate each point individually, or do all points in a single matrix.

First, the coordinates of the triangle $\triangle ABC$ can be written as a **coordinate matrix**. The coordinates of the vertices of a figure are arranged as columns in the matrix.

$$\triangle ABC = \begin{matrix} A & B & C \\ \begin{bmatrix} -1 & 0 & -2 \\ 3 & 2 & 1 \end{bmatrix} \end{matrix}$$

Secondly, translating the triangle 3 units to the right means each x-coordinate increases by 3.

Translating the triangle 1 unit down means that each y-coordinate decreases by 1.

The translation matrix that will do this is $\begin{bmatrix} 3 & 3 & 3 \\ -1 & -1 & -1 \end{bmatrix}$.

Finally, to determine the coordinates of the vertices of the translated triangle $\triangle A'B'C'$ add the translation matrix to the coordinate matrix.

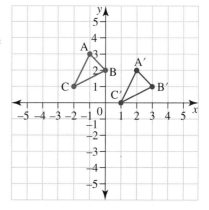

$$\begin{matrix} A & B & C \\ \begin{bmatrix} -1 & 0 & -2 \\ 3 & 2 & 1 \end{bmatrix} \end{matrix} + \begin{bmatrix} 3 & 3 & 3 \\ -1 & -1 & -1 \end{bmatrix} = \begin{matrix} A' & B' & C' \\ \begin{bmatrix} 2 & 3 & 1 \\ 2 & 1 & 0 \end{bmatrix} \end{matrix}$$

The coordinates of the vertices of the translated triangle

$\triangle A'B'C' = \begin{bmatrix} 2 & 3 & 1 \\ 2 & 1 & 0 \end{bmatrix}$ are $A'(2, 2), B'(3, 1)$ and $C'(1, 0)$.

WORKED EXAMPLE 2 Determining a translation matrix

Determine the translation matrix if $\triangle ABC$ with coordinates $A(-1, 3)$, $B(0, 2)$ and $C(-2, 1)$ is translated to $\triangle A'B'C'$ with coordinates $A'(2, 4)$, $B'(3, 3)$ and $C'(1, 2)$.

THINK	WRITE
1. Write the coordinates of $\triangle ABC$ as a coordinate matrix.	The coordinates of the vertices of a figure are arranged as columns in the matrix. $$\triangle ABC = \begin{matrix} A & B & C \\ \begin{bmatrix} -1 & 0 & -2 \\ 3 & 2 & 1 \end{bmatrix} \end{matrix}$$
2. Write the coordinates of the vertices of the translated triangle $\triangle A'B'C'$ as a coordinate matrix.	$$\triangle A'B'C' = \begin{matrix} A' & B' & C' \\ \begin{bmatrix} 2 & 3 & 1 \\ 4 & 3 & 2 \end{bmatrix} \end{matrix}$$
3. Calculate the translation matrix by recalling the matrix equation: $P' = P + T$. Rearrange the equation to make T the subject and apply the rules of matrix subtraction.	$$\begin{matrix} P' & & P \\ \begin{bmatrix} 2 & 3 & 1 \\ 4 & 3 & 2 \end{bmatrix} = \begin{bmatrix} -1 & 0 & -2 \\ 3 & 2 & 1 \end{bmatrix} + T \end{matrix}$$ $$\begin{matrix} & P' & & P \\ T = & \begin{bmatrix} 2 & 3 & 1 \\ 4 & 3 & 2 \end{bmatrix} - \begin{bmatrix} -1 & 0 & -2 \\ 3 & 2 & 1 \end{bmatrix} \end{matrix}$$
4. Translating the triangle 3 units to the right means that each x-coordinate increases by 3. Translating the triangle 1 unit up means that each y-coordinate increases by 1.	The translation matrix is: $$T = \begin{bmatrix} 3 & 3 & 3 \\ 1 & 1 & 1 \end{bmatrix}$$

10.2.4 Invariant properties of translations

Properties which are unchanged by a transformation are called **invariants**. For translations, properties such as length, shape and area are invariant since they are unchanged when a line, curve or shape undergoes a translation.

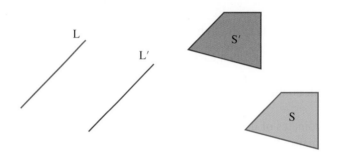

Consider the examples given which show a line L which undergoes a translation to L', and a shape S which undergoes a translation to S'. Notice that the length of the line L is unchanged under the translation, and the length of sides, shape and area of S is unchanged under translation.

10.2.5 Translations of a curve

Note that this content is covered in the mathematical methods course.

A translation of a curve maps every original point (x, y) of the curve onto a new unique and distinct image point (x', y').

Consider the parabola with the equation $y = x^2$.

If the parabola is translated 3 units in the positive direction of the x-axis, what is the image equation and what happens to the coordinates?

As seen from the table of values, each coordinate (x, y) has a new coordinate pair or image point $(x + 3, y)$.

x	y	(x, y)	$x' = x + 3$	$y' = y$	(x', y')
-3	9	$(-3, 9)$	$-3 + 3$	9	$(0, 9)$
-2	4	$(-2, 4)$	$-2 + 3$	4	$(1, 4)$
-1	1	$(-1, 1)$	$-1 + 3$	1	$(2, 1)$
0	0	$(0, 0)$	$0 + 3$	0	$(3, 0)$
1	1	$(1, 1)$	$1 + 3$	1	$(4, 1)$
2	4	$(2, 4)$	$2 + 3$	4	$(5, 4)$
3	9	$(3, 9)$	$3 + 3$	9	$(6, 9)$

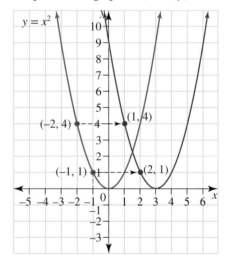

The matrix equation for the translation of any point on the curve $y = x^2$ can be written as:

$$\begin{bmatrix} x' \\ y' \end{bmatrix} = \begin{bmatrix} x \\ y \end{bmatrix} + \begin{bmatrix} 3 \\ 0 \end{bmatrix}$$

The image equations for the two coordinates are $x' = x + 3$ and $y' = y$.

Rearranging the image equations to make the pre-image coordinates the subject, we get

$$x' = x + 3 \leftrightarrow x = x' - 3 \quad \text{and} \quad y = y'.$$

To determine the image equation, substitute the image expressions into the pre-image equation.

$$y = x^2$$
$$y = y' \quad x = x' - 3$$
$$\therefore y' = (x' - 3)^2$$

The equation of the parabola, $y = x^2$, after a translation of 3 units in the positive direction of the x-axis is $y = (x - 3)^2$.

WORKED EXAMPLE 3 Determining a matrix translation of a line

Determine the equation of the image of the line with equation $y = x + 1$ after it is transformed by the translation matrix $T = \begin{bmatrix} 2 \\ 1 \end{bmatrix}$.

THINK

1. Recall the matrix equation for the transformation and apply it to the given transformation.

2. State the image equations for the two coordinates.

3. Rearrange the equations to make the pre-image coordinates x and y the subjects.

4. Substitute the image equations into the pre-image equation to determine the image equation.

5. Graph the image and pre-image equation to verify the translation.

WRITE

$$\begin{bmatrix} x' \\ y' \end{bmatrix} = \begin{bmatrix} x \\ y \end{bmatrix} + \begin{bmatrix} 2 \\ 1 \end{bmatrix}$$

$x' = x + 2$ and $y' = y + 1$

$x = x' - 2$ and $y = y' - 1$

$$y = x + 1$$
$$y' - 1 = (x' - 2) + 1$$
$$y' = x'$$

The image equation is $y = x$.

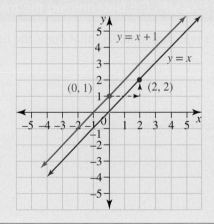

Determine the equation of the image of the parabola with equation $y = x^2$ after it is transformed by the translation matrix $T = \begin{bmatrix} -3 \\ 1 \end{bmatrix}$.

THINK	WRITE
1. Recall the matrix equation for the transformation and apply it to the given translation.	$\begin{bmatrix} x' \\ y' \end{bmatrix} = \begin{bmatrix} x \\ y \end{bmatrix} + \begin{bmatrix} -3 \\ 1 \end{bmatrix}$
2. State the image equations for the two coordinates.	$x' = x - 3$ and $y' = y + 1$
3. Rearrange the equations to make the pre-image coordinates x and y the subjects.	$x = x' + 3$ and $y = y' - 1$
4. Substitute the image expressions into the pre-image equation to determine the image equation.	$y = x^2$ $y' - 1 = (x' + 3)^2$ $y' = (x' + 3)^2 + 1$ The image equation is $y = (x + 3)^2 + 1$.
5. Graph the image and pre-image equation to verify the translation.	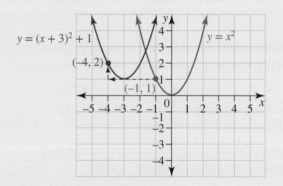

Determine a translation matrix that maps the line with equation $y = x$ onto the line with equation $y = x - 7$.

THINK	WRITE
1. Write a matrix equation for a translation.	$\begin{bmatrix} x' \\ y' \end{bmatrix} = \begin{bmatrix} x \\ y \end{bmatrix} + \begin{bmatrix} a \\ b \end{bmatrix}$ $= \begin{bmatrix} x + a \\ y + b \end{bmatrix}$

2. State the separate equations for the x- and y-coordinates of the image and then rearrange them to make x and y the subjects.

$$x' = x + a \qquad y' = y + b$$
$$x = x' - a \qquad y = y' - b$$

3. Substitute these into the equation of the line $y = x$ and rearrange to make y' the subject.

$$y = x$$
$$y' - b = x' - a$$
$$y' = x' - a + b$$

4. Recall that the equation of the image is $y' = x' - 7$. Determine values of a and b that satisfy this equation.

$$y' = x' - 7$$
$$\therefore -a + b = -7$$
$$a = 7, \ b = 0 \text{ satisfy this equation.}$$

5. State the translation that maps the line $y = x$ onto the line $y = x - 7$.

$$T = \begin{bmatrix} 7 \\ 0 \end{bmatrix}$$

Note: There are infinitely many translations that map the line $y = x$ onto the line $y = x - 7$. Any values of a and b for which $b - a = -7$ will be valid answers e.g. $\begin{bmatrix} 3 \\ 4 \end{bmatrix}$.

Exercise 10.2 Translations

learn on

10.2 Exercise	10.2 Exam questions on

Simple familiar	Complex familiar	Complex unfamiliar
1, 2, 3, 4, 5, 6, 7, 8, 9, 10, 11, 12, 13, 14	15, 16, 17, 18, 19, 20	N/A

These questions are even better in jacPLUS!
• Receive immediate feedback
• Access sample responses
• Track results and progress

Find all this and MORE in jacPLUS ▶

Simple familiar

1. **WE1** A chess player moves his knight 1 square to the right and 2 squares up from position $(2, 5)$. Write the translation matrix and determine the new position of the knight.

2. Determine the image of the point $(-1, 0)$ using the matrix equation for the translation $\begin{bmatrix} x' \\ y' \end{bmatrix} = \begin{bmatrix} x \\ y \end{bmatrix} + \begin{bmatrix} -5 \\ 2 \end{bmatrix}$.

3. Determine the image of the point $(1, 2)$ using the matrix equation for translation $\begin{bmatrix} x' \\ y' \end{bmatrix} = \begin{bmatrix} x \\ y \end{bmatrix} + \begin{bmatrix} 3 \\ -2 \end{bmatrix}$.

4. Determine the image of the point $(3, -4)$ using the matrix equation for translation $\begin{bmatrix} x' \\ y' \end{bmatrix} = \begin{bmatrix} x \\ y \end{bmatrix} + \begin{bmatrix} -1 \\ 2 \end{bmatrix}$.

5. **WE2** Determine the translation matrix if $\triangle ABC$ with coordinates $A(0, 0)$, $B(2, 3)$ and $C(-3, 4)$ is translated to $\triangle A'B'C'$ with coordinates $A'(1, -2)$, $B'(3, 1)$ and $C'(-2, 2)$.

6. Determine the translation matrix if $\triangle ABC$ with coordinates $A(3, 0)$, $B(2, 4)$ and $C(-2, -5)$ is translated to $\triangle A'B'C'$ with coordinates $A'(4, 2)$, $B'(3, 6)$ and $C'(-1, -3)$.

7. The image equations are given by $x' = x + 2$ and $y' = y + 1$. Express the translation in matrix equation form.

8. a. On a Cartesian plane, draw $\triangle ABC = \begin{bmatrix} 0 & 1 & -2 \\ 0 & 3 & 1 \end{bmatrix}$ and

   $\triangle A'B'C' = \begin{bmatrix} 2 & 3 & 0 \\ -1 & 2 & 0 \end{bmatrix}$.

   b. Calculate the translation matrix if $\triangle ABC = \begin{bmatrix} 0 & 1 & -2 \\ 0 & 3 & 1 \end{bmatrix}$ is

   translated to $\triangle A'B'C' = \begin{bmatrix} 2 & 3 & 0 \\ -1 & 2 & 0 \end{bmatrix}$.

9. **WE3** Determine the equation of the image of the line with equation $y = x - 3$ after it is transformed by the translation matrix $T = \begin{bmatrix} -1 \\ 3 \end{bmatrix}$ and graph the image and pre-image equations to verify the translation.

10. Determine the equation of the image of the line with equation $y = x - 1$ after it is transformed by the translation matrix $T = \begin{bmatrix} 3 \\ -2 \end{bmatrix}$ and graph the image and pre-image equations to verify the translation.

11. Determine the equation of the image of the line with equation $y = x + 3$ after it is transformed by the translation matrix $T = \begin{bmatrix} -2 \\ 1 \end{bmatrix}$ and graph the image and pre-image equations to verify the translation.

12. **WE4** Determine the equation of the image of the parabola with equation $y = x^2$ after it is transformed by the translation matrix $T = \begin{bmatrix} 2 \\ -1 \end{bmatrix}$ and graph the image and pre-image equations to verify the translation.

13. Determine the equation of the image of the parabola with equation $y = x^2 + 1$ after it is transformed by the translation matrix $T = \begin{bmatrix} -3 \\ 0 \end{bmatrix}$ and graph the image and pre-image equations to verify the translation.

14. Determine the equation of the image of the parabola with equation $y = x^2 - 2$ after it is transformed by the translation matrix $T = \begin{bmatrix} 7 \\ -4 \end{bmatrix}$ and graph the image and pre-image equations to verify the translation.

Complex familiar

15. **WE5** Determine a translation matrix that maps the line with equation $y = x$ onto the line with equation $y = x + 2$.

16. Determine a translation matrix that maps the parabola with equation $y = x^2$ onto the parabola with equation $y = (x - 7)^2 + 3$.

17. Determine the translation matrix that maps the parabola with equation $y = x^2$ onto the parabola with equation $y = x^2 - 4x + 10$.

18. Determine the translation matrix that maps the circle with equation $x^2 + y^2 = 9$ onto the circle with equation $(x - 1)^2 + y^2 = 9$.

19. Determine the translation matrix that maps the parabola with equation $y = x^2$ onto the parabola with equation $y = (x - a)^2 + b$.

20. Determine the translation matrix that maps the circle with equation $x^2 + y^2 = r^2$ onto the circle with equation $(x - a)^2 + y^2 = r^2$.

Fully worked solutions for this chapter are available online.

LESSON
10.3 Reflections and rotations

SYLLABUS LINKS

- Use basic linear transformations: rotations about the origin and reflection in a line that passes through the origin, and the representations of these transformations by 2×2 matrices.
  - rotation of angle θ anticlockwise about the origin: $\begin{bmatrix} \cos(\theta) & -\sin(\theta) \\ \sin(\theta) & \cos(\theta) \end{bmatrix}$
  - reflection in the line $y = \tan(\theta)$: $\begin{bmatrix} \cos(2\theta) & \sin(2\theta) \\ \sin(2\theta) & -\cos(2\theta) \end{bmatrix}$
- Apply these transformations to points in the plane and polygons.

Source: Specialist Mathematics Senior Syllabus 2024 © State of Queensland (QCAA) 2024; licensed under CC BY 4.0

Matrix multiplication can be used to determine the Cartesian coordinates of a reflected or rotated image when the position and orientation of an object changes without changing its size.

10.3.1 Reflections

A **reflection** is a transformation in a line. This line is called a line of reflection, or the **mediator**, and the image point is a mirror image of the pre-image point.

In a reflection:
- the image, P′, and the pre-image, P, are equidistant from the line of reflection.
- the line of reflection is perpendicular to the line joining the image, P′, and the pre-image, P.
- M is the symbol used for a matrix reflection. The subscript describes the line of reflection.

The following reflections will be considered:
- reflection in the x-axis ($y = 0$)
- reflection in the y-axis ($x = 0$)
- reflections in lines that pass through the origin.
- reflection in the line $y = x \tan(\theta)$.

Line of reflection or mediator

Pre-image   Image

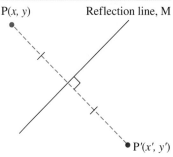

P(x, y)   Reflection line, M

P′(x', y')

10.3.2 Reflection in the x-axis ($y = 0$)

The reflection in the x-axis maps the point P(x, y) onto the point P'(x', y'), giving the image point $(x', y') = (x, -y)$.

The matrix for a reflection in the x-axis is:

$$M_x = \begin{bmatrix} 1 & 0 \\ 0 & -1 \end{bmatrix}$$

In matrix form, the reflection for any point in the x-axis is

$$\begin{matrix} P' & & T & & P \end{matrix}$$

$$\begin{bmatrix} x' \\ y' \end{bmatrix} = \begin{bmatrix} 1 & 0 \\ 0 & -1 \end{bmatrix} \begin{bmatrix} x \\ y \end{bmatrix}$$

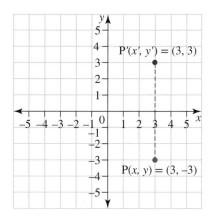

This matrix multiplication means that the x-coordinate stays the same and the y-coordinate changes sign.

10.3.3 Reflection in the y-axis ($x = 0$)

The reflection in the y-axis maps the point P(x, y) onto the point P'(x', y'), giving the image point $(x', y') = (-x, y)$.

The matrix for a reflection in the y-axis is:

$$M_y = \begin{bmatrix} -1 & 0 \\ 0 & 1 \end{bmatrix}$$

In matrix form, the reflection for any point in the y-axis is:

$$\begin{matrix} P' & & T & & P \end{matrix}$$

$$\begin{bmatrix} x' \\ y' \end{bmatrix} = \begin{bmatrix} -1 & 0 \\ 0 & 1 \end{bmatrix} \begin{bmatrix} x \\ y \end{bmatrix}$$

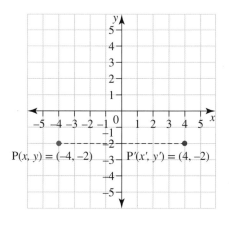

This matrix multiplication means that the x–coordinate changes sign and the y–coordinate stays the same.

WORKED EXAMPLE 6 Determining the reflection of a point using matrices

Determine the image of the point $(-2, 3)$:
a. after a reflection in the x-axis
b. after it is transformed by the matrix $\begin{bmatrix} -1 & 0 \\ 0 & 1 \end{bmatrix}$. Comment on your answer.

THINK	WRITE
a. 1. Recall the reflection matrix for reflection in the x-axis.	a. $M_x = \begin{bmatrix} 1 & 0 \\ 0 & -1 \end{bmatrix}$
2. Apply the matrix equation for reflection in the x-axis matrix, $P' = TP$.	$\begin{bmatrix} x' \\ y' \end{bmatrix} = \begin{bmatrix} 1 & 0 \\ 0 & -1 \end{bmatrix} \begin{bmatrix} x \\ y \end{bmatrix}$
3. Substitute the pre-image point into the matrix equation.	The pre-image point is $(-2, 3)$. $\begin{bmatrix} x' \\ y' \end{bmatrix} = \begin{bmatrix} 1 & 0 \\ 0 & -1 \end{bmatrix} \begin{bmatrix} -2 \\ 3 \end{bmatrix}$

4. Calculate the coordinates of the image point.

$$\begin{bmatrix} x' \\ y' \end{bmatrix} = \begin{bmatrix} 1 & 0 \\ 0 & -1 \end{bmatrix} \begin{bmatrix} -2 \\ 3 \end{bmatrix} = \begin{bmatrix} -2 \\ -3 \end{bmatrix}$$

The image point is $(-2, -3)$.

b. 1. Recall the matrix equation, $P' = TP$.

b.
$$\begin{bmatrix} x' \\ y' \end{bmatrix} = \begin{bmatrix} -1 & 0 \\ 0 & 1 \end{bmatrix} \begin{bmatrix} x \\ y \end{bmatrix}$$

2. Substitute the pre-image point into the matrix equation.

The pre-image point is $(-2, 3)$.

$$\begin{bmatrix} x' \\ y' \end{bmatrix} = \begin{bmatrix} -1 & 0 \\ 0 & 1 \end{bmatrix} \begin{bmatrix} -2 \\ 3 \end{bmatrix}$$

3. Calculate the coordinates of the image point.

$$\begin{bmatrix} x' \\ y' \end{bmatrix} = \begin{bmatrix} -1 & 0 \\ 0 & 1 \end{bmatrix} \begin{bmatrix} -2 \\ 3 \end{bmatrix} = \begin{bmatrix} 2 \\ 3 \end{bmatrix}$$

The image point is $(2, 3)$. It is a reflection in the y-axis from the pre-image point.

TI	THINK	DISPLAY/WRITE	CASIO	THINK	DISPLAY/WRITE
a. 1. On a Calculator page, complete the entry line as: $\begin{bmatrix} 1 & 0 \\ 0 & -1 \end{bmatrix} \times \begin{bmatrix} -2 \\ 3 \end{bmatrix}$ then press ENTER. *Note:* Matrix templates can be found by pressing the templates button.		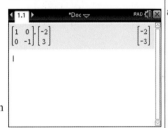	**a. 1.** On a Run-Matrix screen, complete the entry line as: $\begin{bmatrix} 1 & 0 \\ 0 & -1 \end{bmatrix} \times \begin{bmatrix} -2 \\ 3 \end{bmatrix}$ then press EXE. *Note:* Matrix templates can be found by selecting MATH by pressing F4, then selecting MAT/VCT by pressing F1.		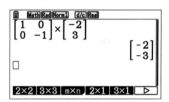
2. The answer appears on the screen.		The image point is $(-2, -3)$.	**2.** The answer appears on the screen.		The image point is $(-2, -3)$.

WORKED EXAMPLE 7 Determining a reflection in the y-axis using matrices

Determine the equation of the image of the graph of $y = (x + 1)^2$ after it is reflected in the y-axis. Verify your solution.

THINK

WRITE

1. Recall the reflection matrix for reflection in the y-axis and apply it to the matrix equation.

$$M_y = \begin{bmatrix} -1 & 0 \\ 0 & 1 \end{bmatrix}$$

$$\begin{bmatrix} x' \\ y' \end{bmatrix} = \begin{bmatrix} -1 & 0 \\ 0 & 1 \end{bmatrix} \begin{bmatrix} x \\ y \end{bmatrix}$$

2. Determine the image coordinates.

$x' = -x$
$y' = y$

3. Rearrange the equations to make the pre-image coordinates x and y the subjects.

$x = -x'$
$y = y'$

4. Substitute the image expressions into the pre-image equation $y = (x + 1)^2$ to determine the equation of the image.

$$y = (x + 1)^2$$
$$y' = (-x' + 1)^2$$
$$\quad = [-(x' - 1)]^2$$
$$y' = (x' - 1)^2$$

The equation of the image is $y = (x - 1)^2$.

5. Graph the image and the pre-image equation to verify the reflection.

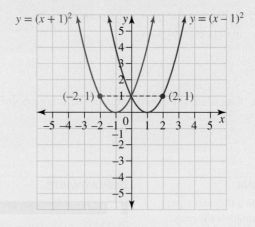

Alternatively:

6. Recall the reflection matrix for reflection in the y-axis and apply it to the matrix equation, $P' = TP$.

$$\begin{bmatrix} x' \\ y' \end{bmatrix} = \begin{bmatrix} -1 & 0 \\ 0 & 1 \end{bmatrix} \begin{bmatrix} x \\ y \end{bmatrix}$$

7. Determine the pre-image coordinates by using the inverse of the transformational matrix, $P = T^{-1}P'$.

$$\begin{bmatrix} x \\ y \end{bmatrix} = -\begin{bmatrix} 1 & 0 \\ 0 & -1 \end{bmatrix} \begin{bmatrix} x' \\ y' \end{bmatrix}$$

8. Multiply and simplify the matrix equation.

$$x = -x'$$
$$y = y'$$

9. Substitute the image expressions into the pre-image equation $y = (x + 1)^2$ to determine the image equation.

$$y = (x + 1)^2$$
$$y' = (-x' + 1)^2$$
$$y' = (x' - 1)^2$$

The equation of the image is $y = (x - 1)^2$.

10.3.4 Reflection in a line that passes through the origin $(0,0)$

Reflection in the line with equation $y = x$

A reflection in the line $y = x$ maps the point P(x, y) onto the point P$'(x', y')$, giving the image point $(x', y') = (y, x)$.

The matrix for a reflection in the line $y = x$ is $M_{y=x} = \begin{bmatrix} 0 & 1 \\ 1 & 0 \end{bmatrix}$.

In matrix form, a reflection in the line $y = x$ is

$$\begin{bmatrix} x' \\ y' \end{bmatrix} = \begin{bmatrix} 0 & 1 \\ 1 & 0 \end{bmatrix} \begin{bmatrix} x \\ y \end{bmatrix}.$$

A reflection in the line $y = -x$ maps the point P(x, y) onto the point P$'(x', y')$, giving the image point $(x', y') = (-y, -x)$.

The matrix for a reflection in the line $y = -x$ is

$$M_{y=-x} = \begin{bmatrix} 0 & -1 \\ -1 & 0 \end{bmatrix}$$

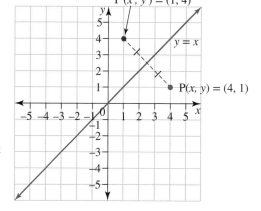

Similarly, in matrix form, a reflection in the line $y = -x$ is $\begin{bmatrix} x' \\ y' \end{bmatrix} = \begin{bmatrix} 0 & -1 \\ -1 & 0 \end{bmatrix} \begin{bmatrix} x \\ y \end{bmatrix}$

This matrix multiplication means that the x and y–coordinates swap positions, and in the case of the reflection in the line $y = -x$, they change signs as well.

WORKED EXAMPLE 8 Determining the reflection in the line $y = x$ using matrices

Consider the triangle A(3, 1), B(4, −1) and C(6, 0). Determine the coordinates of the image of the triangle ABC after a reflection in the line with equation $y = x$.

THINK	WRITE
1. Recall the reflection matrix for reflection in the line $y = x$.	$M_{y=x} = \begin{bmatrix} 0 & 1 \\ 1 & 0 \end{bmatrix}$
2. Use the matrix equation for a reflection about the line with equation $y = x$: P$' = $ TP.	$\begin{bmatrix} x' \\ y' \end{bmatrix} = \begin{bmatrix} 0 & 1 \\ 1 & 0 \end{bmatrix} \begin{bmatrix} x \\ y \end{bmatrix}$
3. Substitute the first pre-image point into the matrix equation.	The first pre-image point is $(3, 1)$. $\begin{bmatrix} x' \\ y' \end{bmatrix} = \begin{bmatrix} 0 & 1 \\ 1 & 0 \end{bmatrix} \begin{bmatrix} 3 \\ 1 \end{bmatrix}$
4. Calculate the coordinates of the image point.	$\begin{bmatrix} x' \\ y' \end{bmatrix} = \begin{bmatrix} 0 & 1 \\ 1 & 0 \end{bmatrix} \begin{bmatrix} 3 \\ 1 \end{bmatrix} = \begin{bmatrix} 1 \\ 3 \end{bmatrix}$ The first image point is $(1, 3)$.
5. Substitute the second pre-image point into the matrix equation.	The second pre-image point is $(4, -1)$ $\begin{bmatrix} x' \\ y' \end{bmatrix} = \begin{bmatrix} 0 & 1 \\ 1 & 0 \end{bmatrix} \begin{bmatrix} 4 \\ -1 \end{bmatrix}$
6. Calculate the coordinates of the second image point.	$\begin{bmatrix} x' \\ y' \end{bmatrix} = \begin{bmatrix} 0 & 1 \\ 1 & 0 \end{bmatrix} \begin{bmatrix} 4 \\ -1 \end{bmatrix} = \begin{bmatrix} -1 \\ 4 \end{bmatrix}$ The second image point is $(-1, 4)$

7. Substitute the third pre-image point into the matrix equation.

The third pre-image point is $(6, 0)$

$$\begin{bmatrix} x' \\ y' \end{bmatrix} = \begin{bmatrix} 0 & 1 \\ 1 & 0 \end{bmatrix} \begin{bmatrix} 6 \\ 0 \end{bmatrix}$$

8. Calculate the coordinates of the third image point.

$$\begin{bmatrix} x' \\ y' \end{bmatrix} = \begin{bmatrix} 0 & 1 \\ 1 & 0 \end{bmatrix} \begin{bmatrix} 6 \\ 0 \end{bmatrix} = \begin{bmatrix} 0 \\ 6 \end{bmatrix}$$

The third image point is $(0, 6)$

9. State the coordinates of all three points.

The coordinates of the image of the triangle are A'$(1, 3)$, B'$(-1, 4)$ and C'$(0, 6)$.

WORKED EXAMPLE 9 Determining the reflection in the line $y = -x$ using matrices

Determine the equation of the image of $y = x^2$ after a reflection in the line $y = -x$. Verify your solution.

THINK	WRITE
1. Recall the matrix for a reflection in the line $y = -x$.	$M_{y = -x} = \begin{bmatrix} 0 & -1 \\ -1 & 0 \end{bmatrix}$
2. Write the matrix equation for a reflection in the line $y = -x$; $P' = TP$.	$\begin{bmatrix} x' \\ y' \end{bmatrix} = \begin{bmatrix} 0 & -1 \\ -1 & 0 \end{bmatrix} \begin{bmatrix} x \\ y \end{bmatrix}$
3. Multiply the matrices.	$\begin{bmatrix} x' \\ y' \end{bmatrix} = \begin{bmatrix} -y \\ -x \end{bmatrix}$
4. Determine the image equations.	$x' = -y$ and $y' = -x$
5. Rearrange the equations to make x and y the subjects.	$y = -x'$ and $x = -y'$
6. Substitute the image expressions into the pre-image equation $y = x^2$ and make y the subject.	$y = x^2$ $-x' = (-y')^2$ $-x' = y'^2$ $\therefore y' = \pm\sqrt{-x'}$
7. Write the answer.	The equation of the image is $y = \pm\sqrt{-x}$.
8. Graph the image and the pre-image to verify the reflection.	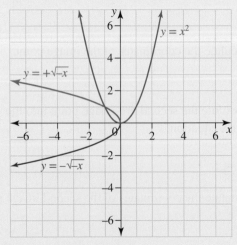

10.3.5 Reflection in the line $y = x\tan(\theta)$

The line $y = x\tan(\theta)$ might be more easily recognised as $y = mx$, where m is the gradient of the line which passes through the origin.

Remember that the gradient $m = \dfrac{y_2 - y_1}{x_2 - x_1}$ and tangent ratio $= \dfrac{\text{rise}}{\text{run}}$.

Therefore the tangent and gradient ratios provide the same information: $\dfrac{\text{rise}}{\text{run}} = \dfrac{y_2 - y_1}{x_2 - x_1}$.

Carefully examine these diagrams that illustrate reflection of the points $(1, 0)$ and $(0, 1)$ in the line $y = x\tan(\theta)$.

Note the following from these diagrams.

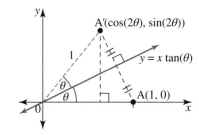

For the point A$(1, 0)$:
1. Point A is reflected to a point equidistant from and perpendicular to, the line $y = x\tan(\theta)$.
2. The angle from the x-axis to A$'$ is 2θ.
3. The x-coordinate of the right-angled triangle is $\cos(2\theta)$.
4. The y-coordinate of this triangle is $\sin(2\theta)$.
5. Hence, point $(1, 0) \to (\cos(2\theta), \sin(2\theta))$.

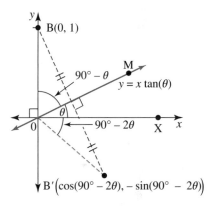

For the point B$(0, 1)$:
1. Point B is reflected to a point equidistant from, and perpendicular to, the line $y = x\tan(\theta)$.
2. $\angle MOB = 90° - \theta$; therefore, $\angle MOB' = 90° - \theta$.
3. Therefore, $\angle XOB' = (90° - \theta) - \theta = 90° - 2\theta$.
4. The x-coordinate $= \cos(90° - 2\theta)$.
5. The y-coordinate $= -\sin(90° - 2\theta)$ because the angle is in the fourth quadrant.
6. Hence, point $(0, 1) \to [\cos(90° - 2\theta), -\sin(90° - 2\theta)]$.
7. Using trigonometric ratios, this simplifies to yield $(\sin(2\theta), -\cos(2\theta))$.
   (Remember that $\sin(30°) = \cos(60°)$, etc.)
   Hence, the general reflection matrix in the line $y = x\tan(\theta)$ is:

$$M_{y = x\tan(\theta)} = \begin{bmatrix} \cos(2\theta) & \sin(2\theta) \\ \sin(2\theta) & -\cos(2\theta) \end{bmatrix}$$

WORKED EXAMPLE 10 Determining the reflection in any line using matrices

Determine the equation of the image of the line $y = -x - 1$ after a reflection in the line $y = \sqrt{3}x$. Verify your solution.

THINK

1. Recall the general reflection matrix for a reflection in the line $y = x\tan(\theta)$.

WRITE

$$M_{y = x\tan(\theta)} = \begin{bmatrix} \cos(2\theta) & \sin(2\theta) \\ \sin(2\theta) & -\cos(2\theta) \end{bmatrix}$$

2. Determine θ by equating the general form of the equation with the given equation and recall the exact value.

$$y = \sqrt{3}x$$
$$y = x\tan(\theta)$$
$$\therefore \tan(\theta) = \sqrt{3}$$

$$\theta = \tan^{-1}\left(\sqrt{3}\right)$$
$$\therefore \theta = \frac{\pi}{3}$$

3. Substitute $\dfrac{\pi}{3}$ for θ into the general reflection matrix for a reflection in the line $y = x\tan(\theta)$.

$$M_{y=x\tan(\theta)} = \begin{bmatrix} \cos(2\theta) & \sin(2\theta) \\ \sin(2\theta) & -\cos(2\theta) \end{bmatrix}$$

$$M_{y=\sqrt{3}x} = \begin{bmatrix} \cos\left(\dfrac{2\pi}{3}\right) & \sin\left(\dfrac{2\pi}{3}\right) \\ \sin\left(\dfrac{2\pi}{3}\right) & -\cos\left(\dfrac{2\pi}{3}\right) \end{bmatrix}$$

4. Simplify the reflection matrix.

$$= \begin{bmatrix} -\dfrac{1}{2} & \dfrac{\sqrt{3}}{2} \\ \dfrac{\sqrt{3}}{2} & \dfrac{1}{2} \end{bmatrix}$$

5. State the matrix equation for a reflection in the line $y = \sqrt{3}x$.

$$\begin{bmatrix} x' \\ y' \end{bmatrix} = M_{y=\sqrt{3}x}\begin{bmatrix} x \\ y \end{bmatrix}$$

$$\begin{bmatrix} x' \\ y' \end{bmatrix} = \begin{bmatrix} -\dfrac{1}{2} & \dfrac{\sqrt{3}}{2} \\ \dfrac{\sqrt{3}}{2} & \dfrac{1}{2} \end{bmatrix}\begin{bmatrix} x \\ y \end{bmatrix}$$

6. Rearrange the matrix equation to make the pre-image point (x, y) the subject.

$$\begin{bmatrix} x \\ y \end{bmatrix} = \begin{bmatrix} -\dfrac{1}{2} & \dfrac{\sqrt{3}}{2} \\ \dfrac{\sqrt{3}}{2} & \dfrac{1}{2} \end{bmatrix}^{-1}\begin{bmatrix} x' \\ y' \end{bmatrix}$$

7. Determine the inverse of $\begin{bmatrix} -\dfrac{1}{2} & \dfrac{\sqrt{3}}{2} \\ \dfrac{\sqrt{3}}{2} & \dfrac{1}{2} \end{bmatrix}$ and simplify the matrix equation.

$$\begin{bmatrix} x \\ y \end{bmatrix} = \frac{1}{-\frac{1}{4}-\frac{3}{4}}\begin{bmatrix} \dfrac{1}{2} & -\dfrac{\sqrt{3}}{2} \\ -\dfrac{\sqrt{3}}{2} & -\dfrac{1}{2} \end{bmatrix}\begin{bmatrix} x' \\ y' \end{bmatrix}$$

$$= \begin{bmatrix} -\dfrac{1}{2} & \dfrac{\sqrt{3}}{2} \\ \dfrac{\sqrt{3}}{2} & \dfrac{1}{2} \end{bmatrix}\begin{bmatrix} x' \\ y' \end{bmatrix}$$

8. Multiply the matrices to determine the pre-image expressions.

$$x = -\frac{1}{2}x' + \frac{\sqrt{3}}{2}y'$$

$$y = \frac{\sqrt{3}}{2}x' + \frac{1}{2}y'$$

9. Substitute the image expressions into the pre-image equation $y = -x - 1$ to determine the equation of the image.

$y = -x - 1$ becomes

$$\frac{\sqrt{3}}{2}x' + \frac{1}{2}y' = \frac{1}{2}x' - \frac{\sqrt{3}}{2}y' - 1$$

$$\frac{1}{2}y' + \frac{\sqrt{3}}{2}y' = \frac{1}{2}x' - \frac{\sqrt{3}}{2}x' - 1$$

10. Make y' the subject and simplify the equations by rationalising the denominator.

$$\frac{1+\sqrt{3}}{2}y' = \frac{1-\sqrt{3}}{2}x' - 1$$

$$y' = \left(\sqrt{3} - 2\right)x' + 1 - \sqrt{3}$$

11. Write the answer.

The equation of the image of the line $y = -x - 1$ is $y = (\sqrt{3} - 2)x + 1 - \sqrt{3}$.

12. Sketch the image and the pre-image graphs to verify the reflection.

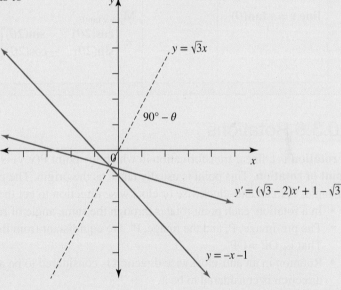

The specific reflection matrices do not need to be memorised. They are similar and easily confused. While shown in the table below for convenience, they can be generated from the general reflection matrix.

Matrix reflections

A summary of the matrices for reflections are shown in the following table.

Reflection in	Value of θ	Matrix	Matrix equation
x-axis	$\theta = 0$	$M_x = \begin{bmatrix} 1 & 0 \\ 0 & -1 \end{bmatrix}$	$\begin{bmatrix} x' \\ y' \end{bmatrix} = \begin{bmatrix} 1 & 0 \\ 0 & -1 \end{bmatrix} \begin{bmatrix} x \\ y \end{bmatrix} = \begin{bmatrix} x \\ -y \end{bmatrix}$
y-axis	$\theta = \dfrac{\pi}{2}$	$M_y = \begin{bmatrix} -1 & 0 \\ 0 & 1 \end{bmatrix}$	$\begin{bmatrix} x' \\ y' \end{bmatrix} = \begin{bmatrix} -1 & 0 \\ 0 & 1 \end{bmatrix} \begin{bmatrix} x \\ y \end{bmatrix} = \begin{bmatrix} -x \\ y \end{bmatrix}$
line $y = x$	$\theta = \dfrac{\pi}{4}$	$M_{y=x} = \begin{bmatrix} 0 & 1 \\ 1 & 0 \end{bmatrix}$	$\begin{bmatrix} x' \\ y' \end{bmatrix} = \begin{bmatrix} 0 & 1 \\ 1 & 0 \end{bmatrix} \begin{bmatrix} x \\ y \end{bmatrix} = \begin{bmatrix} y \\ x \end{bmatrix}$
line $y = -x$	$\theta = \dfrac{3\pi}{4}$	$M_{y=-x} = \begin{bmatrix} 0 & -1 \\ -1 & 0 \end{bmatrix}$	$\begin{bmatrix} x' \\ y' \end{bmatrix} = \begin{bmatrix} 0 & -1 \\ -1 & 0 \end{bmatrix} \begin{bmatrix} x \\ y \end{bmatrix} = \begin{bmatrix} -y \\ -x \end{bmatrix}$
line $y = x\tan(\theta)$		$M_{y=x\tan(\theta)}$ $= \begin{bmatrix} \cos(2\theta) & \sin(2\theta) \\ \sin(2\theta) & -\cos(2\theta) \end{bmatrix}$	$\begin{bmatrix} x' \\ y' \end{bmatrix} = \begin{bmatrix} \cos(2\theta) & \sin(2\theta) \\ \sin(2\theta) & -\cos(2\theta) \end{bmatrix} \begin{bmatrix} x \\ y \end{bmatrix}$

10.3.6 Rotations

A **rotation** is a linear transformation in which the point $P(x, y)$ is rotated about a fixed point called the centre or **point of rotation**. This point is usually taken as the origin. The pre-image point, $P(x, y)$, can be rotated through an angle, θ, in an anti-clockwise or clockwise direction to get the image point, $P'(x', y')$.

- In a rotation, each point rotates through the same angle of rotation, θ.
- The pre-image, P, and the image, P', are equidistant from the origin.
  That is, $OP = OP'$.
- Rotation in an anti-clockwise direction is considered to be a positive rotation, and rotation in a clockwise direction is considered to be a
  negative rotation.
- R_θ is the symbol used for the matrix rotation in an anti-clockwise direction; $R_{-\theta}$ is used for clockwise direction.

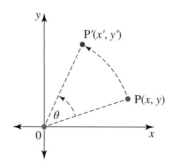

10.3.7 Special rotations in an anticlockwise direction

In this section, we will discuss transformations involving rotations of 90°, 180°, 270° and 360°, as well as general rotations.

Rotation of 90°

Consider the figure shown.

As the plane rotates through $\theta = 90°$ about the origin, point $(1, 0)$ will map to point $(0, 1)$ and point $(0, 1)$ will map to point $(-1, 0)$.

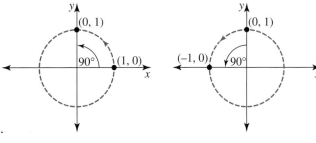

Hence, the identity matrix, I, is altered to

$$\downarrow \quad \downarrow$$
$$\begin{bmatrix} 0 & -1 \\ 1 & 0 \end{bmatrix} \text{ to achieve a rotation of } 90° \text{ about the origin.}$$

It is most important that you recognise the pattern that is displayed by the columns in the matrix and the coordinates of the image points. This concept forms the basis of the next section of work and totally eliminates 'remembering' formulas so that you will be able to understand what is happening to the points.

Hence, $R_{90°} = \begin{bmatrix} 0 & -1 \\ 1 & 0 \end{bmatrix}$ and is the matrix of rotation.

In general terms

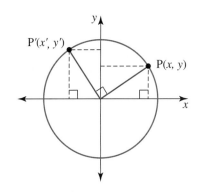

$$(x, y) \rightarrow (-y, x)$$

$$\begin{bmatrix} x' \\ y' \end{bmatrix} = \begin{bmatrix} 0 & -1 \\ 1 & 0 \end{bmatrix} \begin{bmatrix} x \\ y \end{bmatrix}$$
$$x' = -y$$
$$y' = x$$

As mentioned earlier, these rotation matrices should not be learned. They are quite similar and can be too readily confused. Sketch the original $(1, 0)$ and $(0, 1)$ points and then use their images to build the rotation matrices.

Rotation of 180°

In the diagrams below, notice that point $(1, 0)$ is mapped onto point $(-1, 0)$ and point $(0, 1)$ is mapped onto $(0, -1)$.

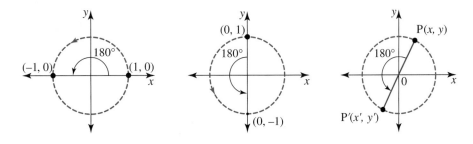

Therefore, $R_{180°} = \begin{bmatrix} -1 & 0 \\ 0 & -1 \end{bmatrix}$ where $(x, y) \rightarrow (-x, -y)$.

Rotation of 270°

In the diagrams below, notice that point $(1, 0)$ is mapped onto point $(0, -1)$ and point $(0, 1)$ is mapped onto point $(1, 0)$.

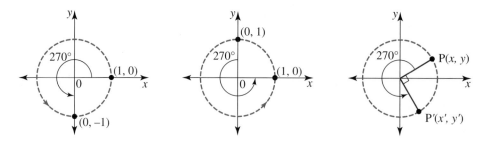

Therefore, $R_{270°} = \begin{bmatrix} 0 & 1 \\ -1 & 0 \end{bmatrix}$ where $(x, y) \to (y, -x)$.

Rotation of 360°

$R_{360°} = \begin{bmatrix} 1 & 0 \\ 0 & 1 \end{bmatrix}$ because $R_{360°}$ essentially leaves the original unchanged (or mapped onto itself).

General rotation of θ

Consider the points $A(1, 0)$ and $B(0, 1)$ that are rotated through angle θ about the origin in an anti-clockwise direction.

Careful examination of the diagram shows that point $A(1, 0)$ is mapped onto point $A'(\cos(\theta), \sin(\theta))$ and point $B(0, 1)$ is mapped onto point $B'(-\sin(\theta), \cos(\theta))$, where

$$\cos(\theta) = x \, (\text{horizontal})$$
$$\text{and } \sin(\theta) = y \, (\text{vertical})$$

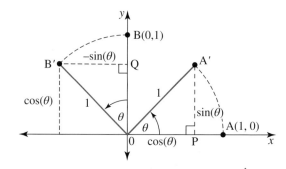

R_θ is the matrix rotation for an anticlockwise rotation through θ about the origin.

$$R_\theta = \begin{bmatrix} \cos(\theta) & -\sin(\theta) \\ \sin(\theta) & \cos(\theta) \end{bmatrix}$$

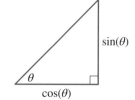

$R_{-\theta}$ is the matrix rotation when θ is taken in a clockwise, negative rotation about the origin as shown.

$$R_{-\theta} = \begin{bmatrix} \cos(-\theta) & -\sin(-\theta) \\ \sin(-\theta) & \cos(-\theta) \end{bmatrix}$$

$$= \begin{bmatrix} \cos(\theta) & \sin(\theta) \\ -\sin(\theta) & \cos(\theta) \end{bmatrix}$$

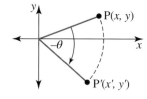

since $\cos(-\theta) = \cos(\theta)$ and $\sin(-\theta) = -\sin(\theta)$.

Both R_θ and $R_{-\theta}$ can be used to confirm the specific cases of R_{90}, R_{180}, and R_{270}.

A summary of the matrices for rotations are shown in the following table.

Matrix rotations

Rotation of	Value of θ	Matrix	Matrix equation
90°	$\theta = \dfrac{\pi}{2}$	$R_{90°} = \begin{bmatrix} 0 & -1 \\ 1 & 0 \end{bmatrix}$	$\begin{bmatrix} x' \\ y' \end{bmatrix} = \begin{bmatrix} 0 & -1 \\ 1 & 0 \end{bmatrix} \begin{bmatrix} x \\ y \end{bmatrix} = \begin{bmatrix} -y \\ x \end{bmatrix}$
180°	$\theta = \pi$	$R_{180°} = \begin{bmatrix} -1 & 0 \\ 0 & -1 \end{bmatrix}$	$\begin{bmatrix} x' \\ y' \end{bmatrix} = \begin{bmatrix} -1 & 0 \\ 0 & -1 \end{bmatrix} \begin{bmatrix} x \\ y \end{bmatrix} = \begin{bmatrix} -x \\ -y \end{bmatrix}$
270°	$\theta = \dfrac{-\pi}{2}$	$R_{270°} = \begin{bmatrix} 0 & 1 \\ -1 & 0 \end{bmatrix}$	$\begin{bmatrix} x' \\ y' \end{bmatrix} = \begin{bmatrix} 0 & 1 \\ -1 & 0 \end{bmatrix} \begin{bmatrix} x \\ y \end{bmatrix} = \begin{bmatrix} y \\ -x \end{bmatrix}$
360°	$\theta = 0$	$R_{360°} = \begin{bmatrix} 1 & 0 \\ 0 & 1 \end{bmatrix}$	$\begin{bmatrix} x' \\ y' \end{bmatrix} = \begin{bmatrix} 1 & 0 \\ 0 & 1 \end{bmatrix} \begin{bmatrix} x \\ y \end{bmatrix} = \begin{bmatrix} x \\ y \end{bmatrix}$
θ		$R_{\theta} = \begin{bmatrix} \cos(\theta) & -\sin(\theta) \\ \sin(\theta) & \cos(\theta) \end{bmatrix}$	$\begin{bmatrix} x' \\ y' \end{bmatrix} = \begin{bmatrix} \cos(\theta) & -\sin(\theta) \\ \sin(\theta) & \cos(\theta) \end{bmatrix} \begin{bmatrix} x \\ y \end{bmatrix}$

WORKED EXAMPLE 11 Determining a rotation of a point using matrices

The point $(2, -2)$ is rotated $\dfrac{\pi}{4}$ about the origin in an anti-clockwise direction. Determine the coordinates of the image of the point after this transformation.

THINK

1. Sketch the point $(2, -2)$ and the angle $\dfrac{\pi}{4}$.

2. Recall the rotation matrix equation for an anti-clockwise rotation.

WRITE

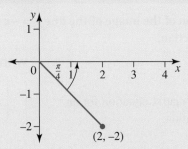

$$\begin{bmatrix} x' \\ y' \end{bmatrix} = \begin{bmatrix} \cos(\theta) & -\sin(\theta) \\ \sin(\theta) & \cos(\theta) \end{bmatrix} \begin{bmatrix} x \\ y \end{bmatrix}$$

3. Substitute the pre-image point, $(2, -2)$, and $\dfrac{\pi}{4}$ for θ and evaluate.

$$\begin{bmatrix} x' \\ y' \end{bmatrix} = \begin{bmatrix} \cos\left(\dfrac{\pi}{4}\right) & -\sin\left(\dfrac{\pi}{4}\right) \\ \sin\left(\dfrac{\pi}{4}\right) & \cos\left(\dfrac{\pi}{4}\right) \end{bmatrix} \begin{bmatrix} 2 \\ -2 \end{bmatrix}$$

$$\begin{bmatrix} x' \\ y' \end{bmatrix} = \begin{bmatrix} \dfrac{1}{\sqrt{2}} & -\dfrac{1}{\sqrt{2}} \\ \dfrac{1}{\sqrt{2}} & \dfrac{1}{\sqrt{2}} \end{bmatrix} \begin{bmatrix} 2 \\ -2 \end{bmatrix}$$

$$= \begin{bmatrix} \dfrac{2}{\sqrt{2}} + \dfrac{2}{\sqrt{2}} \\ \dfrac{2}{\sqrt{2}} - \dfrac{2}{\sqrt{2}} \end{bmatrix}$$

$$= \begin{bmatrix} \dfrac{4}{\sqrt{2}} \\ 0 \end{bmatrix}$$

4. Rationalise the denominator and simplify.

$$= \begin{bmatrix} \dfrac{4}{\sqrt{2}} \times \dfrac{\sqrt{2}}{\sqrt{2}} \\ 0 \end{bmatrix}$$

$$= \begin{bmatrix} 2\sqrt{2} \\ 0 \end{bmatrix}$$

5. State the coordinates of the image point.

The image point is $(2\sqrt{2}, 0)$.

WORKED EXAMPLE 12 Determining a rotation of a line using matrices

Determine the equation of the image of the line $y = -x + 4$ under a rotation of $30°$ about the origin in an anti-clockwise direction.

THINK	WRITE
1. Recall the rotation matrix equation for an anti-clockwise rotation.	$\begin{bmatrix} x' \\ y' \end{bmatrix} = \begin{bmatrix} \cos(\theta) & -\sin(\theta) \\ \sin(\theta) & \cos(\theta) \end{bmatrix} \begin{bmatrix} x \\ y \end{bmatrix}$
2. Substitute $30°$ for θ and evaluate.	$\begin{bmatrix} x' \\ y' \end{bmatrix} = \begin{bmatrix} \cos(30°) & -\sin(30°) \\ \sin(30°) & \cos(30°) \end{bmatrix} \begin{bmatrix} x \\ y \end{bmatrix}$
	$\begin{bmatrix} x' \\ y' \end{bmatrix} = \begin{bmatrix} \dfrac{\sqrt{3}}{2} & -\dfrac{1}{2} \\ \dfrac{1}{2} & \dfrac{\sqrt{3}}{2} \end{bmatrix} \begin{bmatrix} x \\ y \end{bmatrix}$

3. Rearrange the matrix equation to make the pre-image point (x, y) the subject.

$$\begin{bmatrix} x \\ y \end{bmatrix} = \begin{bmatrix} \dfrac{\sqrt{3}}{2} & -\dfrac{1}{2} \\ \dfrac{1}{2} & \dfrac{\sqrt{3}}{2} \end{bmatrix}^{-1} \begin{bmatrix} x' \\ y' \end{bmatrix}$$

4. Determine the inverse of $\begin{bmatrix} \dfrac{\sqrt{3}}{2} & \dfrac{1}{2} \\ -\dfrac{1}{2} & \dfrac{\sqrt{3}}{2} \end{bmatrix}$ and simplify the matrix equation.

$$\begin{bmatrix} x \\ y \end{bmatrix} = \dfrac{1}{\left(\frac{\sqrt{3}}{2} \times \frac{\sqrt{3}}{2} \right) - \left(-\frac{1}{2} \times \frac{1}{2} \right)} \begin{bmatrix} \dfrac{\sqrt{3}}{2} & \dfrac{1}{2} \\ -\dfrac{1}{2} & \dfrac{\sqrt{3}}{2} \end{bmatrix} \begin{bmatrix} x' \\ y' \end{bmatrix}$$

$$\begin{bmatrix} x \\ y \end{bmatrix} = \dfrac{1}{\frac{3}{4} + \frac{1}{4}} \begin{bmatrix} \dfrac{\sqrt{3}}{2} & \dfrac{1}{2} \\ -\dfrac{1}{2} & \dfrac{\sqrt{3}}{2} \end{bmatrix} \begin{bmatrix} x' \\ y' \end{bmatrix}$$

$$\begin{bmatrix} x \\ y \end{bmatrix} = \begin{bmatrix} \dfrac{\sqrt{3}}{2} & \dfrac{1}{2} \\ -\dfrac{1}{2} & \dfrac{\sqrt{3}}{2} \end{bmatrix} \begin{bmatrix} x' \\ y' \end{bmatrix}$$

5. Multiply the matrices to determine the pre-image expressions.

$$x = \frac{\sqrt{3}}{2} x' + \frac{1}{2} y'$$

$$y = -\frac{1}{2} x' + \frac{\sqrt{3}}{2} y'$$

6. Substitute the image expressions into the pre-image equation $y = -x + 4$ to determine the equation of the image.

$$y = -x + 4$$

$$-\frac{1}{2} x' + \frac{\sqrt{3}}{2} y' = -\left(\frac{\sqrt{3}}{2} x' + \frac{1}{2} y' \right) + 4$$

$$-\frac{1}{2} x' + \frac{\sqrt{3}}{2} y' = -\frac{\sqrt{3}}{2} x' - \frac{1}{2} y' + 4$$

7. Make y' the subject.

$$\left(\frac{\sqrt{3}}{2} + \frac{1}{2} \right) y' = \left(\frac{1}{2} - \frac{\sqrt{3}}{2} \right) x' + 4$$

$$y' = \frac{\left(1 - \sqrt{3} \right)}{\sqrt{3} + 1} x' + \frac{8}{\sqrt{3} + 1}$$

8. Simplify the terms by rationalising the denominator.

$$y' = \left(\sqrt{3} - 2 \right) x' + 4 \left(\sqrt{3} - 1 \right)$$

9. Write the answer.

The equation of the image of the line $y = -x + 4$ is
$$y' = \left(\sqrt{3} - 2 \right) x' + 4 \left(\sqrt{3} - 1 \right).$$

10. Use technology to sketch the image and the pre-image graphs to verify the rotation.

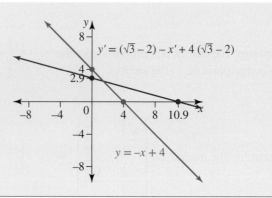

10.3.8 Invariant properties of reflections and rotations

Recall that properties which are unchanged by a transformation are called invariants. For reflections and rotations, invariant properties include length, area and angles between corresponding sides.

Invariant points are points which are unchanged by a transformation. For reflections, any point on the line of reflection is an invariant point, and for rotations, the centre of rotation is the only invariant point (unless the rotation is a multiple of $360°$, in which case all points are invariant points).

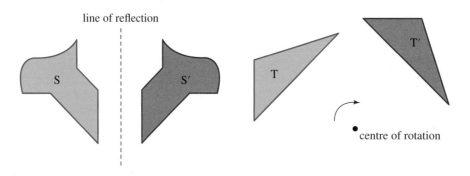

Exercise 10.3 Reflections and rotations

learn on

10.3 Exercise	10.3 Exam questions on

Simple familiar	Complex familiar	Complex unfamiliar
1, 2, 3, 4, 5, 6, 7, 8, 9, 10	11, 12, 13, 14, 15, 16, 17, 18, 19, 20	N/A

These questions are even better in jacPLUS!
- Receive immediate feedback
- Access sample responses
- Track results and progress

Find all this and MORE in jacPLUS

Simple familiar

1. **WE6a** Determine the image of the point $(-3, -1)$ after a reflection in the x-axis.

2. **WE6b** Determine the image of the point $(5, -2)$ after it is transformed by the matrix $\begin{bmatrix} -1 & 0 \\ 0 & 1 \end{bmatrix}$.

3. **WE7** Determine the equation of the image of the graph of $y = (x-2)^2$ after it is reflected in the y-axis.

4. Determine the equation of the image of the graph of $y = x^2 + 1$ after it is reflected in the x-axis.

5. **WE8** Determine the image of the points $(-2, 5)$, $(3, 9)$ and $(8, -5)$ after a reflection in the line with equation $y = x$.

6. Determine the image of the point $(9, -6)$ after a reflection in the line with equation $y = -x$.

7. Determine the equation of the image of $y = x^2$ after a reflection in the line $y = x$.

8. **WE9** Determine the equation of the image of $y = 2x^2 + 1$ after a reflection in the line $y = -x$.

9. The line with equation $y = -x + 3$ is transformed according to the matrix equations given. Determine the equation of the image of the line after this transformation.

   a. $\begin{bmatrix} x' \\ y' \end{bmatrix} = \begin{bmatrix} 1 & 0 \\ 0 & -1 \end{bmatrix} \begin{bmatrix} x \\ y \end{bmatrix}$

   b. $\begin{bmatrix} x' \\ y' \end{bmatrix} = \begin{bmatrix} -1 & 0 \\ 0 & 1 \end{bmatrix} \begin{bmatrix} x \\ y \end{bmatrix}$

10. The parabola with equation $y = x^2 + 2x + 1$ is transformed according to the matrix equations given. Determine the equation of the image of the parabola after this transformation.

   a. $\begin{bmatrix} x' \\ y' \end{bmatrix} = \begin{bmatrix} 1 & 0 \\ 0 & -1 \end{bmatrix} \begin{bmatrix} x \\ y \end{bmatrix}$

   b. $\begin{bmatrix} x' \\ y' \end{bmatrix} = \begin{bmatrix} -1 & 0 \\ 0 & 1 \end{bmatrix} \begin{bmatrix} x \\ y \end{bmatrix}$

Complex familiar

11. **WE10** Determine the equation of the image of the line $y = x + 1$ after a reflection in the line $y = \dfrac{\sqrt{3}}{3}x$.

12. Determine the equation of the image of the line $y = 2x + 3$ after a reflection in the line $y = \sqrt{3}x$.

13. Determine the equation of the image of the line $y = 4 - x$ after a reflection in the line $y = \dfrac{1}{\sqrt{3}}x$.

14. **WE11** The point $(5, 4)$ is rotated $\dfrac{\pi}{3}$ about the origin in a clockwise direction. Determine the coordinates of the image of the point after this transformation.

15. Determine the matrices for the following rotations about the origin.

   a. $90°$ clockwise   b. $180°$ clockwise   c. $45°$ anti-clockwise   d. $\dfrac{\pi}{6}$ anti-clockwise

16. a. Determine the coordinates of the image of the point A$(7, -6)$ rotated $270°$ about the origin in a clockwise direction.
   b. Determine the coordinates of the image of the point A$(7, -6)$ rotated $90°$ about the origin in an anti-clockwise direction.
   c. Show that a clockwise rotation of $270°$ is the same as an anti-clockwise rotation of $90°$ about the origin.

17. **WE12** Determine the equation of the image of the line $y = -3x + 1$ under a rotation of $45°$ about the origin in an anti-clockwise direction.

18. Determine the equation of the image of the line $y = 2x + 1$ under a rotation of $90°$ about the origin in a clockwise direction.

19. Determine the equation of the image of the line $2x + 3y - 6 = 0$ under a rotation of $180°$ about the origin. Verify your solution by sketching the graph.

20. Show that the equation of the image of the line $y = x + 4$ under a rotation of $45°$ about the origin in an anticlockwise direction is the same as the image under a rotation of $315°$ about the origin in a clockwise direction.

Fully worked solutions for this chapter are available online.

LESSON
10.4 Dilations

10.4.1 Dilations from the x- and y-axes

A **dilation** is a linear transformation that changes the size of a figure. The figure is enlarged or reduced parallel to either axis or both axes. A dilation requires a centre point and a scale factor.

A dilation is defined by a scale factor, denoted in general terms by λ.

If λ is greater than 1, the figure is enlarged.

If λ is between 0 and 1, the figure is reduced.

Original Reduction Enlargement

$0 < \lambda < 1$ $\lambda > 1$

Dilation parallel to the x-axis (or from the y-axis)

A dilation (by a factor of a) from the y-axis or parallel to the x-axis is represented by the following matrix equation.

$$\begin{bmatrix} x' \\ y' \end{bmatrix} = \begin{bmatrix} a & 0 \\ 0 & 1 \end{bmatrix} \begin{bmatrix} x \\ y \end{bmatrix} = \begin{bmatrix} a x \\ y \end{bmatrix}$$

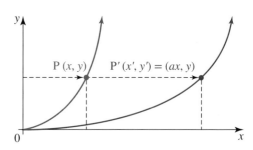

The points (x, y) are transformed onto points with the same y-coordinate.

The point moves away from the y-axis in the direction of the x-axis by a factor of a. This determines the horizontal enlargement of the figure if $a > 1$ or the horizontal compression if $0 < a < 1$.

Dilations parallel to the y-axis (or from the x-axis)

A dilation (by a factor of b) from the x-axis or parallel to the y-axis is represented by the following matrix equation.

$$\begin{bmatrix} x' \\ y' \end{bmatrix} = \begin{bmatrix} 1 & 0 \\ 0 & b \end{bmatrix} \begin{bmatrix} x \\ y \end{bmatrix} = \begin{bmatrix} x \\ by \end{bmatrix}$$

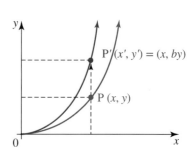

The point moves away from the x-axis in the direction of the y-axis by a factor of b. This determines the vertical enlargement of the figure if $b > 1$ or if $0 < b < 1$, the vertical compression.

WORKED EXAMPLE 13 Determining a dilation of a point using matrices

Determine the coordinates of the image of the point $(3, -4)$ under a dilation of factor $\dfrac{1}{2}$ parallel to the y-axis.

THINK	WRITE
1. Recall the dilation matrix for dilations from the x-axis.	$\begin{bmatrix} 1 & 0 \\ 0 & b \end{bmatrix}$
2. Identify the dilation factor and use the matrix equation for dilation by substituting the value of b.	The dilation factor is $b = \dfrac{1}{2}$. $\begin{bmatrix} x' \\ y' \end{bmatrix} = \begin{bmatrix} 1 & 0 \\ 0 & \dfrac{1}{2} \end{bmatrix} \begin{bmatrix} x \\ y \end{bmatrix}$
3. Substitute the pre-image point into the matrix equation.	The pre-image point is $(3, -4)$. $\begin{bmatrix} x' \\ y' \end{bmatrix} = \begin{bmatrix} 1 & 0 \\ 0 & \dfrac{1}{2} \end{bmatrix} \begin{bmatrix} 3 \\ -4 \end{bmatrix}$
4. Calculate the coordinates of the image point.	$\begin{bmatrix} x' \\ y' \end{bmatrix} = \begin{bmatrix} 3 \\ -2 \end{bmatrix}$ The image point is $(3, -2)$.

TI	THINK	DISPLAY/WRITE	CASIO	THINK	DISPLAY/WRITE
1. On a Calculator page, complete the entry line as: $\begin{bmatrix} 1 & 0 \\ 0 & \dfrac{1}{2} \end{bmatrix} \times \begin{bmatrix} 3 \\ -4 \end{bmatrix}$ then press ENTER. *Note:* Matrix templates can be found by pressing the templates button.	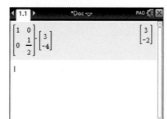	1. On a Run-Matrix screen, complete the entry line as: $\begin{bmatrix} 1 & 0 \\ 0 & \dfrac{1}{2} \end{bmatrix} \times \begin{bmatrix} 3 \\ -4 \end{bmatrix}$ then press EXE. *Note:* Matrix templates can be found by selecting MATH by pressing F4, then selecting MAT/VCT by pressing F1.	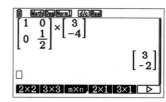
2. The answer appears on the screen.	The image point is $(3, -2)$.	2. The answer appears on the screen.	The image point is $(3, -2)$.

WORKED EXAMPLE 14 Determining a dilation of a curve using matrices

Determine the equation of the image of the parabola with equation $y = x^2$ after it is dilated by a factor of 2 parallel to the x-axis.

THINK	WRITE
1. Recall the matrix equation for dilation.	$\begin{bmatrix} x' \\ y' \end{bmatrix} = \begin{bmatrix} 2 & 0 \\ 0 & 1 \end{bmatrix} \begin{bmatrix} x \\ y \end{bmatrix} = \begin{bmatrix} 2x \\ y \end{bmatrix}$

▶

2. Determine the expressions of the image coordinates in terms of the pre-image coordinates.

$x' = 2x$ and $y' = y$

3. Rearrange the equations to make the pre-image coordinates x and y the subjects.

$x = \dfrac{x'}{2}$ and $y = y'$

4. Substitute the image values into the pre-image equation to determine the image equation.

$y = x^2$

$y' = \left(\dfrac{x'}{2}\right)^2$

$\quad = \dfrac{(x')^2}{4}$

The equation of the image is $y = \dfrac{x^2}{4}$.

5. Graph the image and the pre-image equations to verify the transformation.

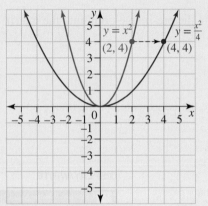

10.4.2 Dilation from both x- and y-axes

A summary of the matrices for dilations are shown in the following table.

Dilation from both axes

Direction	Matrix	Matrix equation
Parallel to the x-axis	$\begin{bmatrix} a & 0 \\ 0 & 1 \end{bmatrix}$	$\begin{bmatrix} x' \\ y' \end{bmatrix} = \begin{bmatrix} a & 0 \\ 0 & 1 \end{bmatrix} \begin{bmatrix} x \\ y \end{bmatrix} = \begin{bmatrix} ax \\ y \end{bmatrix}$
Parallel to the y-axis	$\begin{bmatrix} 1 & 0 \\ 0 & b \end{bmatrix}$	$\begin{bmatrix} x' \\ y' \end{bmatrix} = \begin{bmatrix} 1 & 0 \\ 0 & b \end{bmatrix} \begin{bmatrix} x \\ y \end{bmatrix} = \begin{bmatrix} x \\ by \end{bmatrix}$
Parallel to both axes	$\begin{bmatrix} a & 0 \\ 0 & b \end{bmatrix}$	$\begin{bmatrix} x' \\ y' \end{bmatrix} = \begin{bmatrix} a & 0 \\ 0 & b \end{bmatrix} \begin{bmatrix} x \\ y \end{bmatrix} = \begin{bmatrix} ax \\ by \end{bmatrix}$

where a and b are the dilation factors in the x-axis and y-axis directions respectively.

- When $a \neq b$ the object is skewed.
- When $a = b = \lambda$, the size of the object is enlarged or reduced by the same factor, and the matrix equation is:

$$\begin{bmatrix} x' \\ y' \end{bmatrix} = \begin{bmatrix} \lambda & 0 \\ 0 & \lambda \end{bmatrix} \begin{bmatrix} x \\ y \end{bmatrix}$$

$$= \lambda \begin{bmatrix} 1 & 0 \\ 0 & 1 \end{bmatrix} \begin{bmatrix} x \\ y \end{bmatrix}$$

$$= \lambda I \begin{bmatrix} x \\ y \end{bmatrix}$$

$$= \lambda \begin{bmatrix} x \\ y \end{bmatrix}$$

$$= \begin{bmatrix} \lambda x \\ \lambda y \end{bmatrix} \text{ where } \lambda \text{ is the dilation factor.}$$

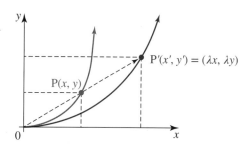

WORKED EXAMPLE 15 Performing dilations in real-world situations using matrices

Jo has fenced a rectangular vegetable patch with fence posts at A(0, 0), B(3, 0), C(3, 4) and D (0, 4).

a. She wants to increase the size of the vegetable patch by a dilation factor of 3 in the *x*-direction and a dilation factor of 1.5 in the *y*-direction. Determine where Jo should relocate the fence posts.

b. Jo has noticed that the vegetable patch in part a is too long and can only increase the vegetable patch size by a dilation factor of 2 in both the *x*-direction and the *y*-direction. Determine where she should relocate the fence posts and if this will give her more area to plant vegetables. Explain.

THINK

a. 1. Draw a diagram to represent this situation.

2. State the coordinates of the vegetable patch as a coordinate matrix.

3. State the dilation matrix.

WRITE

a.

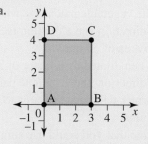

The coordinates of the vegetable patch ABCD can be written as a coordinate matrix.

$$\begin{bmatrix} 0 & 3 & 3 & 0 \\ 0 & 0 & 4 & 4 \end{bmatrix}$$

$$\begin{bmatrix} 3 & 0 \\ 0 & 1.5 \end{bmatrix}$$

4. Multiply the dilation matrix by the coordinate matrix to calculate the new fence post coordinates.

$$\begin{bmatrix} 3 & 0 \\ 0 & 1.5 \end{bmatrix} \begin{bmatrix} 0 & 3 & 3 & 0 \\ 0 & 0 & 4 & 4 \end{bmatrix} = \begin{bmatrix} 0 & 9 & 9 & 0 \\ 0 & 0 & 6 & 6 \end{bmatrix}$$

The new fence posts are located at
$A'(0,0), B'(9,0), C'(9,6)$ and $D'(0,6)$.

b. 1. State the coordinates of the vegetable patch as a coordinate matrix.

b. The coordinates of the vegetable patch $2 \times 2 \times 12 = 48$ units^2 can be written as a coordinate matrix:

$$\begin{bmatrix} 0 & 3 & 3 & 0 \\ 0 & 0 & 4 & 4 \end{bmatrix}$$

2. State the dilation matrix.

The dilation matrix is $\begin{bmatrix} 2 & 0 \\ 0 & 2 \end{bmatrix}$.

3. Calculate the new fence post coordinates $A'B'C'D'$ by multiplying the dilation matrix by the coordinate matrix.

$$\begin{bmatrix} 2 & 0 \\ 0 & 2 \end{bmatrix} \begin{bmatrix} 0 & 3 & 3 & 0 \\ 0 & 0 & 4 & 4 \end{bmatrix} = 2 \begin{bmatrix} 1 & 0 \\ 0 & 1 \end{bmatrix} \begin{bmatrix} 0 & 3 & 3 & 0 \\ 0 & 0 & 4 & 4 \end{bmatrix}$$

$$= \begin{bmatrix} 0 & 6 & 6 & 0 \\ 0 & 0 & 8 & 8 \end{bmatrix}$$

The new fence posts are located at
$A'(0,0), B'(6,0), C'(6,8)$ and $D'(0,8)$.

4. Draw a diagram of the original vegetable patch, and the two transformed vegetable patches on the same Cartesian plane.

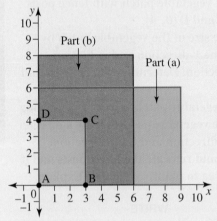

5. Determine the area for each vegetable patch.

The vegetable patch size when dilated by a factor of 3 in the x-direction and a dilation factor of 1.5 in the y-direction gives an area of 54 units^2.

When dilated by a factor of 2 in both the x-direction and the y-direction, the vegetable patch has an area of 48 units^2. The farmer will have less area to plant vegetables in the second option.

Note that you could use simply multiply the original area by the determinant of the dilation matrices.

a. $3 \times 1.5 \times 12 = 54$ units^2

b. $2 \times 2 \times 12 = 48$ units^2

10.4.3 Invariant properties of dilations

Unlike translations, reflections and rotations, dilations generally have few, if any, invariant properties. Lengths of sides, angles and areas are usually all affected by a dilation. For dilations from the x-axis (parallel to the y-axis), points on the x-axis are invariant points, and for dilations from the y-axis (parallel to the x-axis), points on the y-axis are invariant points.

Exercise 10.4 Dilations

learn on

10.4 Exercise	10.4 Exam questions on

Simple familiar	Complex familiar	Complex unfamiliar
1, 2, 3, 4, 5, 6, 7, 8, 9, 10	11, 12, 13, 14, 15, 16, 17, 18, 19, 20	N/A

These questions are even better in jacPLUS!
- Receive immediate feedback
- Access sample responses
- Track results and progress

Find all this and MORE in jacPLUS ▶

Simple familiar

1. **WE13** Determine the coordinates of the image of the point $(2, -1)$ after a dilation of factor 3 parallel to the y-axis.

2. Determine the coordinates of the image of $(-1, 4)$ after a dilation factor of 2 parallel to the y-axis.

3. A man standing in front of a carnival mirror looks like he has been dilated 3 times wider. Write a matrix equation for this situation.

4. Determine the coordinates of the image of the point $(-1, 4)$ after a dilation by $\begin{bmatrix} 2 & 0 \\ 0 & 1 \end{bmatrix}$ parallel to the x-axis.

5. Determine the coordinates of the image of $(2, -5)$ after a dilation of 3 parallel to the x-axis.

6. **WE14** Determine the equation of the image of the parabola with equation $y = x^2$ after it is dilated by a factor of 3 parallel to the x-axis.

7. Determine the equation of the image of the parabola with equation $y = x^2$ after it is dilated by a factor of $\dfrac{1}{2}$ from the x-axis.

8. A transformation T is given by $\begin{bmatrix} x' \\ y' \end{bmatrix} = \begin{bmatrix} 3 & 0 \\ 0 & 2 \end{bmatrix} \begin{bmatrix} x \\ y \end{bmatrix}$.

   a. Determine the coordinates of image of the point $A(-1, 3)$.
   b. Describe the transformation represented by T.

9. Determine the equation of the image of the cubic function with equation $y = x^3$ after it is dilated by a factor of 2 parallel to the x-axis. Verify your solution graphically.

10. Determine the equation of the image of the curve with equation $y = \sqrt{x}$ after it is dilated by a factor of 4 from the x-axis. Verify your solution graphically.

11. **WE15** A farmer has fenced a vegetable patch with fence posts at A(0, 0), B(3, 0), C(3, 4) and D(0, 4). She wants to increase the vegetable patch size by a dilation factor of 1.5 in the x-direction and a dilation factor of 3 in the y-direction. Determine where she should relocate the fence posts.

12. Jack wants to plant flowers on a flower patch with corners at A(2, 1), B(4, 1), C(3, 2) and D(1, 2). He wants to increase the flower patch size by a dilation factor of 2 in both the x-direction and the y-direction. Determine where he should relocate the new corners of the flower patch.

13. Determine the equation of the image of the line with equation $2y + x = 3$ after it is dilated by $\begin{bmatrix} 1 & 0 \\ 0 & 2 \end{bmatrix}$.

14. Determine the equation of the image of the parabola with equation $y = x^2 - 1$ after it is dilated by $\begin{bmatrix} 3 & 0 \\ 0 & 1 \end{bmatrix}$.

15. Determine the equation of the image of the hyperbola with equation $y = \dfrac{1}{x+1}$ after it is dilated by a factor of 2 from the y-axis.

16. The equation $y = 2\sqrt{x}$ is transformed according to $\begin{bmatrix} x' \\ y' \end{bmatrix} = \begin{bmatrix} 2 & 0 \\ 0 & 3 \end{bmatrix} \begin{bmatrix} x \\ y \end{bmatrix}$.

    a. Determine the transformations represented by this matrix equation.
    b. Determine the equation of the image of $y = 2\sqrt{x}$ after this transformation.

17. Determine the equation of the image of the circle with equation $x^2 + y^2 = 4$ after it is transformed according to $\begin{bmatrix} x' \\ y' \end{bmatrix} = \begin{bmatrix} 2 & 0 \\ 0 & 1 \end{bmatrix} \begin{bmatrix} x \\ y \end{bmatrix}$.

18. The coordinates of $\triangle ABC$ can be written as a coordinate matrix $\begin{bmatrix} -2 & -1 & -3 \\ 0 & 3 & 2 \end{bmatrix}$. It has undergone a transformation T given by $\begin{bmatrix} x' \\ y' \end{bmatrix} = \begin{bmatrix} a & 0 \\ 0 & 2 \end{bmatrix} \begin{bmatrix} x \\ y \end{bmatrix}$.

    a. Determine the dilation factor, a, if the image coordinate point A$'$ is $(-3, 0)$.
    b. Calculate the coordinates of the vertices of $\triangle A'B'C'$.

19. Calculate the dilation factor from the y-axis when the graph of $y = \dfrac{1}{x^2}$ maps on to the graph of $y = \dfrac{1}{3x^2}$.

20. a. Determine the equation of the image of $x + 2y = 2$ under a dilation by a factor of 3 parallel to the x-axis.
    b. Determine if there is an invariant point.

Fully worked solutions for this chapter are available online.

LESSON
10.5 Combinations of transformations

SYLLABUS LINKS

- Use basic linear transformations: dilations of the form $(x, y) \rightarrow (ax, by)$, rotations about the origin and reflection in a line that passes through the origin, and the representations of these transformations by 2×2 matrices.
  - dilation of factor a parallel to the x-axis and factor b parallel to the y-axis: $\begin{bmatrix} a & 0 \\ 0 & b \end{bmatrix}$
  - rotation of angle θ anticlockwise about the origin: $\begin{bmatrix} \cos(\theta) & -\sin(\theta) \\ \sin(\theta) & \cos(\theta) \end{bmatrix}$
- reflection in the line $y = \tan(\theta)$: $\begin{bmatrix} \cos(2\theta) & \sin(2\theta) \\ \sin(2\theta) & -\cos(2\theta) \end{bmatrix}$
- Apply these transformations to points in the plane and polygons.
- Understand and use composition of linear transformations and the corresponding matrix products.
- Understand and use inverses of linear transformations and the relationship with the matrix inverse.
- Understand and use the relationship between the determinant and the effect of a linear transformation on area.
- Determine geometric results by matrix multiplications, e.g. showing that the combined effect of two reflections in lines through the origin is a rotation.

Source: Specialist Mathematics Senior Syllabus 2024 © State of Queensland (QCAA) 2024; licensed under CC BY 4.0

10.5.1 Double transformation matrices

A **combined transformation** is made up of two or more transformations.

If a linear transformation T_1 of a plane is followed by a second linear transformation T_2, then the results may be represented by a single transformation matrix T.

When transformation T_1 is applied to the point P(x, y) it results in P′(x', y').

When transformation T_2 is then applied to P′(x', y') it results in P″(x'', y'').

Summarising in matrix form:

$$\begin{bmatrix} x' \\ y' \end{bmatrix} = T_1 \begin{bmatrix} x \\ y \end{bmatrix}$$

$$\begin{bmatrix} x'' \\ y'' \end{bmatrix} = T_2 \begin{bmatrix} x' \\ y' \end{bmatrix}$$

Substituting $T_1 \begin{bmatrix} x \\ y \end{bmatrix}$ for $\begin{bmatrix} x' \\ y' \end{bmatrix}$ into $\begin{bmatrix} x'' \\ y'' \end{bmatrix} = T_2 \begin{bmatrix} x' \\ y' \end{bmatrix}$ results in $\begin{bmatrix} x' \\ y' \end{bmatrix} = T_2 T_1 \begin{bmatrix} x \\ y \end{bmatrix}$.

Combined transformations

To form the single transformation matrix **T**, the first transformation matrix T_1 must be *pre-multiplied* by the second transformation matrix T_2. The order of multiplication is important.

$$T = T_2 T_1$$

Note that if there were a third transformation T_3, then the single transformation matrix would be $T = T_3 T_2 T_1$. The order of operation is important, as matrix multiplication is not commutative. And remember that translation matrices are to be added, not multiplied.

Common transformation matrices used for combinations of transformations	
$M_x = \begin{bmatrix} 1 & 0 \\ 0 & -1 \end{bmatrix}$	Reflection in the x-axis
$M_y = \begin{bmatrix} -1 & 0 \\ 0 & 1 \end{bmatrix}$	Reflection in the y-axis
$M_{y = x\tan(\theta)} = \begin{bmatrix} \cos(2\theta) & \sin(2\theta) \\ \sin(2\theta) & -\cos(2\theta) \end{bmatrix}$	Reflection in the line $y = x\tan(\theta)$
$R_\theta = \begin{bmatrix} \cos(\theta) & -\sin(\theta) \\ \sin(\theta) & \cos(\theta) \end{bmatrix}$	Anti-clockwise (positive) rotation about origin
$D_{a,1} = \begin{bmatrix} a & 0 \\ 0 & 1 \end{bmatrix}$	Dilation in one direction parallel to the x-axis
$D_{1,b} = \begin{bmatrix} 1 & 0 \\ 0 & b \end{bmatrix}$	Dilation in one direction parallel to the y-axis
$D_{a,b} = \begin{bmatrix} a & 0 \\ 0 & b \end{bmatrix}$	Dilation parallel to both the x- and y-axes (a and b are the dilation factors)

Note: Translations are not linear transformations. The combined effect of two translations $\begin{bmatrix} a \\ b \end{bmatrix}$ and $\begin{bmatrix} c \\ d \end{bmatrix}$ is found by addition, $\begin{bmatrix} a+c \\ b+d \end{bmatrix}$.

WORKED EXAMPLE 16 Determining a single transformation matrix

Determine the single transformation matrix T that describes a reflection in the x-axis followed by a dilation of factor 3 from the y-axis.

THINK	WRITE
1. Determine the transformation matrices being used.	$T_1 = $ reflection in the x-axis $T_1 : M_x = \begin{bmatrix} 1 & 0 \\ 0 & -1 \end{bmatrix}$ $T_2 = $ dilation of factor 3 parallel to the x-axis $T_2 : D_{3,1} = \begin{bmatrix} 3 & 0 \\ 0 & 1 \end{bmatrix}$

2. State the combination of transformations matrix and simplify.

$T = T_2 T_1$
$T = D_{3,1} M_x$

$$T = \begin{bmatrix} 3 & 0 \\ 0 & 1 \end{bmatrix} \begin{bmatrix} 1 & 0 \\ 0 & -1 \end{bmatrix}$$

$$= \begin{bmatrix} 3 & 0 \\ 0 & -1 \end{bmatrix}$$

3. State the single transformation matrix.

The single transformation matrix is: $T = \begin{bmatrix} 3 & 0 \\ 0 & -1 \end{bmatrix}$

WORKED EXAMPLE 17 Determining the effect of a combination of transformations

Calculate the coordinates of the image of the point P(2, 3) under a reflection in the y-axis followed by a rotation of 90° about the origin in an anti-clockwise direction.

THINK

1. State the transformation matrices, T_1 and T_2.

WRITE

$T_1 = $ reflection in the y-axis
$T_2 = $ rotation of 90° anti-clockwise

$$T_1 = M_y = \begin{bmatrix} -1 & 0 \\ 0 & 1 \end{bmatrix}$$

$$T_2 = R_{90°} = \begin{bmatrix} \cos(90°) & -\sin(90°) \\ \sin(90°) & \cos(90°) \end{bmatrix}$$

2. Determine the single transformation matrix and simplify.

$T = T_2 T_1$
$T = R_{90} \circ M_y$

$$T = \begin{bmatrix} \cos(90°) & -\sin(90°) \\ \sin(90°) & \cos(90°) \end{bmatrix} \begin{bmatrix} -1 & 0 \\ 0 & 1 \end{bmatrix}$$

$$= \begin{bmatrix} 0 & -1 \\ 1 & 0 \end{bmatrix} \begin{bmatrix} -1 & 0 \\ 0 & 1 \end{bmatrix}$$

$$= \begin{bmatrix} 0 & -1 \\ -1 & 0 \end{bmatrix}$$

The single transformation matrix is $\begin{bmatrix} 0 & -1 \\ -1 & 0 \end{bmatrix}$.

3. State the single transformation matrix equation.

$$\begin{bmatrix} x' \\ y' \end{bmatrix} = \begin{bmatrix} 0 & -1 \\ -1 & 0 \end{bmatrix} \begin{bmatrix} x \\ y \end{bmatrix}$$

4. Substitute the pre-image (2, 3) into the matrix equation.

$$\begin{bmatrix} x' \\ y' \end{bmatrix} = \begin{bmatrix} 0 & -1 \\ -1 & 0 \end{bmatrix} \begin{bmatrix} 2 \\ 3 \end{bmatrix}$$

5. Calculate the coordinates of the image point.

$$\begin{bmatrix} x' \\ y' \end{bmatrix} = \begin{bmatrix} 0 & -1 \\ -1 & 0 \end{bmatrix} \begin{bmatrix} 2 \\ 3 \end{bmatrix}$$

$$\begin{bmatrix} x' \\ y' \end{bmatrix} = \begin{bmatrix} -3 \\ -2 \end{bmatrix}$$

6. State the answer.

The coordinates of the image point are $(-3, -2)$.

TI	THINK	DISPLAY/WRITE

1. Ensure the calculator is in Degree mode.
On a Calculator page, complete the entry line as:
$$\begin{bmatrix} \cos(90) & -\sin(90) \\ \sin(90) & \cos(90) \end{bmatrix} \times \begin{bmatrix} -1 & 0 \\ 0 & 1 \end{bmatrix} \times \begin{bmatrix} 2 \\ 3 \end{bmatrix}$$
then press ENTER.
Note: Matrix templates can be found by pressing the templates button.

2. The answer appears on the screen.

The image point is $(-3, -2)$.

CASIO	THINK	DISPLAY/WRITE

1. On a Run-Matrix screen, complete the entry line as:
$$\begin{bmatrix} \cos(90) & -\sin(90) \\ \sin(90) & \cos(90) \end{bmatrix} \times$$
$$\begin{bmatrix} -1 & 0 \\ 0 & 1 \end{bmatrix} \times \begin{bmatrix} 2 \\ 3 \end{bmatrix}$$
then press EXE.
Note: Matrix templates can be found by selecting MATH by pressing F4, then selecting MAT/VCT by pressing F1.

2. The answer appears on the screen.

The image point is $(-3, -2)$.

WORKED EXAMPLE 18 Determining the effect of two reflections in lines through the origin

Show that the result of a reflection in the line $y = x$, followed by a reflection in the line $y = \sqrt{3}x$ is a rotation.

THINK

1. State the transformation matrices T_1 and T_2.

WRITE

$T_1 = $ reflection in the line $y = x$

$T_2 = $ reflection in the line $y = \sqrt{3}x$

$$T_2 = M_{y = x\tan(\theta)} = \begin{bmatrix} \cos\left(\dfrac{2\pi}{3}\right) & \sin\left(\dfrac{2\pi}{3}\right) \\ \sin\left(\dfrac{2\pi}{3}\right) & -\cos\left(\dfrac{2\pi}{3}\right) \end{bmatrix}$$

$$= \begin{bmatrix} -\dfrac{1}{2} & \dfrac{\sqrt{3}}{2} \\ \dfrac{\sqrt{3}}{2} & \dfrac{1}{2} \end{bmatrix}$$

2. Determine the single transformation matrix and simplify.

$$T = T_2 T_1$$

$$= \begin{bmatrix} -\dfrac{1}{2} & \dfrac{\sqrt{3}}{2} \\ \dfrac{\sqrt{3}}{2} & \dfrac{1}{2} \end{bmatrix} \begin{bmatrix} 0 & 1 \\ 1 & 0 \end{bmatrix}$$

$$= \begin{bmatrix} \dfrac{\sqrt{3}}{2} & -\dfrac{1}{2} \\ \dfrac{1}{2} & \dfrac{\sqrt{3}}{2} \end{bmatrix}$$

$$= \begin{bmatrix} \cos\left(\dfrac{\pi}{6}\right) & -\sin\left(\dfrac{\pi}{6}\right) \\ \sin\left(\dfrac{\pi}{6}\right) & \cos\left(\dfrac{\pi}{6}\right) \end{bmatrix}$$

3. State the single transformation matrix.

The single transformation matrix is a rotation of $\dfrac{\pi}{6}$ in the anticlockwise direction.

10.5.2 Inverse transformation matrices

The inverse of a transformation matrix will transform the image of a point or shape back to its original position.

Note: $\begin{bmatrix} x' \\ y' \end{bmatrix} = T \begin{bmatrix} x \\ y \end{bmatrix}$

Inverse transformation matrices

By determining the inverse of a matrix, we get:

$$\begin{bmatrix} x \\ y \end{bmatrix} = T^{-1} \begin{bmatrix} x' \\ y' \end{bmatrix}$$

WORKED EXAMPLE 19 Determining the inverse transformation of a point

Determine the coordinates of the pre-image point, **P(x, y)**, under a reflection in the *x*-axis followed by a rotation of **90°** about the origin in a clockwise direction of the image point **P′(−3, −2)**.

THINK	WRITE
1. State the transformation matrices, T_1 and T_2.	$T_1 =$ reflection in the x-axis $T_2 =$ rotation of $90°$ clockwise $T_1 = M_x = \begin{bmatrix} 1 & 0 \\ 0 & -1 \end{bmatrix}$ $T_2 = R_{-90°} = \begin{bmatrix} \cos(90°) & \sin(90°) \\ -\sin(90°) & \cos(90°) \end{bmatrix}$
2. Determine the single transformation matrix and simplify.	$T = T_2 T_1$ $T = R_{-90°} M_x$ $T = \begin{bmatrix} \cos(90°) & \sin(90°) \\ -\sin(90°) & \cos(90°) \end{bmatrix} \begin{bmatrix} 1 & 0 \\ 0 & -1 \end{bmatrix}$ $= \begin{bmatrix} 0 & 1 \\ -1 & 0 \end{bmatrix} \begin{bmatrix} 1 & 0 \\ 0 & -1 \end{bmatrix}$ $= \begin{bmatrix} 0 & -1 \\ -1 & 0 \end{bmatrix}$ The single transformation matrix is $\begin{bmatrix} 0 & -1 \\ -1 & 0 \end{bmatrix}$.
3. State the single transformation matrix equation.	$\begin{bmatrix} x' \\ y' \end{bmatrix} = \begin{bmatrix} 0 & -1 \\ -1 & 0 \end{bmatrix} \begin{bmatrix} x \\ y \end{bmatrix}$
4. Substitute the image point $P'(-3, -2)$ into the matrix equation.	$\begin{bmatrix} -3 \\ -2 \end{bmatrix} = \begin{bmatrix} 0 & -1 \\ -1 & 0 \end{bmatrix} \begin{bmatrix} x \\ y \end{bmatrix}$
5. Using $\begin{bmatrix} x \\ y \end{bmatrix} = T^{-1} \begin{bmatrix} x' \\ y' \end{bmatrix}$, rearrange the equation to make the image point (x, y) the subject.	$\begin{bmatrix} x \\ y \end{bmatrix} = \begin{bmatrix} 0 & -1 \\ -1 & 0 \end{bmatrix}^{-1} \begin{bmatrix} -3 \\ -2 \end{bmatrix}$
6. Determine the inverse matrix for the single transformation matrix.	$T^{-1} = \begin{bmatrix} 0 & -1 \\ -1 & 0 \end{bmatrix}$
7. Simplify the matrix equation.	$\begin{bmatrix} x \\ y \end{bmatrix} = \begin{bmatrix} 0 & -1 \\ -1 & 0 \end{bmatrix}^{-1} \begin{bmatrix} -3 \\ -2 \end{bmatrix}$ $= \begin{bmatrix} 0 & -1 \\ -1 & 0 \end{bmatrix} \begin{bmatrix} -3 \\ -2 \end{bmatrix}$ $\therefore \begin{bmatrix} x \\ y \end{bmatrix} = \begin{bmatrix} 2 \\ 3 \end{bmatrix}$
8. State the answer.	The coordinates of the pre-image point are $(2, 3)$.

A triangle ABC is transformed under the transformation matrix $T = \begin{bmatrix} 1 & 2 \\ -3 & 4 \end{bmatrix}$ to give vertices at A′(−1, −7), B′(4, 18) and C′(11, 7). Determine the vertices A, B and C.

THINK	WRITE
1. Write the image vertices, A′(−1, −7), B′(4, 18) and C′(11, 7), of ΔABC as a coordinate matrix.	$\begin{bmatrix} -1 & 4 & 11 \\ -7 & 18 & 7 \end{bmatrix}$
2. Write the pre-image vertices, A, B and C, of ΔABC as a coordinate matrix.	Let $A = (a, b), B = (c, d), C = (e, f)$. $\begin{bmatrix} a & c & e \\ b & d & f \end{bmatrix}$
3. Recall the matrix equation and substitute known values.	$\begin{bmatrix} -1 & 4 & 11 \\ -7 & 18 & 7 \end{bmatrix} = \begin{bmatrix} 1 & 2 \\ -3 & 4 \end{bmatrix} \begin{bmatrix} a & c & e \\ b & d & f \end{bmatrix}$
4. Rearrange the equation to make the image points the subject.	$\begin{bmatrix} a & c & e \\ b & d & f \end{bmatrix} = \begin{bmatrix} 1 & 2 \\ -3 & 4 \end{bmatrix}^{-1} \begin{bmatrix} -1 & 4 & 11 \\ -7 & 18 & 7 \end{bmatrix}$
5. Determine the inverse of the transformation matrix.	$T^{-1} = \dfrac{1}{10} \begin{bmatrix} 4 & -2 \\ 3 & 1 \end{bmatrix}$
6. Simplify the matrix equation.	$\begin{bmatrix} a & c & e \\ b & d & f \end{bmatrix} = \begin{bmatrix} 1 & 2 \\ -3 & 4 \end{bmatrix}^{-1} \begin{bmatrix} -1 & 4 & 11 \\ -7 & 18 & 7 \end{bmatrix}$ $\begin{bmatrix} a & c & e \\ b & d & f \end{bmatrix} = \dfrac{1}{10} \begin{bmatrix} 4 & -2 \\ 3 & 1 \end{bmatrix} \begin{bmatrix} -1 & 4 & 11 \\ -7 & 18 & 7 \end{bmatrix}$ $= \dfrac{1}{10} \begin{bmatrix} 10 & -20 & 30 \\ -10 & 30 & 40 \end{bmatrix}$ $= \begin{bmatrix} 1 & -2 & 3 \\ -1 & 3 & 4 \end{bmatrix}$
7. Write the answer.	The vertices of triangle ABC are A(1, −1), B(−2, 3) and C(3, 4).

TI | THINK	DISPLAY/WRITE	CASIO | THINK	DISPLAY/WRITE
On a Calculator page, complete the entry line as: $\begin{bmatrix} 1 & 2 \\ -3 & 4 \end{bmatrix}^{-1} \times$ $\begin{bmatrix} -1 & 4 & 11 \\ -7 & 18 & 7 \end{bmatrix}$ then press ENTER. *Note:* Matrix templates can be found by pressing the templates button.	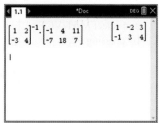 The vertices of triangle ABC are A(1, −1), B(−2, 3) and C(3, 4).	On a Main screen, complete the entry line as: $\begin{bmatrix} 1 & 2 \\ -3 & 4 \end{bmatrix}^{-1} \times$ $\begin{bmatrix} -1 & 4 & 11 \\ -7 & 18 & 7 \end{bmatrix}$ then press EXE.	 The vertices of triangle ABC are A(1, −1), B(−2, 3) and C(3, 4).

10.5.3 Interpreting the determinant of the transformation matrix

A single transformation matrix is represented by T. When a shape is transformed by this transformation matrix, the magnitude of the determinant of matrix T gives the ratio of the image area to the original area.

> **Determinant and area of an image**
>
> **Area of image = |det(T)| × area of object**
>
> **Where |det(T)| represents the area scale factor for the transformation.**
>
> **If det(T) is negative, then the transformation will have involved some reflection.**

Thus, if a polygon of area $6\,\text{cm}^2$ is transformed by a matrix with a determinant of 8 (or -8) for instance, the new polygon will have an area of $48\,\text{cm}^2$.

Recall that the determinant of a 2×2 matrix, $T = \begin{bmatrix} a & b \\ c & d \end{bmatrix}$, is $\det(T) = ad - bc$.

WORKED EXAMPLE 21 Using determinants to determine areas of transformed images

The triangle ABC is mapped by the transformation represented by $T = \begin{bmatrix} 1 & -\sqrt{3} \\ \sqrt{3} & 1 \end{bmatrix}$ onto the triangle A′B′C′. Given that the area of ΔABC is 8 units², calculate the area of A′B′C′.

THINK	WRITE
1 Recall the determinant formula and calculate for the given matrix.	$T = \begin{bmatrix} 1 & -\sqrt{3} \\ \sqrt{3} & 1 \end{bmatrix}$ $\det(T) = 1 + 3 = 4$
2 Analyse the determinant.	The determinant is 4, which means that the area of the image is 4 times the area of the original object.
3 Calculate the area of the image, $\Delta A'B'C'$.	Area of image = $\|\det(T)\| \times$ area of object Area of $\Delta A'B'C' = 4 \times 8$ $\qquad\qquad\qquad = 32$
4 Write the answer.	The area of $\Delta A'B'C'$ is 32 units².

TI | THINK	DISPLAY/WRITE	CASIO | THINK	DISPLAY/WRITE
1. On a Calculator page, press MENU, then select: 7: Matrix & Vector 3: Determinant. Complete the entry line as: $\det\left(\begin{bmatrix} 1 & -\sqrt{3} \\ \sqrt{3} & 1 \end{bmatrix}\right)$ then press ENTER. *Note:* Matrix templates can be found by pressing the templates button.	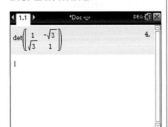	1. On a Run-Matrix screen, press OPTN and select MAT/VCT by pressing F2, then select Det by pressing F3. Complete the entry line as: $\det\left(\begin{bmatrix} 1 & -\sqrt{3} \\ \sqrt{3} & 1 \end{bmatrix}\right)$ then press EXE. *Note:* Matrix templates can be found by selecting MATH by pressing F4, then selecting MAT/VCT by pressing F1.	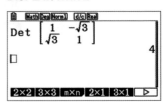

2. Complete the next entry line as: ans × 8 then press ENTER.

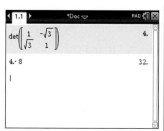

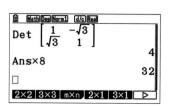

2. The answer appears on the screen.	The area is 32 units².
2. The answer appears on the screen.	The area is 32 units².

Exercise 10.5 Combinations of transformations

learn on

10.5 Exercise	**10.5 Exam questions** on

Simple familiar	Complex familiar	Complex unfamiliar
1, 2, 3, 4, 5, 6, 7, 8, 9, 10, 11, 12	13, 14, 15, 16, 17, 18	19, 20

These questions are even better in jacPLUS!
- Receive immediate feedback
- Access sample responses
- Track results and progress

Find all this and MORE in jacPLUS ▶

Simple familiar

1. **WE16** Determine the single transformation matrix T that describes a reflection in the y-axis followed by a dilation factor of 3 parallel to the y-axis.

2. Determine the single transformation matrix T that describes a reflection in the line $y = x$ followed by a dilation of factor 2 from both the x- and y-axes.

3. **WE17** Calculate the coordinates of the image of the point $P(1, -3)$ under a reflection in the x-axis followed by a rotation of 180° about the origin in an anti-clockwise direction.

4. Calculate the coordinates of the image of the point $P(-2, 2)$ under a reflection in the line $y = x$ followed by a rotation of 45° about the origin in an anti-clockwise direction.

5. Describe fully a sequence of two geometrical transformations represented by $T = \begin{bmatrix} 0 & -1 \\ 1 & 0 \end{bmatrix} \begin{bmatrix} 2 & 0 \\ 0 & 2 \end{bmatrix}$.

6. Determine the equation of the image of $y = x^2$ under a double transformation: a reflection in the x-axis followed by a dilation factor of 2 parallel to both the x- and y-axes.

7. Determine the equation of the image of $y = \sqrt{x}$ under a double transformation: a reflection in the y-axis followed by a dilation of 3 parallel to the x-axis.

8. a. State the transformations that have undergone $T\left(\begin{bmatrix} x' \\ y' \end{bmatrix}\right) = \begin{bmatrix} 1 & 0 \\ 0 & -1 \end{bmatrix} \begin{bmatrix} 0 & 1 \\ 1 & 0 \end{bmatrix} \begin{bmatrix} x \\ y \end{bmatrix}$.

   b. Determine the image of the curve with equation $2x - 3y = 12$.

9. a. State the transformations that have undergone $T\left(\begin{bmatrix} x' \\ y' \end{bmatrix}\right) = \begin{bmatrix} 2 & 0 \\ 0 & -1 \end{bmatrix} \begin{bmatrix} x \\ y \end{bmatrix} + \begin{bmatrix} 1 \\ 2 \end{bmatrix}$.

   b. Determine the image of the curve with equation $y = 2x^2 - 1$.

10. **WE20** A triangle ABC is transformed under the transformation matrix $T = \begin{bmatrix} 3 & 2 \\ 5 & 8 \end{bmatrix}$ to give vertices at $A'(0, 0), B'(4, 18)$ and $C'(9, 15)$. Determine the vertices A, B and C.

11. **WE21** The triangle ABC is mapped by the transformation represented by $T = \begin{bmatrix} 3 & 1 \\ 1 & 2 \end{bmatrix}$ onto the triangle $A'B'C'$. Given that the area of ABC is 4 units^2, calculate the area of $A'B'C'$.

12. A rectangle ABCD is transformed under the transformation matrix $T = \begin{bmatrix} 3 & 2 \\ 5 & 8 \end{bmatrix}$, to give vertices at $A'(0, 0), B'(3, 0), C'(3, 2)$ and $D'(0, 2)$.

   a. Determine the vertices of the square ABCD.

   b. Calculate the area of the figure ABCD.

Complex familiar

13. **WE18** Show that the result of a reflection in the line $y = -x$, followed by a reflection in the line $y = \dfrac{\sqrt{3}}{3}x$ is a rotation.

14. **WE19** Determine the coordinates of the pre-image point, P(x, y), under a reflection in the x-axis followed by a rotation of 90° about the origin in an anti-clockwise direction of the image point $P'(1, -2)$.

15. **a.** Calculate the coordinates of the image of point $P'(x', y')$ when the point P(x, y) undergoes a double transformation: a reflection in the y-axis followed by a translation of 4 units in the positive direction of the x-axis.

   b. Reverse the order of the pair of transformations in part **a** and determine if the image is different.

16. Determine the coordinates of the pre-image point, P(x, y), under a reflection in the y-axis followed by a dilation of factor 3 from the y-axis of the image point $P'(-3, 6)$.

17. The triangle $\triangle ABC$ is mapped by the transformation represented by $T = \begin{bmatrix} 3 & -1 \\ 1 & 2 \end{bmatrix}$ onto the triangle $\triangle A'B'C'$. Given that the area of $\triangle ABC$ is 10 units^2, calculate the area of $\triangle A'B'C'$.

18. A rectangle ABCD with vertices at $A(0, 0), B(2, 0), C(2, 3)$ and $D(0, 3)$ is transformed under the transformation matrix $T = \begin{bmatrix} 2 & -1 \\ 1 & 2 \end{bmatrix}$. Calculate the new area of the transformed rectangle.

Complex unfamiliar

19. If D_λ denotes a dilation factor of λ parallel to both axes, determine what single dilation would be equivalent to D_λ^2.

20. Check whether the transformation 'a reflection in the y-axis followed by a reflection in the line $y = x$' is the same as 'a reflection in the line $y = x$ followed by a reflection in the y-axis'.

Fully worked solutions for this chapter are available online.

LESSON
10.6 Review

10.6.1 Summary

10.6 Exercise

learn on

10.6 Exercise	10.6 Exam questions on

Simple familiar	Complex familiar	Complex unfamiliar
1, 2, 3, 4, 5, 6, 7, 8, 9, 10, 11, 12	13, 14, 15, 16	17, 18, 19, 20

Simple familiar

1. Determine the single transformation matrix for a dilation of factor 7 from the y-axis followed by a reflection in the line $y = x$.

2. The point $(-3, 2)$ is translated by the matrix $\begin{bmatrix} -1 \\ -5 \end{bmatrix}$. Determine the new coordinates of the point.

3. Determine the transformation matrix for a dilation of factor λ parallel to the y-axis followed by a reflection in the line $y = x$, then a reflection in the x-axis.

4. Calculate the coordinates of the image of the points A(0, 1) and C(3, 2) under the transformation defined by $\begin{bmatrix} 2 & 3 \\ -1 & 0 \end{bmatrix}$.

5. State the coordinates of the image of the point P(x, y) under a translation defined by $\begin{bmatrix} 2 \\ -1 \end{bmatrix}$ followed by a reflection in the x axis.

6. Determine the equation of the image of $y = 2x^2$ under the translation defined by $\begin{bmatrix} 2 \\ -1 \end{bmatrix}$.

7. Determine the matrix equation that represents a dilation of factor 3 parallel to the x-axis followed by a translation of 2 units in the negative direction of the x-axis and 5 units in the positive direction of the y-axis.

8. Determine the coordinates of the pre-image point $D(a, b)$ under a transformation defined by $T = \begin{bmatrix} 0 & 2 \\ -1 & 0 \end{bmatrix}$ of the image point $D'(10, 3)$.

9. Determine the equation of the image of the graph of $y = \sqrt{x}$ after a dilation of factor 3 parallel to the x-axis followed by a translation of 2 units in the negative direction of the x-axis and 5 units in the positive direction of the y-axis.

10. Determine the final image point when point $P(2, -1)$ undergoes two reflections. It is firstly reflected in the x-axis and then reflected in the line $y = x$.

11. Calculate the coordinates of the vertices of the image of pentagon ABCDE with $A(2, 5)$, $B(4, 4)$, $C(4, 1)$, $D(2, 0)$ and $E(0, 3)$ after a reflection in the y-axis.

12. A triangle ABC with vertices $A(2, -1)$, $B(-4, 0)$ and $C(5, 2)$ is rotated $45°$ anticlockwise.
    a. Calculate the coordinates of vertices A', B' and C' of the rotated triangle.
    b. Compare the area of the triangle $A'B'C'$ to that of triangle ABC.

Complex familiar

13. The area of the image is 10 times the area of the original shape when it has undergone a transformation $T = \begin{bmatrix} k & 3k \\ 1 & k \end{bmatrix}$. Determine the value(s) of k.

14. The line $y = 2x - 1$ undergoes a succession of translations defined by $T_1 = \begin{bmatrix} 1 \\ -4 \end{bmatrix}$ and $T_2 = \begin{bmatrix} -2 \\ 3 \end{bmatrix}$. Verify that the order in which these translations occur has no effect on the result.

15. Determine the equation of the image of the line $y = -2x + 2$ under a rotation of $\dfrac{\pi}{2}$ about the origin in an anticlockwise direction. Sketch the original and the image.

16. Determine the image of the point $P(1, 2)$ after a reflection in the line $y = \sqrt{3}x$.

Complex unfamiliar

17. A farmer has fenced a vegetable patch with fence posts at $A(0, 0)$, $B(3, 0)$, $C(3, 4)$ and $D(0, 4)$. She wants to increase the size of the vegetable patch by a dilation factor of 1.5 in the x-direction and a dilation factor of 3 in the y-direction. Where should she relocate the fence posts?

18. Avril has plotted all of the locations for a cartoon on the Cartesian plane. The vertices of the outline of a character's house have coordinates $(-3, 5)$, $(-3, 7)$, $(5, 7)$ and $(5, 5)$. She wants to move the points so that they are reflected in the y-axis.
To add more animation to her cartoon, she wants not only to reflect the outline of the house in the y-axis, but also to combine it with a dilation factor of 2 in the x-axis and then a translation of 2 units across in the negative x-direction and 5 units down in the negative y-direction. State the transformation matrix equation and determine the new coordinates of the house after the combinations of transformations.

19. A computer designer wants to animate a marshmallow being squashed. The marshmallow is modelled by the unit circle equation $x^2 + y^2 = 1$. To make it look squashed, it has to undergo the following transformations for the animation:

- a dilation factor of 2 from the y-axis and factor of 3 from the x-axis
- a translation of 1 unit across the x-axis and 2 units up the y-axis.
  Determine the area of the squashed marshmallow.

20. Mark wants to build a deck for entertaining outside. To plan out his deck, he uses a coordinate grid where each square represents 1 metre. The coordinates for the vertices of his deck are $(2, 2)$, $(8, 2)$, $(2, 6)$ and $(8, 6)$.

a. Mark decides to increase the deck by dilating it by a factor of 1.5 from the x-axis.
  Mark ordered enough decking wood to cover an area of $40 \, \text{m}^2$. Justify whether Mark has enough to build his new deck?

b. On the blueprint of Mark's house, the coordinates of the corners of the garage are $(1, 5)$, $(1, 25)$, $(31, 5)$ and $(31, 25)$. On the blueprint, 1 unit represents 1 cm.

  The scale of the blueprint is $\dfrac{1}{30}$ of the actual structure. Mark has two cars. Each car needs a parking area of $15 \, \text{m}^2$ in the garage. Does he have enough space to park both cars in his garage?

c. Mark decides to transform a garden bed in the form of a triangle with vertices at $A(-2, -2)$, $B(0, 2)$ and $C(2, -2)$ using the transformation equations $x' = x + y$ and $y' = x - y$.
  Calculate the scale factor by which the area has been increased.

Fully worked solutions for this chapter are available online.

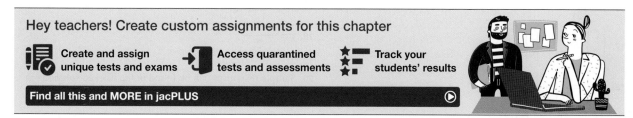

Answers

Chapter 10 Matrices and transformations

10.2 Translations

10.2 Exercise

1. $\begin{bmatrix} x' \\ y' \end{bmatrix} = \begin{bmatrix} 3 \\ 7 \end{bmatrix}, (3, 7)$

2. $\begin{bmatrix} x' \\ y' \end{bmatrix} = \begin{bmatrix} -6 \\ 2 \end{bmatrix}, (-6, 2)$

3. $\begin{bmatrix} x' \\ y' \end{bmatrix} = \begin{bmatrix} 4 \\ 0 \end{bmatrix}, (4, 0)$

4. $\begin{bmatrix} x' \\ y' \end{bmatrix} = \begin{bmatrix} 2 \\ -2 \end{bmatrix}, (2, -2)$

5. $T = \begin{bmatrix} 1 & 1 & 1 \\ -2 & -2 & -2 \end{bmatrix}$

6. $T = \begin{bmatrix} 1 & 1 & 1 \\ 2 & 2 & 2 \end{bmatrix}$

7. $\begin{bmatrix} x' \\ y' \end{bmatrix} = \begin{bmatrix} x \\ y \end{bmatrix} + \begin{bmatrix} 2 \\ 1 \end{bmatrix}$

8. a.

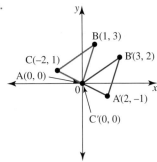

b. $T = \begin{bmatrix} 2 & 2 & 2 \\ -1 & -1 & -1 \end{bmatrix}$

9. $y = x + 1$

10. $y = x - 6$

11. $y = x + 6$

12. $y = (x - 2)^2 - 1$

13. $y = (x + 3)^2 + 1$

14. $y = (x - 7)^2 - 6$

15. $T = \begin{bmatrix} 0 \\ 2 \end{bmatrix}$

16. $T = \begin{bmatrix} 7 \\ 3 \end{bmatrix}$

17. $\begin{bmatrix} x' \\ y' \end{bmatrix} = \begin{bmatrix} x \\ y \end{bmatrix} + \begin{bmatrix} 2 \\ 6 \end{bmatrix}$

18. $\begin{bmatrix} x' \\ y' \end{bmatrix} = \begin{bmatrix} x \\ y \end{bmatrix} + \begin{bmatrix} 1 \\ 0 \end{bmatrix}$

19. $\begin{bmatrix} x' \\ y' \end{bmatrix} = \begin{bmatrix} x \\ y \end{bmatrix} + \begin{bmatrix} a \\ b \end{bmatrix}$

20. $\begin{bmatrix} x' \\ y' \end{bmatrix} = \begin{bmatrix} x \\ y \end{bmatrix} + \begin{bmatrix} a \\ 0 \end{bmatrix}$

10.3 Reflections and rotations

10.3 Exercise

1. $\begin{bmatrix} x' \\ y' \end{bmatrix} = \begin{bmatrix} -3 \\ 1 \end{bmatrix}, (-3, 1)$

2. $\begin{bmatrix} x' \\ y' \end{bmatrix} = \begin{bmatrix} -5 \\ -2 \end{bmatrix}, (-5, -2)$

3. $y = (x + 2)^2$

4. $y = -x^2 - 1$

5. $\begin{bmatrix} x' \\ y' \end{bmatrix} = \begin{bmatrix} 5 \\ -2 \end{bmatrix}, (5, -2)$

   $\begin{bmatrix} x' \\ y' \end{bmatrix} = \begin{bmatrix} 9 \\ 3 \end{bmatrix}, (9, 3)$

   $\begin{bmatrix} x' \\ y' \end{bmatrix} = \begin{bmatrix} -5 \\ 8 \end{bmatrix}, (5, -8)$

6. $\begin{bmatrix} x' \\ y' \end{bmatrix} = \begin{bmatrix} 6 \\ -9 \end{bmatrix}, (6, -9)$

7. $y = \pm \sqrt{x}$

8. $y = \pm \dfrac{\sqrt{-2(x + 1)}}{2}$

9. a. $y = x - 3$ b. $y = x + 3$

10. a. $y = -(x^2 + 2x + 1)$ b. $y = x^2 - 2x + 1$

11. $y = -(\sqrt{3} - 2)x + 1 - \sqrt{3}$

12. $(2 + \sqrt{3})x + (1 - 2\sqrt{3})y - 6 = 0$

13. $y = 4(1 + \sqrt{3}) - (2 + \sqrt{3})x$

14. $\left(\dfrac{5}{2} + 2\sqrt{3}, -\dfrac{5\sqrt{3}}{2} + 2 \right)$

15. a. $\begin{bmatrix} 0 & 1 \\ -1 & 0 \end{bmatrix}$ b. $\begin{bmatrix} -1 & 0 \\ 0 & -1 \end{bmatrix}$

    c. $\dfrac{1}{\sqrt{2}} \begin{bmatrix} 1 & -1 \\ 1 & 1 \end{bmatrix}$ d. $\dfrac{1}{2} \begin{bmatrix} \sqrt{3} & -1 \\ 1 & \sqrt{3} \end{bmatrix}$

16. a. $(6, 7)$

    b. $(6, 7)$

    c. The points are the same.

17. $y = -\dfrac{x}{2} + \dfrac{\sqrt{2}}{4}$

18. $y = \dfrac{-x + 1}{2}$

19. $2x + 3y - 6 = 0$

20. Sample responses can be found in the worked solutions in the online resources.

10.4 Dilations

10.4 Exercise

1. $\begin{bmatrix} x' \\ y' \end{bmatrix} = \begin{bmatrix} 2 \\ -3 \end{bmatrix}$, $(2, -3)$

2. $\begin{bmatrix} x' \\ y' \end{bmatrix} = \begin{bmatrix} -1 \\ 8 \end{bmatrix}$, $(-1, 8)$

3. $\begin{bmatrix} x' \\ y' \end{bmatrix} = \begin{bmatrix} 3 & 0 \\ 0 & 1 \end{bmatrix} \begin{bmatrix} x \\ y \end{bmatrix}$

4. $\begin{bmatrix} x' \\ y' \end{bmatrix} = \begin{bmatrix} -2 \\ 4 \end{bmatrix}$, $(-2, 4)$

5. $\begin{bmatrix} x' \\ y' \end{bmatrix} = \begin{bmatrix} 6 \\ -5 \end{bmatrix}$, $(6, -5)$

6. $y = \left(\dfrac{x}{3}\right)^2 = \dfrac{x^2}{9}$

7. $y = \dfrac{x^2}{2}$

8. a. $\begin{bmatrix} x' \\ y' \end{bmatrix} = \begin{bmatrix} -3 \\ 6 \end{bmatrix}$, $(-3, 6)$

   b. A dilation of 3 parallel to the x-axis and a dilation of 2 parallel to the y-axis.

9. $y = \dfrac{1}{8}x^3$

10. $y = 4\sqrt{x}$

11. $(0, 0)$, $(4.5, 0)$, $(4.5, 12)$, $(0, 12)$

12. $(4, 2)$, $(8, 2)$, $(6, 4)$, $(2, 4)$

13. $x + y = 3$

14. $y = \left(\dfrac{x}{3}\right)^2 - 1 = \dfrac{x^2}{9} - 1$

15. $y = \dfrac{2}{x + 2}$

16. a. A dilation of 2 parallel to the x-axis and a dilation of 3 parallel to the y-axis.

    b. $y = 3\sqrt{2x}$

17. $\dfrac{x^2}{4} + y^2 = 4$

18. a. $a = \dfrac{3}{2}$

    b. $\begin{bmatrix} -3 & -\dfrac{3}{2} & -\dfrac{9}{2} \\ 0 & 6 & 4 \end{bmatrix}$;

    $A'(-3, 0)$, $B'\left(-\dfrac{3}{2}, 6\right)$, $C'\left(-\dfrac{9}{2}, 4\right)$

19. $a = \pm\dfrac{1}{\sqrt{3}}$ or $\pm\dfrac{\sqrt{3}}{3}$

20. a. $y = -\dfrac{x}{6} + 1$

    b. Invariant point is $(0, 1)$.

10.5 Combinations of transformations

10.5 Exercise

1. $T = \begin{bmatrix} -1 & 0 \\ 0 & 3 \end{bmatrix}$

2. $T = \begin{bmatrix} 0 & 2 \\ 2 & 0 \end{bmatrix}$

3. $(-1, -3)$

4. $(2\sqrt{2}, 0)$

5. Dilation of factor 2 parallel to both axes followed by an anticlockwise rotation of 90° about the origin.

6. $y = -\dfrac{x^2}{2}$

7. $y = \sqrt{-\dfrac{x}{3}}$

8. a. Reflection in line $y = x$ followed by a reflection in x-axis.

   b. $3x + 2y = -12$ or $y = -\dfrac{3}{2}x - 6$

9. a. Dilation of factor 2 parallel to the x-axis followed by a reflection in the x-axis and translation of 1 unit in the positive x-direction and 2 units in the positive y-direction.

   b. $y = -\dfrac{1}{2}(x - 1)^2 + 3$

10. $A(0, 0)$, $B\left(\dfrac{-2}{17}, \dfrac{17}{7}\right)$, $C(3, 0)$

11. 20 units^2

12. a. $A(0, 0)$, $B\left(\dfrac{12}{7}, \dfrac{-15}{14}\right)$, $C\left(\dfrac{10}{7}, \dfrac{-9}{14}\right)$, $D\left(\dfrac{-2}{7}, \dfrac{3}{7}\right)$

    b. $\dfrac{3}{7}$ units^2

13. The single transformation matrix is a rotation of $\dfrac{\pi}{3}$ in the clockwise direction.

14. $(-2, 1)$

15. a. $(-x + 4, \ y)$

    b. Yes: $(-x - 4, \ y)$

16. $(1, 6)$

17. 70 units^2

18. 30 units^2

19. D_λ^2 gives a dilation factor of λ^2 parallel to both axes.

20. Not the same.

10.6 Review

10.6 Exercise

1. $\begin{bmatrix} 0 & 1 \\ 7 & 0 \end{bmatrix}$

2. $(-4, -3)$

3. $\begin{bmatrix} 1 & 0 \\ 0 & -1 \end{bmatrix} \begin{bmatrix} 0 & 1 \\ 1 & 0 \end{bmatrix} \begin{bmatrix} 1 & 0 \\ 0 & \lambda \end{bmatrix}$

4. $A'(3, 0)$, $C'(12, -3)$

5. $(x + 2, -y + 1)$ or $(x + 2, -(y - 1))$

6. $y = 2x^2 - 8x + 7$

7. $\begin{bmatrix} x' \\ y' \end{bmatrix} = \begin{bmatrix} 3 & 0 \\ 0 & 1 \end{bmatrix} \begin{bmatrix} x \\ y \end{bmatrix} + \begin{bmatrix} -2 \\ 5 \end{bmatrix}$

8. $(-3, 5)$

9. $y = \dfrac{\sqrt{3(x+2)}}{3} + 5$

10. $(1, 2)$

11. $A'(-2, 5), B'(-4, 5), C'(-4, 1), D'(-2, 0), E'(0, 3)$

12. a. $A'\left(\dfrac{1}{\sqrt{2}}, -\dfrac{3}{\sqrt{2}}\right), B'\left(-2\sqrt{2}, 2\sqrt{2}\right),$
    $C'\left(\dfrac{7}{\sqrt{2}}, -\dfrac{3}{\sqrt{2}}\right)$

    b. The areas are the same.

13. $k = 5$ or $k = -2$

14. Sample responses can be found in the worked solutions in the online resources.

15. $y = \dfrac{x}{2} + 1$

16. $P'\left(\dfrac{2\sqrt{3}-1}{2}, \dfrac{\sqrt{3}+2}{2}\right)$

17. She should relocate the fence to points $(0, 0)$, $\left(\dfrac{9}{2}, 0\right)$,
    $(0, 12)$ and $\left(\dfrac{9}{2}, 12\right)$.

18. $\begin{bmatrix} x' \\ y' \end{bmatrix} = \begin{bmatrix} -2 & 0 \\ 0 & 1 \end{bmatrix} \begin{bmatrix} x \\ y \end{bmatrix} + \begin{bmatrix} -2 \\ -5 \end{bmatrix}$
    $(4, 0), (4, 2), (-12, 2), (-12, 0)$

19. 6π units^2

20. a. Area $= 6 \times 6 = 36\,\text{m}^2$
    Yes, he only needs $36\,\text{m}^2$ and has $4\,\text{m}^2$ to spare.

    b. He needs $30\,\text{m}^2$ for the car and has $54\,\text{m}^2$ available; he has enough space.

    c. Area factor $= 2$

GLOSSARY

absolute value: $|x|$, the magnitude of a number

addition principle: if there are n ways of performing operation A and m ways of performing operation B, then there are $m + n$ ways of performing A *or* B.

additive inverse: when two matrices $A + B = O$, matrix B is an additive inverse of A.

affix: the point corresponding to this complex number $z = a + bi$, that is the point with Cartesian coordinates (a, b)

alternate segment: the chord $\overline{AD}$ divides the circle into two segments. The segment closest to the tangent $\overline{HD}$ contains $\angle$ADH. The alternate segment contains $\angle$AGD and this angle is known as the angle in the alternate segment. $\angle$ADH $= \angle$AGD

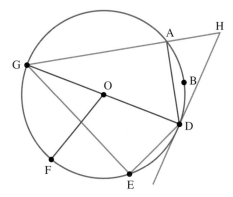

angle at the centre: the angle made by two radii intersecting the centre of a circle. As O is the centre of the circle, $\angle$FOD is called an angle at the centre subtended by the arc $\overset{\frown}{\text{FED}}$.

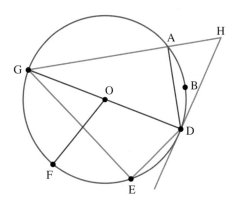

angle at the circumference: the angle at the circumference of a circle. Consider the angle ∠AGD. As G is on the circumference of the circle, then ∠AGD is called an angle at the circumference subtended by the arc $\overset{\frown}{ABD}$. The same angle can also be described as standing on chord $\overline{AD}$.

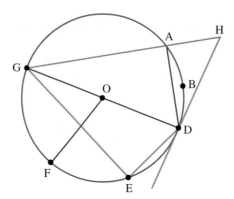

angle difference identities: $\sin(A - B) = \sin(A)\cos(B) - \cos(A)\sin(B)$,
$\cos(A - B) = \cos(A)\cos(B) + \sin(A)\sin(B)$

angle in a semicircle: the angle at the circumference of a semi-circle subtended by the arc of the semicircle. As $\overline{GD}$ is a diameter, ∠GAD is called an angle in a semicircle.

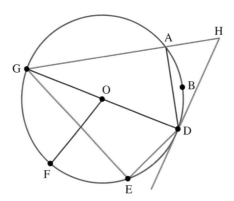

angle sum identities: $\sin(A + B) = \sin(A)\cos(B) + \cos(A)\sin(B)$, $\cos(A + B) = \cos(A)\cos(B) - \sin(A)\sin(B)$

applied forces: forces acting on objects in physical contact with each other; include tensile forces, contact forces, compressive forces

arc: three points on a circle describe an arc. $\overset{\frown}{ABD}$ begins at point A and then passes through B to end at D. A minor arc is smaller than a semicircle, and a major arc is larger than a semicircle.

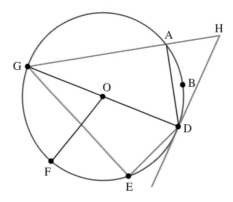

Argand diagram: plotting a complex number $x + yi$ as a point (x, y) on the complex plane with real and imaginary axes (the Argand plane)

Argand plane: geometric representation of complex numbers $x + yi$ with real and imaginary axes. The axes are denoted by 'Re(z)' and 'Im(z)' respectively.

argument of z: $\arg(z)$; the angle measurement anticlockwise of the positive real axis; $\arg(z) = \theta$, $\tan(\theta) = \dfrac{y}{x}$

arrangement: a way of choosing things where order is important; see **permutation**

asymptote: the graphical representation of an undefined function $f(x) = 0$. Graphically it is a line that a graph approaches but never reaches. A horizontal asymptote shows the long-term behaviour as $x \to \infty$; a vertical asymptote may occur where a function is undefined, such as at $x = 0$ for the hyperbola $y = \dfrac{1}{x}$.

axiom: a proposition that is assumed to be true; from Greek, meaning 'agreed starting point'. Every area of mathematics has its own basic axioms.

balanced set of forces: an object at rest will stay at rest or an object in motion will maintain its motion under balanced forces. Objects under balanced forces are in equilibrium.

binomial theorem: a rule for expanding expressions of the form $(x + y)^n$:
$$(x + y)^n = x^n + {}^nC_1 x^{n-1}y + {}^nC_2 x^{n-2}y^2 + \ldots {}^nC_r x^{n-r}y^r + \ldots y^n$$

Cartesian form of a complex number: $x + yi$; consists of an ordered pair of numbers (x, y) which can be plotted on the complex plane or Argand plane

Cartesian form of a vector: from the origin to point (x, y), the Cartesian form of vector $\underset{\sim}{u}$ is $\underset{\sim}{u} = x\hat{\imath} + y\hat{\jmath}$.

centre of the circle: O is the centre of the circle. One of Euclid's postulates states that any circle can be defined by its centre and its radius.

chord: an interval jointing two points on a circle. $\overline{AG}$, $\overline{AD}$ and $\overline{GD}$ are all chords.

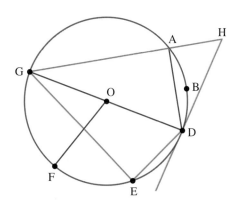

circumference: the distance around a circle (that is, the length of the circle if it was opened up and straightened into a line segment). It may also be used to refer to the edge of the circle. Points on the circumference are really points on the circle.

collinear vectors: parallel vectors that lie along the same line or parallel lines

column matrix: a matrix with only one column; a vector matrix

combination: the number of ways of choosing r things from n distinct things where order is not important; ${}^nC_r = \dfrac{n!}{r!\,(n-r)!}$, also written as $\dbinom{n}{r}$

combined transformation: a combination of two or more transformations

complex conjugate roots: the complex solutions for quadratic equations with $\Delta < 0$ (i.e. those that have no real solutions); expressed in the form $a \pm bi$

complex number: $z = a + bi$; a real number added to a multiple of the imaginary unit i, when a and b are real numbers; $\text{Re}(z) = a$, $\text{Im}(z) = b$

component (of a vector): the components of a vector $\underset{\sim}{u}$ are the two horizontal and vertical vectors (parallel to the unit vectors $\hat{\imath}$, $\hat{\jmath}$) that sum to $\underset{\sim}{u}$.

conjugate of a complex number: the conjugate, $\bar{z}$, of a complex number z has the opposite sign of the imaginary component. Multiplication or addition of a complex number and its conjugate results in a real number.

contradiction: in proof by contradiction, the opposite of what you are trying to prove is assumed to be true. Eventually, the proof will reach a statement that cannot be true, meaning that the initial assumption must be false.

contrapositive: a statement that contains the negation of both terms of the original statement but in the reverse order. For example, the contrapositive of 'if P, then Q' is 'if not Q, then not P'. The contrapositive of a true statement is also true.

converse: the converse of a statement has the same propositions but in reverse order. The converse may be false, even if the original statement is true. For example, the converse of 'if P, then Q' is 'if Q, then P'; symbolically, the converse of $P \Rightarrow Q$ is $Q \Rightarrow P$ or $P \Leftarrow Q$.

coordinate matrix: a matrix that represents the coordinates of points as columns in the matrix

cosecant function: the reciprocal of the sine function; $\operatorname{cosec}(x) = \dfrac{1}{\sin(x)}$, $\sin(x) \neq 0$

cotangent function: the reciprocal of the tangent function; $\cot(x) = \dfrac{1}{\tan(x)} = \dfrac{\cos(x)}{\sin(x)}$, $\sin(x) \neq 0$

counter example: an example which shows a statement is not true

cyclic quadrilateral: a quadrilateral whose vertices all lie on a circle. ADEG is a cyclic quadrilateral.

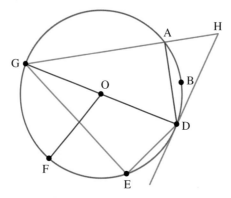

de Moivre's theorem: $(r\operatorname{cis}(\theta))^n = r^n\operatorname{cis}(n\theta)$, where $\operatorname{cis}(\theta) = \cos(\theta) + i\sin(\theta)$

definition: the agreed meaning of a term

dependent: two events are dependent if where one event is followed by the other, the order is important.

determinant of a matrix: a value associated with a square matrix, evaluated by multiplying the elements in the leading diagonal and subtracting the product of the elements in the other diagonal; for a matrix $A = \begin{bmatrix} a & b \\ c & d \end{bmatrix}$, $\det(A) = ad - bc$

diameter: a chord that passes through the centre of a circle. $\overline{GD}$ is a chord and a diameter.

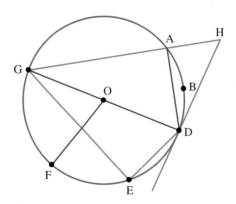

dilation: a linear transformation that enlarges or reduces the size of a figure by a scale factor k parallel to either axis or both

domain: the set of possible x-values for a function; the set of all x-values of the ordered pairs (x, y) that make up a relation

dot product: or scalar product; multiplication of two vectors which results in a scalar

double angle identities: $\sin(2A) = 2\sin(A)\cos(A)$,
$$\cos(2A) = \cos^2(A) - \sin^2(A)$$
$$= 2\cos^2(A) - 1$$
$$= 1 - 2\sin^2(A)$$

elements: the members of a set; $a \in A$ means a is an element of, or belongs to, the set A. If a is not an element of the set A, this is written as $a \notin A$.

elements of a matrix: the numbers in the matrix

equilibrium: a condition when all forces acting on an object are balanced; that is, the net or resultant force is zero

equivalent statements: statements P and Q are equivalent if the converse of $P \Rightarrow Q$ is true; that is, $P \Rightarrow Q$ and $Q \Rightarrow P$; symbol $\Leftrightarrow$.

existential quantifier: there exists; symbolically, $\exists$. For a propositional function, consider if there exists a value for the variable which will make the statement true.

exterior angle to a cyclic quadrilateral: when one side of a cyclic quadrilateral is extended to an exterior point, the exterior angle is created between this extended segment and the cyclic quadrilateral. $\angle DAH$ is an exterior angle.

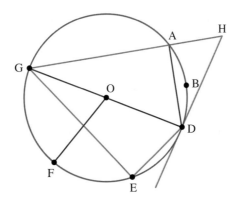

external (to a circle): points outside the circle are external to the circle.

factorial: $n!$, multiplying each of the integers from n down to 1; $0! = 1$

field force: force acting on objects without physical contact, for example gravity

fixed point: a point of the domain of a function which is mapped onto itself after a transformation; an invariant point

force: a push or pull on an object; includes field force and applied force (e.g. contact, tensile)

function: a relation in which the set of ordered pairs (x, y) have each x-coordinate paired to a unique y-coordinate; typically expressed as $y = f(x)$

half-angle identities: $\sin(A) = 2\sin\left(\dfrac{A}{2}\right)\cos\left(\dfrac{A}{2}\right)$,
$$\cos(A) = \cos^2\left(\frac{A}{2}\right) - \sin^2\left(\frac{A}{2}\right)$$
$$= 2\cos^2\left(\frac{A}{2}\right) - 1$$
$$= 1 - 2\sin^2\left(\frac{A}{2}\right)$$

identical vector: a vector with the same magnitude and direction as the original

identity: a relationship that is true for all possible values of the variable or variables

identity matrix: I; a square matrix that has 1s down the leading diagonal and 0s on the other diagonal; for example, the 2×2 identity matrix is $I = \begin{bmatrix} 1 & 0 \\ 0 & 1 \end{bmatrix}$

iff: the condition 'if and only if'; when all stated conditions must apply

image: a point or figure after a transformation

imaginary number: i; the root of the equation $x^2 = -1$; that is, $x = \pm \sqrt{i^2}$

implication: if P, then Q; symbolically, $P \Rightarrow Q$

inclusion–exclusion principle: for two sets S and T: $n(S \cup T) = n(S) + n(T) - n(S \cap T)$.
For three sets S, T and R: $n(S \cup T \cup R) = n(S) + n(T) + n(R) - n(S \cap T) - n(T \cap R) - n(S \cap R) + n(S \cap T \cap R)$.

inertial mass: resistance to acceleration

integers: all of the positive and negative whole numbers and zero: $\ldots -3, -2, -1, 0, 1, 2, 3 \ldots$

interior opposite angle in a cyclic quadrilateral: the angle inside the quadrilateral that does not share any rays with the exterior angle. If the exterior angle is $\angle$DAH, then the interior opposite angle in GADE is $\angle$GED.

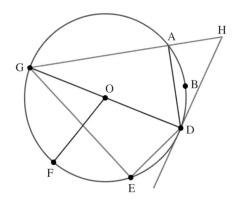

intersection: the set containing the elements common to both A and B; denoted as $A \cap B$

invariant point: a point of the domain of a function which is mapped onto itself after a transformation; a fixed point

invariants: properties which are unchanged by a transformation

irrational numbers: real numbers not including the rational numbers; that is, real numbers that cannot be expressed in the form $\dfrac{a}{b}$, where $b \neq 0$

kinematics: the study of the motion of objects

leading diagonal: the principal diagonal of a square matrix, from upper left to bottom right

linear transformation: a transformation of a vector in which the origin does not move. A linear transformation can be represented by a 2×2 matrix $\begin{bmatrix} a & b \\ c & d \end{bmatrix}$. Linear transformations include rotations around the origin, reflection in lines passing through the origin and dilations. Translations are not linear transformations.

magnitude: $|x|$; the absolute value of a number

magnitude (of a vector): $|u|$ or r; the length of any directed line segment representing the vector $\underset{\sim}{u}$

major arc: an arc that is larger than a semicircle

major segment: segment with an intercepted arc more than the semicircle

mathematical statement: a mathematical sentence that is either true or false; a proposition. For example, '5 is a prime number' is a true mathematical statement.

matrix: a rectangular array used to store and display data

mediator: a line of reflection

minor arc: an arc that is smaller than a semicircle

minor segment: segment with an intercepted arc less than the semicircle

minors: the determinants of a smaller square matrix formed by deleting one row and one column from a larger square matrix

modulus: the absolute value of a number

modulus of a complex number: $|z|$; for a complex number, $z = x + iy$, the length of the line segment from the origin to point z; $|z| = \sqrt{x^2 + y^2}$

multiple angle identities: $\sin(3A) = 3\sin(A) - 4\sin^3(A)$, $\cos(3A) = 4\cos^3(A) - 3\cos(A)$, $\sin(4A) = \cos(A)\left(4\sin(A) - 8\sin^3(A)\right)$, $\cos(4A) = 8\cos^4(A) - 8\cos^2(A) + 1$

multiplication principle: if there are n ways of performing operation A and m ways of performing operation B, then there are $m \times n$ ways of performing A *and* B.

multiplicative inverse: a number multiplied by its multiplicative inverse equals 1. That is, $z^{-1} = \dfrac{1}{z}$, $zz^{-1} = 1$.

multiplicative inverse (of a matrix): A^{-1}; the inverse of a square matrix A such that $AA^{-1} = A^{-1}A = 1$. For a 2×2 matrix $A = \begin{bmatrix} a & b \\ c & d \end{bmatrix}$, $A^{-1} = \dfrac{1}{\det(A)} \begin{bmatrix} d & -b \\ -c & a \end{bmatrix}$, $\det(A) \neq 0$.

natural numbers: the counting numbers 1, 2, 3, ...

negation: the opposite of a mathematical statement. If the statement is P, the negation is 'not P'; symbolically, ¬P.

net force: the vector sum of all real forces acting on an object; resultant force

newton (N): the SI unit of weight, the amount of force required to accelerate 1 kg of mass 1 m/s^2

null matrix: a square matrix that consists entirely of '0' elements; a zero matrix

order of a matrix: the size of a matrix, written as rows × columns

particle: an object taken as a point; used to model an object in Newtonian dynamics as a representation of the object's motion. The original object's shape, change in shape and/or internal movements are irrelevant.

Pascal's triangle: a triangle formed by rows of the binomial coefficients of the terms in the expansion of $(a + b)^n$ with n as the row number

permutation: the number of ways of choosing r things from n distinct things when order is important; $^n\mathrm{P}_r = \dfrac{n!}{(n-r)!} = n \times (n-1) \times (n-2) \times ... \times (n-r+1)$

point of rotation: the fixed point around which a rotation occurs

points on the circle: any point that lies on a circle

polar form (of a complex number): the complex number z expressed in terms of θ; $z = r\,\mathrm{cis}(\theta)$

polar form (of a vector): a vector expressed in the form (r, θ) where r is the magnitude of the vector and $\theta = \tan^{-1}\left(\dfrac{y}{x}\right)$, which is the direction of the vector (the angle the vector makes to the positive direction of the x-axis)

position vector: a vector that defines a point by magnitude and direction relative to the origin

postulate: in geometry, a postulate is a statement that is assumed to be true without proof. (This is usually called an axiom in other areas of mathematics.)

pre-image: the original figure before a transformation

probability: the long-term proportion or relative frequency of the occurrence of an event

projection: a vector v can be defined by a projection acting in the direction of vector u and a projection acting perpendicular to u.

proof: demonstration that a statement is always true using definitions, postulates and previously proven statements in a formal sequence of steps

proposition: a sentence that is either true or false; a mathematical statement. For example, '5 is a prime number' is a true proposition.

propositional function: a proposition that includes variables

Pythagorean identities: for any angle A, $\sin^2(A) + \cos^2(A) = 1$, $\tan^2(A) + 1 = \sec^2(A)$ and $\cot^2(A) + 1 = \mathrm{cosec}^2(A)$.

quantifiers: used with a propositional function to give information about the scope of the function's variables. There are 2 main quantifiers:
- the universal quantifier (for all/for each): ∀
  (consider if the function is true for all possible values of the variable)
- the existential quantifier (there exists): ∃
  (consider if there is a value for the variable that will make the statement true).

radius: any line segment joining a point on the circle to the centre

range: the set of possible *y*-values for a function; the set of all *y*-values of the ordered pairs (*x*, *y*) that make up a relation

rational numbers: real numbers that can be written in the form $\frac{a}{b}$ where *a* and *b* are integers, the only common factor between *a* and *b* is 1, and $b \neq 0$

real numbers: numbers that can be represented on a number line

reciprocal: the multiplicative inverse of the original; 1 divided by the original. The reciprocal of *f*(*x*) is $\frac{1}{f(x)}$.

reflection: a transformation of a point, line or figure defined by a line of reflection, where the image point is a mirror image of the pre-image point

relation: any set of ordered pairs (*x*, *y*)

resistive forces: forces that occur when two objects move or attempt to move relative to one another, for example air drag and friction

restricted domain: a subset of a function's maximal domain, often due to practical limitations on the independent variable in modelling situations

resultant force: the vector sum of all real forces acting on an object; net force

roots of an equation: the solutions of an equation

rotation: a transformation of a point, line or figure where each point is moved a constant amount around a fixed point

row matrix: a matrix with only one row

scalar: a quantity that has only magnitude, no direction

scalar resolute: $\hat{u} \cdot \underset{\sim}{v}$; the magnitude of vector $\underset{\sim}{v}$ acting in the direction of vector $\underset{\sim}{u}$

secant: a line that touches a circle at two distinct points. $\overline{GH}$ is a secant.

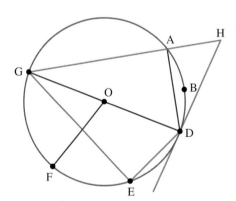

secant function: the reciprocal of the cosine function; $\sec(x) = \frac{1}{\cos(x)}$, $\cos(x) \neq 0$

segment: a chord divides a circle into two regions; the larger region is the major segment and the small region is the minor segment.

selection: the number of ways of choosing things when order is not important; see **combination**

set: a collection of elements

singular matrix: a matrix that has a zero determinant and hence has no multiplicative inverse

square matrix: a matrix with an equal number of rows and columns

standard form (of a complex number): the expression of a complex number as a real number added to a multiple of the imaginary unit i, when a and b are real numbers; $z = a + bi$

standing: $\angle AGD$ is called an angle at the circumference standing on chord $\overline{AD}$ (or subtended by the arc $\overset{\frown}{ABD}$).

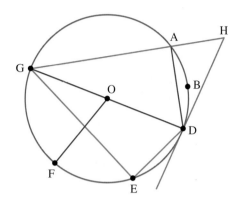

statics: a situation when the net or resultant force acting on an object is zero; the object is in equilibrium

subtend: an arc on a circle subtends the angle at the circumference opposite it. The arc $\overset{\frown}{ABD}$ is said to subtend the angle $\angle AGD$, where A, B, D and G are points on the circle.

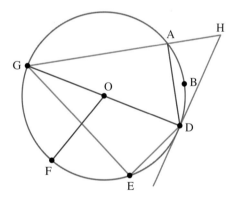

surd: a root of a number that does not have an exact answer. Surds are irrational numbers. Surds themselves are exact numbers, for example $\sqrt{6}$; their decimal approximations are not.

tangent: a line that touches a circle at one point

theorem: a mathematical statement that can be shown to be true using a proof

thrust: the force required to move an object forward

trace of a matrix: $\text{tr}(A)$ for matrix A; the sum of the leading diagonal elements

transformation: a geometric operation that may change the shape and/or the position of a point or set of points

translation: a transformation of a point, line or figure where each point in the plane is moved a given distance in a horizontal or vertical direction

translation matrix: the representation of a translation by a matrix

trigonometric function: sine (sin), cosine (cos) or tangent (tan). On a unit circle, $\cos(\theta)$ is the x-coordinate of the trigonometric point $[\theta]$; $\sin(\theta)$ is the y-coordinate of the trigonometric point $[\theta]$; and $\tan(\theta)$ is the length of the intercept that the line through the origin and the trigonometric point $[\theta]$ cuts off on the tangent drawn to the unit circle at $(1, 0)$.

unbalanced forces: forces acting on an object to cause a change in motion

undefined: a mathematical expression that does not have a meaning, such as division by zero. Division of a number by zero cannot be undone by multiplication by zero, $\dfrac{x}{0} \times 0 \neq x$; hence, division by zero is undefined.

union: the set of all elements in any one set and in a combination of sets; $A \cup B$ contains the elements in A or in B or in both A and B

unit circle: a circle of radius 1 about the origin, from which trigonometric ratios can be considered

unit vectors: the vectors $\hat{\imath}$ and $\hat{\jmath}$ with magnitudes of 1 unit, which allow a vector to be resolved into its components; $\underset{\sim}{u} = x\hat{\imath} + y\hat{\jmath}$

universal quantifier: for all; symbolically, $\forall$. For a propositional function, consider if the function is true for all possible values of the variable.

universal set: the complete set of objects being considered; usually represented by ξ

vector: a quantity that has both magnitude and direction

vector resolute: the direction of vector $\underset{\sim}{v}$ acting parallel or perpendicular to the direction of vector $\underset{\sim}{u}$;
$$\underset{\sim}{v}_{\parallel} = \underset{\sim}{u} - (\underset{\sim}{u} \cdot \hat{v})\,\hat{v}; \; \underset{\sim}{v}_{\perp} = (\underset{\sim}{u} \cdot \hat{v})\,\hat{v}$$

vertical line test: a test that determines whether a graph is that of a function; any vertical line that cuts the graph of a function does so exactly once

weight: a vector quantity equal to the mass of an object multiplied by the acceleration due to gravity; $\underset{\sim}{W} = m\underset{\sim}{g}$; weight of mass m in gravitational field $\underset{\sim}{g}$

zero matrix: a square matrix that consists entirely of '0' elements